Masses and Ground State electron configurations of the elements

Closed shell

s block elements **p block elements**

d block elements – Transition metals

Post transition metals

f block elements

periods	1	2	3	4	5	6	7	8	9	10	11	12	13	14	15	16	17	18	shell
1	1 H 1.0079 $1s^1$																	2 He 4.003 $1s^2$	K
2	3 Li 6.941 $2s^1$	4 Be 9.012 $2s^2$											5 B 10.81 $2p^1$	6 C 12.01 $2p^2$	7 N 14.01 $2p^3$	8 O 16.00 $2p^4$	9 F 19.00 $2p^5$	10 Ne 20.18 $2p^6$	L
3	11 Na 22.99 $3s^1$	12 Mg 24.30 $3s^2$											13 Al 26.98 $3p^1$	14 Si 28.09 $3p^2$	15 P 30.97 $3p^3$	16 S 32.07 $3p^4$	17 Cl 35.45 $3p^5$	18 Ar 39.95 $3p^6$	M
4	19 K 39.10 $4s^1$	20 Ca 40.08 $4s^2$	21 Sc 44.96 $4s^2 3d^1$	22 Ti 47.87 $4s^2 3d^2$	23 V 50.94 $4s^2 3d^3$	24 Cr 52.00 $4s^1 3d^5$	25 Mn 54.94 $4s^2 3d^5$	26 Fe 55.85 $4s^2 3d^6$	27 Co 58.93 $4s^2 3d^7$	28 Ni 58.69 $4s^2 3d^8$	29 Cu 63.55 $4s^1 3d^{10}$	30 Zn 65.39 $4s^2 3d^{10}$	31 Ga 69.72 $4p^1$	32 Ge 72.61 $4p^2$	33 As 74.92 $4p^3$	34 Se 78.96 $4p^4$	35 Br 79.90 $4p^5$	36 Kr 83.80 $4p^6$	N
5	37 Rb 85.47 $5s^1$	38 Sr 87.62 $5s^2$	39 Y 88.91 $5s^2 4d^1$	40 Zr 91.22 $5s^2 4d^2$	41 Nb 92.91 $5s^2 4d^3$	42 Mo 95.94 $5s^1 4d^5$	43 Tc 98.91 $5s^2 4d^5$	44 Ru 101.1 $5s^1 4d^7$	45 Rh 102.9 $5s^1 4d^8$	46 Pd 106.4 $5s^0 4d^{10}$	47 Ag 107.9 $5s^1 4d^{10}$	48 Cd 112.4 $5s^2 4d^{10}$	49 In 114.8 $5p^1$	50 Sn 118.7 $5p^2$	51 Sb 121.8 $5p^3$	52 Te 127.6 $5p^4$	53 I 126.9 $5p^5$	54 Xe 131.3 $5p^6$	O
6	55 Cs 132.9 $6s^1$	56 Ba 137.3 $6s^2$	57 La 138.9 $6s^2 5d^1$	72 Hf 178.5 $6s^2 5d^2$	73 Ta 180.9 $6s^2 5d^3$	74 W 183.8 $6s^2 5d^4$	75 Re 186.2 $6s^2 5d^5$	76 Os 190.2 $6s^2 5d^6$	77 Ir 192.2 $6s^2 5d^7$	78 Pt 195.1 $6s^1 5d^9$	79 Au 197.0 $6s^1 5d^{10}$	80 Hg 200.6 $6s^2 5d^{10}$	81 Tl 204.4 $6p^1$	82 Pb 207.2 $6p^2$	83 Bi 209.0 $6p^3$	84 Po 209 $6p^4$	85 At 210 $6p^5$	86 Rn 222 $6p^6$	P
7	87 Fr 223.0 $7s^1$	88 Ra 226.0 $7s^2$	89 Ac 227.0 $7s^2 6d^1$	104 Rf 261 $7s^2 6d^2$	105 Db 262 $7s^2 6d^3$	106 Sg 266 $7s^2 6d^4$	107 Bh 264 $7s^2 6d^5$	108 Hs 269 $7s^2 6d^6$	109 Mt 268 $7s^2 6d^7$	110 Ds 271 $7s^2 6d^8$	111 Rg 272 $7s^1 6d^{10}$	112 Cn 277 $7s^2 6d^{10}$	113 Uut	114 Fl 289 $7p^2$	115 Uup	116 Lv 293 $7p^4$	117 Uus	118 Uuo	Q

Lanthanoid series ($6s^2$)

58 Ce	59 Pr	60 Nd	61 Pm	62 Sm	63 Eu	64 Gd	65 Tb	66 Dy	67 Ho	68 Er	69 Tm	70 Yb	71 Lu
140.1 $5d^0 4f^2$	140.9 $5d^0 4f^3$	144.2 $5d^0 4f^4$	145 $5d^0 4f^5$	150.4 $5d^0 4f^6$	152.0 $5d^0 4f^7$	157.2 $5d^1 4f^7$	158.9 $5d^0 4f^9$	162.5 $5d^0 4f^{10}$	164.9 $5d^0 4f^{11}$	167.3 $5d^0 4f^{12}$	168.9 $5d^0 4f^{13}$	173.0 $5d^0 4f^{14}$	175.0 $5d^1 4f^{14}$

Actinoid series ($7s^2$)

90 Th	91 Pa	92 U	93 Np	94 Pu	95 Am	96 Cm	97 Bk	98 Cf	99 Es	100 Fm	101 Md	102 No	103 Lw
232.0 $6d^2 5f^0$	231.0 $6d^1 5f^2$	238.0 $6d^1 5f^3$	237.0 $6d^1 5f^4$	244 $6d^0 5f^6$	243 $6d^0 5f^7$	247 $6d^1 5f^7$	247 $6d^0 5f^9$	251 $6d^0 5f^{10}$	252 $6d^0 5f^{11}$	257 $6d^0 5f^{12}$	258 $6d^0 5f^{13}$	259 $6d^0 5f^{14}$	262 $6d^1 5f^{14}$

f block elements

Fundamental Constants and Useful Unit Conversion

Fundamental Constants

Newton $(N) = \mathrm{kg\, m\, s^{-2}} = \mathrm{J\, m^{-1}}$

Joule $(J) = 10^7\, \mathrm{erg} = 1$ newton meter $(N\, m) = 1\, \mathrm{kg\, m^2\, s^{-2}}$

Watt $(W) = \mathrm{J\, s^{-1}}$

Charge on an electron $(e^-) = -1.6022 \times 10^{-19}$ coulomb (C)

Charge of the proton $= 1.6022 \times 10^{-19}$ C

Electron mass, $m_e = 9.109384 \times 10^{-31}$ kg

Proton mass, $m_p = 1.672623 \times 10^{-27}$ kg

Neutron mass, $m_n = 1.674928 \times 10^{-27}$ kg

Atomic mass unit, $u = 1.660539 \times 10^{-27}$ kg

Avogadro's number, $N_A = 6.022 \times 10^{23}$ atoms $\mathrm{mol^{-1}}$

Boltzmann's constant, $k = 1.38065 \times 10^{-23}\, \mathrm{J\, °K^{-1}}$

Gas constant, $R = k(N_A) = 8.314\, \mathrm{J\, mol^{-1}\, °K^{-1}}$

$\qquad\qquad\qquad = 0.082057\, \mathrm{L\, atm\, mol^{-1}\, °K^{-1}}$

$\qquad\qquad\qquad = 1.987\, \mathrm{cal\, mol^{-1}\, °K^{-1}}$

Faraday, $F = e\, (N_A) = 9.64853 \times 10^4\, \mathrm{C\, mol^{-1}}$

Planck constant, $h = 6.627 \times 10^{-34}\, \mathrm{J\, s}$

Velocity of light, $c = 2.99792458 \times 10^8\, \mathrm{m\, s^{-1}}$

Vacuum permittivity, $\varepsilon_0 = 8.854188 \times 10^{-12}\, \mathrm{J^{-1}\, C^2\, m^{-1}}$

Conversion Factors

$1\, \mathrm{eV} = 1.602189 \times 10^{-19}\, \mathrm{J} = 96.485\, \mathrm{kJ\, mol^{-1}} = 8065.5\, \mathrm{cm^{-1}}$ ($\bar{v} = \mathrm{cm^{-1}}$ or Kaiser, K)

$1\, \mathrm{cm^{-1}} = 1.986 \times 10^{-23}\, \mathrm{J} = 11.96\, \mathrm{J\, mol^{-1}} = 0.1240\, \mathrm{meV}$

1 calorie (cal) $= 4.184\, \mathrm{J}$

1 Å (angstrom) $= 10^{-8}\, \mathrm{cm} = 10^{-10}\, \mathrm{m} = 100\, \mathrm{pm} = 0.100\, \mathrm{nm}$

Standard pressure, $P = 1\, \mathrm{atm} = 760\, \mathrm{mm\, Hg}$ (torr) $= 101.325\, \mathrm{kPa} = 1.01325 \times 10^5\, \mathrm{N\, m^{-2}}$

$1\, \mathrm{bar} = 10^5\, \mathrm{Pa}$

At $T = 298.15\, \mathrm{°K}$, $RT = 2.4788\, \mathrm{kJ\, mol^{-1}}$ and $RT/F = 25.691\, \mathrm{mV}$

Prefixes

a	f	p	n	μ	m	c	d	k	M	G	T	P
atto	femto	pico	nano	micro	milli	centi	deci	kilo	mega	giga	tera	peta
10^{-18}	10^{-15}	10^{-12}	10^{-9}	10^{-6}	10^{-3}	10^{-2}	10^{-1}	10^3	10^6	10^9	10^{12}	10^{15}

Inorganic Chemistry for Geochemistry and Environmental Sciences

Inorganic Chemistry for Geochemistry and Environmental Sciences

Fundamentals and Applications

GEORGE W. LUTHER, III

*School of Marine Science & Policy,
University of Delaware, USA*

WILEY

Library of Congress Cataloging-in-Publication Data

Names: Luther III, George W.
Title: Inorganic chemistry for geochemistry and environmental sciences :
    fundamentals and applications / George W. Luther, III.
Description: Chichester, West Sussex : John Wiley & Sons, Inc., 2016. |
    Includes bibliographical references and index.
Identifiers: LCCN 2015047266 | ISBN 9781118851371 (cloth) | ISBN 9781118851401 (epdf) | ISBN 9781118851418 (epub)
Subjects: LCSH: Chemistry, Inorganic. | Geochemistry. |
    Bioinorganic chemistry. | Transition metals–Environmental aspects. |
    Sulfides–Environmental aspects. | Chemical ecology.
Classification: LCC QH541.15.C44 L88 2016 | DDC 577/.14–dc23 LC record available at http://lccn.loc.gov/2015047266

A catalogue record for this book is available from the British Library.

Set in 9/11pt, TimesLTStd by SPi Global, Chennai, India.

Printed in the UK

To B.J., Gregory and Stephanie

Contents

About the Author

Professor George W. Luther, III, School of Marine Science and Policy, University of Delaware, USA

Professor George W. Luther, III, has joint appointments in the Department of Chemistry and Biochemistry, Department of Civil and Environmental Engineering and the Department of Plant and Soil Science at the University of Delaware, USA.

He taught an American Chemical Society accredited course on advanced inorganic chemistry from 1973 to 1986 to senior undergraduate students. As he moved into environmental and marine chemistry, he began using environmental examples in inorganic chemistry. In 1988, he started a similar course titled "Marine Inorganic Chemistry," which is being taught biannually at the University of Delaware, attracting students in Chemical Oceanography, Chemistry and Biochemistry, Geology/Geochemistry, Civil and Environmental Engineering, and Plant and Soil Science. In 2004, he was awarded the Clair C. Patterson Award from the Geochemical Society for outstanding contributions to environmental geochemistry.

In 2013, he was awarded the Geochemistry Division Medal by the American Chemical Society for his wide-ranging contributions to aqueous geochemistry. He is recognized for the application of physical inorganic chemistry to the transfer of electrons between chemical compounds in the environment, and also the development of chemical sensors for quantifying the presence of elements and compounds in natural waters.

He was named a fellow of the American Association for the Advancement of Science in 2011, the American Geophysical Union in 2012, the Geochemical Society in 2014, and the American Chemical Society in 2015.

Preface

For the past 25 years, I have been teaching an inorganic chemistry course primarily to graduate students in chemistry, chemical oceanography, geochemistry, soil and plant science, and civil and environmental engineering. My goal in the course has been to use a physical inorganic chemistry approach with many chemical examples from geochemistry and environmental and marine chemistry so the students could gain better understanding of environmental processes at the molecular level. Frequently, as the students performed their own research, they encountered some puzzling aspects, which they wished to better understand or explain. I wish to thank all my former students, postdoctoral students, and colleagues both at the University of Delaware and elsewhere for encouraging me in this endeavor. I note only a few including those who provided valuable input or information for some of the chapters: Herbert Allen, Rachael Austin, Alison Butler, Thomas Church, Dominic DiToro, Alyssa Findlay, Amy Gartman, Chin-Pao Huang, Rob Mason, Frank Millero, James Morgan, Véronique Oldham, Ann Ploskonka, David Rickard, Charles Riordan, Tim Rozan, Timothy Shaw, Donald Sparks, Werner Stumm, Martial Taillefert, Adam Wallace, and Jessica Wallick. Of course, any errors are due to my carelessness or lack of attention to detail.

Although inorganic chemists study all the elements of the periodic table, those studying inorganic chemistry in an environmental setting must sometimes do it at trace or ultra-trace level concentrations. In this book, the concepts of physical inorganic chemistry are used to study natural chemical processes occurring in the ocean, water, soil, sediment, and atmosphere as well as those related to anthropogenic activities. A couple of relevant examples of inorganic chemistry on other planets such as Mars and in interstellar space are also provided. Understanding the principles of inorganic chemistry including chemical bonding, one and two electron transfer processes in oxidation–reduction chemistry (redox), acid–base chemistry, transition metal ligand complexes, metal catalysis including enzyme catalysis, and more are essential to describing earth processes over all time scales ranging from ~1 nsec to geologic time (Gyr). The fields of geochemistry and environmental chemistry depend on the principles of physical inorganic chemistry. I hope the student will understand the relationship between these fields by using the fundamental concepts from thermodynamics, kinetics, and a detailed understanding of electronic structure. To aid in visualizing orbitals and molecular structures, I have used the most recent version (8.0.10) of the HyperChem™ program from Hypercube, Inc. (Gainesville, FL). Still, students should be able to "draw" orbitals and structures so the book uses both idealized drawings and computer-generated models throughout.

Broadly speaking, the book has three sections. Chapter 1 discusses the distribution of the elements through the cosmos and on earth with emphasis on large-scale chemical processes that occur on earth, which is profoundly influenced by the presence of water as solvent. The other chapters reference many of the processes in Chapter 1. Chapters 2 through 9 give the foundations of inorganic chemistry with traditional examples that most inorganic chemists would be familiar with; in addition, there are many geochemical and environmental reactions and processes given as examples to introduce or to explain concepts. In the last four chapters, the concepts from Chapters 2 through 9 are used to describe a host of geochemical, environmental and bioinorganic chemistry examples, which are also cross-referenced to processes in Chapter 1.

Chapter 2 introduces the thermodynamics of redox chemistry, describes the oxidation state of important elements in nature, and emphasizes one and two electron transfer step reactions; data and concepts from this chapter are used throughout the text. Chapter 3 describes atomic theory, the buildup of the periodic table, and the periodic properties of the elements. Chapter 4 describes molecules using the principles of symmetry and group theory, and I decided to have a separate chapter rather than intersperse these topics into other chapters. Chapter 5 introduces bonding theories for nonmetals, and the frontier molecular orbital approach is used to describe numerous examples of chemical reactivity for small molecules of geochemical and environmental interest. The frontier molecular orbital theory approach is used often in subsequent chapters to gain understanding and predict chemical reactivity. Although this approach is well used in inorganic and organic chemistry, it is less used by scientists studying the environment. Chapter 6 continues the description of covalent bonding in metals and semiconductors, and then proceeds with the ionic bonding model including the importance of nanoparticles in inorganic chemistry. Chapter 7 reviews acid–base chemistry and leads directly into transition metal chemistry, which is described in detail in Chapters 8 and 9. Chapter 8 provides the basics of transition metal chemistry including bonding theories (e.g., valence bond theory, crystal field theory, and molecular orbital theory), and the spectroscopy and magnetic properties of metal ligand complexes. Chapter 9 gives details on the thermodynamics and the kinetics of metal ligand complexes and their substitution electron transfer reactions while introducing concepts of transition metal catalysis.

Chapters 10 through 13 give many examples of transition metal chemistry in the environment. Chapter 10 describes the chemistry of metals with molecular oxygen including the oxidation of reduced iron and manganese, the reversible binding of oxygen in reduced iron and copper for transport in blood, the use of O_2 and enzyme systems to oxidize hydrocarbons and ammonium ion, and the photochemical formation of O_2 in the oxygen-evolving complex. Chapter 11 describes the chemistry of the dissolution of manganese and iron oxides with hydrogen sulfide and the oxidation of pyrite by O_2 and soluble Fe(III). The formation of metal sulfide nanoparticles and particles is described in Chapter 12, which ends with a discussion on FeS phases as a catalytic source for the origin of life and with the ability of ferredoxins to activate small molecules such as carbon dioxide. Chapter 13 describes the uptake of metals primarily in single-celled organisms, and uses information on stability constants of metal–ligand complexes and their kinetics from Chapter 9 to provide a quantitative description.

In this book, attempts are made to show the interrelationship between topics in different chapters so that the reader can better understand the principles of physical inorganic chemistry. The discovery and use of these relationships by the student should further our knowledge of environmental processes from the molecular level to the global level.

George W. Luther, III
Lewes, DE
2016

Cover art: To convey chemistry from the molecular to the macroscopic level, the background is a photo of a black smoker hydrothermal vent spewing black iron sulfide and pyrite (nano)particles. The superimposed chemical models show a representation for the contact of hydrogen sulfide with the surface of FeS nanoparticles to form pyrite (FeS_2) nanoparticles that can then aggregate to form microscopic and larger particles. The two red dots in the photo are 10 cm apart.

Photo credit: Image courtesy George Luther, Univ. of Delaware/NSF/ROV Jason 2012©Woods Hole Oceanographic Institution

SEM credit: Image from collaborative work of Amy Gartman and George Luther

Companion Website

This book is accompanied by a companion website:

www.wiley.com/go/luther/inorganic

The website includes:

- A comprehensive set of PowerPoint slides for use by lecturers
- Exercises for students, to accompany each chapter in the book
- Solutions to the exercises

1

Inorganic Chemistry and the Environment

1.1 Introduction

Understanding the atomic structure of the atom is important to understanding how atoms combine to build molecules including minerals, aqueous materials, and gases. However, nature builds the atoms of the elements starting with the simplest elements, hydrogen and helium, which are the most abundant in the cosmos. As described in the Big Bang Theory [1–6 and references therein], the origin of the universe (and the periodic table) started with elementary particles under enormous gravitational attraction concentrated in an extremely densely packed point, which exploded causing an expanding universe and the release of enormous energy and fundamental particles. Within 0.01 s of the Big Bang, temperatures have been predicted to be in the range of 10^9–10^{11} °K. After 100 s and on cooling to ~10^9 °K, the elementary particles began to combine under the force of **gravity**. Here, positively charged protons and neutral neutrons start to combine to form the lighter elements (**nucleosynthesis**) and their isotopes. Their combination in the nucleus occurs by the **strong force**, which is the short-range (10^{-15} m) attractive force between protons and neutrons that binds these particles in the nucleus while overcoming the repulsive force of the protons with each other. At this time and under these conditions, the electrons are totally ionized from the nucleus and cannot combine with the elements until cooling occurs at about 10^6 °K. At this lower temperature, the **electromagnetic force** begins to take effect and the combination of the electrons with the positive nuclei to form neutral atoms occurs. Once there is a buildup of neutral atoms, chemical processes can occur that eventually lead to life and biological processes.

1.1.1 Energetics of Processes

To understand the energies associated with a wide range of processes at temperatures from absolute zero to these extreme Big Bang temperatures, the Boltzmann energy-temperature relationship (Equation 1.1) provides perspective:

$$E \, (\text{Joule}) = kT \qquad (1.1)$$

where $k = 1.38065 \times 10^{-23}$ J °K^{-1}; multiplying k by 6.2415×10^{18} eV J^{-1} (as $1 \, \text{eV} = 1.602189 \times 10^{-19}$ J) gives E in units of eV (electron volts). Figure 1.1 is a plot of E (eV) versus T(°K) that gives the temperature and corresponding energy at which several well-known processes occur. Multiplying k by Avogadro's number ($A = 6.022 \times 10^{23}$ atoms mol^{-1}) provides R, the gas constant, 8.314 J mol^{-1} and E in units of J mol^{-1} °K^{-1}.

Inorganic Chemistry for Geochemistry and Environmental Sciences: Fundamentals and Applications, First Edition. George W. Luther, III.
© 2016 John Wiley & Sons, Ltd. Published 2016 by John Wiley & Sons, Ltd.
Companion Website: www.wiley.com/go/luther/inorganic

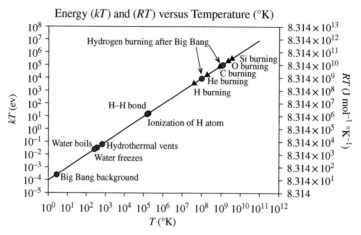

Figure 1.1 *Log–log plot of energy versus temperature; circles include the temperature at which some familiar chemical and physical processes including hydrothermal vents (~360°C) found on deep ocean ridges occur. Triangles indicate the T and E parameters for nucleosynthesis in a sun that has 10–20 times the mass of our sun*

Before continuing with the process of nucleosynthesis, it is necessary to define the general symbol used for nuclides, which includes their nuclear and charged properties, as $_{Z}^{A}El^{x\pm}$ where El is the element symbol, A = atomic mass (total number of protons and neutrons or total **nucleons**), Z = atomic number (number of protons) and $x^{\pm}$ is the charge due to loss or gain of electrons. The difference of $A - Z$ equals the number of neutrons (N). Isotopes of an element have different atomic masses and the same atomic number due to a different number of neutrons in the nucleus.

Immediately after the Big Bang, the buildup of He (and other light elements) from protons and neutrons occurred through several multistep nuclear processes. Equations 1.2–1.5 show one example (positive charges are omitted for simplicity after Equation 1.2). The free neutron has a half-life of 13 min so the formation of hydrogen (Equation 1.2) occurs rapidly with formation of the electron (e^{-} or β^{-}) and one of the neutrinos $v^{(+)}$ (another radiation component; see Table 1.1). The first nuclear reaction in this sequence is between the proton ($_{1}^{1}H^{+}$ or p^{+}) and the neutron to form positively charged deuterium (deuteron, $_{1}^{2}H^{+}$), and the buildup of $_{1}^{2}H^{+}$ eventually leads to positively charged tritium ($_{1}^{3}H^{+}$) and positively charged helium ($_{2}^{3}He^{2+}$) formation. For example, continued reaction of the proton with the deuteron produces the doubly charged $_{2}^{3}He^{2+}$ (Helium-3). Under these extreme temperatures ($\sim 10^{8-9}$ °K), the repulsive forces of the positively charged particles can be overcome so that the charged particles combine in the nucleus, which has a size on the order of 10^{-15} m diameter. (At higher temperatures, the $_{1}^{1}H^{+}$ can decay due to photodissociation.) $_{2}^{3}He^{2+}$ can then combine with another neutron to form $_{2}^{4}He^{2+}$ (also known as the alpha particle; Helium-4) where γ indicates gamma rays that are at the high energy region of the electromagnetic spectrum (Figure 1.2). The energy released for Equation 1.3 and subsequent reactions is substantial, and maintains or increases the initial temperature. Because of these extreme temperatures, the elements were actually in a plasma state.

$$_{0}^{1}n \rightarrow \, _{1}^{1}H^{+} + e^{-} + v^{(+)} \tag{1.2}$$

$$_{1}^{1}H + \, _{0}^{1}n \rightarrow \, _{1}^{2}H + \gamma \tag{1.3}$$

$$_{1}^{2}H + \, _{1}^{1}H \rightarrow \, _{2}^{3}He + \gamma \tag{1.4}$$

$$_{2}^{3}He + \, _{0}^{1}n \rightarrow \, _{2}^{4}He + \gamma \tag{1.5}$$

Table 1.1 *Some important atomic and subatomic particles. One atomic mass unit (amu) equals 1.6606 × 10⁻²⁷ kg (the atomic mass constant) and the elementary charge is 1.602 × 10⁻¹⁹ coulomb (C). Spin is in units of h/(2π) or ħ (h = Planck's constant). β⁻ and β⁺ are ejected from the nucleus with β⁺ formally an antiparticle to β⁻; the reaction of β⁻ with β⁺ leads to their annihilation and release of γ-ray energy. The electron neutrino, v_e or $v⁻$, and the positron neutrino, v_e or $v⁺$ (also known as the antineutrino), account for excess energy release during nuclear reactions (there are two other neutrinos and antineutrinos that are not important to this discussion)*

Symbol	Particle	Mass (amu) [7, 8]	Mass #	Charge	Spin
β⁻ or e⁻	beta particle	5.486×10^{-4}	0	−1	+1/2
β⁺ or e⁺	Positron	5.486×10^{-4}	0	+1	+1/2
$^1_1H^+$	Proton	1.007277	1	+1	+1/2
^1_1H	H atom	1.007825	1	0	+1/2
^1_0n	Neutron	1.008665	1	0	+1/2
^4_2He or α	alpha particle	4.00120	4	+2	0
γ-ray	gamma ray	0	0	0	1
v_e or $v⁻$	Neutrino	$\ll 5.486 \times 10^{-4}$	0	0	+1/2
$\bar{v}_e$ or $v⁺$	Antineutrino	$\ll 5.486 \times 10^{-4}$	0	0	+1/2

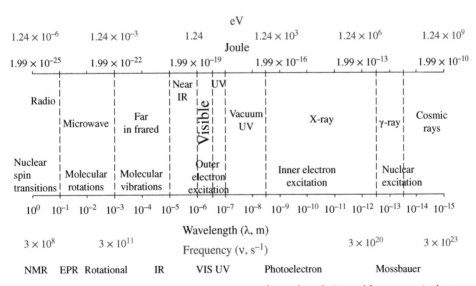

Figure 1.2 *The electromagnetic spectrum is given in terms of wavelength (λ) and frequency (ν, bottom axis) and energy in eV and Joule per atom (top axis; see Equation 3.2; recall c = νλ). 1 eV = 1.602 × 10⁻¹⁹ J. The types of spectroscopic techniques for these energy regions are at the bottom*

1.2 Neutron–Proton Conversion

The proton and the neutron interconvert in atoms to increase nuclear stability via the **weak nuclear force** that operates at 10^{-18} m [4]. Equation 1.2 is an example of spontaneous beta emission (decay) of the neutron. Other examples of neutron conversion to a proton via reaction with a particle and an electron neutrino are given in Equations 1.6a and 1.6b, respectively. The proton is stable to decay so requires another particle

(Equation 1.6c) or energy (antineutrino; Equation 1.6d) to transform it to a neutron.

$$\tfrac{1}{0}n + \beta^+ \rightarrow \tfrac{1}{1}H^+ + \upsilon^{(+)} \tag{1.6a}$$

$$\tfrac{1}{0}n + \upsilon^{(-)} \rightarrow \tfrac{1}{1}H^+ + \beta^- \tag{1.6b}$$

$$\tfrac{1}{1}H^+ + \beta^- \rightarrow \tfrac{1}{0}n + \upsilon^{(-)} \tag{1.6c}$$

$$\tfrac{1}{1}H^+ + \upsilon^{(+)} \rightarrow \tfrac{1}{0}n + \beta^+ \tag{1.6d}$$

1.3   Element Burning Reactions – Buildup of Larger Elements

Further buildup of the elements occurs in the stars and during stellar explosions (supernovae), and the energy released for those reactions above and subsequent reactions is substantial (Section 1.4). The initial process in stellar nuclear synthesis is **hydrogen burning** (our sun is undergoing this process), which is described with the following set of equations (1.7a–1.7d) and which occurs at temperatures of $1 - 3 \times 10^7\,°K$.

$$2(\tfrac{1}{1}H + \tfrac{1}{1}H \rightarrow \tfrac{2}{1}H + \beta^+ + \upsilon^{(-)}) \tag{1.7a}$$

β^+ is the positron (also given as e^+) and $\upsilon^{(+)}$ is one of the neutrinos.

$$2(\tfrac{2}{1}H + \tfrac{1}{1}H \rightarrow \tfrac{3}{2}He + \gamma) \tag{1.7b}$$

$$\tfrac{3}{2}He + \tfrac{3}{2}He \rightarrow \tfrac{4}{2}He + 2\,\tfrac{1}{1}H \tag{1.7c}$$

The sum of Equations 1.7a–1.7c gives the overall nuclear reaction to produce the alpha particle, $\tfrac{4}{2}He$ (Equation 1.7d), and is one of three **proton–proton chain sequences** [6].

$$4\,\tfrac{1}{1}H \rightarrow \tfrac{4}{2}He + 2\,\beta^+ + 2\,\gamma + 2\,\upsilon^{(-)} \tag{1.7d}$$

Once helium is produced, the core of the star continues to collapse and gravitational forces drive the temperature up to $1 - 3 \times 10^8\,°K$ where $\tfrac{4}{2}He$ fuses with itself in a reaction sequence (Equations 1.8a–1.8d) to form $\tfrac{8}{4}Be$ (half-life of 2×10^{-16} s), which rapidly fuses with another $\tfrac{4}{2}He$ to form carbon (the triple α process), which can react further with helium. This sequence is one of the **helium burning** reaction sequences.

$$3\,\tfrac{4}{2}He \rightarrow \tfrac{12}{6}C + \gamma \tag{1.8a}$$

$$\tfrac{12}{6}C + \tfrac{4}{2}He \rightarrow \tfrac{16}{8}O + \gamma \tag{1.8b}$$

$$\tfrac{16}{8}O + \tfrac{4}{2}He \rightarrow \tfrac{20}{10}Ne + \gamma \tag{1.8c}$$

$$\tfrac{20}{10}Ne + \tfrac{4}{2}He \rightarrow \tfrac{24}{12}Mg + \gamma \tag{1.8d}$$

Another hydrogen burning nuclear reaction sequence that occurs is the **carbon–nitrogen–oxygen cycle** where these elements act as catalysts to convert protons into helium as in Equations 1.9a–1.9g.

$$\tfrac{12}{6}C + \tfrac{1}{1}H \rightarrow \tfrac{13}{7}N + \gamma \tag{1.9a}$$

$$\tfrac{13}{7}N \rightarrow \tfrac{13}{6}C + \beta^+ + \upsilon^{(-)} \tag{1.9b}$$

$$\tfrac{13}{6}C + \tfrac{1}{1}H \rightarrow \tfrac{14}{7}N + \gamma \tag{1.9c}$$

$$^{14}_{7}\text{N} + ^{1}_{1}\text{H} \rightarrow ^{15}_{8}\text{O} + \gamma \tag{1.9d}$$

$$^{15}_{8}\text{O} \rightarrow ^{15}_{7}\text{N} + \beta^+ + v^{(-)} \tag{1.9e}$$

$$^{15}_{7}\text{N} + ^{1}_{1}\text{H} \rightarrow ^{12}_{6}\text{C} + ^{4}_{2}\text{He} \tag{1.9f}$$

with the net sum of these reactions being equal to

$$4^{1}_{1}\text{H} \rightarrow ^{4}_{2}\text{He} + 2\,\beta^+ + 3\,\gamma + 2\,v^{(-)} \tag{1.9g}$$

When carbon and oxygen become abundant, carbon and oxygen burning continues the buildup of the heavier elements, for example, Equations 1.10a–1.10d. These **carbon burning** and **oxygen burning** processes occur inside stars that have a mass of 1.4 or greater than the mass of our sun and lead to production of $^{28}_{14}\text{Si}$.

$$^{12}_{6}\text{C} + ^{12}_{6}\text{C} \rightarrow ^{23}_{11}\text{Na} + ^{1}_{1}\text{H} + \gamma \tag{1.10a}$$

$$^{12}_{6}\text{C} + ^{12}_{6}\text{C} \rightarrow ^{23}_{12}\text{Mg} + ^{1}_{0}\text{n} + \gamma \tag{1.10b}$$

$$^{16}_{8}\text{O} + ^{16}_{8}\text{O} \rightarrow ^{28}_{14}\text{Si} + ^{4}_{2}\text{He} + \gamma \tag{1.10c}$$

$$^{16}_{8}\text{O} + ^{16}_{8}\text{O} \rightarrow ^{31}_{15}\text{P} + ^{1}_{1}\text{H} + \gamma \tag{1.10d}$$

Silicon burning is the addition of seven alpha particles starting with $^{28}_{14}\text{Si}$ in a series of reactions similar to Equations 1.8a–1.8d to form the successive elements until $^{56}_{28}\text{Ni}$ is formed. $^{56}_{28}\text{Ni}$ then decays to Co and stable $^{56}_{26}\text{Fe}$ via β^+ decay (Equation 1.11). A **stable nucleus** is defined as one that does not emit particles and/or radiation.

$$^{56}_{28}\text{Ni} \rightarrow ^{56}_{27}\text{Co} + \beta^+ \rightarrow ^{56}_{26}\text{Fe} + \beta^+ \tag{1.11}$$

1.4   Nuclear Stability and Binding Energy

Nuclear stability reaches a maximum at $^{56}_{26}\text{Fe}$ and can be calculated for each nuclide using the Einstein mass–energy relationship of Equation 1.12 where c is the velocity of light ($^{58}_{26}\text{Fe}$ and $^{62}_{28}\text{Ni}$ are slightly more stable than $^{56}_{26}\text{Fe}$ by 2 and 4 keV, respectively, but are not found in abundance on earth). Here, the sum of the individual particle masses of the neutron and proton show a loss in mass compared to the actual atomic mass (after nucleosynthesis); thus, the mass loss can be converted into energy, which is known as the binding energy of the nucleus. Figure 1.3 shows a plot of the binding energy per nucleon versus atomic number. The binding energy increases substantially from the proton to carbon, and then increases gradually to iron.

$$E\,(\text{eV}) = \Delta m\,c^2 \tag{1.12}$$

where $\Delta m = m_{\text{nucleons}} - m_{\text{nucleus}}$ in units of amu. For $c = 2.9979 \times 10^8 \text{ m s}^{-1}$ and the atomic mass constant of a nuclide $= 1.6606 \times 10^{-27}$ kg amu^{-1}, the energy unit is kg m^2 s^{-2} (or Joule) where 1 J $= 6.2415 \times 10^{18}$ eV (or 1 eV $= 1.602189 \times 10^{-19}$ J). Thus, the energy associated with a mass loss of 1 amu (the proton mass) is 1.492×10^{-10} J or 9.315×10^8 eV (0.9315 GeV), and on a molar basis, it is 8.9875×10^{13} J mol^{-1} °K^{-1}. For a mass loss of 5.486×10^{-4} amu (the electron mass), the energy in Joules is 8.187×10^{-14} or 5.11×10^5 eV (0.511 MeV) and on a molar basis, it is 4.93×10^{10} J mol^{-1} °K^{-1}. Frequently, masses for nuclear particles are given in units of energy (eV).

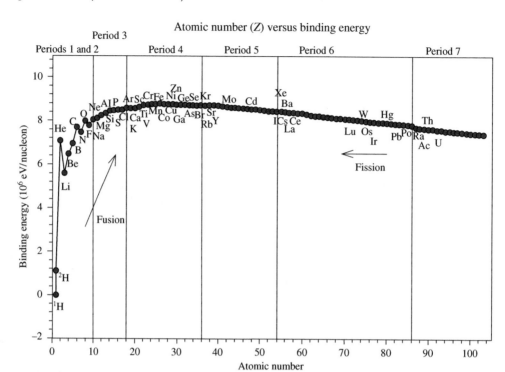

Figure 1.3 *Plot of the binding energy per nucleon for the most stable isotope of each atom [data from 7, 8]. Although the binding energy is normally plotted as a positive entity, the energy released is an exothermic process and has a negative value*

1.4.1   The "r" and "s" Processes

The binding energy decreases with increasing atomic number after iron, and nucleosynthesis of the elements after $^{56}_{26}$Fe becomes more difficult for two highly charged nuclei to overcome the coulombic repulsion to get close enough for fusion to occur. At this point, neutron capture is the primary pathway for nucleosynthesis as the neutron can penetrate a positively charged nucleus thereby increasing the mass of the nucleus. There are two neutron capture processes that result in the production of new elements. First, slow neutron capture or the "**s**" **process** occurs with the addition of one neutron to the nucleus, which is then followed by beta emission (β^-); this occurs in stars with a mass of 0.6–10 times that of the sun and the process terminates at the most stable massive nucleus, $^{209}_{83}$Bi. The "s" process results in an increase in the proton to neutron ratio and a new element, as in the formation of technetium (Equation 1.13), which does not occur as a natural element on earth.

$$^{98}_{42}\text{Mo} + ^{1}_{0}\text{n} \rightarrow ^{99}_{42}\text{Mo} + \gamma \rightarrow ^{99}_{43}\text{Tc} + \beta^- + \upsilon^{(+)} \tag{1.13}$$

The second neutron capture process is rapid neutron capture or the "**r**" **process**, which occurs in environments with high neutron density (e.g., core collapse supernovae; $\sim 10^9\,^\circ$K), so that several neutrons may be captured by a heavy seed nucleus such as $^{56}_{26}$Fe on the order of a second or less before β^- decay occurs, resulting in a new element as in Equation 1.14.

$$^{56}_{26}\text{Fe} + 3^{1}_{0}\text{n} \rightarrow ^{59}_{26}\text{Fe} \rightarrow ^{59}_{27}\text{Co} + \beta^- \tag{1.14}$$

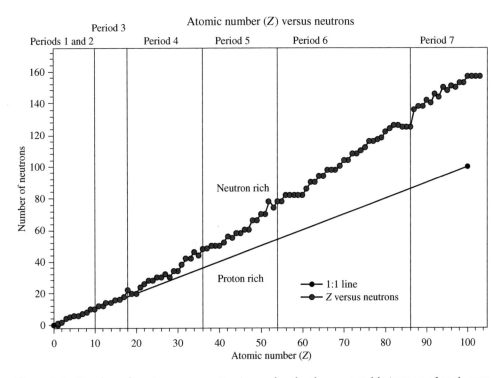

Figure 1.4 *Number of neutrons versus atomic number for the most stable isotope of each atom*

The "r" process occurs at higher temperatures than the "s" process. It is energetic enough to allow for the formation of nuclei larger than $^{209}_{83}\text{Bi}$, and results in a faster increase of atomic mass in the nucleus than the "s" process. The result of both neutron capture processes is an increase in the number of neutrons and protons in the nucleus with increasing atomic number so that the heavier elements have neutron-rich nuclei as shown in Figure 1.4.

For the elements lighter than Ne, there is a tendency for the number of protons to equal the number of neutrons. The study of empirical binding energies and the development of theory for nuclear processes have led to the concept of nuclear shells that are complete when one of the following magic numbers is obtained for the proton or neutron: 2, 8, 20, 28, 50, 82 (also 114 for the proton and 126 and 184 for the neutron; recently 34 has been shown to be a magic number) [9]. When the number of protons or neutrons equals a magic number in the nucleus, the nuclei are highly stable to nuclear decay processes. When both have one of the magic numbers, the nucleus is considered to have "double magic" [10]. Examples of particularly stable elements are $^{4}_{2}\text{He}$, $^{16}_{8}\text{O}$, $^{40}_{20}\text{Ca}$, $^{48}_{28}\text{Ni}$ and, $^{208}_{82}\text{Pb}$. These elements have even numbers of protons (Z), neutrons (N), and thus atomic mass (A). The number of stable nuclides with the four possibilities of even and odd character of Z, N, and A are as follows:

Even Z, even N (even A)   164 nuclei

Even Z, odd N (odd A)   55 nuclei

Odd Z, even N (odd A)   50 nuclei

Odd Z, odd N (even A)   4 nuclei ($^{2}_{1}\text{H}$, $^{6}_{3}\text{Li}$, $^{10}_{5}\text{B}$, $^{14}_{7}\text{N}$)

1.5 Nuclear Stability (Radioactive Decay)

Inspection of Figures 1.3 and 1.4 also indicates that the heavier elements are unstable to radioactive decay; they release alpha particles and beta particles with the formation of stable atoms. Figure 1.5 shows the natural radioactive decay (first-order kinetics) along with the half-life ($t_{1/2}$) for each decay process for three different atomic series; uranium-238, thorium-232, and uranium-235 with the end result being the formation of stable lead isotopes lead-206, lead-208, and lead-207, respectively. The use of the uranium series for geochronology or dating purposes resulted in the determination of the age of the earth (4.55×10^9 year) by Patterson [12] – but not until he was able to build a trace metal clean facility to avoid Pb contamination as lead was used commonly in gasoline, paints, plumbing, pesticides and other uses before and after World War II. Several isotopes in these series are presently used for other dating purposes and to provide reliable information on earth and ocean processes (e.g., Th is used to track the transport of organic carbon from the ocean surface through the water column to the sediments).

Of course neutron capture by some of the heavier elements leads to fission or the splitting of the heavier elements (see Figure 1.3) into two intermediate elements. Fission of uranium also results in significant energy release and produces enough energy to synthesize heavier nuclides (e.g., $^{254}_{98}Cf$).

1.6 Atmospheric Synthesis of Elements

In addition to the formation of the elements in stars and supernovae, which is ongoing, cosmic rays have sufficient energy to induce neutron capture followed by proton emission as in the reaction to form carbon-14 from nitrogen-14 present in earth's atmosphere (Equation 1.15a). This reaction results in a relatively constant supply of carbon-14 ($t_{1/2} = 5730$ year) on the earth. C-14 is also fixed into organic matter along with C-12 and C-13 during photosynthesis, and once an organism dies, the C-14 decays with first-order kinetics and beta particle release (Equation 1.15b) allowing radio-dating of the dead material.

$$^{14}_{7}N + {}^{1}_{0}n + \text{cosmic rays} \rightarrow {}^{14}_{6}C + {}^{1}_{1}H \tag{1.15a}$$

$$^{14}_{6}C \rightarrow {}^{14}_{7}N + \beta^- + \upsilon^{(+)} \tag{1.15b}$$

1.7 Abundance of the Elements

The abundance of the elements is directly related to nucleosynthesis, and this section provides information on their abundance in the cosmos and on earth, then on their abundance in the atmosphere, the oceans, and the human body. Discussion of the transport of the elements between land, atmosphere, rivers, and the oceans couples physical, chemical, and biological processes. The incorporation of the elements into the hard (e.g., bone, shell) and soft (tissue) parts of organisms profoundly impacts their distribution in the environment.

1.7.1 The Cosmos and the Earth's Lithosphere

It should be no surprise that the most abundant elements in the cosmos are hydrogen and helium as shown in Figure 1.6, which shows a general decrease in elemental abundance with atomic number increase. Oxygen, silicon, and iron are also significant in abundance relative to the elements near them. The abundance of the elements on earth (lithosphere only) shows a similar trend (Figure 1.6). The earth consists of three zones, which are the **continental crust** (to about 36 km from the surface), the **mantle** (from 36 to 2900 km), and

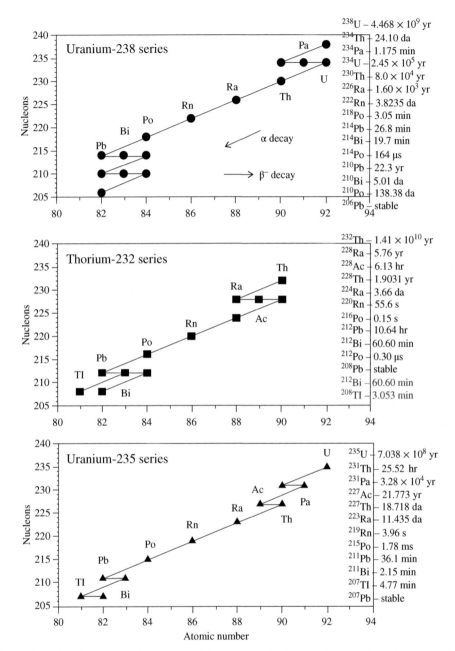

Figure 1.5   *Radioactive decay series for U-238, Th-232, and U-235 to produce stable Pb isotopes. The half-lives ($t_{1/2}$) for each nuclide are given on the right (Source: Data from [11])*

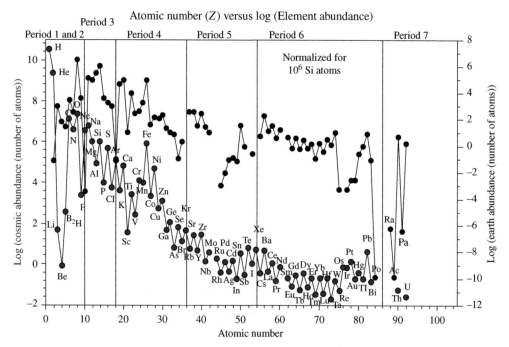

Figure 1.6   *Log plot of the Cosmic and Earth (continental crust or lithosphere) relative elemental abundances , which are normalized to* Si $= 10^6$ *atoms in each case (Source: Data from [13–15])*

the **core** (from 2900 km to the core), and the elements vary in concentration between these three zones or compartments. The crust is the most available for study and for use as a source of the elements and their compounds.

Goldschmidt [16] gave a primary differentiation for the elements on the basis of the types of chemical phases that they could be found in these three zones. He classified the elements into four groups; **siderophilic or iron loving** (e.g., iron, cobalt, nickel), **chalcophilic or copper loving** but principally associated with sulfide (copper, silver, zinc, cadmium, mercury, lead), **lithophilic or stone loving** based on silicate minerals (oxygen, alkaline, and alkaline earth metals) and **atmophilic or vapor loving** (elements mainly as gases such as hydrogen, nitrogen, oxygen, and the noble gases). Obviously, many elements are found in each of the four groups; for example, iron is found in the siderophilic, chalcophilic, and lithophilic groups with the latter groups being considered secondary.

The earth data in Figure 1.6 do not include the composition of water in the oceans and the composition of the elements in the ocean (see Section 1.7.2.1). However, the interaction of the elements and their transport between the atmosphere, hydrosphere, and lithosphere are important to understanding environmental and geochemical processes. The next sections will present information to give context for how the earth functions globally.

1.7.2   Elemental Abundance (Atmosphere, Oceans, and Human Body)

Figure 1.7 shows the major components of the atmosphere, the oceans, and the adult human body. Oxygen is a major component of the atmosphere as a gas (O_2) and as water (H_2O) in the oceans combined with hydrogen. The human body is also mainly composed of water followed by organic carbon. Seventy percent

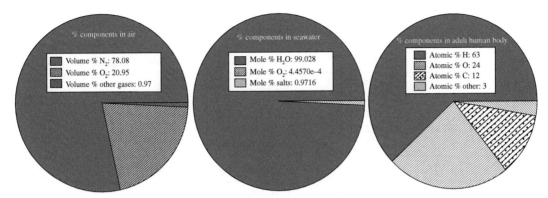

Figure 1.7   *Pie charts showing the major constituents in the atmosphere, seawater, and the human body. Note that O_2 in seawater is imperceptible in the plot*

of the surface of the earth is covered by water in the oceans, which have a mean water depth of 4 km. Water is key to many processes on earth so this text will emphasize reactions that occur in water. There are several **master variables** (ionic strength (I), pH, redox potential, and light to name prominent ones) that are helpful in describing the chemical speciation of the elements in the environment and thus their reactivity. These will be noted throughout the book. Ionic strength varies from freshwater to the oceans to brines, which have 10 times the salinity of seawater and where sodium chloride precipitates. The pH of the environment can range from zero in acid mine drains to near 14 in alkaline lakes. The redox potential (or oxidation–reduction state of a system determined by O_2 as the ultimate oxidant and H_2 as the ultimate reductant; see Sections 2.4 and 2.7) affects the element's oxidation state, which in turn affects other chemical processes. Although the Earth's crust is a dynamic system as a result of plate tectonics, a geochemical interest in this text is the sedimentary system where minerals and nanoparticles form and dissolve. Many of the reactions in the environment occur in the absence of light so it is important to make a distinction between **thermal and photochemical reactions** when discussing chemical reactivity. Photochemistry is important in surface waters as well as in the atmosphere. Using thermodynamics, kinetics, and quantum mechanics, it is possible to explain and predict important environmental reactions.

Freshwater systems include lakes and rivers, which flow into estuaries where salinity varies from essentially zero salinity to full ocean salinity of 35 g dissolved salt per kg of seawater (ppt). The major ions in seawater are given in Table 1.2. The ionic strength of seawater with a salinity of 35 $g\,kg^{-1}$ is 0.7; surface seawater has a pH of 8.1. The major ions in seawater are those that are unreactive or conservative as will be shown in Figure 2.7 for chloride ion in pH $-$ eH plots.

1.7.2.1   Distribution of the Elements in the Water Column of the Ocean

The following discussion describes first the distribution of the elements in the ocean, second the physical transport pathways of the elements into and through the ocean, and then the incorporation of the elements into phytoplankton via photosynthesis at the ocean surface. Lastly, as phytoplankton decay and sink to the sediments, the elements are released back into the water column and in the sediments through the biological pump (see Figure 1.12).

Marine chemists have measured the total concentration of almost every known element in the periodic table from the surface of the ocean to its depths. To measure low or ultra-trace levels (e.g., nanomolar to femtomolar) has required the fabrication of trace metal or element clean facilities as well as instrumentation

Table 1.2   *Major anions and cations in seawater. The last two columns are mass/mass and molar/molar ratios of various ions to chloride ion*

ANIONS	g kg^{-1} (‰)	Molal	Charge (eq. kg^{-1})	Ion/Cl$^-$(g)	Ion/Cl$^-$(M)
Cl$^-$	19.354	0.54591	0.54591	1	1
SO$_4^{2-}$	2.712	0.02825	0.05650	0.1401	0.05175
Br$^-$	0.0673	0.000842	0.000842	0.00384	0.00154
F$^-$	0.0013	0.000068	0.000068	6.7×10^{-5}	1.2×10^{-4}
HCO$_3^-$	0.142	0.00230	0.00230	0.00734	0.00421
CATIONS	**g kg^{-1} (‰)**	**Molal**	**Charge (eq. kg^{-1})**	**Ion/Cl$^-$(g)**	**Ion/Cl$^-$(M)**
Na$^+$	10.770	0.46850	0.46850	0.5565	0.8582
Mg^{2+}	1.290	0.05308	0.10616	0.6665	0.09723
Ca^{2+}	0.4121	0.01028	0.02056	0.02129	0.01883
K$^+$	0.3990	0.01021	0.01021	0.02062	0.01870
Sr^{2+}	0.0079	0.00009	0.00018	4.1×10^{-4}	1.6×10^{-4}

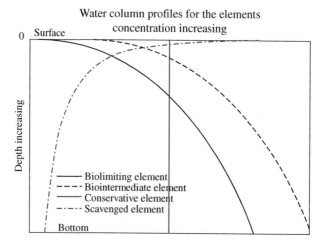

Figure 1.8   *Four types of water column profiles for dissolved species of the elements in the ocean. Dissolved is operationally defined as that material which passes through a 0.4 μm filter*

that can measure these trace levels; here radiochemistry has been an important tool along with advances in electrochemistry and inductively coupled plasma–mass spectrometry (ICP-MS). As a result of this great body of work, chemical oceanographers have produced a classification of the dissolved elements in the ocean with water column profiles (plots of element concentration versus ocean depth; see the GEOTRACES website at http://www.geotraces.org/). Figure 1.8 shows idealized profiles for **conservative**, **recycled** (*biolimiting* and *biointermediate*), and **scavenged** elements.

Conservative elements show no significant change in concentration with depth in the ocean as these are largely unreactive over geologic time. Sodium and chloride ions are the predominant conservative species in seawater. Although there is no significant change in chloride concentration in the ocean, trace amounts of chloride can undergo oxidation to molecular chlorine and eventually form organic chlorine compounds, which can be released to the atmosphere. The recycled elements are those elements that are taken up by phytoplankton into both hard (e.g., bone, shells) and soft (e.g., muscle, tissue, skin) parts. The recycled elements can show a couple of diverse trends: one for biolimiting elements such as hydrogen, phosphorus, silicon, and iron, and the other for biointermediate elements such as calcium. The hard parts are composed of silica or calcium carbonate. Many elements, particularly metals, can be included into the hard part as a co-precipitate or as a **defect** in the crystal structure (Section 6.3.8). In addition to carbon, hydrogen, oxygen, nitrogen, sulfur, and phosphorus, the soft parts contain many of the metals used in enzyme systems such as nitrite reductase that contains iron. Recycled elements show a low concentration in the surface of the ocean due to uptake by algae (phytoplankton) and an increase with depth in the ocean, which is related to the decay of phytoplankton as they die and fall to the bottom of the ocean. The ratio of a recycled element's concentration in the deep Pacific Ocean to its concentration in the deep Atlantic Ocean is greater than one and is related to the mixing time of the ocean. As an example, Si as silicate increases in concentration with time and depth as diatoms sink to the deep ocean and dissolve. Because cold ocean water low in Si content forms in the North Atlantic Ocean and travels south around Antarctica and then into the Indian and Pacific oceans (the **ocean conveyor belt** is a useful first order approximation of ocean water movement; see http://oceanservice.noaa.gov/facts/conveyor.html and http://www.mbari.org/chemsensor/pteo.htm), these older waters have more dissolved silicate. Elements that are under the scavenged classification show a high concentration in the surface ocean and a decreasing concentration with depth. Manganese, aluminum, and lead are among the scavenged elements. Manganese is used in photosystem center II and is in the soft tissue of the organisms. There is more manganese in the surface than other metals that are used in enzyme systems, but manganese is in excess to what the organisms actually need so is not biolimiting. The ratio of a scavenged element's concentration in the deep Pacific Ocean to its concentration in the deep Atlantic Ocean is less than one.

1.7.2.2 Residence Time of the Elements in the Ocean

Figure 1.9a shows an idealized plot of the concentrations versus mean oceanic residence time of many of the elements in the ocean. The oceanic residence time can be calculated using Equation 1.16:

$$\text{Residence time } (\tau) = \frac{[X]\text{ocean} * \text{Ocean volume}}{[X]\text{river} * \text{Annual river flux}} \tag{1.16}$$

where $[X]$ is the concentration of the element concerned, and the ocean volume is 1.37×10^{21} L and the annual river flux is 3.6×10^{16} L per year. The major elements ($>10^{-4}$ M) have long residence times that approach 10% of the age of the earth. The minor elements (10^{-4}–10^{-6} M), including the macronutrients nitrogen and phosphorus, have intermediate residence times. The trace elements (10^{-6} M–10^{-9} M), which include many of the transition metals used in enzyme systems, have residence times less than 1000 years, and finally the ultra-trace elements ($<10^{-9}$ M) have the smallest residence times in the ocean. The residence time of an element is a function of that element's chemical reactivity, and the first hydrolysis constant (pK_H, Equation 1.17, Section 7.3) of the "free" metal ion $[M(H_2O)_x]^{y+}$ is a reasonable measure as shown in Figure 1.9b. As the tendency for an element to form a hydroxo species increases and thus becoming more reactive, the residence time of the element decreases. Figure 1.9a and b shows that the concentration of Fe, the pK_H of Fe(III), and the residence time for Fe are low (small), indicating that oceanic Fe chemistry is governed predominantly by Fe(III) and not Fe(II) (note the 10^8 difference between Fe(III) and Fe(II) in pK_H). This observation of reactivity is true for many other elements that can exist in higher oxidation states (e.g., Ti, Mn, Hg, Th).

$$[\text{Fe}(H_2O)_6]^{3+} \rightarrow [\text{Fe}(H_2O)_5(OH)]^{2+} + H^+ \tag{1.17}$$

The residence time for particles in **sediments** is similar to that for Na^+ and Cl^-. The chemical residence time of a water molecule in the ocean is greater than 10,000 years whereas the physical oceanic mixing or stirring time is 1000 years. Oceanic mixing is described by the conveyor belt model noted above.

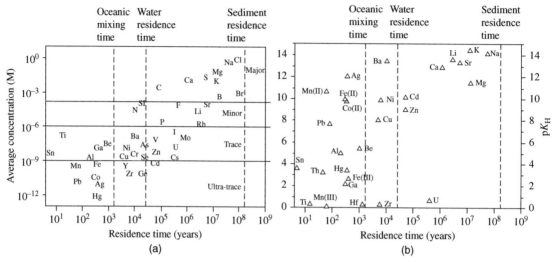

Figure 1.9 (a) The average concentration. (b) First hydrolysis constant, pK_H, versus residence time of many important elements in the ocean

1.7.2.3   Box Model Describing Element Flux to the Ocean and its Sediments

Figure 1.10 shows a six box model where the atmosphere and the rivers provide input of the elements to the surface layer of the ocean. The **weathering of the continents** through the action of rain and oxygen brings elements primarily in their dissolved form into the rivers and then into the estuary (some elements precipitate and get trapped in the estuary as shown in Figure 1.11) and finally in the ocean. The surface layer is dominated by the solubility of the elements and the photosynthesis/respiration cycle, as the elements partition between dissolved and particulate phases. In the center of the ocean, **the atmosphere** provides transport of the elements in dust particles and nanoparticles, which are deposited in the surface ocean by dry and wet deposition (rain) and which are solubilized to some extent so that organisms can uptake nutrients and metals (Si, Fe, Mn, etc.). Particles from dust that are not solubilized and particles from organisms formed in the surface layer settle into the deep layer where they can dissolve and release elements as shown in the classification of elements (Figure 1.8). Eventually particles not dissolved in the deep layer reach the sediment–water interface where they undergo continued dissolution and oxidation. Particles not dissolved at the sediment–water interface become part of the sediments and add to the net sedimentation rate. Oceanic waters also undergo downwelling and upwelling, which allow for dissolved elemental species to be transported between boxes. Upwelling of dissolved biolimiting and biointermediate elements (Figure 1.8) into the surface ocean via storm activity enhances primary productivity, much as river and atmospheric transport of these elements to surface waters does. Hydrothermal vents and seeps are important advective flux components exporting material (dissolved, nanoparticulate, and particulate states) from the crust to the lower ocean. This box model can also be applied to lakes.

1.7.2.4   Elemental Distribution at the Land–Ocean Boundary (Estuaries and Coasts)

Although the atmosphere is the major pathway to transport elements to the center of the surface ocean, the rivers are a major entry for many elements to the coastal ocean. The rivers mix with the coastal ocean waters via

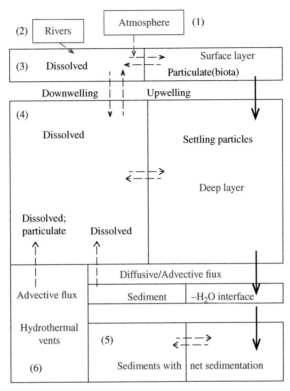

Figure 1.10   *Six box model of the ocean including the atmosphere and rivers as inputs into the surface ocean. Hydrothermal vents and sediment flux provide input to the deep ocean. Numbers indicate the six boxes*

the estuary system, and mixing can result in a homogeneous distribution of chemical species (such as sodium and chloride ions with depth at a given location) or in vertical stratification, which shows an increase in the distribution of chemical species with depth at a given location. Figure 1.11 shows three major classifications or types of estuaries, which are determined by the ratio of tidal flow from seawater to river flow [17]. The salt wedge estuary (Figure 1.11a) is an extreme case of stratification at a given location where the concentration of an element is constant over some surface depth and then increases dramatically with depth as seawater intrudes from below. The Bosporus Strait, which enters into the Black Sea, is an excellent example of this classification, and the ratio of tidal to river flow is ≤1. The partially mixed estuary (Figure 1.11b) shows a regular increase in concentration of an element with depth at a given location; that is, there is some stratification with depth, and the ratio of tidal to river flow ranges from 10 to 1000. The northern Chesapeake Bay above the Potomac River is an example of this type of system. The vertically homogeneous estuary (Figure 1.11c) shows no change in concentration of an element with depth at a given location; that is, there is no stratification with depth, and the ratio of tidal flow to river flow is >1000. The Delaware Bay is an example of a homogeneous estuary in the absence of severe storms that results in freshwater runoff from rain (during storms the freshwater remains on the surface and does not mix readily with the seawater coming in with the tides).

Figure 1.11d shows an example of a property–property plot for the mixing of a conservative element at high concentration and low salinity (terrestrial source) with an oceanic water mass that has a low concentration

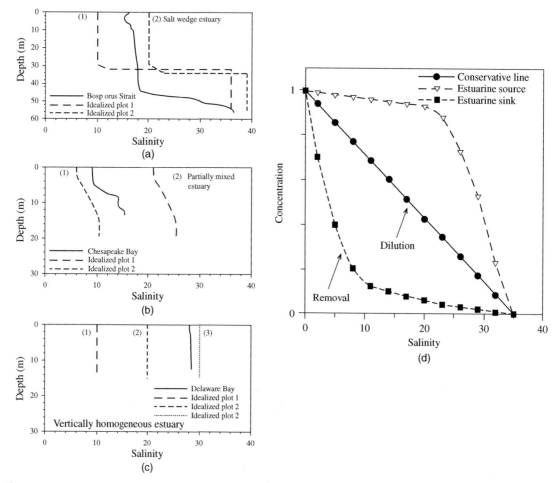

Figure 1.11 *(a–c) Three classifications for the types of estuaries based upon the mixing of river water with oceanic water. In each figure, the solid line shows an actual profile of salinity with depth whereas the dashed lines show idealized plots for each classification. (d) Idealized property–property plot of a property (e.g., concentration of a chemical species) versus salinity*

of the element. The straight line shows dilution or conservative behavior on mixing of freshwater with ocean water. For the vertically homogeneous Delaware Bay (Figure 1.11c), silicate shows conservative behavior. The curved dashed line (triangles, concave downward) above the straight line is an example of an element that shows estuarine source behavior. For the Delaware Bay, phosphate shows source behavior due to sediment resuspension and dissolution. The curved dashed line (squares, concave upward) below the straight line is an example of an element that shows removal behavior (also termed an estuarine sink). In the Delaware Bay, iron, manganese, cobalt, and cadmium show removal behavior whereas copper, nickel, and nitrate show a gradual removal. For iron, oxidation of Fe(II) to Fe(III) results in precipitation of iron oxyhydroxides, which can adsorb many of the other metals, to the sediment. An element with an oceanic source (Na, Cl) would have an inverse plot.

1.8   Scope of Inorganic Chemistry in Geochemistry and the Environment

The above geochemical discussion describes the distribution of the elements and how they are affected primarily by physical forcing. As noted, the atmosphere and the rivers provide the elements to the surface of the ocean far from land and to the coastal ocean, respectively. For example, the atmosphere provides dust particles including sand (SiO_2) and aluminosilicates from deserts that are rich in metals such as Al and Si and that contain trace elements such as P, Fe, Mn, Zn, and Cu. These particles are partially soluble in seawater, and the elements released can be taken up by phytoplankton during photosynthesis for various biochemical and structural functions. Table 1.3 shows some elements essential for life.

1.8.1   Elemental Distribution Based on Photosynthesis and Chemosynthesis

Environmental inorganic chemistry can be broadly broken into two subject areas, that which is natural and that which is anthropogenic. Many natural processes on earth (often referred to as **(bio)geochemistry**) are driven by the **carbon cycle** via **photosynthesis** that requires macronutrients (such as nitrate and phosphate) and micronutrients (such as the transition metals), which are used in a host of enzyme systems. In photosynthesis, the transformation of carbon dioxide into organic matter requires light to initiate electron and hydrogen ion transfer to eventually produce phytoplankton and sea grasses in the ocean and grasses and forests on land. A molecular understanding of CO_2 (Sections 5.7.2 and 7.7.3.1) is important as it is a poor electron acceptor, which affects its dissolution from the atmosphere to surface waters (known as the solubility pump in Figure 1.12) and its reactivity with other molecules (e.g., Section 12.6.3).

During photosynthesis, the elements are combined in surface waters to form organic matter in the form of phytoplankton. Figure 1.12 shows the biological pump for lakes and the ocean. Here, carbon dioxide, the major nutrients (nitrate, phosphate, and N_2), and trace elements used for enzymes along with the major elements, Ca and Si, used to form the hard parts of organisms are taken up by phytoplankton with the production and release of oxygen to the water column and the atmosphere. Much of the phytoplankton organic carbon produced becomes part of the ocean food web, and some organic compounds are released as soluble entities to the water column. Prior to the 1970s, transition metals dissolved in the oceanic water column were considered

Table 1.3   *Selected elements and their uses in biochemistry and bioinorganic chemistry. The first five are considered major and minor elements as shown in Figure 1.9*

Element	Some biological uses
Nitrogen	Protein and nucleic acid synthesis
Phosphorus	Nucleic acids, teeth, bones, shells
Sulfur	Protein synthesis, cell division, electron transport
Silicon	Diatom hard parts (SiO_2)
Calcium	Foraminifera shells, coral, bones, teeth, electrical conduction
Iron	Electron transport, N_2 fixation, O_2 activation, storage and transport
Copper	Electron transport, O_2 activation, storage and transport
Manganese	Photosystem center II – the O_2 evolving complex (OEC), oxidative stress
Zinc	Carbonic anhydrase, nucleic acid regulation
Vanadium	Haloperoxidases, N_2 fixation
Molybdenum	N_2 fixation

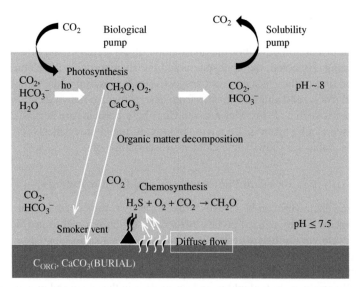

Figure 1.12   *The biological pump (also white downward arrows) shows carbon transfer from the ocean surface to its sediments. Burial of organic carbon and calcium carbonate, which are produced via photosynthesis, occurs in sediments. Emissions from hydrothermal vents release metals and other chemicals to the deep ocean*

to exist as only inorganic compounds such as chloride (halide), hydroxide, and carbonate complexes before they precipitated as oxides and carbonates. In reducing sediments, many metals then reacted with hydrogen sulfide to form metal sulfides. Now, most transition metals are known to complex with organic chelates so that greater than 99% of the metal is organically complexed [18]; this includes metals known to hydrolyze and precipitate as oxides or hydroxides such as Fe(III).

On death of the phytoplankton, the soft or fleshy parts are decomposed by oxygen back to bicarbonate and carbon dioxide, and this process is known as **respiration**. Additional chemical constituents in organic compounds are also released during decomposition (Figure 1.8) and are available for reaction with other compounds. Although much respiration occurs in the water column, the hard parts of dead phytoplankton (silica and calcium carbonate) have a density greater than that of seawater, which permits them to sink to the sediments at the bottom of the ocean where further organic matter decomposition occurs. The entire process of organic matter production in the surface and its transport to the sediments is known as **the biological pump**.

Chemosynthesis occurs in the dark as inorganic chemical reactions (e.g., the oxidation of H_2S with O_2) mediated by chemolithic autotrophic microbes fuel the energy, hydrogen ions, and electrons needed to fix carbon dioxide into organic matter. [**Chemolithic** means stone or inorganic chemicals that microbes use to fix CO_2 (**Autotrophic** process)] Chemosynthesis is found primarily near hydrothermal vents on ocean ridges; here, H_2S, H_2, CO, and other reduced materials are released from the crust to the ocean.

The decomposition of dead organic matter (C_{ORG}) occurs with a variety of oxidizing agents and leads to the partial regeneration of carbon dioxide and bicarbonate (HCO_3^-). However, a significant amount of organic matter results in the production and burial of oil and natural gas, which are used as raw materials to produce a wealth of materials beneficial to society. The transformation of natural products from photosynthesis and chemosynthesis frequently occurs with the use of inorganic reagents, many of which are catalysts. Chemical reactions that are performed in the laboratory can be used to gain understanding about reaction kinetics and pathways that occur in nature. For example, the Fischer–Tropsch reaction (Section 9.13.4), which is a catalytic process for organic compound formation from simple molecules with metals as catalysts, is now being studied as a possible reaction that occurs in the chimneys of hydrothermal vents [19]. Thus, it is important to demonstrate the synergism between laboratory and field studies. In order to better understand environmental chemistry and geochemistry, researchers can benefit from an understanding of the principles of physical inorganic chemistry, which can explain and predict fundamental molecular interactions.

1.8.2 Stratified Waters and Sediments – the Degradation of Organic Matter by Alternate Electron Acceptors

On death, the soft parts from phytoplankton undergo oxidation via bacterial mediation. The oxidation of organic matter both in the water column and sediments is represented by the following seven equations (1.18–1.23) where $C_{106}H_{263}O_{110}N_{16}P$ is the Redfield ratio for carbon, hydrogen, oxygen, nitrogen, and phosphorus [20, 21]. Here, amino acids, peptides, proteins, DNA, and RNA can be decomposed with the release of NH_4^+ and phosphate to the water column as shown in Figure 1.8 for biolimiting and biointermediate elements. NH_4^+ is oxidized to NO_3^- by O_2. The first oxidant used to decompose organic matter is oxygen, which provides the maximum free energy for the oxidation of organic carbon. Once oxygen is consumed, nitrate is used as an oxidant followed by manganese dioxide and then iron(III) oxyhydroxides and sulfate. Methanogenesis finally occurs when no other natural oxidants are available. For all these reactions, phosphate is a product; however, the nitrogen product changes from nitrate (Equation 1.18) to N_2 (Equations 1.19–1.21) and ammonium ion (Equations 1.22–1.24) as shown in Section 12.7.

OXYGEN REDUCTION/RESPIRATION (reverse of photosynthesis)

$$138O_2 + C_{106}H_{263}O_{110}N_{16}P \rightarrow 106HCO_3^- + 16NO_3^- + HPO_4^{2-} + 16H_2O + 124H^+$$

$$\Delta G = -3190(kJ\,mol^{-1}) \tag{1.18}$$

NITRATE REDUCTION

$$94.4NO_3^- + C_{106}H_{263}O_{110}N_{16}P \rightarrow 106HCO_3^- + 55.2N_2 + HPO_4^{2-} + 71.2H_2O + 13.6H^+$$

$$\Delta G = -3030(kJ\,mol^{-1}) \tag{1.19}$$

IODATE REDUCTION

$$78.7IO_3^- + C_{106}H_{263}O_{110}N_{16}P \rightarrow 106HCO_3^- + 8N_2 + HPO_4^{2-} + 78.7I^- + 24H_2O + 108H^+$$

$$\Delta G = -2868(kJ\,mol^{-1}) \tag{1.20}$$

MnO$_2$ REDUCTION

$$236MnO_2 + C_{106}H_{263}O_{110}N_{16}P + 364H^+ \rightarrow 8N_2 + 106HCO_3^- + HPO_4^{2-} + 236Mn^{2+} + 636H_2O$$

$$\Delta G = -3090(kJ\,mol^{-1}) \tag{1.21}$$

FeOOH REDUCTION

$$424FeOOH + C_{106}H_{263}O_{110}N_{16}P + 756H^+ \rightarrow 106HCO_3^- + 16NH_4^+ + HPO_4^{2-} + 424Fe^{2+} + 636H_2O$$

$$\Delta G = -1330(kJ\,mol^{-1}) \tag{1.22}$$

SULFATE REDUCTION

$$53SO_4^{2-} + C_{106}H_{263}O_{110}N_{16}P \rightarrow 106HCO_3^- + 16NH_4^+ + HPO_4^{2-} + 53HS^- + 36H^+$$

$$\Delta G = -380(kJ\,mol^{-1}) \tag{1.23}$$

METHANOGENESIS

$$C_{106}H_{263}O_{110}N_{16}P + 53H_2O \rightarrow 53HCO_3^- + 53CH_4 + 16NH_4^+ + HPO_4^{2-} + 39H^+$$

$$\Delta G = -350(kJ\,mol^{-1}) \tag{1.24}$$

Figure 1.13a provides a schematic representation of pore water vertical profiles for dissolved oxygen, nitrate, manganese(II), iron(II), and hydrogen sulfide as determined by microelectrode profiling [21] along with the profile for metal oxide phases. Vertically stratified or stagnant water columns such as the Black Sea and stratified lakes, which are maintained by physical forcing, also give similar chemical profiles. Although the oceanic water column is expected to be oxygenated (**oxic**), there are several large ocean basins (e.g., the Arabian Sea and the tropical Eastern Pacific Ocean off Peru) that have significant consumption of O_2 below the **photic** (or light) penetration zone for 1000 m of a 4000 m water column. This O_2 consumption zone is called the oxygen minimum zone (OMZ) as O_2 concentrations are as low as 10 nM [22]. Here, nitrogen processes and reactions become more important with release of N_2 and NH_4^+ and consumption of NO_3^-. The initial increase in nitrate in these zones is due to O_2 oxidizing the NH_4^+ from peptides in organic matter. When both O_2 and H_2S are not detectable in the OMZ, this zone is often termed **suboxic,** indicating that other elements are more important in maintaining the redox state of the zone. When O_2 is absent and H_2S is present, the zone is termed **anoxic**. The flow of chemical species and/or electrons at the interfaces of these zones allows for catalytic cycles between the elements that facilitate important chemical transformations that have had a major impact on earth processes. These catalytic cycles are composed of many reactions, some of which are purely inorganic (or **abiotic**) chemical reactions that occur in conjunction or in competition with **biotic** reactions mediated by microbes, which can use the free energy from the reactions to grow and multiply. For example, some iron oxidizing bacteria (FeOB) live at the interface where there is a decreasing oxygen gradient from above and a decreasing Fe^{2+} gradient from below as in Figure 1.13b. At the interface of these opposing gradients, the chemical kinetics of the reaction between Fe^{2+} and O_2 are slow due to the low concentration (ranging from nanomolar to a few micromolar) of each reactant. These FeOB facilitate and can use the energy of the reaction between Fe^{2+} and O_2 at low concentrations to fix carbon dioxide into organic matter. The major waste byproduct is the formation of Fe(III) minerals or solids; some researchers believe FeOB were major contributors to some of the banded iron formations (BIFs) that formed over 2 billion years ago [23]. Figure 1.13c shows an extreme mixing pattern of reduced vent fluids with oxygenated seawater right above the vent orifice. A range of chemical reactions occurs including oxidation and precipitation of reduced metals to form metal (oxy)hydroxides. Similar to vertical mixing in Figure 1.13c, horizontal mixing of Fe^{2+} and O_2 also occurs when reduced ground (or seep) waters from aquifers are ejected or mixed with coastal oxygenated waters, for example, in the beach zone and on the continental shelf (submarine groundwater discharge, SGD) [24].

These organic matter decomposition reactions as well as the secondary abiotic and biotic reactions (e.g., Fe^{2+} oxidation) are exceptionally important for geochemistry and environmental chemistry. Many of the element transformations are multielectron transfer processes, which undergo one or two electron transfers per reaction. In particular, O_2 frequently undergoes four separate one-electron reductions to water (forming **reactive oxygen species or ROS** as intermediates), manganese dioxide undergoes two-electron reduction to manganese(II) [sometimes via Mn(III)], sulfate undergoes eight-electron reduction to hydrogen sulfide (forming a host of S intermediate oxidation states including elemental sulfur, polysulfides, thiosulfate, and

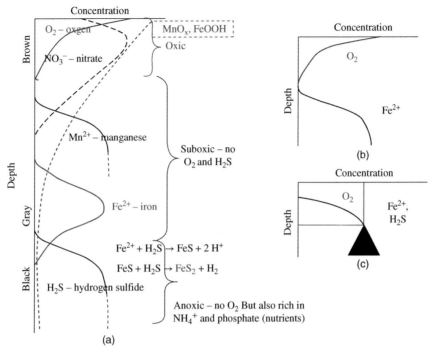

Figure 1.13   *(a) Idealized vertical porewater profiles (concentration versus depth) of major dissolved redox chemical species in sediments and stagnant water basins based on Equations 1.18 through 1.24. The colors on the left indicate the metal oxide (brown), pyrite (gray), and FeS zones (black). (b) Overlap of O_2 and Fe^{2+} in lakes. (c) Mixing of reduced Fe^{2+} and H_2S from vents with oxygenated ocean waters. All figures represent one-dimensional profiles*

sulfite before forming sulfate) and reduced nitrogen in peptides undergoes an eight-electron transfer to nitrate (forming several intermediate oxidation states) as reduced carbon is oxidized to carbon dioxide. During these multielectron reactions, many of the formed intermediates provide unique oxidation–reduction chemical species or redox properties that cannot be readily reproduced in the laboratory. In many instances, these reactions are facilitated by microbial enzymes that are actively being investigated by bioinorganic chemists, geochemists, and environmental chemists.

1.9   Summary

This chapter showed that the distribution of elements in the cosmos and on earth is first a function of physical processes. The synthesis of the elements occurs via high-energy nuclear processes. The broad distribution of the elements on earth is a result of geophysical, hydrological, and atmospheric processes. As the elements undergo dynamic physical transport, chemical transformations occur. The emphasis has been on natural biogeochemical processes; however, humans have manufactured a variety of chemicals that have entered the atmosphere and the hydrosphere. Some have resulted in harmful effects as in the case of halocarbon emissions resulting in depletion of ozone in the stratosphere. Chemicals entering the atmosphere are transported rather quickly to all regions of the earth whereas chemicals entering rivers and other local waterways are transported primarily to the coastal ocean. Finally chemicals also percolate through surface soils and enter aquifers potentially contaminating ground waters.

1.9.1   Environmental Inorganic Chemistry

Some general questions and items of relevance in studying (geo)chemical and environmental chemical reactions include the following:

1. Why is a particular reaction kinetically slow although thermodynamically favorable?
2. What are the exact chemical species of two elements that lead to a chemical reaction?
3. How do master variables such as pH, ionic strength, and redox state influence how chemical speciation varies in the environment?

   Some of the specific examples of geochemical and environmental relevance, which focus on electron transfer reactions in the environment, include the following:

1. Kinetic slowness of H_2S oxidation by O_2. Why then is O_2 a "weak" oxidant that requires microbial intervention to speed up the reaction?
2. Biogeochemical reactions including activation of small molecules (H_2, O_2, CO, NO, N_2, I_2, H_2O, H_2S) with metals as catalysts; synthesis of organic molecules, and the origins of life (e.g., hydrothermal vents).
3. Manganese(II) and iron(II) oxidation by O_2 (implications regarding banded iron formations and oxygen transport carriers in organisms).
4. Mn(III, IV) and Fe(III) mineral dissolution by H_2S and organic ligands (siderophores and other ligands solubilize and stabilize these higher oxidation states in solution).
5. Mechanisms of pyrite, FeS_2, oxidation [Fe(III) versus O_2 reactivity].
6. Pyrite formation via different sulfur reactants (FeS with polysulfides or H_2S).
7. Metal complexation and centers in enzymes, metal uptake by organisms (thermodynamics and kinetics of metal acquisition by organisms).
8. Metalloenzymes capable of oxidizing CH_4 and NH_4^+ and of reducing N_2 and CO_2.
9. Properties of manganese in the oxygen-evolving complex found in photosystem center II.

References

1. Boesgard, A. M. and Steigman, G. (1985) Big Bang Nucleosynthesis: theories and observations. *Annual Review of Astronomy and Astrophysics*, **23**, 319–378.
2. Choppin, G., Rydberg, J., Liljenzin, J.-O., and Ekberg, C. (2013) *Radiochemistry and Nuclear Chemistry* (4th ed.), Academic Press.
3. Hix, W. R. and Thielemann, F.-K. (1999) Computational methods for nucleosynthesis and nuclear energy generation. *Journal of Computational and Applied Mathematics*, **109**, 321–351.
4. Olive, K. A., Steigman, G. and Walker, T. P. (2000) Primordial nucleosynthesis: theory and observations. *Physics Reports*, **333–334**, 389–407.
5. Steigman, G. (2007) Primordial nucleosynthesis in the precision cosmology era. *Annual Review of Nuclear and Particle Science*, **57**, 463–491.
6. Truran, J. W. (1984). Nucleosynthesis. *Annual Review of Nuclear and Particle Science*, **34**, 53–97.

7. Audi, G., Wapstra, A. H. and Thibault, C. (2003) The AME2003 atomic mass evaluation (II). Tables, graphs and references. *Nuclear Physics A*, **729**, 337–676.

8. Wapstra, A. H., Audi, G. and Thibault, C. (2003) The AME2003 atomic mass evaluation (I). Evaluation of input data, adjustment procedures. *Nuclear Physics A*, **729**, 129–336.

9. Steppenbeck, D. and others. (2013) Evidence for a new nuclear 'magic number' from the level structure of ^{54}Ca. *Nature*, **502**, 207–210.

10. Janssens, R. V. F. (2009) Unexpected doubly magic nucleus. *Nature*, **459**, 1069–1070.

11. Friedlander, G., Kennedy, J. W., Macias, E. S. and Miller, J. M. (1981) *Nuclear and Radiochemistry* (3rd ed.), John Wiley and Sons, NY.

12. Patterson, C. (1956) Age of meteorites and the earth. *Geochimica Cosmochimica Acta*, **10**, 230–237.

13. Cameron, A. G. W. (1973) Abundances of the elements in the solar system. *Space Science Reviews*, **15**, 121–146.

14. Urey, H. C. (1964) A review of atomic abundances in chondrites and the origin of meteorites. *Reviews of Geophysics*, **2**, 1–34.

15. Anders, E. and Grevesse, N. (1989) Abundances of the elements: Meteoritic and solar. *Geochimica Cosmochimica Acta*, **53**, 197–214.

16. Goldschmidt, V. M. (1954) *Geochemistry* (ed A. Muir) Clarendon Press, Oxford. pp. 730.

17. Knauss, J. A. (1997) *Introduction to Physical Oceanography* (2nd ed.), Waveland Press, Inc., Long Grove, IL.

18. Gledhill, M. and Buck, K. N. (2012) The organic complexation of iron in the marine environment: a review. *Frontiers in Microbiology*, 3, article 69. doi: 10.3389/fmicb.2012.00069.

19. McCollum, T. M. and Seewald, J. S. (2007) Abiotic synthesis of organic compounds in deep-sea hydrothermal environments. *Chemical Reviews*, **107**, 382–401.

20. Froelich, P. N., Klinkhammer, G. P., Bender, M. L., Luedtke, N. A., Heath, G. R., Cullen, D. et al. (1979) Early oxidation of organic matter in pelagic sediments of the eastern equatorial Atlantic: Suboxic diagenesis. *Geochimica Cosmochimica Acta*, **43**, 1075–1090.

21. Luther, G. W., III, Sundby, B., Lewis, B. L., Brendel, P. J. and Silverberg, N. (1997) Interactions of manganese with the nitrogen cycle: alternative pathways to dinitrogen. *Geochimica Cosmochimica Acta*, **61**, 4043–4052.

22. Revsbech, N. P., Larsen, L. H., Gundersen, J., Dalsgaard, T., Ulloa, O. and Thamdrup, B. (2009) Determination of ultra-low oxygen concentrations in oxygen minimum zones by the STOX sensor. *Limnology and Oceanography Methods*, **7**, 371–381.

23. Koehler, I., Konhauser, K. and Kappler, A. (2010) Role of microorganisms in banded iron formations, In *Geomicrobiology: Molecular and Environmental Perspective* (eds L. L. Barton, M. Mandl and A. Loy). Springer, Dordrecht, pp. 309–324.

24. Swarzenski, P. W. (2007) U/Th radionuclides as coastal groundwater tracers. *Chemical Reviews*, **107**, 663–674.

2

Oxidation–Reduction Reactions (Redox)

2.1 Introduction

A countless number of metal and non-metal reactions occur that undergo electron transfer. **Oxidation** involves the loss of electrons, and **reduction** involves the gain of electrons. The **oxidation state** of an element in a compound or ion is positive when electrons are lost and negative when electrons are gained relative to the element in its standard state, which has an oxidation state or number of zero. In a large number of compounds, the oxidation state of hydrogen is typically +1 (when bound to metals it is normally −1) and that of oxygen is −2.

2.1.1 Energetics of Half Reactions

In a complete redox reaction, there are two half-reactions where one chemical species must accept electrons (the oxidizing reagent, oxidant, or electron acceptor) and another chemical species must donate the electrons (the reducing agent, reductant, or electron donor). Redox reactions can occur between two chemical species in the same physical state (solution) or in different physical states (e.g., solid FeOOH reacting with aqueous hydrogen sulfide, Section 11.2). There are many important redox reactions that are exploited for their production of energy in voltaic or galvanic cells (battery technology). Equation 2.1 shows the relationship between ΔG_r^0 (the Gibbs free energy), E^0 (the standard reduction potential), and K (the equilibrium constant) of a redox reaction. T is the temperature in Kelvin, and standard state conditions are 25 °C (298.16 °K) and 1 atm with all reactants at unit activity (molarity is often used). Positive E^0 values indicate a **spontaneous** or thermodynamically favorable reaction (**exergonic** reaction) whereas negative E^0 values indicate an unfavorable reaction (**endergonic** reaction). Equation 2.1 is applicable to half-reactions and full reactions. Although ΔG_r^0 values for half-reactions are additive, the number of electrons (n) must be considered when adding E^0 values (see **Frost diagrams** below).

$$\Delta G_r^0 = -nFE^0 = -RT \ln K \qquad (2.1)$$

Figure 2.1 shows experimental setups for the determination of E^0 for a redox reaction and for half-reactions. Figure 2.1a shows that Zn metal (oxidation state of 0) loses two electrons to dissolve and form Zn^{2+} (oxidation state of +2) as Cu^{2+} in solution gains two electrons and is deposited on the original Cu electrode, which grows

Inorganic Chemistry for Geochemistry and Environmental Sciences: Fundamentals and Applications, First Edition. George W. Luther, III.
© 2016 John Wiley & Sons, Ltd. Published 2016 by John Wiley & Sons, Ltd.
Companion Website: www.wiley.com/go/luther/inorganic

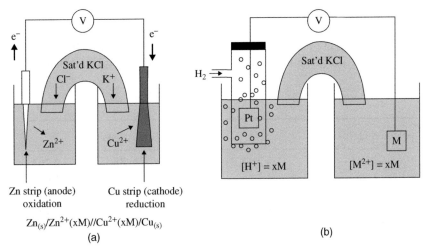

Figure 2.1 *(a) Galvanic cell with a KCl salt bridge for the internal circuit and a voltmeter (V). (b) Measurement of potential for half-reactions using the SHE. At standard state, all concentrations of chemical species are at unit activity (~1 atm or bar; 1 M)*

in size. The shorthand notation for metals in aqueous solution is given as Zn^{2+} or M^{x+}, which is a simplification of the free metal ion that has water molecules attached to it, $[Zn(H_2O)_6]^{2+}$ or $[M(H_2O)_6]^{x+}$ (see Section 7.3). When the circuit is closed, the reaction produces a voltage of +1.100 V. These two half-reactions and the overall reaction are written as

$$Zn_s \rightarrow Zn^{2+} + 2e^- \quad \text{plus} \quad Cu^{2+} + 2e^- \rightarrow Cu_s \quad \text{equals} \quad Cu^{2+} + Zn_s \rightarrow Cu_s + Zn^{2+}$$

The voltage for each half-reaction is determined with a standard hydrogen electrode (SHE), which is defined by convention with a voltage of 0.000 V.

$$H^+_{(aq)} + e^- \rightarrow \frac{1}{2} H_{2(g)}$$

From the SHE voltage, the standard reduction potential for the reduction half-reactions of the metal ions is determined as in Figure 2.1b; the values and notation for the Cu and Zn reduction couples are as follows:

$$Cu^{2+} + 2e^- \rightarrow Cu_s \quad E^0_{red}(Cu^{2+}, Cu) = +0.337 \text{ V}; E^0_{ox} = -0.337 \text{ V}$$

$$Zn^{2+} + 2e^- \rightarrow Zn_s \quad E^0_{red}(Zn^{2+}, Zn) = -0.763 \text{ V}; E^0_{ox} = +0.763 \text{ V}$$

The half-reaction with the largest positive value is the **cathode** where reduction occurs (e.g., Cu^{2+}, the oxidized species, is reduced) and the least positive value is the **anode** where oxidation occurs (Zn, the reduced species, is oxidized). Zn metal is commonly used as a sacrificial anode attached to the hulls of boats. To determine a spontaneous redox reaction, the reduction potential of the anode is subtracted from the reduction potential of the cathode as follows (similar to adding the E^0 for the reduction reaction to the E^0 for the oxidation reaction).

$$E^0_{cell} = E^0_{cat} - E^0_{an} = E^0_{red} + E^0_{ox} = +0.337 - (-0.763) = +1.100 \text{ V}$$

From these potential measurements, the electrochemical series of the elements and of their compounds is produced. Oxidants have a positive standard reduction potential, and reductants have a negative standard reduction potential. The following half-reactions and standard potentials at 25 °C demonstrate the series. Standard potentials are typically given in 1 M acid (pH = 0) or 1 M base (pH = 14) at unity activity of all chemical species.

Selected Oxidants	V	Selected Reductants	V
$F_2(g) + 2e^- \rightarrow 2F^-$	+2.87	$Zn^{2+} + 2e^- \rightarrow Zn_s$	−0.763
$Cl_2(g) + 2e^- \rightarrow 2Cl^-$	+1.36	$Al^{3+} + 3e^- \rightarrow Al_s$	−1.68
$Br_2(l) + 2e^- \rightarrow 2Br^-$	+1.065	$Ca^{2+} + 2e^- \rightarrow Ca_s$	−2.84
$I_2(s) + 2e^- \rightarrow 2I^-$	+0.535	$Li^+ + e^- \rightarrow Li_s$	−3.04

2.1.2   Standard Potential and the Stability of a Chemical Species of an Element

The standard potential for each chemical species of an element can be used to predict stability of species in solution. Here, the standard reduction potentials for the eight equations representing six manganese species below in 1 M acid are used as the example.

$$Mn(H_2O)_6^{2+} + 2e^- \rightarrow Mn + 6H_2O \qquad E^0 = -1.18 \text{ V}$$

$$Mn(H_2O)_6^{3+} + e^- \rightarrow Mn(H_2O)_6^{2+} \qquad E^0 = +1.50 \text{ V}$$

$$MnO_2 + 4H^+ + e^- + 6H_2O \rightarrow Mn(H_2O)_6^{3+} + 2H_2O \qquad E^0 = +0.95 \text{ V}$$

$$MnO_4^{3-} + 4H^+ + e^- \rightarrow MnO_2 + 2H_2O \qquad E^0 = +4.27 \text{ V}$$

$$MnO_4^{2-} + e^- \rightarrow MnO_4^{3-} \qquad E^0 = +0.27 \text{ V}$$

$$MnO_4^- + e^- \rightarrow MnO_4^{2-} \qquad E^0 = +0.56 \text{ V}$$

$$MnO_2 + 4H^+ + 2e^- + 6H_2O \rightarrow Mn(H_2O)_6^{2+} + 2H_2O \qquad E^0 = +1.23 \text{ V}$$

$$MnO_4^- + 4H^+ + 3e^- \rightarrow MnO_2 + 2H_2O \qquad E^0 = +1.70 \text{ V}$$

$$MnO_4^- + 8H^+ + 5e^- + 6H_2O \rightarrow Mn(H_2O)_6^{2+} + 4H_2O \qquad E^0 = +1.51 \text{ V}$$

2.1.2.1   Latimer Diagrams

These electrochemical reactions can be summarized in a **Latimer** diagram (Figure 2.2), which shows the reduction potentials of the above reactions by indicating the most oxidized species on the left and the most reduced on the right of the diagram while omitting the H_2O, H^+, and OH^- needed to balance the equations. Summing the three reduction potentials for the one electron transfers from MnO_4^- to MnO_2 $(0.56 + 0.27 + 4.27)$ gives a value of $E^0 = 5.1$; likewise, the three electron reduction of MnO_4^- to MnO_2 gives a value of $nE^0 = (3(+1.70)) = 5.1$. This is an example of **Hess's law**, which indicates that the total energy change during the course of a chemical reaction is the same whether the reaction progresses in one step or several steps; i.e., the energy change is independent of the path taken from the initial to the final state. This is a statement of the principle of the conservation of energy.

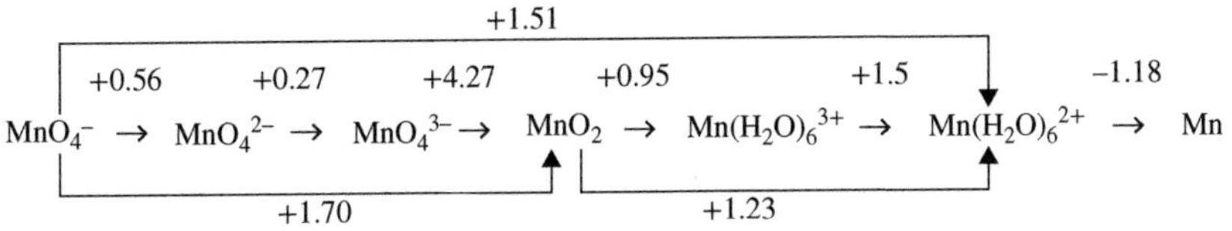

$$\text{MnO}_4^- \xrightarrow{+0.56} \text{MnO}_4^{2-} \xrightarrow{+0.27} \text{MnO}_4^{3-} \xrightarrow{+4.27} \text{MnO}_2 \xrightarrow{+0.95} \text{Mn(H}_2\text{O})_6^{3+} \xrightarrow{+1.5} \text{Mn(H}_2\text{O})_6^{2+} \xrightarrow{-1.18} \text{Mn}$$

with spanning values: +1.51, +1.70, +1.23

Figure 2.2 *Latimer diagram for Mn species at pH = 0. E^0 data from [1] (Source: Data from [1])*

$$\text{Ag}^{2+} \xrightarrow{+1.989} \text{Ag}^+ \xrightarrow{+0.799} \text{Ag}_{(s)}$$

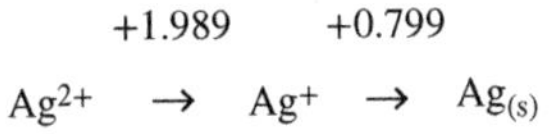

Figure 2.3 *Latimer diagram for Ag species at pH = 0. E^0 data from [1] (Source: Data from [1])*

A species tends to **disproportionate** into its two neighbors if the standard reduction potential on the right is greater than that on the left of the Latimer diagram as in the case of $\text{Mn(H}_2\text{O})_6^{3+}$.

$$2\text{Mn(III)} \rightarrow \text{MnO}_2 + \text{Mn(II)} \quad \text{or} \quad 2\text{Mn(III)} \rightarrow \text{Mn(IV)} + \text{Mn(II)}$$

In this case, the reduction of $\text{Mn(H}_2\text{O})_6^{3+}$ to $\text{Mn(H}_2\text{O})_6^{2+}$ has $E^0{}_{right} = +1.50$ V whereas the reduction of MnO_2 to $\text{Mn(H}_2\text{O})_6^{3+}$ has $E^0{}_{left} = +0.95$ V. The difference between these two one-electron step reactions is thermodynamically favorable.

$$E^0{}_{right} - E^0{}_{left} = 1.50 \text{ V} - (0.95 \text{ V}) = +0.55 \text{ V}$$

where the negative sign in front of $E^0{}_{left}$ indicates an oxidation reaction for $\text{Mn(H}_2\text{O})_6^{3+}$.

The reverse of disproportionation is **comproportionation** where two species of the same element with different oxidation states form a product that has an intermediate oxidation state.

The comproportionation reaction of Ag^{2+} with $\text{Ag}_{(s)}$ leads to formation of Ag^+.

$$\text{Ag}^{2+} + \text{Ag}_{(s)} \rightarrow 2\text{Ag}^+$$

The relevant reactions and standard potentials are given below and Figure 2.3 (above) is the Latimer diagram.

$$\text{Ag}^{2+} + e^- \rightarrow \text{Ag}^+ \quad E^0{}_{left} = +1.989 \text{ V}; \quad \text{Ag}^+ + e^- \rightarrow \text{Ag}_{(s)} \quad E^0{}_{right} = +0.799$$

$$E^0{}_{comp} = E^0{}_{left} - E^0{}_{right} = +1.989 - (+0.799) = +1.190 \text{ V}$$

Comproportionation reactions can also occur in multielectron transfer reactions. The reaction of MnO_4^- with *excess* $\text{Mn(H}_2\text{O})_6^{2+}$ and H^+ forms $\text{Mn(H}_2\text{O})_6^{3+}$, which can disproportionate over time as above. However, the stoichiometric reaction of MnO_4^- with $\text{Mn(H}_2\text{O})_6^{2+}$ forms stable MnO_2; in this case, $E^0{}_{left} > E^0{}_{right}$ in the Latimer diagram for the Mn species above.

2.1.2.2 *Frost Diagrams*

A Frost diagram uses the same equations and data as in the Latimer diagram but explicitly accounts for the number of electrons transferred between each chemical couple for an element based on Equation 2.1. The plot is nE^0 (or $-\Delta G_r^0 / F$) in volts versus the oxidation state of the element and is formally a plot of the Gibbs free energy of the different redox species versus their oxidation states. An element in its standard state has a

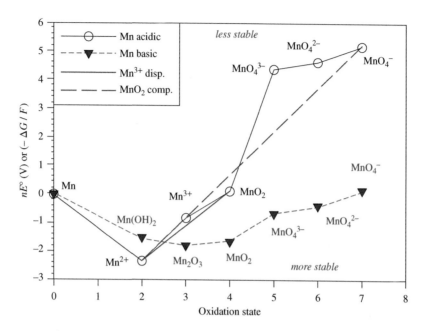

Figure 2.4   *Frost diagrams for Mn species at* pH = 0 *and* pH = 14

value of $nE^0 = 0$ by definition as $\Delta G_f^0 = 0$. The acidic Mn half-reactions above yield the black line in the Frost diagram (Figure 2.4). The plot starts with Mn(0) with a value of $nE^0 = 0$; then Mn^{2+} with nE^0 value of -2.36 from $[(2*(-1.18)) + (0)]$; then Mn^{3+} with nE^0 value of -0.86 from $[(1*(+1.50)) + (-2.36)]$, and so on. There are five consecutive one-electron transfer half-reactions from MnO_4^- to Mn^{2+}.

The slopes of the lines give an indication of the direction of a reaction. For Mn(0) oxidation to Mn^{2+} in acidic solution, there is a negative slope, which indicates that Mn should be a reducing agent and should be oxidized. For couples with positive slopes, the redox species with the highest nE^0 value should be readily reduced. In a Frost diagram, stability of an oxidation state increases from top to bottom as in Figure 2.4, which shows that Mn^{2+} is the most stable species under acidic conditions. A chemical species that tends to disproportionate will have a nE^0 value above the line between its two nearest neighbors as shown for Mn^{3+} above the dark blue line between Mn^{2+} and MnO_2. A chemical species that will form due to comproportionation will have an nE^0 value below the line between its two nearest neighbors (this is the case for Ag^+). The dashed blue line in Figure 2.4 shows the comproportionation reaction to form MnO_2 from the stoichiometric reaction of MnO_4^- with Mn^{2+}. Inspection of Figure 2.4 (and Figure 2.5) shows that high oxidation state redox species (dashed blue line Figure 2.4) are also more stable in basic solution than in acidic solution indicating that the central atom can undergo oxidation in base. At pH = 14, $Mn(OH)_2$, Mn_2O_3, and MnO_2 are the most stable species.

Figure 2.5 shows the Frost diagram for chlorine species (most half-reactions are two electron transfers). In this example, the most stable redox species is the chloride ion under both acidic and basic conditions. Thus, the formation of any other chlorine species under natural conditions should lead to their reduction and eventual formation of Cl^-, which is the dominant form in nature over all pH. In Chapter 1 (Section 1.7.1), Cl^- was shown to be a conservative species in seawater. The dashed and solid blue lines indicate that $HClO_2(ClO_2^-)$ will disproportionate to HClO or HOCl (OCl^-) and ClO_3^-; this is an example of forming a higher oxidation state chemical species without the need for an additional oxidant.

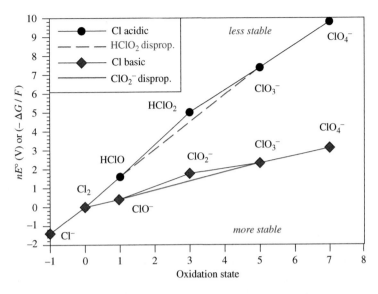

Figure 2.5 *Frost diagrams for Cl species at* pH = 0 *and* pH = 14

2.2 Variation of Standard Potential with pH (the Nernst Equation)

Reactants and products are not normally at unit activity so the voltage of a redox reaction (Equation 2.2) is dependent on the concentration of each species. The voltage is then expressed using the **Nernst** equation (Equation 2.3) where Q is the reaction quotient (Equation 2.4) (or the equilibrium constant K at equilibrium). At the standard state, the $\ln Q$ term in Equation 2.3 yields the simplified log expression in Equation 2.4. The numbers 1 and 2 indicate the chemical species of elements 1 and 2, whereas the letters indicate the stoichiometry of the reaction. For solids, the activity is defined as 1; however, this may not be true for nanoparticles.

$$a\mathrm{Ox}_1 + b\mathrm{Red}_2 \rightarrow c\mathrm{Red}_1 + d\mathrm{Ox}_2 \tag{2.2}$$

$$E = E^0 - \frac{RT}{nF} \ln Q = E^0 - \frac{0.059}{n} \log Q \tag{2.3}$$

$$Q = \frac{[\mathrm{Red}]_1{}^c [\mathrm{Ox}]_2{}^d}{[\mathrm{Ox}]_1{}^a [\mathrm{Red}]_2{}^b} \tag{2.4}$$

For an initial concentration of Zn^{2+} of 1 M and Cu^{2+} of 0.1 M, Q equals 10; thus, $E = +1.100 - 0.030 = +1.070$ V. The change in voltage across a membrane in a pH electrode is the basis for determining the H^+ concentration or activity of the solution.

2.3 Thermodynamic Calculations and pH Dependence

An excellent way to demonstrate the thermodynamics and redox state of a chemical species or chemical environment over pH is to use Equation 2.3 in the following manner to calculate the E_h (the potential of a chemical system) and the negative log of the electron activity $\{\mathrm{e}^-\}$ (Equation 2.5) where the brackets { }

indicate the activity of a species. Calculations are performed for aqueous species

$$pe = -\log\{e^-\} \tag{2.5}$$

over the pH range of 1–14, which covers the environmental span of acid mine drainage areas to photosynthetically active and alkali lake areas. Thermodynamic data in $kJ\,mol^{-1}$ at 25 °C and 1 atm are from three main sources [1–3; Appendix 2.1]. The value used for the Gibbs free energy for $Fe^{2+}(-90.53\ kJ\,mol^{-1})$ is that discussed in [4]. The basic mathematical approach has been fully developed in standard textbooks and publications [5, 6]. First, a reduction half-reaction for each redox couple is written as for the case of aqueous oxygen reduction to water in Equation 2.6a. From the known Gibbs free energies or standard redox potentials, $pe°$ $(= \log K)$ is calculated at the standard state conditions for each half-reaction (Equations 2.6b and 2.6c), which is normalized to a one-electron reaction (e.g., Equation 2.6a becomes Equation 2.6d).

$$\begin{array}{cccc} O_{2(aq)} + 4H^+ + 4e^- \rightarrow & 2H_2O \\ 16.32 \quad 0 \quad 0 & 2(-237.18) \quad \Delta G^0_f \text{ in } (kJ\,mol^{-1}) \end{array} \tag{2.6a}$$

The standard state ΔG^0_f for the reaction = -490.68 $kJ/2$ mol H_2O or 4 mol of electrons. The quotient now referred to as the equilibrium constant (K^0_4) is given in Equation 2.6b where $\{\}$ indicates activity for each chemical species. *Log* $K^0_4 = \Delta G/[(-2.303)RT] = 86.00$ for 4 mol of electrons or 21.50 for 1 mol of electrons where

$$K^0_4 = \frac{\{H_2O\}^2}{\{O_2\}\,\{H^+\}^4\,\{e^-\}^4} \quad \text{so } \log K^0_4 = 86.00 = \log \frac{\{H_2O\}^2}{\{O_2\}\,\{H^+\}^4\,\{e^-\}^4} \tag{2.6b}$$

or

$$\log K^0_4 = -\log\{O_{2(aq)}\} - \log\{H^+\}^4 - \log\{e^-\}^4 \tag{2.6c}$$

For an one-electron reaction, Equation 2.6a becomes Equation 2.6d.

$$\tfrac{1}{4} O_{2(aq)} + H^+ + e^- \rightarrow \tfrac{1}{2} H_2O \tag{2.6d}$$

Similarly, Equation 2.6c then becomes Equation 2.6e, which is rearranged to Equation 2.6f.

$$\tfrac{1}{4}\log K^0_4 = -\tfrac{1}{4}\log\{O_{2(aq)}\} - \log\{H^+\} - \log\{e^-\} = -\tfrac{1}{4}\log\{O_{2(aq)}\} + pH + pe \tag{2.6e}$$

$$pe = \tfrac{1}{4}\log K^0_4 + \tfrac{1}{4}\log\{O_{2(aq)}\} - pH \tag{2.6f}$$

From the Nernst equation, $pe° = \tfrac{1}{4}\log K^0_4 = 21.50$ (standard state value) and on substituting into Equation 2.6f becomes Equation 2.6g.

$$pe = pe° + \tfrac{1}{4}\log\{O_{2\,(aq)}\} - pH = 21.50 + \tfrac{1}{4}\log\{O_{2\,(aq)}\} - pH \tag{2.6g}$$

At 250 μM O_2 (250×10^{-6} M; 100% air saturation at ~ 25 °C), this expression becomes Equation 2.6h.

$$pe = 21.50 + \tfrac{1}{4}\log\{250 \times 10^{-6}\ M\} - pH = 20.60 - pH \tag{2.6h}$$

At 1 μM O_2 (10^{-6} M), this expression becomes Equation 2.6i.

$$pe = 21.50 + \tfrac{1}{4}\log\{10^{-6}\ M\} - pH = 20.00 - pH \tag{2.6i}$$

At unit activity for all reagents $p\varepsilon = p\varepsilon°$. At unit activity of all reagents other than the H^+, the approximate pH dependence is calculated for all reactions. In the $O_{2\,(aq)}$ example, Equation 2.6g becomes Equation 2.6j, which will be used for all aqueous solution calculations unless stated otherwise.

$$p\varepsilon = p\varepsilon° - pH = 21.50 - pH \qquad (2.6j)$$

The same process can be performed for the H_2O/H_2 gas couple using Equation 2.7a or 2.7b and assuming the activity of hydrogen or $\{H_2\}$ is 1. The final result is Equation 2.8.

$$2H_2O + 2e^- \rightarrow H_2 + 2OH^- \qquad (2.7a)$$

$$H^+_{(aq)} + e^- \rightarrow \tfrac{1}{2}H_{2(g)} \qquad (2.7b)$$

$$p\varepsilon = -pH \qquad (2.8)$$

2.4 Stability Field of Aqueous Chemical Species

Plots of $p\varepsilon$ versus pH (Figure 2.6 or the **Pourbaix** diagram) for Equations 2.6j and 2.8 show the redox stability field of H_2O between the two diagonal lines. E_h and $p\varepsilon$ are related through Equations 2.1 and 2.3 to give Equation 2.9 ($n = 1$ for H^+ reduction). E_h is plotted on the right of Figure 2.6.

$$(0.059)p\varepsilon = E_h \qquad (2.9)$$

In a similar manner, these calculations can be used to show what ions and compounds of the elements are stable within the stability field of water. Thus, the two **master variables**, pH and $p\varepsilon$ (E_h), together can predict thermodynamic stability in the natural environment. In Chapter 1 (Section 1.7.1), Cl^- and other simple ions were shown to have conservative behavior in seawater. Using the equations in Figure 2.7, the $p\varepsilon°$ versus pH diagram for four chlorine species (Cl^-, Cl_2, $HOCl$, OCl^-) in water shows that only Cl^- is predicted to exist in water under standard state conditions. The calculation assumes that O_2 is the strongest or only

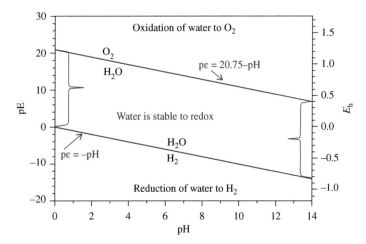

Figure 2.6 *The $p\varepsilon°$ versus pH diagram for water stability. The solid blue lines are calculated from Equations 2.6j and 2.8 and represent the stability field of water*

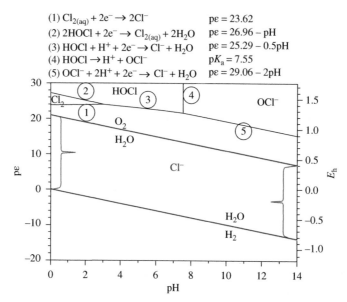

(1) $Cl_{2(aq)} + 2e^- \rightarrow 2Cl^-$ $p\varepsilon = 23.62$
(2) $2HOCl + 2e^- \rightarrow Cl_{2(aq)} + 2H_2O$ $p\varepsilon = 26.96 - pH$
(3) $HOCl + H^+ + 2e^- \rightarrow Cl^- + H_2O$ $p\varepsilon = 25.29 - 0.5pH$
(4) $HOCl \rightarrow H^+ + OCl^-$ $pK_a = 7.55$
(5) $OCl^- + 2H^+ + 2e^- \rightarrow Cl^- + H_2O$ $p\varepsilon = 29.06 - 2pH$

Figure 2.7 *Pourbaix ($p\varepsilon°$ versus pH) diagram for low oxidation state chlorine species. The area between the blue solid lines represents the stability field of water. The numbers indicate the lines calculated from the five chlorine reactions*

oxidant in nature. For stronger oxidants such as H_2O_2, O_3, and the OH radical (•OH), which are at nanomolar concentrations (or transient species) in the environment, the blue stability lines would shift and HOCl and OCl^- could exist (see Figure 2.10). Although chlorate can form from disproportionation of ClO_2^-, chlorate and perchlorate are more oxidized species of chlorine; thus, they are prepared via stronger oxidants and/or by electrolysis, photochemical, electrical discharge, or radiolysis.

2.5 Natural Environments

Figure 2.8 shows where several natural waters exist within the stability field of water. These are approximate as they can vary pending the O_2 concentration found in the environment. For example, several regions of the ocean have oxygen minimum zones (OMZs) as discussed in Chapter 1 (Section 1.8.2) so that the stability field of ocean water can extend from pH 7.5 to 8.1, and from pE of fully oxygenated to non-detectable O_2 (note the dashed vertical lines in Figure 2.8).

2.6 Calculations to Predict Favorable Chemical Reactions

A pe(pH) for a half-reaction at a given pH can be calculated for any chemical species and some common ones of environmental relevance are presented in Table 2.1. These reactions are multistep reactions normalized to one electron so that the thermodynamic favorability of a reaction can be calculated (see Equation 2.1). The calculated value for each half-reaction is given as a function of pH in the lower part of Table 2.1, and these can easily be entered into spreadsheets for quick calculations of full reactions from two half-reactions

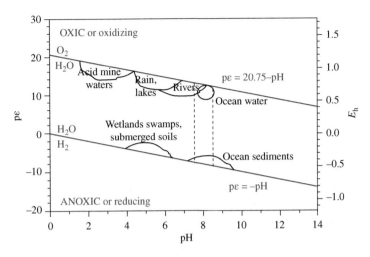

Figure 2.8 *Selected natural waters within the stability field of water*

Table 2.1 *Important half-reactions for Fe^{2+}, Mn^{2+}, O_2, NH_4^+, and NO_3^- calculated as in the text. The $p\varepsilon^\circ$ (log K^0) is the standard state condition (activity = 1 in all species including H^+). Multiplying the calculated $p\varepsilon$ by 0.059 V gives the reduction potential, E, at 25 °C*

Selected half-reactions and $p\varepsilon(pH) = p\varepsilon^\circ(\log K^\circ) - n\,pH$

(A)	$1/4 O_{2(aq)} + H^+ + e^- \rightarrow 1/2 H_2O$	$p\varepsilon = p\varepsilon^\circ - pH = 21.50 - pH$
(B)	$1/5 NO_3^- + 6/5 H^+ + e^- \rightarrow 1/10 N_2 + 3/5 H_2O$	$p\varepsilon = p\varepsilon^\circ - 1.2\,pH = 21.05 - 1.2\,pH$
(C)	$1/8 NO_3^- + 5/4 H^+ + e^- \rightarrow 1/8 NH_4^+ + 3/8 H_2O$	$p\varepsilon = p\varepsilon^\circ - 1.25\,pH = 14.9 - 1.25\,pH$
(D)	$1/6 N_2 + 4/3 H^+ + e^- \rightarrow 1/3 NH_4^+$	$p\varepsilon = p\varepsilon^\circ - 1.33\,pH = 4.65 - 1.33\,pH$
(E)	$1/2 MnO_{2(s)} + 2H^+ + e^- \rightarrow 1/2 Mn^{2+} + H_2O$	$p\varepsilon = p\varepsilon^\circ - 2\,pH = 20.8 - 2\,pH$
(F)	$FeOOH_{(s)} + 3H^+ + e^- \rightarrow Fe^2 + 2H_2O$	$p\varepsilon = p\varepsilon^\circ - 3\,pH = 13.37 - 3\,pH$

$p\varepsilon(pH) = \text{Log } K(pH)$ for half-reactions A through F at the selected pH values

pH	$p\varepsilon[A]$	$p\varepsilon[B]$	$p\varepsilon[C]$	$p\varepsilon[D]$	$p\varepsilon[E]$	$p\varepsilon[F]$
0	21.50	21.05	14.9	4.65	20.8	13.37
1	20.50	19.85	13.65	3.32	18.8	10.37
2	19.50	18.65	12.4	1.98	16.8	7.37
3	18.50	17.45	11.15	0.65	14.8	4.37
4	17.50	16.25	9.90	−0.68	12.8	1.37
5	16.50	15.05	8.65	−2.02	10.8	−1.63
6	15.50	13.85	7.4	−3.35	8.8	−4.63
7	14.50	12.65	6.15	−4.7	6.8	−7.63
8	13.50	11.45	4.9	−6.02	4.8	−10.63
9	12.50	10.25	3.65	−7.35	2.8	−13.63
10	11.50	9.05	2.4	−6.7	0.8	−16.63

(see next section). When H^+ or OH^- is not in a balanced equation for a half-reaction, there is no pH dependence on the half-reaction. The pε calculated is termed pε(pH), which provides log K for each half-reaction at a given pH. Concentration dependence for the other reactants is not considered in the calculation; thus, these are considered standard state calculations. Note that Equations 2.6h–2.6j show a 1.50 log change for an O_2 concentration range from 1 μM to unity activity so the calculations could vary an order of magnitude or more in either direction when concentration dependence is included. To test that a reaction is thermodynamically favorable, calculations are performed by combining different half-reactions at a given pH.

2.6.1   Coupling Half-Reactions

Using the data from Table 2.1, the reaction of O_2 (Equation A) with Mn^{2+} (Equation F) to form H_2O and MnO_2 at pH = 7 is used as an example of coupling two half-reactions to determine whether a reaction is favorable. In this case, log K [for the complete reaction] = pε(O_2) – pε(Mn^{2+}) = 14.50 – (6.8) = 7.70. The negative sign in front of pε(Mn^{2+}) indicates that the half-reaction is now an oxidation half-reaction. This calculation is for the complete reduction of O_2 to water (a four-electron process).

Figure 2.9a shows the oxidation of Fe^{2+} and Mn^{2+} with O_2 to form various Fe(II,III) and Mn(II, III,IV) (oxy)hydroxide solids over the pH range 0–14. From this analysis, the reactions are predicted to occur over all

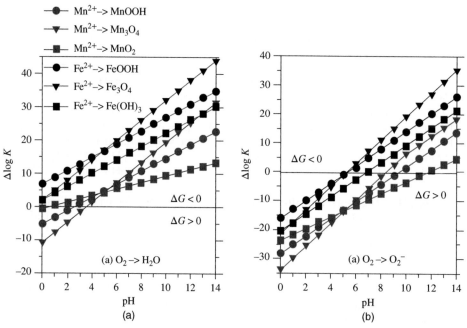

Figure 2.9   *(a) Thermodynamic calculations for the multielectron reactions of Fe^{2+} and Mn^{2+} with O_2 to form H_2O from Table 2.2 (Reaction O1 coupled with reactions Mn1–Mn3 and Fe1–Fe3). Positive values for Δlog K on the y-axis indicate a favorable forward reaction. A negative value for Δ log K indicates an unfavorable forward reaction; however, the reverse reaction is favorable. (b) Thermodynamic calculations for the one-electron transfer reactions of Fe^{2+} and Mn^{2+} with O_2 to form O_2^-. Reaction O6 coupled with reactions Mn1–Mn3 and Fe1–Fe3 (Source: From [5], Figures 1A, 4A. Reproduced with permission from Springer)*

pH for Fe^{2+} and over most pH values for Mn^{2+}. However, this analysis is in contrast to the known reactivity and kinetics of these reactions that indicate that Mn^{2+} is mainly unreactive below a pH of 9, and Fe^{2+} is unreactive below a pH of 5 [3; see Section 10.4.1].

To make better predictions of reactivity, calculations for the thermodynamics of one-electron or two-electron transfers need to be made as most chemical reactions proceed in one or two electron steps at most [7]. Table 2.2 gives a list of the pε(pH) equations for O, Mn, Fe, Cu, N, S, N, and halogen species for relevant one- and two-electron transfer reactions [5, 6] to be considered throughout this text. For dissolved Fe^{2+} and Mn^{2+}, only the $M(H_2O)_6^{2+}$ species are provided for simplicity. The hydroxo (or hydrolysis) complexes $M(OH)(H_2O)_5^+$ or $M(OH)_2(H_2O)_4^0$ start to become important as pH increases.

2.6.2 One-Electron Oxygen Transformations with Fe^{2+} and Mn^{2+} to Form O_2^-

Stumm and Morgan [3] noted that the one-electron reduction of O_2 by Fe^{2+} leads to the reactive oxygen species, superoxide (O_2^-), which is unstable (see reaction O6 in Table 2.2). In their analysis, they used the iron species Fe^{2+} at pH = 1, $Fe(OH)^+$ at pH = 4, and $Fe(OH)_2$ at pH = 7. Their data showed that the first electron transfer from Fe^{2+} to O_2 is a key step and controls the reaction progress and rate at a pH > 5.

Figure 2.9b shows that the same predictions are made over the pH range from 1 to 14 but without specifying the Fe(II) and Mn(II) hydroxo species. In fact, Figure 2.9b shows that the one-electron oxidation of Fe^{2+} by

Table 2.2 *Reduction equations for relevant O, Mn, Fe, Cu, N, S, C, and halogen reactions normalized to one electron – all soluble species are aqueous forms. Activities of all reactants (except H^+) are unity. Multiplying the calculated pε by 0.059 V gives the reduction potential E, at 25 °C and the desired pH*

OXYGEN AND HYDROGEN REACTIONS

Four-electron reaction normalized to one electron

$1/4 O_{2(aq)} + H^+ + e^- \rightarrow 1/2 H_2O$	$p\varepsilon = p\varepsilon^\circ - pH = 21.50 - pH$	(O1)

Two-electron reactions normalized to one electron

$1/2 O_{2(aq)} + H^+ + e^- \rightarrow 1/2 H_2O_2$	$p\varepsilon = p\varepsilon^\circ - pH = 13.18 - pH$	(O2)
$1/2 H_2O_2 + H^+ + e^- \rightarrow 1 H_2O$	$p\varepsilon = p\varepsilon^\circ - pH = 29.82 - pH$	(O3)
$1/2 O_3 + H^+ + e^- \rightarrow 1/2 O_{2(aq)} + 1/2 H_2O$	$p\varepsilon = p\varepsilon^\circ - pH = 34.64 - pH$	(O4)
$1/2\,^1O_{2(aq)} + H^+ + e^- \rightarrow 1/2 H_2O_2$	$p\varepsilon = p\varepsilon^\circ - pH = 21.57 - pH$	(O5)
$H^+ + e^- \rightarrow 1/2 H_2$	$p\varepsilon = p\varepsilon^\circ - pH = 0 - pH = -pH$	(H1)

One-electron transfer reactions only

$O_{2(aq)} + e^- \rightarrow O_{2\,(aq)}^-$	$p\varepsilon = p\varepsilon^\circ = -2.72$	(O6)
$1/2 O_{2\,(aq)}^- + 2 H^+ + e^- \rightarrow 1/2 H_2O_2$	$p\varepsilon = p\varepsilon^\circ - 2\,pH = 29.08 - 2\,pH$	(O7)
$H_2O_2 + H^+ + e^- \rightarrow H_2O + \bullet OH$	$p\varepsilon = p\varepsilon^\circ - pH = 16.71 - pH$	(O8)
$\bullet OH + e^- \rightarrow OH^-$	$p\varepsilon = p\varepsilon^\circ = 28.92 + pOH$	(O9)
$\bullet OH + H^+ + e^- \rightarrow H_2O$	$p\varepsilon = p\varepsilon^\circ = 42.92 - pH$	(O9a)
$^1O_{2\,(aq)} + e^- \rightarrow O_{2\,(aq)}^-$	$p\varepsilon = p\varepsilon^\circ = 14.04$	(O10)
$O_3 + e^- \rightarrow O_3^-$	$p\varepsilon = p\varepsilon^\circ = 17.08$	(O11)

(continued overleaf)

Table 2.2 *(continued)*

MANGANESE REACTIONS

Two-electron reactions normalized to one electron

$1/2MnO_2 + 2H^+ + e^- \rightarrow 1/2\,Mn^{2+} + H_2O$	$p\varepsilon = p\varepsilon^o - 2\,pH = 20.80 - 2\,pH$	(Mn1)
$1/2\,Mn_3O_4 + 4H^+ + e^- \rightarrow 3/2Mn^{2+} + 2\,H_2O$	$p\varepsilon = p\varepsilon^0 - 4\,pH = 30.82 - 4\,pH$	(Mn2)

One-electron transfer reaction only

$MnOOH + 3\,H^+ + e^- \rightarrow Mn^{2+} + 2\,H_2O$	$p\varepsilon = p\varepsilon^o - 3\,pH = 25.35 - 3\,pH$	(Mn3)
$MnO_2 + H^+ + e^- \rightarrow MnOOH$	$p\varepsilon = p\varepsilon^o - pH = 16.22 - pH$	(Mn4)

IRON REACTIONS

Two-electron reaction normalized to one electron

$1/2Fe_3O_4 + 4H^+ + e^- \rightarrow 3/2Fe^{2+} + 2H_2O$	$p\varepsilon = p\varepsilon^o - 4\,pH = 18.20 - 4\,pH$	(Fe1)

One-electron transfer reactions only

$FeOOH + 3H^+ + e^- \rightarrow Fe^{2+} + 2H_2O$	$p\varepsilon = p\varepsilon^o - 3\,pH = 13.37 - 3\,pH$	(Fe2)
$Fe(OH)_3 + 3H^+ + e^- \rightarrow Fe^{2+} + 3H_2O$	$p\varepsilon = p\varepsilon^o - 3\,pH = 18.03 - 3\,pH$	(Fe3)

IRON-SULFIDE REACTIONS

Two-electron reaction normalized to one electron

$1/2Fe^{2+} + 1/16S_8 + e^- \rightarrow 1/2\,FeS_{mack}$	$p\varepsilon = p\varepsilon^o = 1.34$	(FeS1)
$Fe^{3+} + HS^- + e^- \rightarrow FeS_{mack} + H^+$	$p\varepsilon = p\varepsilon^o = 18.50 + pH$	(FeS2)

Cu²⁺ REDUCTION

$Cu^{2+} + e^- \rightarrow Cu^+$	$p\varepsilon = p\varepsilon^o = 2.72$	(Cu1)

SULFIDE REACTIONS

Two-electron reactions normalized to one electron

$1/2S + H^+ + e^- \rightarrow 1/2H_2S$	$p\varepsilon = p\varepsilon^o - pH = 2.44 - pH$	(S1)
$1/2S + 1/2\,H^+ + e^- \rightarrow 1/2HS^-$	$p\varepsilon = p\varepsilon^o - 0.5\,pH = -1.06 - 0.5\,pH$	(S2)

One-electron transfer reactions only

$HS\bullet + e^- \rightarrow HS^-$	$p\varepsilon = p\varepsilon^o = 18.26$	(S3)
$HS\bullet + H^+ + e^- \rightarrow H_2S$	$p\varepsilon = p\varepsilon^o - pH = 25.21 - pH$	(S4)
$S + H^+ + e^- \rightarrow HS\bullet$	$p\varepsilon = p\varepsilon^o - pH = -20.33 - pH$	(S5)

Eight-electron reactions normalized to one electron

$1/8SO_4^{2-} + 5/4H^+ + e^- \rightarrow 1/8H_2S + 1/2\,H_2O$	$p\varepsilon = p\varepsilon^o - 1.25\,pH = 18.03 - 1.25\,pH$	(S6)

Table 2.2 *(continued)*

NITROGEN REACTIONS

Two-electron reactions (per N) normalized to one electron

$^1/_2\,NO_3^- + H^+ + e^- \rightarrow\ ^1/_2\,NO_2^- + ^1/_2 H_2O$	$p\varepsilon = p\varepsilon^o - pH = 14.15 - pH$	(N1)
$^1/_2\,NO_2^- + H^+ + e^- \rightarrow\ ^1/_4\,N_2O_2^{2-} + ^1/_2\,H_2O$	$p\varepsilon = p\varepsilon^o - pH = 10.78 - pH$	(N2)
$^1/_2\,NO_2^- + 3/2\,H^+ + e^- \rightarrow\ ^1/_4\,N_2O + 3/4\,H_2O$	$p\varepsilon = p\varepsilon^o - 1.5\,pH = 24.97 - 1.5\,pH$	(N3)
$^1/_2\,N_2O + H^+ + e^- \rightarrow\ ^1/_2\,N_2 + ^1/_2\,H_2O$	$p\varepsilon = p\varepsilon^o - pH = 29.91 - pH$	(N4)
$^1/_2\,N_2H_5^+ + ^3/_2\,H^+ + e^- \rightarrow NH_4^+$	$p\varepsilon = p\varepsilon^o - 1.5\,pH = 21.56 - 1.5\,pH$	(N5a)
$^1/_2\,N_2H_4 + 2H^+ + e^- \rightarrow NH_4^+$	$p\varepsilon = p\varepsilon^o - 2\,pH = 25.51 - 2\,pH$	(N5b)
$^1/_2\,N_2H_4 + H^+ + e^- \rightarrow NH_3$	$p\varepsilon = p\varepsilon^o - pH = 15.86 - pH$	(N5c)
$^1/_2\,NH_3OH^+ + H^+ + e^- \rightarrow\ ^1/_2\,NH_4^+ + ^1/_2\,H_2O$	$p\varepsilon = p\varepsilon^o - pH = 22.77 - pH$	(N6a)
$^1/_2\,NH_2OH + H^+ + e^- \rightarrow\ ^1/_2\,NH_3 + ^1/_2\,H_2O$	$p\varepsilon = p\varepsilon^o - pH = 25.16 - pH$	(N6b)
$^1/_2\,NH_2OH + 1.5\,H^+ + e^- \rightarrow\ ^1/_2\,NH_4^+ + ^1/_2\,H_2O$	$p\varepsilon = p\varepsilon^o - 1.5\,pH = 29.77 - 1.5\,pH$	(N6c)
$^1/_4\,N_2 + ^5/_4\,H^+ + e^- \rightarrow\ ^1/_4\,N_2H_5^+$	$p\varepsilon = p\varepsilon^o - 1.25\,pH = -3.89 - 1.25\,pH$	(N7a)
$^1/_4\,N_2 + H^+ + e^- \rightarrow\ ^1/_4\,N_2H_4$	$p\varepsilon = p\varepsilon^o - pH = -5.80 - pH$	(N7b)
$^1/_2\,N_2O + ^3/_2\,H^+ + e^- \rightarrow NH_3OH^+$	$p\varepsilon = p\varepsilon^o - 1.5\,pH = -0.92 - 1.5\,pH$	(N8)
$NH_3OH^+ + ^1/_2\,H^+ + e^- \rightarrow\ ^1/_2\,N_2H_5^+ + H_2O$	$p\varepsilon = p\varepsilon^o - 0.5\,pH = 24.02 - 0.5\,pH$	(N9)
$^1/_2\,N_2 + H_2O + H^+ + e^- \rightarrow NH_2OH$	$p\varepsilon = p\varepsilon^o - pH = -22.82 - pH$	(N10)

One-electron reactions (per N) transfer reactions only

$NO_3^- + 2\,H^+ + e^- \rightarrow NO_2 + H_2O$	$p\varepsilon = p\varepsilon^o - 2\,pH = 13.07 - 2\,pH$	(N11)
$NO_2 + e^- \rightarrow NO_2^-$	$p\varepsilon = p\varepsilon^o = 15.6$	(N12)
$NO_2^- + 2\,H^+ + e^- \rightarrow NO + H_2O$	$p\varepsilon = p\varepsilon^o - 2\,pH = 19.87 - 2\,pH$	(N13)
$2NO + 2\,e^- \rightarrow N_2O_2^{2-}$	$p\varepsilon = p\varepsilon^o = 3.04$	(N14)
$NO + H^+ + e^- \rightarrow\ ^1/_2\,N_2O + ^1/_2\,H_2O$	$p\varepsilon = p\varepsilon^o - pH = 26.82 - pH$	(N15)

Hydrolysis decomposition reaction to form N_2O

$N_2O_2^{2-} + 2\,H^+ \rightarrow N_2O + H_2O$	$p\varepsilon = p\varepsilon^o - 2\,pH = 25.13 - 2\,pH$	(N16)

Three-electron reactions normalized to one electron

$^1/_3NO + H^+ + e^- \rightarrow\ ^1/_3NH_2OH$	$p\varepsilon = p\varepsilon^o - pH = 3.69 - pH$	(N17)

Four-electron reactions (per N) normalized to one electron

$^1/_4\,N_2O + ^1/_4\,H_2O + H^+ + e^- \rightarrow\ ^1/_2NH_2OH$	$p\varepsilon = p\varepsilon^o - pH = -7.87 - pH$	(N18)
$^1/_4NO_2^- + ^5/_4H^+ + e^- \rightarrow\ ^1/_4NH_2OH + ^1/_4H_2O$	$p\varepsilon = p\varepsilon^o - 1.25\,pH = 7.73 - 1.25\,pH$	(N19)
$^1/_4\,NO + H^+ + e^- \rightarrow\ ^1/_8N_2H_4 + ^1/_4\,H_2O$	$p\varepsilon = p\varepsilon^o - pH = 11.27 - pH$	(N20)

Six-electron reactions (per N) normalized to one electron (NH_3 product)

$^1/_6N_2 + H^+ + e^- \rightarrow\ ^1/_3NH_3$	$p\varepsilon = p\varepsilon^o - pH = 1.55 - pH$	(N21)
$^1/_6N_2 + ^4/_3H^+ + e^- \rightarrow\ ^1/_3NH_4^+$	$p\varepsilon = p\varepsilon^o - 1.33\,pH = 4.65 - 1.33\,pH$	(N22)
$^1/_6NO_2^- + ^4/_3H^+ + e^- \rightarrow\ ^1/_6NH_4^+ + ^1/_3H_2O$	$p\varepsilon = p\varepsilon^o - 1.33\,pH = 15.08 - 1.33\,pH$	(N23)

Eight-electron reactions normalized to one electron (NH_3 product)

$^1/_8NO_3^- + ^9/_8H^+ + e^- \rightarrow\ ^1/_8NH_3 + ^3/_8H_2O$	$p\varepsilon = p\varepsilon^o - 1.125\,pH = 13.72 - 1.125\,pH$	(N24)

(continued overleaf)

Table 2.2   (continued)

CARBON REACTIONS

$\frac{1}{2}CH_3OH + H^+ + e^- \rightarrow \frac{1}{2}CH_4 + \frac{1}{2}H_2O$	$p\varepsilon = p\varepsilon^\circ - pH = 9.94 - pH$	(C1)
$\frac{1}{2}CO_2 + H^+ + e^- \rightarrow \frac{1}{2}CO + \frac{1}{2}H_2O$	$p\varepsilon = p\varepsilon^\circ - pH = -1.74 - pH$	(C2)
$\frac{1}{2}CO_2 + H^+ + e^- \rightarrow \frac{1}{2}HCOOH$	$p\varepsilon = p\varepsilon^\circ - pH = -1.93 - pH$	(C3)
$CO_2 + e^- \rightarrow \frac{1}{2}C_2O_4^{2-}$	$p\varepsilon = p\varepsilon^\circ = -10.04$	(C4)
$\frac{1}{2}CO + H^+ + e^- \rightarrow \frac{1}{2}CH_2O$	$p\varepsilon = p\varepsilon^\circ = -0.663 - pH$	(C5)

HALOGEN REACTIONS

Two-electron reactions normalized to one electron

$\frac{1}{2}F_2 + e^- \rightarrow F^-$	$p\varepsilon = p\varepsilon^\circ = 48.51$	(F1)
$\frac{1}{2}Cl_2 + e^- \rightarrow Cl^-$	$p\varepsilon = p\varepsilon^\circ = 23.62$	(CL1)
$\frac{1}{2}Br_2 + e^- \rightarrow Br^-$	$p\varepsilon = p\varepsilon^\circ = 18.58$	(Br1)
$\frac{1}{2}I_2 + e^- \rightarrow I^-$	$p\varepsilon = p\varepsilon^\circ = 10.50$	(IO1)
$\frac{1}{2}HOCl + \frac{1}{2}H^+ + e^- \rightarrow \frac{1}{2}Cl^- + \frac{1}{2}H_2O$	$p\varepsilon = p\varepsilon^\circ - 0.5pH = 25.29 - 0.5pH$	(CL2)
$HOCl + H^+ + e^- \rightarrow \frac{1}{2}Cl_2 + H_2O$	$p\varepsilon = p\varepsilon^\circ - pH = 26.96 - pH$	(CL2a)
$\frac{1}{2}OCl^- + H^+ + e^- \rightarrow \frac{1}{2}Cl^- + \frac{1}{2}H_2O$	$p\varepsilon = p\varepsilon^\circ - pH = 29.06 - pH$	(CL2b)
$\frac{1}{2}HOBr + \frac{1}{2}H^+ + e^- \rightarrow \frac{1}{2}Br^- + \frac{1}{2}H_2O$	$p\varepsilon = p\varepsilon^\circ - 0.5pH = 20.00 - 0.5pH$	(Br2)
$\frac{1}{2}HOI + \frac{1}{2}H^+ + e^- \rightarrow \frac{1}{2}I^- + \frac{1}{2}H_2O$	$p\varepsilon = p\varepsilon^\circ - 0.5pH = 16.66 - 0.5pH$	(IO2)

One-electron transfer reactions only

$Cl\bullet + e^- \rightarrow Cl^-$	$p\varepsilon = p\varepsilon^\circ = 40.76$	(CL3)
$Br\bullet + e^- \rightarrow Br^-$	$p\varepsilon = p\varepsilon^\circ = 32.47$	(Br3)
$I\bullet + e^- \rightarrow I^-$	$p\varepsilon = p\varepsilon^\circ = 22.49$	(IO3)

Source: Reproduced and updated with permission from Ref. [5].

O_2 begins to occur at a pH value as low as 5 depending on the Fe(III) species formed whereas the one-electron oxidation of Mn^{2+} by O_2 begins to occur at a pH value as low as 8 depending on the Mn(III) species formed. Kinetic data [8, 9] indicate that oxidation of Mn^{2+} by O_2 is very slow at a pH near 8, the pH of surface seawater. These calculations indicate that the oxidation reactions for Mn^{2+} and Fe^{2+} are controlled by the first electron accepted by O_2 to form O_2^-, a thermodynamically unfavorable species (Section 10.4.1).

Plots such as those in Figure 2.9a and b predict the direction in which a reaction is favorable. For example, a negative value for $\Delta\log K$ indicates an unfavorable forward reaction; however, the reverse reaction is favorable. In Figure 2.9b, although the oxidation of soluble Mn(II) species by O_2 is unfavorable, the reduction of metal oxides by O_2^- in acid forms the Mn(II) species.

2.7   Highly Oxidizing Conditions

2.7.1   Ozonolysis Reactions

Table 2.2 gives the reactions and redox equations for the aqueous reactive oxygen species (ROS are O_2^-, H_2O_2, OH• and O_3), which can be produced in the atmosphere and in natural waters (at nanomolar levels).

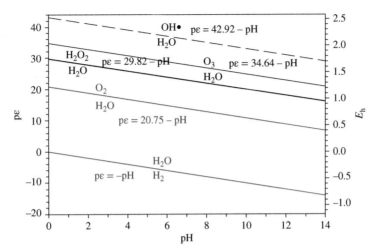

Figure 2.10 *Pourbaix diagram shows the H_2O to O_2, H_2O_2, O_3, and OH• stability lines (ROS reduce to H_2O). H_2O_2, O_3, and OH• are capable of oxidizing Cl^- to OCl^-*

The Pourbaix diagram in Figure 2.10 shows the oxidative strength of aqueous hydrogen peroxide, ozone, and hydroxyl radicals compared to O_2 in water. To decompose organic compounds and to kill bacteria and viruses, ozone is used to treat drinking water, and often with UV radiation. On oxidation of organic compounds, ozone decomposes to O_2 and O atoms, which insert into C–H and C–C bonds to eventually form CO_2 and carbonate species. **O atom transfer** is formally a two-electron transfer reaction. Hydroxyl radicals, one-electron transfer oxidants, are also formed in ozonolysis reactions, and they also quickly oxidize organic compounds to CO_2 species and inorganic compounds. Dissolved manganese(II) and iron(II) are readily oxidized by ozone to form oxide (MnO_2) or (oxy)hydroxide (FeOOH) precipitates that are easily removed from solution. Metals bound to organic ligands such as humic acids are released from the organic ligands, which are also oxidized.

A potential problem with O_3 (as well as Cl_2 and HOCl as disinfectants) occurs when chloride and bromide are present in source water containing natural organic compounds from plant degradation. When the halides oxidize to OX^- (Figure 2.11 for Cl^-), they react with organic matter to form R–X (organohalogen) compounds as HOX is a source of positive halogen ($HO^- X^+$) whereas gaseous HOX leads to radicals (HO + X). The R–X by-products can be carcinogenic.

2.7.2 Atmospheric Redox Reactions

2.7.2.1 Ozone Formation and Destruction

In the atmosphere, UV radiation dissociates O_2 to form O atoms ($|\underline{O}|$ in Equation 2.10), which can react with O_2 to reform O_3 (Equation 2.11; Table 2.2, O4) and with H_2O to form another potent oxidant, the hydroxyl radical (OH• or OH; Table 2.2, O9) (Equation 2.12). M is a third neutral body absorbing the kinetic energy of O atoms colliding with O_2. O atoms are potent oxidants (see Section 5.7.3) when in the **excited state** (^1D; all electrons paired; Section 3.8.2), which occurs at wavelengths < 242 nm.

$$O_2 + h\upsilon(< 242 \text{ nm}) \rightarrow 2|\underline{O}| \tag{2.10}$$

$$|\underline{O}| + O_2 + M \rightarrow O_3 + M^* \tag{2.11}$$

$$|\underline{O}| + H_2O \rightarrow 2OH• \tag{2.12}$$

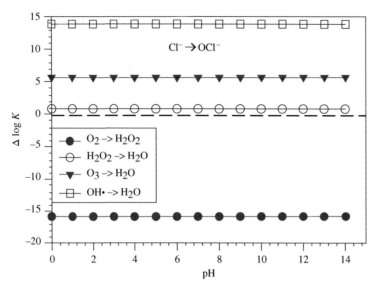

Figure 2.11 *Thermodynamic calculations for the two-electron transfer reactions of Cl⁻ (reaction CL2b) with O_2, H_2O_2, and O_3 (reactions O2, O3, O4; Table 2.2) to form OCl⁻ and the one-electron transfer of Cl⁻ with OH• (O9; Table 2.2). Positive values for $\Delta \log K$ on the y-axis indicate a favorable forward reaction; negative values indicate an unfavorable forward reaction*

Ozone is a major component of the upper stratosphere that is formed continuously, and it is essential in preventing UV electromagnetic radiation (Figure 1.2) from the sun reaching the earth's surface as it absorbs UVB (280–230 nm) and UVC (< 280 nm) radiation. O_3 is very reactive and can recombine with O atoms to form O_2 (Equation 2.13). Unfortunately, the escape of natural and in particular man-made organohalogen (R–X; C–X bonds) compounds including freons to the upper atmosphere also undergo UV bond breakage to form X atoms (Equations 2.14 and 2.15), which react with O_3 (Equation 2.16) and create an **ozone hole**, which is monitored closely by several agencies, over Antarctica in the austral spring.

$$O + O_3 \rightarrow 2\,O_2 \tag{2.13}$$

$$CFCl_3 + h\upsilon \rightarrow CFCl_2 + Cl \tag{2.14}$$

$$CF_2Cl_2 + h\upsilon \rightarrow CF_2Cl + Cl \tag{2.15}$$

$$X + O_3 \rightarrow XO + O_2 \tag{2.16}$$

Although X atoms do not readily form in solution, they are a common species in the atmosphere, and atmospheric reactions are typically not complicated by water and H^+ transfers or formation. In addition to photochemical decomposition of anthropogenic gaseous R–X compounds, the formation of halogen atoms in the atmosphere is also related to major natural processes involving plants and algae in marine and other natural waters, which facilitate volatile R–X and R–S formation and subsequent photochemical decomposition via homolytic bond cleavage of the C–X bond as in Equations 2.14 and 2.15. For example, aquatic macroalgae and microalgae form X_2 or HOX as in Equations 2.17–2.19 [10]. Also, iodide in the interstitial fluids of the brown kelp, *Laminaria digitata*, and other plants can be released and used as an antioxidant to react with H_2O_2 and O_3 during oxidative stress [11]. Oxidation of iodide above or out of the water leads to formation of I_2 (Equation 2.19), and I_2 formation can be three orders of magnitude larger than R–I and other R–X compound formation.

$$2X^- + H_2O_2 + 2H^+ \rightarrow X_2 + 2H_2O \qquad (2.17)$$

$$X^- + H_2O_2 + H^+ \rightarrow HOX + H_2O \qquad (2.18)$$

$$2X^- + O_3 + 2H^+ \rightarrow X_2 + O_2 + H_2O \qquad (2.19)$$

Once X_2 and HOX are formed (Equations 2.17–2.19), they can react with natural organic matter in surface waters to form R–X compounds, which are released to the atmosphere. Then, X_2, HOX, and R–X chemical species can be photochemically decomposed to X atoms, R radicals, and OH radical as in Equations 2.20–2.22. Because bonds of iodine with other atoms are weaker than the corresponding Cl and Br analogs, iodine reactions are more facile.

$$I_2 + h\upsilon \rightarrow 2I \qquad (2.20)$$

$$RI + h\upsilon \rightarrow R + I \qquad (2.21)$$

$$HOI + h\upsilon \rightarrow OH + I \qquad (2.22)$$

Upon formation, X atoms can undergo a variety of thermodynamically favorable **O atom transfer** reactions with ROS and NO (Equations 2.23–2.25) [12]. The destruction of O_3 is a main reaction that leads to XO radicals that can react with NO to form NO_2.

$$X + O_3 \rightarrow XO + O_2 \qquad (2.23)$$

$$XO + NO \rightarrow X + NO_2 \qquad (2.24)$$

XO can also react with HO_2 to abstract an H· atom to form HOX (Equation 2.25), which is photochemically sensitive and regenerates to OH (Equation 2.22).

$$XO + HO_2 \rightarrow HOX + O_2 \qquad (2.25)$$

Both OH and HO_2 (protonated superoxide) also result in O_3 destruction (Equations 2.26 and 2.27) in a similar manner to the halogen reactions above. Thus, decreasing halogen emissions to the atmosphere is of critical importance.

$$O_3 + OH \rightarrow HO_2 + O_2 \qquad (2.26)$$

$$O_3 + HO_2 \rightarrow OH + 2O_2 \qquad (2.27)$$

2.7.2.2   *Gas Phase Reactions Leading to the Formation of Nitrate, Sulfate, and Perchlorate*

Unfortunately, the release of anthropogenic and natural organohalogen compounds to the atmosphere is not the only major cause for the destruction of O_3. The burning of fossil fuels leads to the production of NO_x compounds such as NO_2 and NO_3. There are three possible ways to form nitric acid (HNO_3) in the atmosphere [13]. First, there is the direct reaction of NO_2 with OH• as in Equation 2.28. Second, NO_2 reacts with O_3 followed by proton abstraction from the hydrocarbon such as dimethyl sulfide (DMS; Equations 2.29 and 2.30). Third, NO_2 and NO_3 can react to form N_2O_5, which then reacts with water to form nitric acid (Equations 2.31 and 2.32).

$$NO_2 + OH\bullet + M \rightarrow HNO_3 + M \tag{2.28}$$

$$NO_2 + O_3 \rightarrow NO_3 + O_2 \tag{2.29}$$

$$NO_3 + HC, DMS \rightarrow HNO_3 + products \tag{2.30}$$

$$NO_2 + NO_3 \leftrightarrow N_2O_5 \tag{2.31}$$

$$N_2O_5 + H_2O + surf_{(aq)} \rightarrow 2HNO_3 \tag{2.32}$$

The burning of fossil fuels enriched in sulfur also leads to SO_2 formation and release to the atmosphere. The emissions are in addition to those from **volcanoes**, which are a natural source of SO_2, HOCl, and HCl to the atmosphere. SO_2 can react to form sulfuric acid (H_2SO_4) in the atmosphere as in Equations 2.33–2.35. Although HO_2 is produced in Equation 2.34, further reaction of the products can occur (see Equation 2.25).

$$SO_2 + OH \rightarrow HSO_3 \tag{2.33}$$

$$HSO_3 + O_2 \rightarrow SO_3 + HO_2 \tag{2.34}$$

$$SO_3 + H_2O \rightarrow H_2SO_4 \tag{2.35}$$

Perchlorate, ClO_4^-, has been found on the surface of Mars and on Earth's deserts [14]. The release of HCl to the atmosphere leads to the formation of ClO_4^- from the reaction of ClO_3^- and ClO_2 with reactive oxygen species. The exact pathways for the formation of ClO_4^- are not known, but several pathways have been proposed, and two are noted. First, the reaction of BrO with ClO in an O atom transfer reaction leads to formation of ClO_2, which then reacts with O_3 to form ClO_3 (Equations 2.36 and 2.37), which further reacts with O_3 or OH· (Equation 2.38) to form $HClO_4$ [15].

$$BrO + ClO \rightarrow Br + ClO_2 \tag{2.36}$$

$$ClO_2 + O_3 \rightarrow ClO_3 + O_2 \tag{2.37}$$

$$ClO_3 + OH\bullet + M \rightarrow HClO_4 + M \tag{2.38}$$

The direct reaction of Cl atoms with O_3 has been proposed to form ClO_3 (Equation 2.39), which then reacts as in Equation 2.38 to form ClO_4^-. Of course, chlorine atoms can react with reactive oxygen species in three separate oxygen transfer reactions to form ClO_3.

$$Cl + O_3 + M \rightarrow ClO_3 \tag{2.39}$$

APPENDIX 2.1

Gibbs free energies of formation for selected environmentally important chemical materials.

$kJ\,mol^{-1}$		$kJ\,mol^{-1}$		$kJ\,mol^{-1}$		$kJ\,mol^{-1}$	
$Br_{2\,(aq)}$	4.18	$Cu^+_{(aq)}$	50.0	$NO_3^-{}_{(aq)}$	−111.3	$O_{2\,(aq)}$	16.32
$Br^-_{(aq)}$	−104.0	$Cu^{2+}_{(aq)}$	65.5	$NO_2^-{}_{(aq)}$	−37.2	$O_2^-{}_{(aq)}$	31.84
$HOBr_{(aq)}$	−82.2			$NO_{2\,(g)}$	51.3	$^1O_{2\,(aq)}$	112
$Br{\bullet}_{(aq)}$	81.3	$F^-_{(aq)}$	−278.8	$NO_{(g)}$	86.57	$O_{3\,(g)}$	163.2
		$Fe^{2+}_{(aq)}$	−90.53	$N_2O_{(g)}$	104.2	O_3^-	65.7
$Cl_{2\,(aq)}$	6.90	$Fe^{3+}_{(aq)}$	−4.60	$N_2O_2^{2-}{}_{(aq)}$	139	$H_2O_{2\,(aq)}$	−134.1
$Cl^-_{(aq)}$	−131.3	$\alpha\text{-FeOOH}$	−488.6	$NH_2OH_{(aq)}$	23.4	$H_2O_{(aq)}$	−237.18
$HOCl_{(aq)}$	−79.9	$Fe(OH)_3$	−699	$NH_3OH^+_{(aq)}$	−56.55	${\bullet}OH_{(g)}$	34.22
$OCl^-_{(aq)}$	−36.8	Fe_3O_4	−1012.6	$N_2H_{4\,(aq)}$	132.57	${\bullet}OH_{(aq)}$	7.74
$Cl{\bullet}_{(aq)}$	101	$FeS_{mackinawite}$	−98.2	$N_2H_{4\,(g)}$	159.3	$OH^-_{(aq)}$	−157.29
		$FeS_{2,\,pyrite}$	−160.2	$N_2H_5^+$	82.4		
CO	−137.27			$NH_{3(g)}$	−16.5	$H_2S_{(aq)}$	−27.87
$CO_{2\,(g)}$	−394.37	$I_{2\,(aq)}$	16.43	$NH_{3\,(aq)}$	−26.57	$HS^-_{(aq)}$	12.05
$CH_2O_{(aq)}$	−129.7	$I^-_{(aq)}$	−51.59	$NH_4^+{}_{(aq)}$	−79.37	$HS{\bullet}$	116.2
$HCOOH$	−372.3	HOI	−99.2			$SO_4^{2-}{}_{(aq)}$	−744.6
CH_3OH	−175.4	$I{\bullet}_{(aq)}$	76.8			$SO_{2\,(g)}$	−300.2
$CO_3^{2-}{}_{(aq)}$	−527.9						
$HCO_3^-{}_{(aq)}$	−586.8	Mn^{2+}	−228.0				
$CH_{4\,(aq)}$	−34.39	Mn_3O_4	−1281				
$CH_{4(g)}$	−50.79	MnO_2	−465.1				
$C_2O_4^{2-}{}_{(aq)}$	674.04	$MnOOH$	−557.7				

References

1. Bard, A. J., Parsons, R. and Jordan, J. (eds) (1985) *Standard Potentials in Aqueous Solution* (1st ed.), Marcel Dekker, New York, pp. 834.
2. Stanbury, D. (1989) Reduction potentials involving inorganic free radicals in aqueous solution, In *Advances in Inorganic Chemistry*, vol. 33 (ed A. G. Sykes). Academic Press, New York, pp. 69–138.
3. Stumm, W. and Morgan, J. J. (1996) *Aquatic Chemistry* (3rd ed.), John Wiley, New York.
4. Rickard, D. and Luther III, G. W. (2007) Chemistry of iron sulfides. *Chemical Reviews*, **107**, 514–562.
5. Luther III, G. W. (2010) The role of one and two electron transfer reactions in forming thermodynamically unstable intermediates as barriers in multi-electron redox reactions. *Aquatic Geochemistry*, **16**, 395–420.

6. Luther III, G. W. (2011) Thermodynamic redox calculations for one and two electron transfer steps: Implications for halide oxidation and halogen environmental cycling, In *Aquatic Redox Chemistry* (eds P. G. Tratnyek, T. J. Grundl and S. B. Haderlein). American Chemical Society (ACS) books, pp. 15–35.

7. Basolo, F. and Pearson R. G. (1967) *Mechanisms of Inorganic Reactions*, Wiley, New York, pp. 701.

8. Von Langen, P. J., Johnson, K. S., Coale, K. H., and Elrod, V. A. (1997) Oxidation kinetics of manganese(II) in seawater at nanomolar concentration. *Geochimica et Cosmochimica Acta*, **61**, 4945–4954.

9. Morgan, J. J. (2005) Kinetics of reaction between O_2 and Mn(II) species in aqueous solution. *Geochimica et Cosmochimica Acta*, **69**, 35–48.

10. Hughes, C., Malin, G.; Nightingale, P. D. and Liss, P. S. (2006) The effect of light stress on the release of volatile iodocarbons by three species of marine microalgae. *Limnology and Oceanography*, **51**, 2849–2854.

11. Küpper, F. C., Carpenter, L. J., McFiggans, G. B., Palmer, C. J., Waite, T. J., Boneberg, E.-M., Woitsch, S., Weiller, M., Abela, R., Grolimund, D., Potin, P., Butler, A., Luther, III, G. W., Kroneck, P. M. H., Meyer-Klaucke, W. and Feiters, M. C. (2008) Iodide accumulation provides kelp with an inorganic antioxidant impacting atmospheric chemistry. *Proceedings of the National Academy of Sciences U. S. A.*, **105**, 6954–6958.

12. Whaley, L. K., Furneaux, K. L.; Goddard, A., Lee, J. D., Mahajan, A., Oetjen, H., Read, K. A., Kaaden, N., Carpenter, L. J., Lewis, A. C., Plane, J. M. C., Saltzman, E. S., Wiedensohler, A. and Heard, D. E. (2010) The chemistry of OH and HO_2 radicals in the boundary layer over the tropical Atlantic Ocean. *Atmospheric Chemistry and Physics*, **10**, 1555–1576.

13. Catling, D. C., Claire, M. W., Zahnle, K. J., Quinn, R. C., Clark, B. C., Hecht, M. H. and Kounaves, S. (2010) Atmospheric origins of perchlorate on Mars and in the Atacama. *Journal of Geophysical Research*, **115**, e00e11, doi: 10.1029/2009JE003425, 15 pages.

14. Smith, M. L., Claire, M. W., Catling, D. C. and Zahnle, K. J. (2014) The formation of sulfate, nitrate and perchlorate salts in the Martian atmosphere. *Icarus*, **231**, 51–64.

15. Rao, B., Anderson, T. A., Reeder, A. and Jackson, W. A. (2010) Perchlorate Formation by Ozone Oxidation of Aqueous Chlorine/ Oxy-Chlorine Species: Role of ClxOy Radicals. *Environmental Science and Technology*, **44**, 2961–2967.

3

Atomic Structure

3.1 History

The development of atomic theory was one of the most successful endeavors in the history of science as it has led to major advances in physics, chemistry, and biology benefitting the human race. This chapter reviews many major findings that are important in understanding atomic properties that are codified in the periodic table of elements. Once these properties are understood, it is possible to understand chemical bonding, redox properties, and ultimately chemical reactivity. At the heart of these developments is the use of absorption and emission spectroscopy (UV, Vis, X-ray) as a tool to provide data to understand the structure of the atom. Later, electron spectroscopy for chemical analysis (ESCA) or X-ray photoelectron spectroscopy (XPS) was important in providing further information including accurate orbital energies for atoms and molecules. Historical articles [1–3] have been written celebrating the 100th anniversary of the Bohr atom and the results leading up to it.

When atoms are energetically stimulated by heating or electrical discharge across two electrodes in a glass or quartz tube, colors are frequently observed. The study of the wavelength of spectral lines once atoms were energetically stimulated was well developed in the 19th century, and emission spectroscopy even led to the discovery of helium during a solar eclipse in the 1860s. On earth, helium was found to emanate from the uranium ore, cleveite, in the 1890s, and large reserves were later found in natural gas fields in the United States. The wavelength, λ, and the frequency, v (in Hz, s^{-1}), of light are related through the velocity of light, c ($c = \lambda v$), which is $2.99792458 \times 10^8 \ m\,s^{-1}$ ($\sim 3.0 \times 10^8 \ m\,s^{-1}$).

During the 1880s, Balmer studied the visible region of the hydrogen atom (H•) spectrum and found an empirical relation, which Rydberg modified for all transitions of the H• spectrum (Equation 3.1). (H atoms are created via electrical discharge of H_2 at low pressure to break the H–H bond).

$$\frac{1}{\lambda} = \bar{v} = \mathcal{R} \left[\frac{1}{n_i^2} - \frac{1}{n_{i+1}^2} \right] \tag{3.1}$$

Here, $\bar{v}$ is the wavenumber (units of cm^{-1} or Kaiser, K) and λ in centimeters is the wavelength of the spectral line, $\mathcal{R}$ is the Rydberg constant of $109,677 \ cm^{-1}$, and n_i and n_{i+1} are simple integers indicating the energy state (i = initial or reference state). For the H•, $n_i = 1$ corresponds to the UV series of spectral lines (Lynam lines),

Inorganic Chemistry for Geochemistry and Environmental Sciences: Fundamentals and Applications, First Edition. George W. Luther, III.
© 2016 John Wiley & Sons, Ltd. Published 2016 by John Wiley & Sons, Ltd.
Companion Website: www.wiley.com/go/luther/inorganic

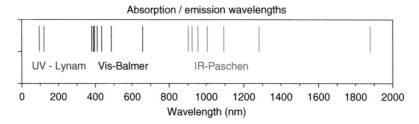

Figure 3.1   *Spectral lines for the Lynam, Balmer, and Paschen series for the hydrogen atom*

$n_i = 2$ corresponds to the visible series of spectral lines (Balmer), and $n_i = 3$ corresponds to the infrared series of spectral lines (Paschen). Figure 3.1 shows the wavelengths for these transitions, and Figure 1.2 shows that they occur near the center of the electromagnetic spectrum.

In 1901, Planck related the energy of a given state, n, to the frequency of light such that electromagnetic energy behaved as both wave and particle (Equation 3.2). Here $h\upsilon$ is termed a photon or particle of light and the frequency is a multiple of an elementary unit ($h = 6.627 \times 10^{-34}$ J s); that is, it is quantized.

$$E = h\upsilon = h\frac{c}{\lambda} = hc\bar{\upsilon} \tag{3.2}$$

Einstein used Planck's work to show that electrons are emitted with a kinetic energy once they absorb enough light to overcome the work function or the binding (potential) energy of the electron (Equation 3.3); that is, the photoelectric effect.

$$E \text{ (kinetic energy)} = h\upsilon - w \text{ (potential energy)} = h\upsilon - w \text{ (orbital energy)} \tag{3.3}$$

The photoelectric effect led to the development of electron spectroscopy for chemical analysis (ESCA) by Siegbahn et al. [4] in the 1960s. Here X-rays of known energy are used to bombard an atom and the kinetic energy of the ejected electron is then used to calculate the work function or orbital energy (w; Equation 3.3). ESCA data will be used throughout this book to give orbital energies for atoms and molecules.

3.2   The Bohr Atom

In 1909, Ernest Rutherford's group demonstrated that atoms were composed of a dense nuclear core (10^{-14}–10^{-15} m) containing neutrons and Z protons (where Z is the atomic number) with electrons revolving around the core. The overall size of an atom is about 1 Å (10^{-8} cm, 100 pm, 10^{-10} m). To show that the electrons would not collapse into the nucleus and thus remain in discrete orbits or energy states around the nucleus, Bohr used all these developments to develop a model of the hydrogen atom, H•. There are three main requirements in Bohr's atomic theory:

1.   The electron revolves around the nucleus in an orbit or stationary state so that the centripetal force equals the electrostatic force of attraction between the electron and the nucleus.
2.   The angular momentum is quantized; that is, $m\upsilon r = nh/(2\pi)$ where m is the mass of the electron, $\upsilon =$ its velocity, r is the radius, and $n =$ an integer for each state (principal quantum number).
3.   An electron moves from one state to another when excited such that $\Delta E = E_2 - E_1 = h\nu$.

With these insights in 1913, Bohr derived the expression (3.4) for the potential energy of a given electron orbit or stationary state for the hydrogen atom and other one-electron ions such that $E \propto -Z^2/n^2$ where e is the electronic charge, n is the principal quantum number, and m is the mass of the electron.

$$E_n = \left[\frac{-2\pi^2 m \, e^4 \, Z^2}{h^2 \, n_i^2} \right] = -\frac{R \, Z^2}{n_i^2} \tag{3.4}$$

The expression for r, the radius in picometers, is Equation 3.5 where $a_0 = 52.9$ pm, the radius of the hydrogen atom.

$$r = \frac{a_0 n^2}{Z} \tag{3.5}$$

On computing the constants $(2\pi^2 m e^4 / h^2)$ in Equation 3.4, the value for R, the Rydberg constant, was found [3]. Thus the experimental data for the H• spectrum were reconciled with theory. The Rydberg expression can be rewritten for all one-electron atoms and ions as Equation 3.6 where $Z = $ atomic number.

$$\frac{1}{\lambda} = \bar{v} = R Z^2 \left[\frac{1}{n_i^2} - \frac{1}{n_{i+1}^2} \right] \tag{3.6}$$

Figure 3.2 shows the H• and He$^+$ energy states from Equation 3.6 plotted versus n state and how the energies of the states become increasingly more positive (less stable) with increasing n value. Because these states are related to orbital energies as shown by Bohr in Equation 3.4, the negative values indicate the orbital energy of each state. As an electron goes from a state n_i to states with increasing n, the ΔE value increases, but the λ of the transition approaches a lower limit. The **ionization** energy or potential (IP) is the energy required to remove the electron from the atom and in the case of H atom it is -13.6 eV, the energy of the $n = 1$ state. For the helium cation, He$^+$, the ionization energy is four times that of the H atom because of the Z^2 term in Equation 3.4.

3.3   The Schrodinger Wave Equation

In 1923, DeBroglie suggested that electron motion has a wave aspect and derived Equation 3.7:

$$\lambda = \frac{h}{mv} \tag{3.7}$$

where λ is the wavelength of the electron and v is the velocity of the electron; at 40 kV, $\lambda = 5$ pm. Thus electrons can behave as waves and particles, and this is known as the "wave particle duality." This expression was confirmed shortly thereafter and is the basis for electron microscopy. This duality led Heisenberg to state his uncertainty principle, which indicates that the electron's position and momentum cannot be both known with certainty (Equation 3.8),

$$\Delta_{mv} \, \Delta_x \geq \frac{1}{2} \left(\frac{h}{2\pi} \right) \tag{3.8}$$

Considering this wave particle expression, Schrodinger postulated his equation for the electron in the hydrogen atom and instituted quantum or wave mechanics. Here the wave function, ψ, is a mathematical description of the atom and corresponds to the amplitude of a wave. The function can be a real function (with $+$, $-$ signs) or complex. The sign of ψ is important for bonding considerations.

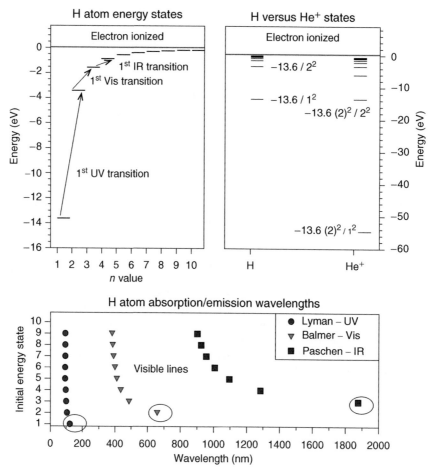

Figure 3.2 *Hydrogen atom and helium cation, He$^+$, energy states. Bottom figure shows that the high energy or low wavelength lines approach a limit; the first transition in each series is circled*

The wave function has the following limitations imposed on it:

1.   It must be finite; if not, a solution cannot be found.
2.   It must be continuous as all waves are.
3.   It must be single valued; that is, two possible values at the same point in space are impossible.

Here, $N^2 \int \psi\psi \, d\tau = 1$ where τ means over all space and N is the normalization constant. Normalization indicates that finding the electron is certainty. Thus, ψ^2 is proportional to the probability of finding the electron over all space; Figure 3.3 shows a plot of the cosine function and its squared plot, indicating that ψ^2 is positive. This is similar to the intensity of light being proportional to the amplitude squared. This is the Born interpretation of ψ.

As demonstrated in Figure 3.4, combining of waves leads to constructive and destructive interference, which will be important when discussing bonding between atoms. When a wave traverses from a positive to a negative value, it crosses through zero, which is called a node.

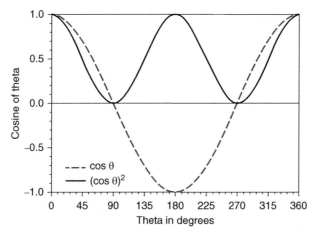

Figure 3.3 *A plot of the cosine function (as shown below in Table 3.2, this is the function for the p_z orbital; $n = 2$; $l = 1$; $m_l = 0$) and the cosine squared*

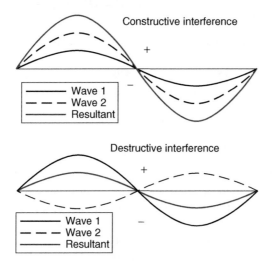

Figure 3.4 *The constructive and destructive interference of waves for a sine function*

The Schrodinger equation for hydrogen-like atoms is given in Equation 3.9

$$Ж\Psi = E\Psi \tag{3.9}$$

where Ж is the Hamiltonian operator. The result is that the operation times the wave function gives the wave function times its energy. The energy of the atom is computed by performing the following mathematical operation (3.10) where an electron's energy is estimated and recomputed until it reaches a minimum energy using Hartree–Fock **self-consistent field theory** (SCF, [5]).

$$E = \frac{\int \psi\, Ж\, \psi\, d\tau}{\int \psi\psi d\tau} \tag{3.10}$$

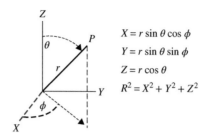

$$X = r \sin \theta \cos \phi$$
$$Y = r \sin \theta \sin \phi$$
$$Z = r \cos \theta$$
$$R^2 = X^2 + Y^2 + Z^2$$

Figure 3.5   *Variables of the polar coordinates for the Schrodinger equation*

The electron can be described in three dimensions as in Figure 3.5 where P is the electron; θ is the angle between the Z axis and r, the distance of the electron from the nucleus; ϕ is the angle between the x axis and the projection of r into the XY plane.

The Schrodinger equation for the motion of a particle in three dimensions is given as

$$\frac{\delta^2 \psi}{\delta^2 x} + \frac{\delta^2 \psi}{\delta^2 y} + \frac{\delta^2 \psi}{\delta^2 z} + \frac{8 \pi^2 m}{h^2}(E - V)\psi = 0$$

and in polar coordinates as (ε_0 is the vacuum permittivity $= 8.854 \times 10^{-12} \, C^2 \, m^{-1} \, J^{-1}$)

$$\frac{\delta}{r^2 \delta r}\left[\frac{r^2 \, \delta_\psi}{\delta r}\right] + \frac{\delta}{r^2 \sin \delta \, \theta}\left[\frac{\sin \theta \, \delta \psi}{\delta \theta}\right] + \frac{\delta^2 \psi}{r^2 \sin^2 \theta \, \delta \phi^2} + \frac{8 \pi^2 m}{h^2}\left[E + \frac{Z \, e^2}{(4 \, \pi \varepsilon_0 r)}\right]\psi = 0$$

The wavefunction, ψ_{nlml}, consists of three components, one for each dimension in space.

$$\psi_{nlml} = N \, R_{nl}(r) * \Theta_{lml}(\theta) * \Phi \phi_{lml}(\phi) = R_{nl}(r) * Y_{lml} \, (\theta, \phi)$$

where $R(r)$ is the radial distribution function and $Y_{lml}(\theta, \phi)$ the spherical harmonics. The three quantum numbers n, l, m_1 result from the solution of the Schrodinger equation, where "n" is related to r and gives information on the *main energy level* that the electron resides in as well as its size or the distance of the electron from the nucleus (this is analogous to n in the Bohr atom); "l" is related to r, θ, and ϕ and gives information on the number of *sublevels* in a given n and on the shapes of the orbitals; and "m_1" is related to θ and ϕ and indicates the number of orbitals in a given l and their *orientation* in space. The wave function for a given combination of n, l and m_1 value is termed an orbital. A relativistic correction to ψ was made by Dirac, resulting in a fourth quantum number (the spin quantum number, m_s) to define the electron's intrinsic magnetic moment in a magnetic field. The values allowed for m_s are $+1/2$ ($\uparrow$, anticlockwise or parallel with an applied magnetic field) and $-1/2$ ($\downarrow$, clockwise or opposed to an applied magnetic field).

3.4   Components of the Wave Function

3.4.1   Radial Part of the Wave Function, $R(r)$

The radial distribution function, $R(r)$, is first discussed, followed by the spherical harmonics, $Y_{lml}(\theta, \phi)$. Hydrogen-like atom wave functions for $R(r)$ and $Y_{lml}(\theta, \phi)$ are provided in Tables 3.1 and 3.2, respectively. The radial functions are products of an exponential function with a polynomial described in the variable r/a_0.

Table 3.1   *Radial distribution functions for energy levels n = 1, 2, 3 for the hydrogen atom. For multielectron atoms, replace r by Zr and multiply each function by (Z)$^{3/2}$*

$n = 1, l = 0$ $R_{10}(r) = 2 \left(\dfrac{1}{a_0} \right)^{3/2} e^{-r/a_0}$

$n = 2, l = 0$ $R_{20}(r) = \dfrac{1}{\sqrt{8}} \left(\dfrac{1}{a_0} \right)^{3/2} \left(2 - \dfrac{r}{a_0} \right) e^{-r/2a_0}$

$n = 2, l = 1$ $R_{21}(r) = \left(\dfrac{1}{\sqrt{24}} \right) \left(\dfrac{1}{a_0} \right)^{3/2} \dfrac{r}{a_0} e^{-r/2a_0}$

$n = 3, l = 0$ $R_{30}(r) = \left(\dfrac{2}{81\sqrt{3}} \right) \left(\dfrac{1}{a_0} \right)^{3/2} \left(27 - 18\dfrac{r}{a_0} + 2\dfrac{r^2}{a_0^2} \right) e^{-r/3a_0}$

$n = 3, l = 1$ $R_{31}(r) = \left(\dfrac{4}{81\sqrt{6}} \right) \left(\dfrac{1}{a_0} \right)^{3/2} \left(6\dfrac{r}{a_0} - \dfrac{r^2}{a_0^2} \right) e^{-r/3a_0}$

$n = 3, l = 2$ $R_{32}(r) = \dfrac{4}{81\sqrt{30}} \left(\dfrac{1}{a_0} \right)^{3/2} \dfrac{r^2}{a_0^2} e^{-r/3a_0}$

Plots of $R(r)$ show typical decay type curves. In Figure 3.6a for the hydrogen atom 1s orbital, the black line shows the decay curve for $R(r)$ ($n = 1$; $l = 0$ from above). Of most importance to chemists for bonding considerations is the square of $R(r)$, which Max Born attributed to the probability of finding the electron or the electron density of a given orbital. The radial probability function $R(r)^2$ and the radial distribution function $R(r)$ are related via the volume of the atom as a spherical atom can be considered as composed of layers (Figure 3.6b) like an onion skin. Thus, the probability of finding the electron in a given spherical layer from r to $r + dr$ can be calculated. Recalling that the volume of sphere is

$$V = \dfrac{4}{3}\pi r^3 \text{ then}$$

$$dV = 4\pi r^2 dr \text{ and } R^2 dV = 4\pi r^2 R^2 dr$$

The blue line in Figure 3.6a shows the contribution of $4\pi r^2$. The black dashed line in Figure 3.6a shows the contributions of $R(r)$ and $4\pi r^2$ to the probability ($\psi^2 = 4\pi r^2 R^2$) with ψ or $R(r)$ decaying quickly to zero and $4\pi r^2$ increasing as the radius or distance from the nucleus increases. At $r = 0$ and at large values of r, the probability of finding the electron is zero, but at intermediate values of r, it reaches a maximum.

Figure 3.7 shows plots of the radial distribution functions and radial probability functions for $n = 1, 2, 3$ and all l values for each n level of the hydrogen atom (Table 3.1). There are several features of the radial distribution functions and the radial probability functions that are important for understanding the properties of atoms, chemical bonding, and electron transfer processes.

The radial distribution functions show that as n increases, the radius increases as the decay is more spread out. There are also nodes where the $R(r)$ function is zero at some distance from the nucleus such that for s orbitals there are $n - 1$ nodes where the function changes sign; for p orbitals there are $n - 2$ nodes, etc.

For the radial probability functions $[R(r)]^2 r^2$, all values are positive as a result of squaring $R(r)$. Electrons in s orbitals have electron density close to the nucleus regardless of the n value whereas p and d electrons

Table 3.2   The $(\theta\phi)$ equations for the hydrogen atom

1	0	0	1s	$\Theta\phi = \dfrac{1}{2\sqrt{\pi}}$
2	0	0	2s	$\Theta\phi = \dfrac{1}{2\sqrt{\pi}}$
2	1	(± 1)	$2p_x$	$\Theta\phi = \dfrac{\sqrt{3}(\sin\Theta\,\cos\phi)}{2\sqrt{\pi}}$
2	1	0	$2p_z$	$\Theta\phi = \dfrac{\sqrt{3}(\cos\Theta)}{2\sqrt{\pi}}$
2	1	(± 1)	$2p_y$	$\Theta\phi = \dfrac{\sqrt{3}(\sin\Theta\,\sin\phi)}{2\sqrt{\pi}}$
3	0	0	3s	$\Theta\phi = \dfrac{1}{2\sqrt{\pi}}$
3	1	(± 1)	$3p_x$	$\Theta\phi = \dfrac{\sqrt{3}(\sin\Theta\,\cos\phi)}{2\sqrt{\pi}}$
3	1	0	$3p_z$	$\Theta\phi = \dfrac{\sqrt{3}(\cos\Theta)}{2\sqrt{\pi}}$
3	1	(± 1)	$3p_y$	$\Theta\phi = \dfrac{\sqrt{3}(\sin\Theta\,\sin\phi)}{2\sqrt{\pi}}$
3	2	(± 2)	$3d_{x^2-y^2}$	$\Theta\phi = \dfrac{\sqrt{15}[\sin^2\Theta\,(\cos^2\phi - \sin^2\Theta)]}{4\sqrt{\pi}}$
3	2	(± 1)	$3d_{xz}$	$\Theta\phi = \dfrac{\sqrt{30}(\sin\Theta\,\cos\Theta\,\cos\phi)}{2\sqrt{2\pi}}$
3	2	0	$3d_{z^2}$	$\Theta\phi = \dfrac{\sqrt{5}(3\cos^2\Theta - 1)}{4\sqrt{\pi}}$
3	2	(± 1)	$3d_{yz}$	$\Theta\phi = \dfrac{\sqrt{30}(\sin\Theta\,\cos\Theta\,\sin\phi)}{2\sqrt{2\pi}}$
3	2	(± 2)	$3d_{xy}$	$\Theta\phi = \dfrac{\sqrt{15}(\sin^2\Theta\,\sin\phi\,\cos\phi)}{2\sqrt{\pi}}$

are found further away from the nucleus. Note that inner electrons can shield the nuclear charge from the outer electrons, and shielding will be discussed when discussing atomic properties (Section 3.8.1). These radial probability functions will be useful for several chemical bonding and electron transfer processes in subsequent chapters.

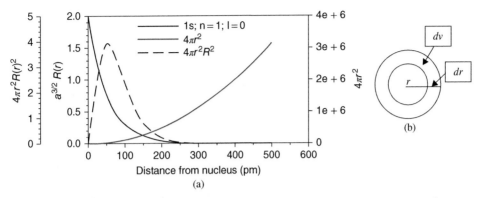

Figure 3.6 *(a) Plots of the 1s orbital (n = 1; l = 0) H atom wave function R(r) from Table 3.1, $4\pi r^2$, and $4\pi r^2 R(r)^2$. (b) Volume of a sphere (shell) of thickness dr*

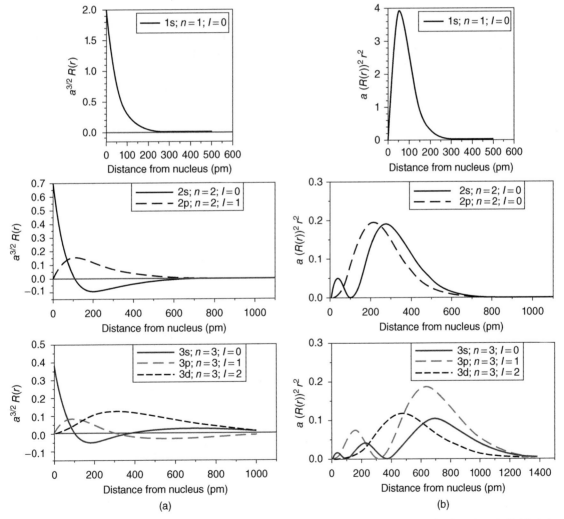

Figure 3.7 *(a) Plots of the hydrogen atom radial wave function versus r or $R_{nl}(r)$ versus r for functions in Table 3.1; (b) Plots of the hydrogen atom probability function or density versus r or $[R_{nl}(r)]^2$ versus r*

3.4.2   Angular Part of the Wavefunction $Y_{lml}(\theta, \phi)$ and Atomic Orbitals

The angular part of the wave function gives information on the number of sublevels and the orientation or shape of the electron in space. Table 3.2 on page 52 gives the hydrogen-like atom spherical harmonic wave functions, $Y_{lml}(\theta, \phi)$.

Figures 3.8 and 3.9 show plots of the $(\theta\phi)$ functions and of their squared functions $(\theta\phi)^2$ for the 1s and $2p_z$ orbitals in two dimensions, respectively. The 1s orbital is spherical over all space. However, the p_z orbital has positive and negative contributions from the cosine function (see also Figure 3.3) so has ungerade symmetry with respect to the center of inversion (see Section 4.2.3.2) through the nucleus. The angular probability plots $(\theta\phi)^2$ for the 1s and $2p_z$ orbitals are shown in Figure 3.9.

For the 1s orbital, these latter plots show no significant change in the shape but for p as well as d and f orbitals, the plots become elongated. The angular probability plots $(\theta\phi)^2$ are a picture or representation of what chemists describe as orbitals, and Figure 3.10 shows idealized angular probability plots or the orbital shapes for the s, p, and d orbitals. The signs of the *original* parts of the wave function, which are important for bonding considerations, for each set of orbitals are also shown. With respect to an inversion center through the nucleus of the atom at the center of the coordinate system, s orbitals are symmetric to inversion so are termed **gerade** (German for even) as the sign is the same everywhere. The p orbitals are not symmetric as the sign changes from "+" to "−" on going from one location to an equivalent location on the opposite side through the center of the nucleus; thus, they are termed **ungerade** (German for uneven). Figure 3.3 shows the change in sign for the cosine function representing the p_z orbital. Likewise, d orbitals are gerade and f orbitals are ungerade. The reader can also generate orbitals from freeware programs at the following web sites (https://www.wolframalpha.com/; http://www.winportal.com/winplot; https://www.wavefun.com/products/spartan.html).

Figure 3.11 shows the general set representation for the angular probability plots of the f orbitals. The cubic set (not drawn) would have the same $f_{5z^3-3zr^2}$ orbital (f_{z^3} for short) plus an f_{x^3} and an f_{y^3} orbital with the same shape on the x and y-axis, respectively. The f_{xyz} is identical in both the general and cubic sets. However, in the

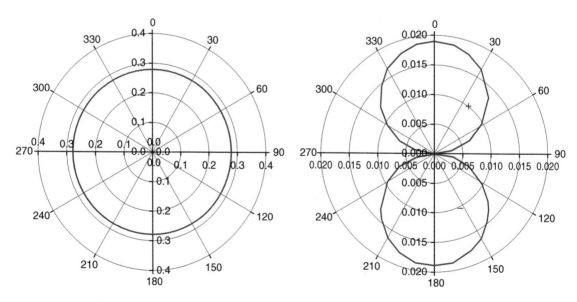

Figure 3.8   *Two-dimensional plots of the $\theta\varphi$ functions for the 1s and $2p_z$ orbitals*

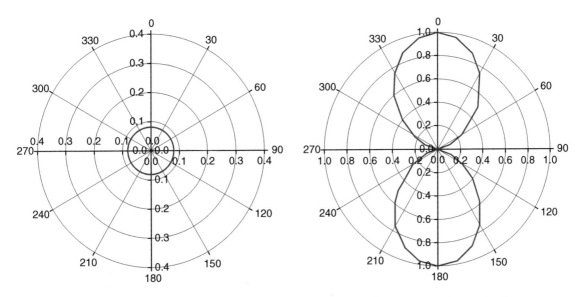

Figure 3.9   *Two-dimensional angular probability plots* $(\theta\phi)^2$ *for the 1s and 2p$_z$ orbitals*

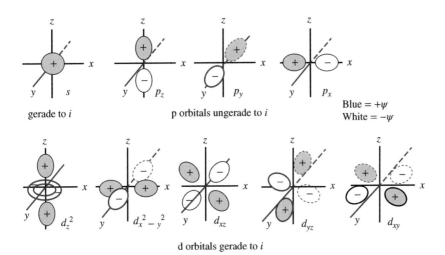

Figure 3.10   *Shapes of the s, p, and d orbitals from angular probability* $(\theta\phi)^2$ *plots. For this figure and Figure 3.11, blue color or shading indicates a "+" sign for the original function* $(\theta\phi)$ *and white shading indicates a "−" sign for the original function* $(\theta\phi)$. *Dashed lines indicate that the object is in the plane of the paper and bold blue lines indicate the object coming out of the plane of the paper*

cubic set, the remaining three orbitals are identical to the f_{xyz} orbital but are rotated 45° about the x, y, and z axes, respectively.

Methods for showing the total probability of finding the electron (multiplying the radial probability function by the angular probability function) anywhere in space are more quantitative descriptions and use shading or contour plots as in Figure 3.12. In these plots, there is an approach to a maximum electron density some

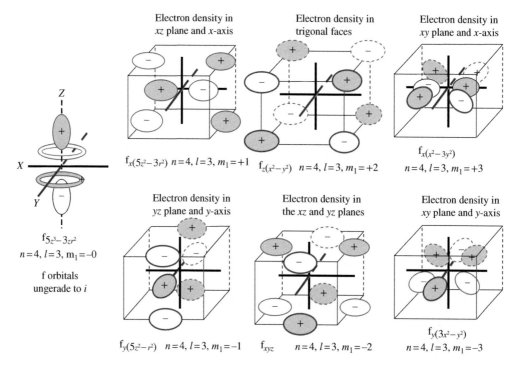

Figure 3.11   *The general set for the shapes of the f orbitals from angular probability $(\theta\phi)^2$ plots. The Cartesian coordinate axes are as in Figure 3.10 and the $f_{5z^3-3zr^2}$ or f_{z^3} orbital on the left*

distance from the nucleus followed by a decrease in electron density on moving farther away from the nucleus. The contour plots are relative to the maximum electron density, which is assigned a value of 1.

3.5   The Four Quantum Numbers

As noted above, when the Schrodinger equation is applied to the hydrogen atom or to any system with one electron and one nucleus, the solution gives the three quantum numbers n, l, and m_l. The **"principal" quantum number n** indicates the main energy level that an electron resides in and may have any integral value from 1 to infinity.

$$n = 1, 2, 3, \ldots, \infty$$

The value of n also indicates the number of sublevels or types of orbitals in the main energy level (for $n = 2$, there are "s" and "p" orbitals or sublevels) and $2n^2$ indicates the maximum number of electrons in the level n for $n = 2$; $2n^2 = 8$). The lower the n value, the more stable the orbital.

The **"azimuthal" or "orbital angular momentum" quantum number** l determines the shape of the orbital and indicates the sublevels in an energy level, n, and may have any integral value from zero to $n - 1$:

$$l = 0, 1, 2, \ldots, (n - 1)$$

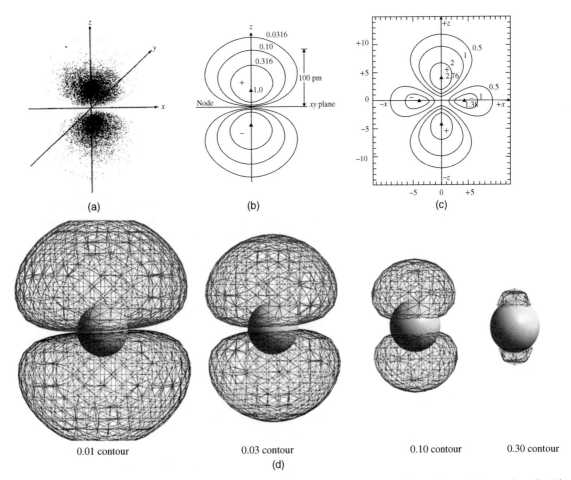

(a)   (b)   (c)

0.01 contour   0.03 contour   0.10 contour   0.30 contour

(d)

Figure 3.12   *(a) Pictorial representation of electron density (Source: From Ref. [6], Figure 5. Reproduced with permission from American Chemical Society). (b) Electron density contours for the hydrogen-like 2p orbital of carbon (Source: From Ref. [7], Figure 8. Reproduced with permission from American Chemical Society). (c)* d_{z^2} *orbital contour (Source: From Ref. [8], Figure 3. Reproduced with permission from American Chemical Society). The signs indicate the sign of the original angular function* ($\theta\phi$)*. (d) Hyperchem™ (version 8.0.10 from Hypercube Inc.) three-dimensional isosurface rendition encompassing the total electron density for the carbon* p_z *orbital at four different contour levels (blue is the "+" part of* ψ *and black is the "−" part of* ψ*); from left to right, the distance to the end of one lobe from the atom's center is 171, 144, 100, and 61 pm, respectively*

For historical reasons related to spectroscopy, l is not normally specified by these integers, but rather by the letters s, p, d, f, g, ... (continuing alphabetically), which correspond to $l = 0, 1, 2, 3, 4, ...$, respectively, where $(2l + 1)$ gives the number of orbitals in a given sublevel. These orbitals have the same energy; that is, they are said to be degenerate in the absence of an electric or magnetic field. The n and l values of an electron are often designated by the notation nl, in which the value of l is indicated by the appropriate letter. Thus a $2s$ electron has $n = 2$ and $l = 0$; a 2p electron has $n = 2$ and $l = 1$; a 3d electron has $n = 3$, $l = 2$.

The **"magnetic" quantum number,** m_l, is related to the component of the angular momentum on a given axis (z) upon application of a magnetic field, which removes degeneracy, and describes the orbital's orientation in space. It may have an integral value ranging from $-l$ to $+l$:

$$m_l = -l, -(l - 1), \ldots, -1, 0, +1, +2, \ldots, l - 1, +l$$

For the p orbitals where $l = 1$, m_l has three values 1, 0, -1 for a total of three orbitals. For the d orbitals where $l = 2$, there are five values, and for the f orbitals where $l = 3$, there are seven values. The s orbital is spherically symmetric around the nucleus and Unsöld's theorem indicates that the sum of the electron density for the set of three p orbitals, set of five d orbitals and set of seven f orbitals is also spherical.

Because an electron has spin, and consequently a magnetic moment that can be oriented either up or down in an applied magnetic field, a fourth quantum number, **the "spin" quantum number** m_s, must be specified. The permissible values of m_s are $\pm 1/2$. The values allowed for m_s are $+1/2$ (↑ or parallel with an applied magnetic field) and $-1/2$ (↓ or opposed to an applied magnetic field). The magnitude of the magnetic moment, μ, in Bohr magnetons $[eh/(4\pi m)]$ with units of 9.27×10^{-24} J T^{-1} is given in Equations 3.11 and 3.12.

$$\mu = 2[S(S + 1)]^{\frac{1}{2}} \text{ where } S = \sum m_s \tag{3.11}$$

or

$$\mu = [n(n + 2)]^{\frac{1}{2}} \text{ where } n = \text{\# of unpaired electrons} \tag{3.12}$$

When $S = 0$, all the electrons in the atom are paired as the individual spins of the electrons in the atom cancel. The atom is termed **diamagnetic** and the electrons are repelled slightly in a magnetic field. When $S > 0$, there are unpaired electrons in the atom, which are attracted in a magnetic field and the atom is called **paramagnetic**.

As a consequence of the restrictions on the quantum numbers, the electron of a hydrogen atom may be assigned only certain combinations of quantum numbers. These permissible combinations for $n = 1, 2, 3, 4$, and 5 are summarized below. Each allowed combination of n, l, and m_l corresponds to an atomic "orbital." Chemists indicate that the electron may be "put into" or "assigned to" a particular orbital or region of space as shown in Table 3.3. Of course, in any orbital, the m_s quantum number may be either $+1/2$ or $-1/2$.

3.6   The Polyelectronic Atoms and the Filling of Orbitals for the Atoms of the Elements

The Schrodinger wave functions for polyelectronic atoms are more complex but several techniques are now available to obtain relatively precise energies and probability densities. Not only are there multiple electrons but the additional nuclear charge, Z, is a principal difference in describing the electrons and their energies. The increased nuclear charge contracts all the orbitals and these adjusted orbitals are called hydrogen-like orbitals [for the wave functions in Table 3.1, replace r by Zr and multiply each function by $(Z)^{3/2}$]. Spectroscopic methods have provided the data leading to the ground state electron configuration of all the elements in order to build up the periodic table of the elements. In a given energy level, the stability of the atomic orbital energies (recall that E is negative in Equation 3.4) tends to increase as s > p > d > f (see Figure 3.15a). At higher n values, the energies of orbitals with different n and l values overlap such that a staggering occurs; for example, 4s < 3d for Ca so the 4s orbital is filled. The filling of the outermost electrons into the orbitals is best described by the following order (part of the **Aufbau** principle, Section 3.7) 1s < 2s < 2p < 3s < 3p < 4s <

Table 3.3 *Some allowed values of the hydrogenic atom quantum numbers to give orbitals and distributions of electrons*

n	l	m_l	m_s	Number of electrons by *orbital* (*l*)	Number of electrons by *n* level	*nl* orbital designation
1	0	0	$\pm 1/2$	2	2	1s
2	0	0	$\pm 1/2$	2	8	2s
2	1	−1, 0, +1	$\pm 1/2$	6		2p
3	0	0	$\pm 1/2$	2	18	3s
3	1	−1, 0, +1	$\pm 1/2$	6		3p
3	2	−2, −1, 0, +1, +2	$\pm 1/2$	10		3d
4	0	0	$\pm 1/2$	2	32	4s
4	1	−1, 0, +1	$\pm 1/2$	6		4p
4	2	−2, −1, 0, +1, +2	$\pm 1/2$	10		4d
4	3	−3, −2, −1, 0, +1, +2, +3	$\pm 1/2$	14		4f
5	0	0	$\pm 1/2$	2	50	5s
5	1	−1, 0, +1	$\pm 1/2$	6		5p
5	2	−2, −1, 0, +1, +2	$\pm 1/2$	10		5d
5	3	−3, −2, −1, 0, +1, +2, +3	$\pm 1/2$	14		5f
5	4	−4, −3, −2, −1, 0, +1, +2, +3, +4	$\pm 1/2$	18		5g

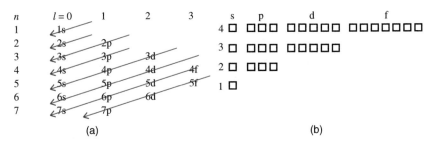

Figure 3.13 *(a) Schematic for the filling of the electrons into the known orbitals of atoms to give their ground state electronic configurations. (b) The number of sublevels and orbitals for a given sublevel from n = 1 through 4*

3d < 4p < 5s < 4d < 5p < 6s < 5d $\simeq$ 4f < 6p < 7s < 6d < 5f. The blue arrows (Figure 3.13) show the order of filling for orbitals, which are known to be filled or partially filled with electrons. Although this order is not fully accurate, it is very useful. The staggering of orbitals is a consequence of the shielding of the outermost electrons by the nuclear charge so that the effective nuclear charge (Z_{eff} or Z^*) is given by $Z_{eff} = Z - S$, where S is a shielding or screening constant (see Section 3.8.1).

The entire periodic table with the ground state electronic configuration based on known orbital energies is given in Figure 3.14. The complete electron configuration is normally given by using the noble gas configuration and then the outermost electrons. For example, Fe ($Z = 26$) has a complete electron configuration of $1s^2\, 2s^2\, sp^6\, 3s^2\, 3p^6\, 4s^2\, 3d^6$. As the argon configuration is $1s^2\, 2s^2\, sp^6\, 3s^2\, 3p^6$, Fe would have the shorthand notation: Ar $4s^3\, 3d^6$. In Figure 3.14, only the outermost or **valence** electrons are given, and the noble gas electron configurations are assumed.

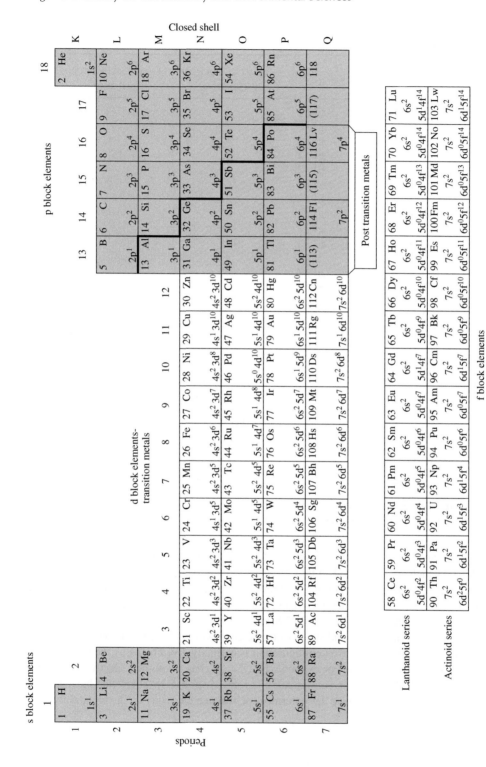

Figure 3.14   *The ground state electron configuration of the elements*

There are seven periods or rows of the known 118 elements, which are also given the notation K through Q. Of the first 92 elements, 91 are found in nature (Tc has been made in nuclear reactions).

From left to right, there are 18 columns or families that can be divided into four blocks (s, p, d, and f) based on the last orbitals filled by electrons. Elements in the same family or group of elements are also called **congeners** and have similar atomic or chemical properties.

Groups 1 (alkali metals) and 2 (alkaline earth metals) have the ns^1 and ns^2 electron configurations; thus they are called the "s" block elements. Group 1 and 2 elements readily lose one or two electrons, respectively so that they exist as ions with $+1$ or $+2$ oxidation states.

The "p" block elements are groups 13–18 with the $ns^2 \, np^n$ electron configuration. Group 16 elements are the chalcogens, Group 17 elements are the halogens, and Group 18 elements are the noble gases. The "s" and "p" block elements are also called the **main group** elements.

The "d" block elements are Groups 3–12 and are called **transition metals** with the general electron configuration $ns^2(n-1)d^n$.

The "f" block elements have the $ns^2(n-1)d^1(n-2)f^n$ electron configuration. Their most common oxidation state is $+3$ for the removal of the three $ns^2(n-1)d^1$ electrons. The **lanthanides** start after lanthanum and the **actinides** start after actinium.

3.7 Aufbau Principle

The Aufbau or building up principle gives the method for determining the ground state electron configuration by taking the above orbital ordering and combining it with Hund's rule and the Pauli Exclusion Principle. **Hund's rule** is the rule of maximum multiplicity, which states that every orbital in a subshell or sublevel is occupied by a single electron before any orbital is doubly occupied and all electrons in singly occupied orbitals have their spins in the same direction or aligned with the magnetic field. This rule takes note of the fact that electrons are of negative charge, and repulsion of electrons in the same region of space is minimized by placing them in different orbitals of the same sublevel. In addition, these electrons with aligned spins shield each other less from the nucleus. The **Pauli Exclusion Principle** states that no two electrons of the same atom can have the same four quantum numbers (i.e., the same energy); that is, two electrons cannot occupy the same region of space with the same spin. They can occupy the same region of space (orbital) if they have a different magnetic moment (m_s). The following are the allowed ground states for the last electron from H through N showing these rules.

^{1}H $= 1\,s^1$	$1, 0, 0, +\tfrac{1}{2}$	
^{2}He $= 1\,s^2$	$1, 0, 0, -\tfrac{1}{2}$	
$_3$Li $= 2\,s^1$	$2, 0, 0, +\tfrac{1}{2}$	
$_4$Be $= 2\,s^2$	$2, 0, 0, -\tfrac{1}{2}$	
$_5$B $= 2\,s^2 2p^1$	$2, 1, +1, +\tfrac{1}{2}$	$p_x \, l = 1;\, m_l = +1$
$_6$C $= 2\,s^2 2p^2$	$2, 1, 0, +\tfrac{1}{2}$	$p_z \, l = 1;\, m_l = 0$ as the p electrons are filled below
$_7$N $= 2\,s^2 2p^3$	$2, 1, -1, +\tfrac{1}{2}$	$p_y \, l = 1;\, m_l = -1$

For carbon, the ground electronic state of the 2p orbitals is represented on the left below:

$$
\begin{array}{ccc}
\uparrow \quad \uparrow \quad \underline{} & \uparrow \quad \downarrow \quad \underline{} & \uparrow\downarrow \quad \underline{} \quad \underline{} \\
2p_x \ 2p_z \ 2p_y & 2p_x \ 2p_z \ 2p_y & 2p_x \ 2p_z \ 2p_y \\
m_l = +1 \quad 0 \quad -1 & +1 \quad 0 \quad -1 & +1 \quad 0 \quad -1
\end{array}
$$

Ground state   Excited states

and its full electron configuration can be represented as $1s^2 \ 2s^2 \ 2p_x{}^1 \ 2p_z{}^1$. The $2p_z$ orbital is assigned $m_l = 0$; this is true for the $3d_{z^2}$ orbital as the z axis is chosen as the axis of the applied magnetic field. Two excited energy states (also called **microstates**) for carbon are on the right of the ground state, and are not in accordance with Hund's rule.

The filling of the d orbitals for the transition metals and of the f orbitals for the lanthanides and actinides occurs in a similar manner. For Fe ($Z = 26$), the electronic configuration is Ar $4s^2 \ 3d^6$ where four of the d orbitals are singly occupied with electrons. For Cr and Mo, the outermost electron configurations are s^1d^5, which gives maximum multiplicity of six unpaired electrons, but W has the s^2d^4 configuration. Cu, Ag, and Au all show the s^1d^{10} configuration. The filling of the d orbitals with half-filled s orbitals for these five elements gives a symmetrical and energetically favorable arrangement of electrons around these atoms.

3.8   Atomic Properties

3.8.1   Orbitals Energies and Shielding

The energy of a polyelectronic atom is a function of Z^2/n^2 (Equation 3.4) but Z increases faster than n as electrons are added. On adding electrons around the nucleus, the inner electrons now shield the outer electrons from the nuclear charge so the energy of an electron is now given as Equation 3.13.

$$
E_n \propto -\left[\frac{(Z-S)^2}{n_i{}^2}\right] = \frac{-Z_{eff}{}^2}{n_i{}^2} \tag{3.13}
$$

where Z_{eff} or $Z^* = Z - S$. Z_{eff} or Z^* is the effective nuclear charge and S is the shielding or screening constant (sometimes given as σ). This update to the energy of the electron (Equation 3.4) is obviously necessary when one compares the ionization potential (IP) of the H atom with the lithium atom. For H, $Z^2/n^2 = 1$ whereas for Li, $Z^2/n^2 = 9/4 = 2.25$. However, the IP of H is 13.6 eV and almost twice that of Li, which has an IP of 5.39 eV. A principal reason for this lowering of the IP with increase in the principal quantum number can be seen with the probability density plots (Figure 3.7) for the 1s and 2s electrons. Here, the 1s electrons penetrate closer to the nucleus than does the 2s electron, which is now shielded by the 1s electrons. Also, the 1s electrons will repel the 2s electron and this combination makes the 2s electron in Li easier to remove. Thus, Z_{eff} must be some number between 1 and 2 to account for the lower IP of Li.

Further inspection of the probability density plots (Figure 3.7) indicates that s electrons are not shielded well by p or d electrons. Electrons from d and f orbitals (higher l values) are poor shielders as they penetrate less to the nucleus. These effects have been taken into account by Slater who provided a set of empirical rules that estimated a value for the shielding constant, S. These rules are listed here:

1. Write out the electronic configuration of the element as follows:
    (1s) (2s, 2p) (3s, 3p) (3d) (4s, 4p) (4d) (4f) (5s, 5p) etc.
    This is different from the Aufbau principle.

2. Electrons in any group to the right of a (ns, np) group contribute zero to S. This rule neglects that the 4s and higher ns electrons can penetrate to the nucleus and shield 3d and ($n-1$)d electrons.
3. All of the other electrons in an (ns, np) group contribute 0.35 to S.
4. All electrons in the $n-1$ level contribute 0.85 to S.
5. All electrons in the $n-2$ level are considered to shield the outer electron completely and contribute 1.00 to S.
6. If the outermost electron being considered is an nd or nf electron, rules 4 and 5 now become: all electrons in the groups to the left of the nd and nf group contribute 1.00 to S.

For the C atom ($Z = 6$), the orbital grouping would be $(1s)^2 (2s, 2p)^4$. $S = (2 \times 0.85) + (3 \times 0.35) = 2.75$. Thus, $Z - S = 6 - 2.75 = 3.25$.

For transition metals, we need to consider both s and d electrons. For Cu ($Z = 29$), the orbital grouping would be $(1s)^2 (2s, 2p)^8 (3s, 3p)^8 (3d)^9 (4s)^1$.

For the 4s electron, $S = (10 \times 1.00) + (17 \times 0.85) = 24.45$ so $Z - S = 29 - 24.45 = 4.55$.
For a 3d electron, $S = (18 \times 1.00) + (8 \times 0.35) = 20.8$ so $Z - S = 29 - 20.8 = 8.2$.

Using Cu as an example, the electrons in d and f orbitals ($S = 1.00$) are screened much better than s and p electrons ($S = 0.85$) by the electrons immediately below them in n value. Also ns and np electrons shield nd and nf effectively ($S = 1.0$). These contributions are consistent with Figure 3.7. However, the 4s orbital is ignored even though it has a finite probability to be near the nucleus. These calculations still allow us to predict that the 4s electron will be removed before a 3d electron in agreement with the known chemistry of the transition metals. Because transition metals have the general electron configuration of $ns^2(n-1)d^n$, the +2 oxidation state where the ns^2 electrons are lost first is very common. Thus, M^{2+} have the $(n-1)d^n$ configuration.

However, this removal of outermost electrons *is in contrast to the Aufbau principle*, which has the 3d orbitals being occupied last. When the energies of all the electrons are summed, the loss or ionization of the 4s electrons is favored as the 3d orbitals are more diffuse and electron–electron repulsions are less than electron–electron repulsions in the 4s orbitals. As Z and Z_{eff} increase across a row, the polyelectronic orbitals behave much as hydrogen orbitals with all orbitals in a given n level residing at a lower or more stable energy than the orbitals in the $n+1$ level. This ordering of 3d versus 4s and 4p orbitals is observed in ESCA spectra, and Figure 3.15a shows this ordering for Kr.

Since Slater proposed these rules, better information on shielding constants has been obtained and Table 3.4 shows values of the effective nuclear charge for the atoms. The trend for Z_{eff} is to increase across a row in the periodic table and decrease for the outermost orbital on descending a family (given n value; e.g., 2s for Li versus 3s for Na) in the periodic table. The periodic properties of the elements, which are described shortly, are directly related to Z_{eff}.

3.8.2 Term Symbols: Coupling of Spin and Orbital Angular Momentum

Figure 3.15a shows the ESCA spectra for the noble gases. Neon has single peaks for the 1s, 2s, and 2p orbitals, but Ar, Kr, and Xe show a splitting of peaks in some instances. For Ar, there are peaks for electrons occupying 2s, 2p, 3s, and 3p orbitals. Note the significant splitting of the peak due to the 2p orbitals; the splitting of the valence p orbitals for Ar (3p), Kr (4p), and Xe (5p) is smaller and only resolved with low energy radiation (Figure 3.15b). The splitting of these peaks is related to the coupling of the orbital angular momentum (L, which is related to rotation of the electron cloud) and the spin angular momentum of the electron (S) to give the

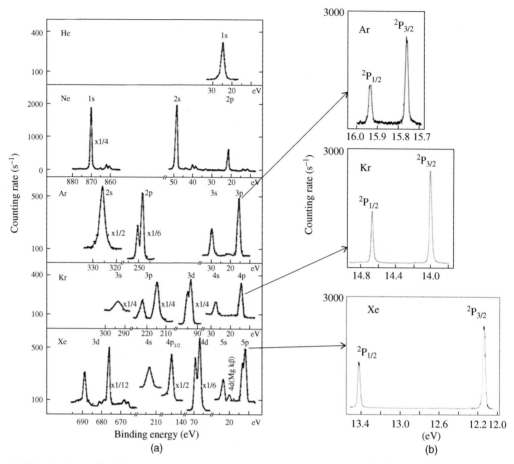

Figure 3.15   *(a) X-ray photoelectron (XPS or ESCA) spectra for He, Ne, Ar, Kr, and Xe excited by Mg Kα X-radiation (Source: From Ref. [4], Figure 4.1. Reproduced with permission). Mg Kα X-radiation (1.254 KeV) is not energetic enough (1) to ionize the 1s electrons for Ar to Xe and not monochromatic enough to resolve the $3p_{1/2}$ and $3p_{3/2}$ peaks for Ar; (2) to ionize the 1s and 2s electrons and not monochromatic enough to resolve the $4p_{1/2}$ and $4p_{3/2}$ peaks for Kr; (3) to ionize the 1s, 2s, and 3s electrons and not monochromatic enough to resolve the $4p_{1/2}$ and $4p_{3/2}$ peaks for Xe. (b) Ar, Kr, and Xe UV photoelectron spectra excited with the low energy He emission of 21.22 eV resolves the 3p, 4p, and 5p peaks of Ar, Kr, and Xe, respectively (Source: From Ref. [9], Figures 3.1–3.3. Reproduced with permission from Wiley, Inc.)*

total angular momentum of the electron (**spin–orbit or Russell–Saunders coupling**), which is represented by the spectroscopic term symbol, J (given as the absolute value of L and S as described in the next paragraph). This coupling is due to the electron–electron repulsions that occur as orbitals are filled because there is more than one option for assigning the electrons to degenerate orbitals that are in different regions of space (as noted for the Carbon 2p orbitals in Section 3.7). The various assignments are called **microstates** and have different energies, which can be resolved spectroscopically into different peaks. This type of coupling is valid for the elements with $Z \leq 30$ and for the examples given throughout this book including the noble gases (Figure 3.15).

Table 3.4 Effective nuclear charges (Z_{eff}) for all occupied orbitals of the first 36 elements

	H		Be	B	C	N	O	F	He
1s	1.00								1.69

	Li	Be	B	C	N	O	F	Ne
1s	2.69	3.68	468	5.67	6.66	7.66	8.65	9.64
2s	1.28	1.91	2.58	3.22	3.85	4.49	5.13	5.76
2p			2.42	3.14	3.83	4.45	5.10	5.76

	Na	Mg	Al	Si	P	S	Cl	Ar
1s	10.63	11.61	12.59	13.57	14.56	15.54	16.52	17.51
2s	6.57	7.39	8.21	9.02	9.82	10.63	11.43	12.23
2p	6.80	7.83	8.96	9.94	10.96	11.98	12.99	14.01
3s	2.51	3.31	4.12	4.90	5.64	6.37	7.07	7.76
3p			4.07	4.29	4.89	5.48	6.12	6.76

	K	Ca	Ga	Ge	As	Se	Br	Kr
1s	18.49	19.47	30.31	31.29	32.27	33.26	34.24	35.23
2s	13.01	13.77	22.60	23.36	24.12	24.88	25.64	26.39
2p	15.02	16.04	27.09	28.08	29.07	30.06	31.05	32.04
3s	8.68	9.60	16.99	17.76	18.59	19.40	20.21	21.03
3p	7.73	8.66	16.20	17.01	17.85	18.70	19.57	20.43
4s	3.49	4.39	7.06	8.04	8.94	9.75	10.55	11.31
3d			15.09	16.25	17.37	18.47	19.55	20.62
4p			6.22	6.78	7.44	8.28	9.02	9.77

	Sc	Ti	V	Cr	Mn	Fe	Co	Ni	Cu	Zn
1s	20.45	21.44	22.42	23.41	24.39	25.38	26.36	27.35	28.33	29.32
2s	14.57	15.37	16.18	16.98	17.79	18.60	19.40	20.12	21.02	21.82
2p	7.06	18.06	19.07	20.07	21.08	22.09	23.09	24.09	25.09	26.09
3s	10.34	11.03	11.70	12.36	13.02	13.67	14.32	14.96	15.59	16.21
3p	9.40	10.10	10.78	11.46	12.11	12.77	13.43	14.08	14.73	15.36
4s	4.63	4.82	4.98	5.13	5.28	5.43	5.57	5.71	5.85	5.96
3d	7.12	8.14	8.98	9.75	10.52	11.18	11.85	12.53	13.20	13.87

The value of $L = \sum m_l$ and $S = \sum m_s$; $J = |L+S|, |L+S|-1, \ldots, |L-S|$ in integer steps. A value of $m_l = 1$ and -1 corresponds to p_x and p_y orbitals, respectively, and the value of $m_l = 0$ corresponds to a p_z orbital when an electric or magnetic field is applied. For the removal of a single p electron from the noble gases to form the M^+ ion as occurs during an ESCA experiment, the five p electrons can be arranged in six separate microstates (Figure 3.16).

For the **top left or ground state**, $L = \sum m_l = 2(1) + 2(0) + 1(-1) = 1$, and $S = \frac{1}{2}$; for the bottom left, $L = \sum m_l = 2(1) + 2(0) + 1(-1) = 1$, and $S = -\frac{1}{2}$. For the top center configuration, $L = \sum m_l = 1(1) + 2(0) + 2(-1) = -1$, and $S = \frac{1}{2}$; for the bottom center, $L = \sum m_l = 1(1) + 2(0) + 2(-1) = -1$, and $S = -\frac{1}{2}$. For the top right configuration, $L = \sum m_l = 2(1) + 1(0) + 2(-1) = 0$, and $S = \frac{1}{2}$; for the bottom right, $L = \sum m_l = 2(1) + 1(0) + 2(-1) = 0$, and $S = -\frac{1}{2}$. These six individual m_1 and m_s values are assembled into the matrix table on the right of Figure 3.16.

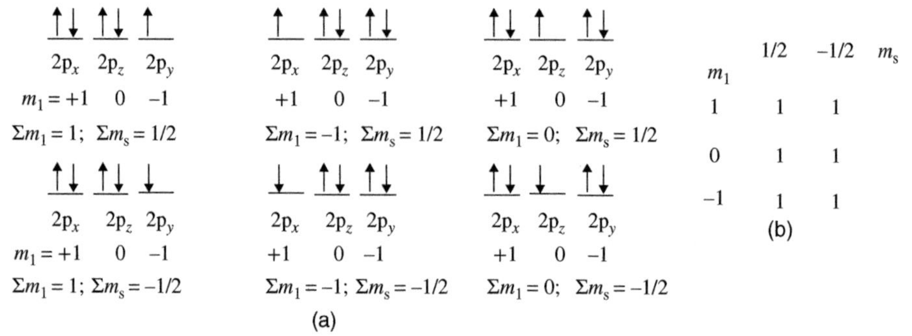

Figure 3.16   *(a) The six microstates of different energy for the filling of five p electrons into the p_x, p_y, and p_z orbitals. (b) The m_1 and m_s matrix comprising the ^2P term symbol*

For a p^1 electronic configuration, the values for L and S are the same as for the p^5 case (a concept known as **hole formalism** as one electron in one of the three p orbitals is similar to a vacancy or missing an electron in one of the three p orbitals as the other positions are occupied). The vector sum of L and S gives the following J values.

$$J = \frac{3}{2} = |L + S| = \left|1 + \frac{1}{2}\right| = \left|-1 - \frac{1}{2}\right|$$

$$J = \frac{1}{2} = |L + S| - 1 = \left|1 + \frac{1}{2}\right| - 1 = \left|-1 - \frac{1}{2}\right| - 1$$

As a result of this spin–orbit coupling, the electron states or levels are now given as spectroscopic term symbols of the form

$$n^{(2S+1)}L_J$$

When $L \geq S$, the multiplicity term $(2S + 1)$ indicates the total number of J states for a given L and S where $L = 0$ is an S state; $L = 1$ is P; $L = 2$ is D, $L = 3$ is F, then continuing alphabetically but omitting J as J is the combination of L and S.

The most stable energy state is the one with maximum multiplicity $(2S + 1)$, then the state with highest L. For **atoms** in the ground state and with sublevels less than half filled, the minimum J value is the lowest energy state, and for sublevels more than half filled, the maximum J value is the lowest energy state (for ions the J values reverse). For Ar^+ in Figure 3.15b, the coupling of electron spin with the angular momentum of the 3p orbitals results in two energy states $3^2P_{1/2}$ (representing the p_z orbital) and $3^2P_{3/2}$ (representing the p_x and p_y orbitals), which are observed in a 1:2 ratio, respectively. The term ^2P is called a doublet P state as the multiplicity is 2; for a multiplicity of 1, the term is singlet; for a multiplicity of 3, the term is triplet. Kr and Xe also have two energy states, $^2P_{1/2}$ and $^2P_{3/2}$, for the 4p and 5p orbitals, respectively.

The s orbitals have only one value of m_1, which is 0, so there is no coupling and no splitting of the peak for 1s orbitals as shown in Figure 3.15a for ionized helium $(1s^1)$, which has a single term symbol or energy state, $^2S_{1/2}$ (the electron spin can be aligned with or against an applied magnetic field).

$$J = \frac{1}{2} = |L + S| = |L - S| = \left|0 + \frac{1}{2}\right| = \left|0 - \frac{1}{2}\right|$$

Russell–Saunders coupling will be discussed in more detail in the discussion on transition metals and the spectroscopy of their complexes (Section 8.8.2.2; the determination of the spectroscopic terms for the d^2

state is explicitly detailed). In Section 2.7.2.1, the dissociation of O_3 led to O_2 and O atoms. On dissociation, the O atom can be in the ground state ($p_x^2 p_z^1 p_y^1$ or the ^3P state) or in an excited state ($p_x^2 p_z^2 p_y^0$ or the ^1D state). As the excited state has an empty orbital, it is more reactive. O_2 also has a ground state $\left(^3\Sigma_g^- \right)$ and an excited state ($^1\Delta_g$) as discussed in Section 5.4.8.1. Dissociation leads to **spin conservation** so both products must be triple or singlet.

3.8.3 Periodic Properties – Atomic Radius

There are several atomic properties that follow from understanding the orbital energies of the atom. The first is the atomic radius, r, for which Equation 3.5 is now modified for a multielectron atom to the relationship in Equation 3.14.

$$r \propto \frac{a_0 n^2}{Z_{\text{eff}}} \tag{3.14}$$

where $a_0 = 52.9\,\text{pm}$. This function indicates a trend that the radius will increase on going down a column or family in the periodic table and will decrease from left to right on going across a row of the periodic table.

Here, the atomic radius is defined for nonmetallic and metallic elements as follows. For nonmetallic elements, the atomic radius is defined as half of the bond or internuclear distance between adjacent atoms of the same element in a molecule (preferably with single bonds). For metals, the metallic radius is defined as half of the experimentally determined distance between the centers of adjacent atoms in the solid. However, different metals have a varying number of adjacent or nearest neighbors so the atomic radius can be normalized to a maximum coordination number of 12 nearest neighbors (Section 6.2.7.1). With these caveats, Table 3.5 provides information on atomic radii.

The table of atomic radii shows the regular increase in radius going down the periodic table signifying the increase in the value of n and the regular decrease on going across the periodic table signifying the increase of the atomic number Z. Of particular note is the first filling of the f orbitals after lanthanum. The atomic radius for the transition metal atoms in row or period 6 is almost identical to the atomic radius of the transition metal atoms in row 5. The similarity in atomic radii is called the **lanthanide contraction** and is a consequence of the poor shielding of the nuclear charge by the electrons in 4f orbitals, which are filled before the 5d orbitals. A similar contraction is also found for gallium and germanium when compared to aluminum and silicon, respectively. This is a consequence of the first filling of the 3d orbitals and is called the scandide contraction. Figure 3.17 provides the atomic radii as Z varies.

For cations of a given element, the ionic radius is smaller than the atomic radius as a result of the increased charge attracting the outermost electrons. For anions of a given element, the ionic radius is larger than the atomic radius as the outermost electrons are not as easily attracted by the nuclear charge. More information on ionic radii is provided in Table 6.4 during the discussion of ionic solids and bonding.

3.8.4 Periodic Properties – Ionization Potential (IP)

The ionization potential (IP) is defined as the energy required to remove an electron from an atom so the value of the IP is endergonic and has a positive sign. It can be described by the following chemical equation:

$$A_{(g)} \rightarrow A_{(g)}^+ + e^-_{(g)} \quad \Delta E_{\text{IP}} = -E_n = +\text{value}$$

Note that the energy of the orbital is negative as it is stable; thus, the IP is defined as $\Delta E_{\text{IP}} = -E_n$. First ionization energies of the elements are given in Table 3.6. Figure 3.18 also shows that the ionization potential

Table 3.5   Atomic radii in picometers from [10]. Values in blue show the "lanthanide contraction"

1	2	3	4	5	6	7	8	9	10	11	12	13	14	15	16	17	18
H 53																	He
Li 157	Be 112											B 88	C 77	N 74	O 73	F 71	Ne
Na 191	Mg 160											Al 143	Si 118	P 110	S 104	Cl 99	Ar
K 235	Ca 197	Sc 164	Ti 147	V 135	Cr 129	Mn 137	Fe 126	Co 125	Ni 125	Cu 128	Zn 137	Ga 140	Ge 122	As 122	Se 117	Br 114	Kr
Rb 250	Sr 215	Y 182	Zr 160	Nb 147	Mo 140	Tc 135	Ru 134	Rh 134	Pd 137	Ag 144	Cd 152	In 150	Sn 140	Sb 141	Te 135	I 133	Xe
Cs 272	Ba 224	La 188	Hf 159	Ta 147	W 141	Re 137	Os 135	Ir 136	Pt 139	Au 144	Hg 155	Tl 155	Pb 154	Bi 152	Po	At	Rn

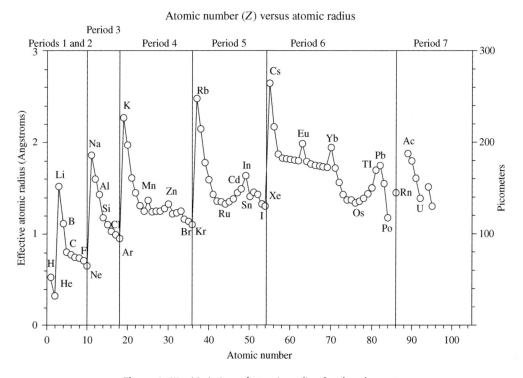

Figure 3.17   *Variation of atomic radius for the elements*

generally decreases on going down a family of the periodic table and increases on going across a row in the periodic table. This trend is almost an exact replica of what was seen for atomic radius, but in reverse order as the equations for atomic radius and ionization potential are inverse in both Z and n. The shielding effect is also observed for the ionization potential of the elements after the first filling of the "f" orbitals (the **lanthanide contraction**). The elements with 5d electrons (from Hf through Pb) have higher IP values than the family members above them. In nature, Tl exists in the +1 (not +3) oxidation state, and Sn and Pb exist in the +2 (not the +4) oxidation state; this is called the **inert "s" pair** effect as the s electrons are not readily removed. Ga and Ge also have values of the IP that are close to Al and Si, respectively.

The ionization potential is useful for describing other properties of the atom and in providing energies of the electrons in their respective orbitals. Figure 3.19 shows the orbital energies, which are the negative value of the ionization energies, and the electron configurations for the first 11 elements in the periodic table. Note that there is a decrease in orbital stability with an increase in n on going down the periodic table (e.g., hydrogen to lithium to sodium). Also, there is an increase in orbital stability on increasing Z on going across the second row of the periodic table, but there are some slight deviations in orbital energies as different orbitals fill up across the periodic table. For boron, it is easier to remove a 2p electron than one of the 2s electrons in beryllium. Recall that s electrons have a finite probability of being found near the nucleus whereas electrons in p, d, and f orbitals do not (see Figure 3.7). For oxygen, there is the first pairing of two electrons in one of the 2p orbitals. As noted before, placing two electrons of opposite spin in the same orbital is a way of balancing the repulsive nature of two electrons that are negative in charge. Thus, it is easier to remove the first electron from oxygen relative to that of nitrogen, which has a filled sub level. Note that the energy gap between the 2s and 2p orbitals increases with the nuclear charge on going across the second row.

Table 3.6 First ionization energy and electron affinity values in electron volts. IE and EA values in blue show the effects of the lanthanide contraction

1	2	3	4	5	6	7	8	9	10	11	12	13	14	15	16	17	18
H 13.6 / 0.754																	**He** 24.6 / −0.50
Li 5.39 / 0.618	**Be** 9.32 / −0.50				IP EA							**B** 8.30 / 0.277	**C** 11.6 / 1.26	**N** 14.5 / −0.07	**O** 13.6 / 1.46	**F** 17.4 / 3.39	**Ne** 21.6 / −1.20
Na 5.14 / 0.548	**Mg** 7.6 / −0.40											**Al** 5.98 / 0.441	**Si** 8.15 / 1.385	**P** 10.5 / 0.747	**S** 10.3 / 2.077	**Cl** 12.9 / 3.617	**Ar** 15.7 / −1.00
K 4.34 / 0.520	**Ca** 6.11 / −0.30	**Sc** 6.54 / –	**Ti** 6.82 / 0.200	**V** 6.74 / 0.500	**Cr** 6.76 / 0.653	**Mn** 7.43 / 0	**Fe** 7.87 / 0.162	**Co** 7.86 / 0.660	**Ni** 7.63 / 1.15	**Cu** 7.72 / 1.27	**Zn** 9.39 / 0	**Ga** 5.99 / 0.30	**Ge** 7.89 / 1.20	**As** 9.81 / 0.81	**Se** 9.75 / 2.02	**Br** 11.8 / 3.36	**Kr** 13.9 / −1.00
Rb 4.17 / 0.486	**Sr** 5.69 / 0	**Y** 6.38 / 0.306	**Zr** 6.84 / 0.425	**Nb** 6.88 / 0.892	**Mo** 7.09 / 0.99	**Tc** 7.28 / 0.70	**Ru** 7.37 / 1.10	**Rh** 7.46 / 1.20	**Pd** 8.34 / 0.60	**Ag** 7.57 / 1.30	**Cd** 8.99 / –	**In** 5.78 / 0.30	**Sn** 7.34 / 1.20	**Sb** 8.64 / 1.07	**Te** 9.00 / 1.97	**I** 10.4 / 3.06	**Xe** 12.1 / −0.80
Cs 3.89 / 0.471	**Ba** 5.21	**La** 5.57 / 0.50	**Hf** 7.0	**Ta** 7.89 / 0.60	**W** 7.98 / 0.60	**Re** 7.88 / 0.15	**Os** 8.7 / 1.10	**Ir** 9.1 / 1.16	**Pt** 9.0 / 2.12	**Au** 9.22 / 2.30	**Hg** 10.4 / 0	**Tl** 6.10 / 0.518	**Pb** 7.41 / 1.03	**Bi** 7.28 / 1.03	**Po** 8.42 / 1.90	**At** 2.8	**Rn** 10.7 / −0.7

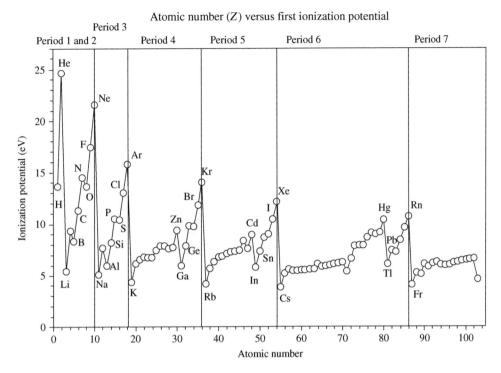

Atomic number (Z) versus first ionization potential

Figure 3.18 *Variation in first ionization potential (eV) for the elements (Source: From Ref. [11])*

The first eight ionization potentials of the first 36 elements are given in Table 3.7. Inspection of these values provides information on the loss of electrons from different levels (n) and sublevels (l) as also shown in Figure 3.19. For example, the second ionization energy of Li is significant as there is removal of a 1s electron after removing the 2s electron. These energies may be used to speculate on the oxidation state or chemical form of an element on another planet such as Mars. Analysis of surface soils from the Martian surface showed that perchlorate and not chloride is the dominant chlorine species unlike Earth [12] (Section 2.7.2.2). The total energy for the first seven electrons to ionize Cl to Cl^{7+} is 409 eV. As the Martian atmosphere is 0.6% of the Earth (as Mars lost its early atmosphere), 409 eV is easily accessible via X-ray and cosmic radiation, which bombards the Martian surface and its atmosphere, allowing formation of a variety of oxidants. Any oxygen or water present would permit stabilization of Cl^{7+} as perchlorate. Similar calculations can be done for ferrate (FeO_4^{2-}) and permanganate (MnO_4^-) and indicate that much of the oxygen on the surface of Mars is locked up with higher nonmetal and metal oxidation states.

3.8.5 Periodic Properties – Electron Affinity (EA)

The electron affinity (EA) is defined as the energy gained or released when an atom accepts an electron into its valence shell. Typically, the energy for this reaction is exergonic for the first electron accepted and has a negative value; thus, EA = $-\Delta E$ to provide a positive number for EA (Table 3.6). There are two equations that can describe the process, and the first is

$$A_{(g)} + e^-_{(g)} \rightarrow A_{(g)}^- \qquad \Delta E_{EA} = -\text{usually}$$

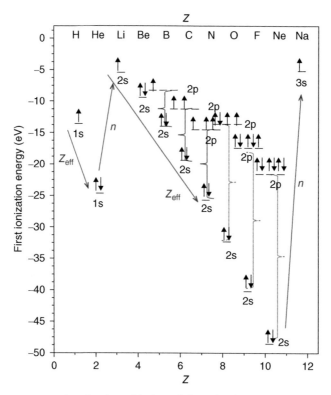

Figure 3.19 *The electron occupation for the orbitals and the orbital energies for the first eleven atoms in the periodic table. Brackets show the 2s, 2p orbitals for a given atom*

The second way to express electron affinity is as the energy required to remove an electron from a negative ion, which is the reverse process of the above. The energy for this process has a positive sign as does the EA value.

$$A_{(g)}^- \rightarrow A_{(g)} + e^-_{(g)} \qquad \Delta E_{EA} = +\text{usually}$$

Removal of the second electron to produce the dianion of an atom (e.g., O and S) results in negative values for the second EA values (endergonic process) as it is energetically difficult for the nuclear charge to overcome the added electron repulsion effects. In fact, the sum of both EA values for O (-6.25 eV) and S (-2.65 eV) [13] to become O^{2-} and S^{2-} shows an endergonic process indicating that these anions are not stable unless stabilized in a crystal (chapter 6) or in a hydrolysis product. Figure 3.20 shows a plot of atomic number versus the first electron affinity for many of the elements. On descending a family of the periodic table, the electron affinity decreases whereas when going across a row of the periodic table there is a tendency for the electron affinity to increase. Table 3.6 also shows that the metals in the second transition metal series frequently have larger EA values than the elements above them in the first transition series whereas the metals in the third transition metal series have similar to or higher EA values than the metals in the second transition metal series. This tendency to attract electrons is related to the filling of the "d" orbitals and is important in metal catalysis (Chapters 9 and 10).

In many instances electron affinity can be negative, indicating that there is no tendency for the atom to accept an electron. For example, helium ($1s^2$) would have to add an electron to a 2s orbital, which is far

Table 3.7   *Ionization energies (eV) for the first 36 elements [11]. Energy required to remove each electron sequentially. To obtain values in MJ mol^{-1} divide by 10.364. (1 eV atom^{-1} = 96.4869 kJ mol^{-1})*

Z	element	I	II	III	IV	V	VII	VII	VIII
1	H	13.598							
2	He	24.587	54.416						
3	Li	5.392	75.638	122.451					
4	Be	9.322	18.211	153.893	217.713				
5	B	8.298	25.154	37.930	259.368	340.217			
6	C	11.260	24.383	47.887	64.492	392.077	489.981		
7	N	14.534	29.601	47.448	77.472	97.888	552.057	667.029	
8	O	13.618	35.116	54.934	77.412	113.896	138.116	739.315	871.387
9	F	17.422	34.970	62.707	87.138	114.240	157.161	185.182	953.886
10	Ne	21.564	40.962	63.45	97.11	126.21	157.93	207.27	239.09
11	Na	5.139	47.286	71.64	98.91	138.39	172.15	208.47	264.18
12	Mg	7.646	15.035	80.143	109.24	141.26	186.50	224.94	265.90
13	Al	5.986	18.828	28.44	119.99	153.71	190.47	241.43	284.59
14	Si	8.151	16.345	33.492	45.141	166.77	205.05	246.52	303.17
15	P	10.486	19.725	30.18	51.37	65.023	220.43	263.22	309.41
16	S	10.360	23.33	34.83	47.30	72.68	88.049	280.93	328.23
17	Cl	12.967	23.81	39.61	53.46	67.8	97.03	114.193	348.28
18	Ar	15.759	27.629	40.74	59.81	75.02	91.007	124.319	143.456
19	K	4.341	31.625	45.72	60.91	82.66	100.0	117.56	154.86
20	Ca	6.113	11.871	50.908	67.10	84.41	108.78	127.7	147.24
21	Sc	6.54	12.80	24.76	73.47	91.66	111.0	138.0	158.7
22	Ti	6.82	13.58	27.491	43.266	99.22	119.36	140.8	168.5
23	V	6.74	14.65	29.310	46.707	65.23	128.12	150.17	173.7
24	Cr	6.766	15.50	30.96	49.1	69.3	90.56	161.1	184.7
25	Mn	7.435	15.640	33.667	51.2	72.4	95	119.27	196.46
26	Fe	7.870	16.18	30.651	54.8	75.0	99	125	151.06
27	Co	7.86	17.06	33.50	51.3	79.5	102	129	157
28	Ni	7.635	18.168	35.17	54.9	75.5	108	133	162
29	Cu	7.726	20.292	36.83	55.2	79.9	103	139	166
30	Zn	9.394	17.964	39.722	59.4	82.6	108	134	174
31	Ga	5.999	20.51	30.71	64				
32	Ge	7.899	15.934	34.22	45.71	93.5			
33	As	9.81	18.633	28.351	50.13	62.63	127.6		
34	Se	9.752	21.19	30.820	42.944	68.3	81.70	155.4	
35	Br	11.814	21.8	36	47.3	59.7	88.6	103.0	192.8
36	Kr	13.999	24.359	36.95	52.5	64.7	78.5	111.0	126

removed from the nuclear charge. This is true of all the noble gases. An electron would also have to be added to the 2p orbital of beryllium ($2s^2$), which is also farther removed from the nucleus; Mg, Ca, Sr, and Ba (beryllium's family members) show similar tendencies. The nitrogen atom would have to add an electron so that it would have the oxygen atom configuration where two electrons would occupy the same p orbital; the electron pairing cannot be stabilized easily by nitrogen's nuclear charge.

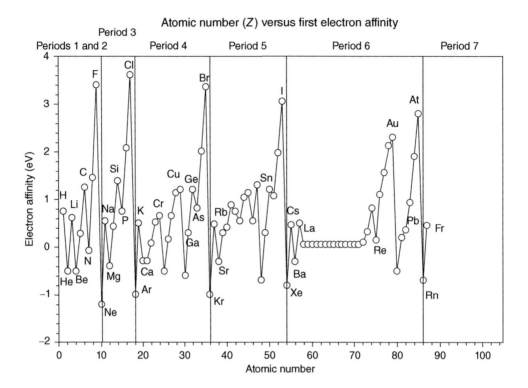

Figure 3.20   *Variation in the first electron affinity (eV) for the elements (Source: From Refs. [14–16])*

3.8.6   Periodic Properties – Electronegativity (χ)

An important atomic parameter is the electronegativity (χ), which is defined as the ability of an atom to draw electrons to itself in a chemical bond. The general trends for electronegativity are an increase across a row in the periodic table and a general decrease down a group in the periodic table, following the general trends for both ionization potential and electron affinity. The value of an atom's electronegativity has been defined in several ways. Linus Pauling produced the first electronegativity values based on bond energies (Equation 5.2). These have been the most frequently used data so all other scales of electronegativity are benchmarked to Pauling's values (χ_P).

Allred and Rochow defined electronegativity (χ_{AR}) based on the effective nuclear charge and atomic radius (Equation 3.15) so that these data are related to the atomic properties as discussed above. This scale formally describes the electric field at the surface of the atom.

$$\chi_{AR} = \frac{Z_{eff}}{r^2} \tag{3.15}$$

and in Pauling units (where r is in picometer) is

$$\chi_P = 0.744 + 35.90 \frac{Z_{eff}}{r^2} \tag{3.16}$$

Mulliken defined electronegativity as the average value of the ionization potential and the electron affinity (Equation 3.17)

$$\chi_M = \frac{IP + EA}{2} \tag{3.17}$$

Table 3.8 *Electronegativity values of the main group elements [17–21]. Allred–Rochow and Mulliken units are given in Pauling units calculated with Equations 3.16 and 3.18, respectively*

H 2.20 / 3.06 / 2.20		Pauling / Mulliken / A-R					He 5.5
Li 0.98 / 1.28 / 0.97	Be 1.57 / 1.99 / 1.47	B 2.04 / 1.83 / 2.01	C 2.55 / 2.67 / 2.50	N 3.04 / 3.08 / 3.07	O 3.44 / 3.22 / 3.50	F 3.98 / 4.43 / 4.10	Ne – / 4.60 / 5.10
Na 0.93 / 1.21 / 1.01	Mg 1.31 / 1.63 / 1.23	Al 1.61 / 1.37 / 1.47	Si 1.90 / 2.03 / 1.74	P 2.19 / 2.39 / 2.06	S 2.58 / 2.65 / 2.44	Cl 3.16 / 3.54 / 2.83	Ar – / 3.36 / 3.30
K 0.82 / 1.03 / 0.91	Ca 1.00 / 1.30 / 1.04	Ga 1.81 / 1.34 / 1.82	Ge 2.01 / 1.95 / 2.02	As 2.18 / 2.26 / 2.20	Se 2.55 / 2.51 / 2.48	Br 2.96 / 3.24 / 2.74	Kr 3.0 / 2.98 / 3.10
Rb 0.82 / 0.99 / 0.89	Sr 0.95 / 1.21 / 0.99	In 1.78 / 1.30 / 1.49	Sn 1.96 / 1.83 / 1.72	Sb 2.05 / 2.06 / 1.82	Te 2.10 / 2.34 / 2.01	I 2.66 / 2.88 / 2.21	Xe 2.6 / 2.59 / 2.40
Cs 0.79 / 0.70 / 0.86	Ba 0.89 / 0.90 / 0.97	Tl 2.04 / 1.80 / 1.44	Pb 2.33 / 1.90 / 1.55	Bi 2.02 / 1.90 / 1.67	Po 2.0 / 2.48 / 1.76	At 2.2 / 2.85 / 1.90	Rn – / 2.12 / 2.06

and in Pauling units is

$$\chi_P = 1.35\,\chi_M^{1/2} - 1.37 \tag{3.18}$$

where the electron affinity is a measure of an atom's tendency to accept an electron and the ionization potential is a measure of an atom's tendency to hold onto an electron as well as its tendency to give up an electron. This scale is considered an absolute scale as it actually uses ionization potential and electron affinity data for the atom in the **valence state**; that is, for electron configurations of the atom when they are part of the molecule; for example carbon would be $2s^1\,2p_x^1\,2p_y^1\,2p_z^1$ for CH_4. Electronegativity values for these three scales are provided in Table 3.8 for comparison and reveal some significant differences in some cases (e.g., Si, Sn, Ge, Pb). However, the general trends on descending a group and going across a row are similar.

3.8.7   Periodic Properties – Hardness (η)

Another important atomic parameter is hardness, which indicates that an atom's or ion's electron cloud is not easily distorted. Pearson [22] defined hardness (η) with Equation 3.19.

$$\eta = \frac{IP - EA}{2} \tag{3.19}$$

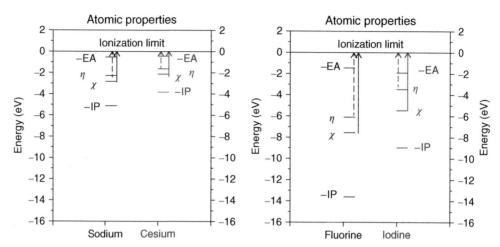

Figure 3.21   *Plots demonstrating the relationship of IP and EA to electronegativity (solid arrows) and hardness (dashed arrows) for Na, Cs, F, and I atoms*

In this case, a larger value for the hardness indicates that the atom is harder or the electron cloud is not easily distorted by another atom or ion. Because IP and EA are used, it is considered an absolute electronegativity scale. Polarizability (α), which is also called softness, is the inverse of hardness so that $\alpha = \eta^{-1}$. A polarizable ion or atom can have its electron cloud or distribution easily distorted. More polarizable species tend to be covalent. Fajans summarized the factors causing polarization into three rules.

1. Small and highly charged cations (hard species) have polarizing ability.
2. Large and highly charged anions (soft species) are easily polarized.
3. Cations that do not have a noble gas configuration such as the transition metals (soft species; see Section 7.8.3) are easily polarized.

A good way to show the relationship of electronegativity and hardness to compare different elements is to plot their values calculated from the energy for the electron affinity and the ionization potential according to Equations 3.17 and 3.19. Figure 3.21 shows a comparison of the ionization potential energy, electron affinity, electronegativity, and the hardness for sodium atom versus cesium atom (a), and fluorine atom versus iodine atom (b); note that the larger the gap between IP and EA, the harder the atom. Here, the electronegativity and the hardness of sodium are slightly larger than those of cesium as expected. Clearly fluorine is much more electronegative and harder than iodine (also sodium and cesium). In addition, iodine is more electronegative (and harder) than sodium and cesium. Lastly, iodine and cesium can be considered softer or more polarizable atoms in their respective families.

Once sodium gives up an electron from its 3s orbital and becomes sodium ion, it has the neon configuration. Removing an extra electron from the second energy level (IP = 47.286 eV) is particularly difficult, and the EA is the first ionization potential so the sodium ion is harder than the sodium atom by an order of magnitude $\eta = (47.286 - 5.139)/2 = 42.147$. Likewise, when the fluorine atom accepts an electron, it obtains a full shell or the neon configuration but does not easily give up its extra electron. Thus, the fluoride ion is also considered a hard anion. Atoms and cations with a large value for hardness have great polarizing capability or ability to distort another atom or ion's electron cloud.

Although ionization potential, electron affinity, electronegativity, and hardness have been provided mainly for atoms, values are known for molecules and ions. In subsequent chapters, the ionization potential will be related to the highest occupied molecular orbital energy (HOMO), and the electron affinity will be related to the lowest unoccupied molecular orbital energy (LUMO). These orbital energies can be determined experimentally by ESCA and other techniques.

References

1. Eugen Schwarz, W. H. (2013) 100th anniversary of Bohr's model of the atom. *Angewandte Chemie international edition in English*, **52**, 12228–12238.
2. Kragh, H. (2013) Niels Bohr between physics and chemistry. *Physics Today*, **66**, 36–41.
3. Clary, D. C. (2013) 100 years of atomic theory. *Science*, **341**, 244–245.
4. Siegbahn, K. et al. (1969) *ESCA Applied to Free Molecules*, North-Holland Publishing Co., Amsterdam. Also, Nordling, C. (1971) ESCA studies of core and valence electrons in gases and solids. *Journal de Physique Colloques*, 32 (C4), pp. C4-254-C4-263. doi: 10.1051/jphyscol:1971447.
5. Engel, T. (2013) *Quantum Chemistry and Spectroscopy* (3rd ed.), Pearson, Boston, MA. pp. 507.
6. Cromer, D. T. (1968) Stereo plots of hydrogen-like electron densities. *Journal of Chemical Education*, **45** (10), 626–631.
7. Ogryzlo, E. A. and Porter, G. B. (1963) Contour surfaces for Atomic and Molecular Orbitals. *Journal of Chemical Education*, **40**, 256–261.
8. Perlmutter-Hayman, B. 1969. The graphical representation of hydrogen-like wave functions. *Journal of Chemical Education*, **46**, 428–430.
9. Turner, D. W., Baker, C., Baker, A. D. and Brundle, C. R. (1970) Molecular photoelectron spectroscopy, John Wiley & Sons Ltd. pp. 386.
10. Sutton, L. E. (ed.) (1958) *Tables of interatomic distances and configuration in molecules and ions*. Supplement 1956–1959. The Chemical Society Special Publication No. 11. pp. 384.; also (1965) No. 18. London.
11. Moore, C. E. (1970) *Ionization potentials and ionization limits from the analyses of optical spectra*, National Standard and Reference Data Series - National Bureau of Standards (NSRDS-NBS) 34, pp. 22. Also see http://www.nist.gov/pml/data/asd.cfm and http://www.nist.gov/pml/data/ion_energy.cfm.
12. Catling, D. C., Clire, M. W., Zahnke K. J., Quinn, R. C., Clark B. C., Hecht, M. H. and Kounaves, S. P. (2010) Atmospheric origins of perchlorate on Mars and in the Atacama. *Journal of Geophysical Research*, **115**, E00E11. doi: 10.1029/2009JE003425.
13. Pearson, R. G. (1991) Negative electron affinities of nonmetallic elements. *Inorganic Chemistry* **30**, 2856–2858.
14. Hotop, H. and Lineberger, W. C. (1975) Binding energies in atomic negative ions. *Journal of Physical and Chemical Reference Data*, **4**, 539.
15. Hotop, H. and Lineberger, W. C. (1985) Binding energies in atomic negative ions:II. *Journal of Physical and Chemical Reference Data*, **14**, 731–750.
16. Andersen, T., Haugen, H. K. and Hotop, H. (1999) Binding energies in atomic negative ions:III. *Journal of Physical and Chemical Reference Data*, **29**, 1511–1533.
17. Allred, A. L. (1961) Electronegativity values form thermochemical data. *Journal of Inorganic and Nuclear Chemistry*, **17**, 215–221.

18. Allred, A. L. and Rochow, E. G. (1958) A scale of electronegativity based on electrostatic force. *Journal of Inorganic and Nuclear Chemistry*, **5**, 264–268.

19. Pauling, L. (1960) *The Nature of the Chemical Bond* (3rd ed.), Cornell University Press, Ithaca, NY. pp. 644.

20. Mulliken, R. S. (1934) A new electroaffinity scale; together with data on valence states and on valence ionization potentials and electron affinities. *Journal of Chemical Physics*, **2,** 782–793.

21. Bratsch, S. G. (1988) Revised Mulliken Electronegativities 1. Calculation and conversion to Pauling Units. *Journal of Chemical Education*, **65**, 34–41.

22. Pearson, R. G. (1988) Absolute electronegativity and hardness: application to Inorganic Chemistry. *Inorganic Chemistry*, **27**, 734–740.

4

Symmetry

4.1 Introduction

A non-mathematical definition of **symmetry** is that symmetry is regularity of form, periodicity, or harmonious arrangement. Objects that demonstrate symmetry are everywhere around us and range from nanomaterials to the cosmos. Nature produces a variety of minerals such as pyrite, which exists in several morphologies and symmetrical forms (Figure 4.1a and b), and biominerals, which are the hard parts of organisms as shown in Figure 4.1c and d for diatoms (SiO_2).

There are many reasons to study symmetry as symmetry helps us classify types of molecules. Once classified, the application of symmetry arguments in the form of group theory (and its application to molecular orbitals and compound reactivity) is one of the most powerful theoretical tools of the chemist. There have been many texts covering symmetry and group theory [2–5] and the following discussion is derived from them.

4.2 Symmetry Concepts

There are two important symmetry concepts, which should not be confused. Both concepts depend on determining the center of the molecule, which may or may not be occupied by an atom.

4.2.1 Symmetry Operation

A symmetry operation is an operation that, when performed on a molecule, results in a new orientation of the object that is indistinguishable from and *superimposable* on the original. Thus, a symmetry operation takes a body into an equivalent configuration.

4.2.2 Symmetry Element

A symmetry element is a geometrical entity (a line, a point, or a plane), which is related to one or more symmetry operations that may be carried out.

Inorganic Chemistry for Geochemistry and Environmental Sciences: Fundamentals and Applications, First Edition. George W. Luther, III.
© 2016 John Wiley & Sons, Ltd. Published 2016 by John Wiley & Sons, Ltd.
Companion Website: www.wiley.com/go/luther/inorganic

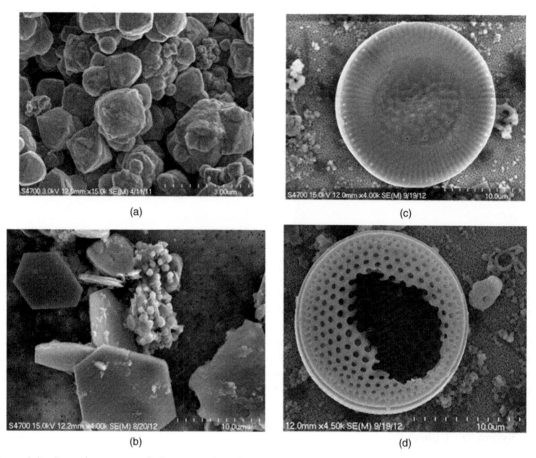

Figure 4.1 *Several symmetrical objects produced in nature. (a) Pyrite (FeS_2) particles (framboids, octahedra). (b) pyrite hexagons (Source: Reproduced from [1], Figure 9b, with permission from Elsevier. Note the micrometer marking at the lower right of each figure; other SEM photos from the collaborative work of Alyssa Findlay, Amy Gartman, and George Luther). (c) Silica (SiO_2) diatom. (d) Partially dissolved diatom frustule*

Symmetry elements and symmetry operations (Table 4.1) are so closely interrelated because the operation can only be defined with respect to the element and at the same time the symmetry element can be demonstrated only by showing that the appropriate symmetry operation exists. Thus, it is necessary to discuss the element in terms of the operation and vice versa.

4.2.3 Symmetry Elements and Operations

4.2.3.1 *Rotation about a Symmetry Axis*

If rotation of a molecule about some axis by some angle (the operation) results in an orientation of the molecule that is **superimposable** on the original, the axis is called a rotational axis (the symmetry element). Figure 4.2 shows the C_2 operation. The drawings note the actual hydrogen atoms before and after performing the operation, which will be important later when using group theory. Molecular models can also be used so drawings and/or models will be used interchangeably in the text.

Table 4.1 *Symmetry elements and their symmetry operations*

Symmetry element – symbol	Symmetry element – description	Symmetry operation
E	Identity	No change
σ	Plane of Symmetry	Reflection through a plane
i	Center of symmetry	Inversion through the center of symmetry
C_n	Axis of symmetry	Rotation about the axis of symmetry by $360/r$ degrees
S_n	Rotation–reflection axis of symmetry (Improper rotation)	Rotation about the axis of symmetry by $360/r$ degrees followed by reflection through a plane perpendicular to the rotation axis

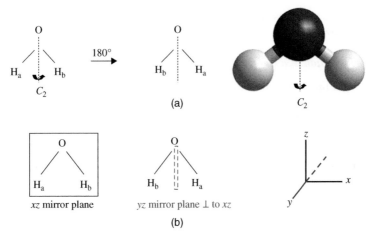

Figure 4.2   *(a) Rotation 180° about the z axis bisecting the H–O–H bond angle (C_2 operation), which is in the xz plane. (b) Mirror planes (σ) for the water molecule discussed below. Model produced with the HyperChem™ program version 8.0.10*

The symbol for the rotation of the water molecule is C_2 for its axis of symmetry. The general symbol is C_n. The value of n is arrived at in the following manner: $n = 360°/r$, where r is the angle of rotation to obtain a superimposable form of the molecule.

The Cartesian axes are used in determining the rotational axes of molecules. For uniformity, the z axis is usually the principal rotational axis and this reference axis must be determined in a regular manner.

General rules for the determination of the Cartesian coordinate axes of a molecule:

1. Place the center of gravity of the molecule at the origin of the coordinate system.
2. **For the assignment of the z axis,** follow these rules [3]:
   A. If there is only one rotational axis in the molecule, this axis is the z axis.
   B. If the molecule has several rotational axes, the one of the highest order is the z axis.
   C. If there are several axes of the highest order, the axis that passes through the greatest number of atoms is the z axis.

3. **For assignment of the x axis, three cases need to be considered:** two apply to planar molecules and the third to nonplanar molecules.

    A. If the molecule is planar and the z axis lies in this plane, the x axis is chosen to the axis ***normal*** to the plane. It is important to specify the Cartesian coordinates in a figure (see Figure 4.2) and remain consistent. Because water is a bent molecule, the x and y axes could be interchanged. For consistency the coordinate system in Figure 4.2 will be used for triatomic molecules and ions that have C_{2v} symmetry (as defined later).

    B. If the molecule is planar and the z axis is perpendicular to the plane, the x axis (which must lie in the plane) is chosen to pass through the largest number of atoms.

    C. If nonplanar molecules have one plane containing a larger number of atoms than any other plane, they are treated as if they were planar and as if this preferred plane were the plane of the molecule. If a decision cannot be made on this basis, it will usually be immaterial as to how the assignments of the x and y axes are made.

For practice, determine the positions of the coordinate axes for these molecules: boron trifluoride; ethylene; trans-dichloroethylene; benzene (Table 4.2 in Section 4.3 has these as examples).

4.2.3.2   *Inversion or Reflection at a Center of Symmetry*

If, in a molecule, a straight line is drawn from every atom through the center of the molecule and continued in the same direction, the line encounters an equivalent atom equidistant from the center (operation), then the molecule possesses a center of symmetry (the element) designated as i for inversion. The inversion process requires that every point (x, y, z) is converted to the position $(-x, -y, -z)$. In some molecules $x = 0$ and therefore $-x = 0$.

For practice, determine that the following chemical species have centers of inversion. These are also in Table 4.2 (Section 4.3).

1. $PtCl_4{}^{2-}$ (square planar)
2. Benzene
3. Nitrogen molecule
4. The staggered form of ethane (number the hydrogens consecutively)

4.2.3.3   *Reflection at a Plane of Symmetry*

If a molecule is bisected by a plane and each atom in one half of the molecule is reflected through the plane and encounters a similar atom in the other half (operation), the molecule is said to possess a mirror plane (symmetry element). The mirror plane is designated as σ. For the bent or angular water molecule in Figure 4.2, both planes are vertical and are designated σ_v. The xz mirror plane has no change and the yz mirror plane reflects the hydrogen atoms through the plane. A horizontal plane (σ_h) is perpendicular to the z axis.

4.2.3.4   *Rotation About an Axis, Followed by Reflection at a Plane Normal to this Axis*

Assume that a molecule is *rotated* around an axis and the resulting orientation is reflected in a plane *perpendicular* to the axis (operation); if the resulting orientation is superimposable on the original molecule, the molecule is said to possess a rotation–reflection axis (the element). The symbol for the rotation–reflection axis is S_n where n is as defined above. Figure 4.3 shows an S_2 improper rotation. Dashed bonds are behind the plane of the paper and bold bonds are in front of the plane of the paper. Here, the z axis is the C=C bond

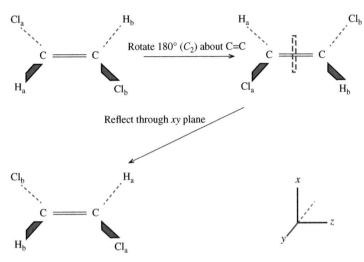

Figure 4.3 C_2 *rotation on the z axis followed by reflection through the xy plane (S_2 operation) for trans-dichloroethane. The dashed and heavy bolded blue lines indicate bonds perpendicular to the plane of the paper*

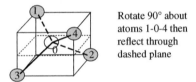

Rotate 90° about atoms 1-0-4 then reflect through dashed plane

Figure 4.4   *Using this figure, show that S_4 and S_4' exist*

axis as this axis goes through the greatest number of atoms. Note that the S_2 improper axis is also an inversion through a center of symmetry.

S_4 would indicate 90° clockwise rotations, while S_4' indicates counterclockwise rotations. For practice, show that AB_4 tetrahedral molecules have an S_4 symmetry axis (Figure 4.4).

The rotation reflection axis, S_n, is a complex symmetry element consisting of a series of two operations, neither of which alone is necessarily a symmetry operation but which together represent a symmetry operation. The same operation can be broken down into rotation and inversion and this complex operation is sometimes called rotation–inversion. Both rotation–reflection and rotation–inversion axes are frequently called improper axes as distinguished from rotational axes, which are called proper axes.

4.2.3.5   *Identity (E)*

The identity could be regarded as a trivial element of symmetry since anything goes into itself if left unchanged. However, the identity operation is important in arriving at a favorable combination of orbitals by the application of the Group Theory. Often it is necessary to carry out two or more symmetry operations consecutively. If the final result can be expressed as equivalent to another symmetry operation, the latter is identified. Rotation of the square planar $PtCl_4^{2-}$ ion by 90° about the C_4 axis gives an equivalent arrangement, but if the individual Cl atoms are tracked, the two arrangements are not identical as the chloride ions have moved in space. Four successive rotations of 90° each would be the same as the identity operation. Thus, there are three 90° rotations ($C_4, C_4^2 = C_2, C_4^3$). Because $C_4^2 = C_2$, it is identical to another symmetry element. Thus, there are two C_4 separate operations to indicate two separate rotations that do not give E; the number 2 indicates the number of symmetry elements in a class of operations (C_4).

Note: Figure 4.2 for the water molecule as the first C_2 does not give an identity operation when each atom is identified. The second rotation of 180° is a symmetry operation that can be defined as $C_2^2 \equiv E$. For water, there is only one C_2 axis of rotation.

4.3   Point Groups

It is possible to compile a list of all of the symmetry elements possessed by a given molecule and then list all of the symmetry operations generated by each of these elements. It can be demonstrated that such a complete list of symmetry operations satisfies the criteria for a mathematical group. Once this is shown, it is necessary to consider what kinds of symmetry groups will be obtained from various possible collections of symmetry operations. The symmetry groups are frequently called **point groups**, because all symmetry elements in a molecule will intersect at a **common point**, which is not shifted by any of the symmetry operations. The point groups are represented by **Schoenflies symbols** such as C_n, D_n, C_{nh}, and C_{nv} as shown in Table 4.2.

Table 4.2   *Examples of important point groups and of molecular species belonging to these point groups. Summing the coefficients for each symmetry element gives the total number of elements in a point group (or h, the order of the group). Dihedral (vertical) planes in D symmetry are labeled σ_d*

Point group	Symmetry elements	Examples
C_1	E	CH_3-CHClBr
C_2	E, C_2	H_2O_2 (cis-nonplanar)
C_i	E, i	CH_3-CHCl-CHCl-CH_3 (trans)
C_s	E, σ_h	NOCl, HOBr
C_{2v}	$E, C_2, \sigma_v(xz), \sigma_v(yz)$	H_2O, CH_2Cl_2, H_2O_2 (cis-planar)
C_{3v}	$E, 2C_3, 3\sigma_v$	NH_3, H_3CCl
$C_{\infty v}$	$E, C_\infty, \infty\sigma_v$	HF, HCN, OCS, CO
C_{2h}	E, C_2, σ_h, i	ClHC=CHCl (trans), H_2O_2 (trans-planar)
D_{2d}	$E, 3C_2$ (mutually $\perp$), $2S_4$ (coincident with one of the C_2), $2\sigma_d$ (thru the S_4 axis)	$H_2C=C=CH_2$
D_{3d}	$E, 2C_3, 3C_2$ ($\perp C_3$), $2S_6$ (coincident with the C_3), $i, 3\sigma_d$	C_2H_6, Si_2H_6 (both staggered)
D_{2h} or V_h	$E, 3C_2$ (mutually $\perp$ and exclusive), 3σ (mutually $\perp$ and exclusive), i	$H_2C=CH$
D_{3h}	$E, 2C_3, 3C_2$ ($\perp C_3$ axis), $3\sigma_v, \sigma_h, 2S_3$	BCl_3, PF_5
D_{4h}	$E, 2C_6, 3C_2'$ and $3C_2''$ ($\perp C_6$), $3\sigma_v, 3\sigma_d, \sigma_h, C_2$ and $2C_3$, $2S_6$ (both coincident with C_6), i	$[PtCl_4]^{2-}$
D_{6h}	$E, 2C_6, 3C_2'$ and $3C_2''$ ($\perp C_6$), $\sigma_v, 3\sigma_d, \sigma_h, C_2$ and $2C_3$, $2S_6$ (both coincident with C_6), i	C_6H_6 (benzene), $Cr(C_6H_6)_2$
$D_{\infty h}$	$E, C_\infty, \infty C_2$ ($\perp C_\infty$), $\infty\sigma_v, \sigma_h, i$	H_2, CO_2, HC≡CH, $[Ag(CN)_2]^-$, XeF_2
T_d	$E, 3C_2$ (mutually $\perp$ $8C_3, 6S_4$ (coincident with C_2), $6\sigma_d$	CH_4, SiF_4, $[Cd(NH_3)_4]^{2+}$, $NiCl_4^{2-}$
O_h	$E, 6C_4$ (mutually $\perp$), $8C_3, 6S_4$ and $3C_2$ (=C_4^2 coincident with C_4), $6C_2, 3\sigma_h, 6\sigma_d, 8S_6$ (coincident with C_3), i	SF_6, $[PtCl_6]^{2-}$, $[Co(NH_3)_6]^{3+}$, $[CoF_6]^{3-}$
I_h	$E, 12C_5, 12C_5^2, 20C_3, 15C_2, i, 12S_{10}, 12S_{10}^3, 20S_6, 15\sigma$	$B_{12}H_{12}^{2-}$, C_{60}

Overall, a point group is a short hand method of classifying a molecule that indicates all of the symmetry elements that it contains.

There are also symmetry groups called *space groups*, which contain operations involving translatory motions. These are helpful in describing crystal lattice types, but are not considered here.

4.3.1 Special Groups and Platonic Solids/Polyhedra

Before showing a scheme to classify molecules, it is important to note the special groups, which are linear molecules, and the regular polyhedral (Platonic solids). Linear molecules are either $C_{\infty v}$ or $D_{\infty h}$ because of the presence or absence of σ_h or i, respectively.

The Platonic solids are those of high symmetry and are readily recognized. They are the **tetrahedron** (T_d), **octahedron and cube** (O_h) and the **icosahedron and dodecahedron** (I_h). The drawn structures in Figure 4.5 and the model structures in Figure 4.6 show the relationship of the cube to the tetrahedron and the octahedron. In the tetrahedron, there are four vertices, four triangular faces (equilateral triangles) and six edges. When comparing the tetrahedron to the cube, alternate vertices of the cube are occupied; the tetrahedron has no σ_h or i. One of the eight C_3 axes goes from the bottom left occupied position through the center of the cube and to the base of the pyramid created by the other three occupied positions.

The cube has 6 square faces, 8 vertices and 12 edges whereas the octahedron has 8 triangular faces, 6 vertices and 12 edges. The six vertices of the octahedron occupy the six square faces of the cube. Another important way to look at the octahedron is to view it down one of the eight C_3 axes so that three positions are out of the plane of the paper and three are below the plane of the paper. This representation is called a trigonal antiprism and will be discussed further in the sections on transition metal complexes.

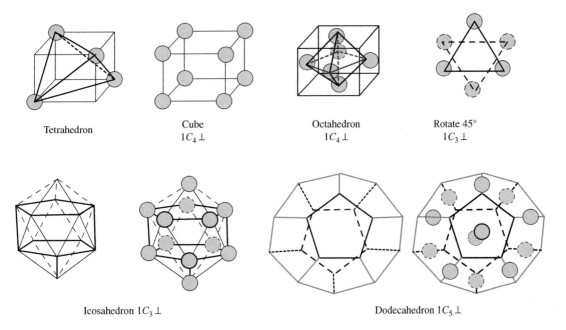

Tetrahedron

Cube
$1C_4\perp$

Octahedron
$1C_4\perp$

Rotate 45°
$1C_3\perp$

Icosahedron $1C_3\perp$

Dodecahedron $1C_5\perp$

Figure 4.5 *The five Platonic solids. Note the spatial relation of the top structures (tetrahedron, cube, and octahedron) to each other and the bottom structures (icosahedron and dodecahedron) to each other. The C_n axes given are those normal to the plane of the paper*

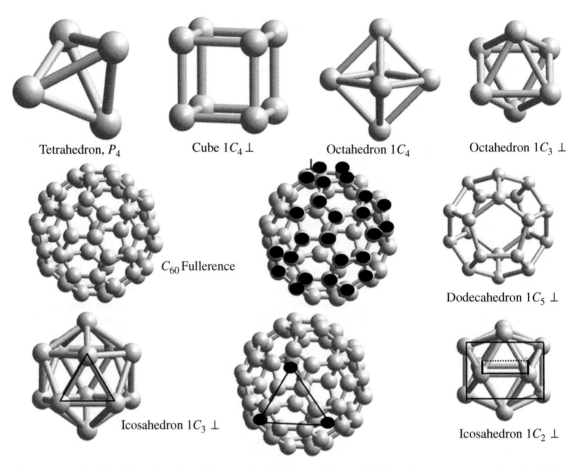

Tetrahedron, P_4 Cube $1C_4 \perp$ Octahedron $1C_4$ Octahedron $1C_3 \perp$

C_{60} Fullerence

Dodecahedron $1C_5 \perp$

Icosahedron $1C_3 \perp$

Icosahedron $1C_2 \perp$

Figure 4.6   *Models of the Platonic solids including fullerene (C_{60}), which has I_h symmetry, produced with the Hyperchem™ program (version 8.0.10 from Hypercube, Inc.). The $\perp$ symbol indicates an axis perpendicular to the plane of the paper. The blackened atoms on C_{60} show the pentagonal faces similar to the dodecahedron, and the triangle on C_{60} shows the similarity with the icosahedron on a C_3 axis. The lower right icosahedral representation shows four atoms above and below the four atoms in one of two square planes; when the upper and lower atoms also form square planes antiprismatic to the central square plane, the structure is termed cuboctahedral*

The icosahedron has 20 vertices, 12 triangular faces, and 30 edges. The structure on the left shows one of the $12C_5$ axes in the plane of the paper from top to bottom with two pentagons in antiprismatic arrangement between the top and bottom vertices. The structure with the 12 vertices occupied shows one of the $20S_6$ axes (this one is perpendicular to the plane of the paper). Note the 2 triangular faces (anti-prismatic) at the top and bottom of the structure and the 6 positions in a hexagon between the triangular faces; this is also one of the $20\ C_3$ axes (see below for coordination number).

The dodecahedron has 12 pentagonal faces, 20 vertices, and 30 edges. The center of the 12 pentagonal faces coincides with the 12 vertices of the icosahedron. The structure with the center of the 12 pentagonal faces occupied is similar to looking down one of the C_5 axes of the icosahedron. Sandwiched between the top

Table 4.3 *Maximum coordination number for an atom at the center of a given polyhedron. The coordination numbers are seen in Figures 4.5 and 4.6*

Polyhedron	Coordination number
Tetrahedron	4 (vertices); 4 (triangular faces)
Cube	8 (vertices); 6 (square faces)
Octahedron	6 (vertices); 8 (triangular faces)
Icosahedron	12 (vertices); 20 (triangular faces)
Dodecahedron	20 (vertices); 12 (pentagonal faces)

and bottom vertices are two antiprismatic pentagons (one has positions with solid lined circles and the other with dashed lined circles).

Lastly, C_{60} **fullerene** is a special case of the icosahedron in Figure 4.6. There are 20 hexagonal faces that form 10 pairs of opposite faces and 12 pentagonal faces that form 6 pairs of opposite faces. The centers of the pentagonal faces are located at the vertices of an icosahedron. There are a total of 60 pentagonal edges and 30 hexagonal edges.

These high symmetry point groups are frequently found in molecules and solids as shown in Table 4.2, and understanding their major symmetry elements is very helpful in assigning the point group to molecules that have lower symmetry. Table 4.3 summarizes the maximum **coordination number** (C.N.) for an atom that is at the center of the polyhedron. The C.N. is typically defined when the vertices or the center of the faces of a polyhedron are occupied by the coordinating atoms. A coordination number of 12 is observed in many metals (see Figures 6.8 and 6.9) and is demonstrated with the icosahedron (look down a S_6/C_3 axis) where the central atom is surrounded in a plane by six atoms and by three atoms from a triangular face above the hexagon and another three atoms from a triangular face below the hexagon.

There are several systematic ways to decide on how to classify molecules into point groups, and Figure 4.7 shows a common one that has six major decision steps. Note that linear molecules and the Platonic solids are easy to discern.

Directions for using the scheme for point group selections are as follows.

1. Linear molecules are considered first (**step 1**) and are designated either $C_{\infty v}$ or $D_{\infty h}$. In order to distinguish between them, look for the presence of a C_∞ axis on the bond axis and $\infty C_2 \perp C_\infty$. If these conditions are verified, the molecule belongs to the point group $D_{\infty h}$; if not, it belongs to the point group $C_{\infty v}$.

2. Molecules with especially high symmetry are also placed in **step 2** of the scheme. These are easily recognized: tetrahedron, T_d ; octahedron, O_h; and icosahedron, I_h, symmetries.

3. If none of the above are found, look to **step 3** of the diagram and search for the proper rotation axis of highest order (C_n, n = maximum). If there is no such axis, then the molecule is of very low symmetry, that is, C_s, C_i or C (**step 6**). Any plane of symmetry or the presence of an inversion center can be used to distinguish between them.

4. In **step 3**, if there are one or more C_n axes, select the one of highest order. Inspect the molecule for nC_2 axes perpendicular to the major C_n axis ($nC_2 \perp C_n$) [**steps 4 and 5**]. Lack of such a set implies a C_{nh}, C_{nv}, or C_n species (**step 4**). Distinction can be made by looking FIRST for a σ_h (indicating C_{nh} symmetry) or

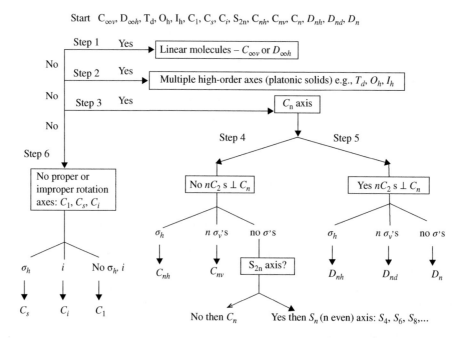

Start $C_{\infty v}$, $D_{\infty h}$, T_d, O_h, I_h, C_1, C_s, C_i, S_{2n}, C_{nh}, C_{nv}, C_n, D_{nh}, D_{nd}, D_n

Figure 4.7   *Scheme for the symmetry classification of molecules*

NEXT for n σ_v (indicating C_{nv} symmetry). If no plane of symmetry is observed, the point group is C_n. If the molecule possesses ONLY an S_{2n} axis, with or without an inversion center, then the point group designation is $S_n (n = \text{even})$.

5. If the condition of $nC_2 \perp C_n$ (**step 5**) is satisfied, the molecule belongs to one of the point groups: D_{nh}, D_{nd} or D_n. These are confirmed by detecting either σ_h or $n\sigma_d$ or no σ, respectively.

4.3.2   Examples of the Use of the Scheme for Determining Point Groups

1. *Methyl chloride (Figure 4.8):* Although methane is a T_d molecule, the presence of a substituted halogen lowers the symmetry. Therefore, look to the bottom of the scheme (steps 3, 4, and 5). There exists only one rotation axis, a C_3 axis collinear with the C–Cl bond. Thus, the molecule must belong to the C or D class. *D* symmetries can easily be eliminated as there are no C_2 axes. The only planes of symmetry are the $3\sigma_v$ lying in the three Cl–C–H planes; thus, the point group designation is C_{3v}. NH_3 and PCl_3O are pyramidal also and have the same point group.
2. *Ethylene (C_2H_4; Figure 4.9):* Inspection of the geometry of ethylene excludes any of the special high symmetry point groups; so go to **step 3**. There are three mutually perpendicular C_2 axes. The major axis may be arbitrarily chosen perpendicular to the molecular plane, but the C=C bond axis has the greatest number of atoms, so is the z axis. The remaining C_2 axes can represent $2C_2 \perp C_2$ and establish a *D* type point group (**step 5**). Finally, there is a σ_h (the xy plane) perpendicular to the C=C bond axis indicating D_{2h} symmetry.

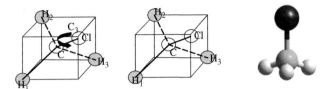

Figure 4.8 *Methyl chloride, CH$_3$Cl. The molecular model shows the C$_3$ axis in the plane of the paper. Model produced with the HyperChem™ program version 8.0.10*

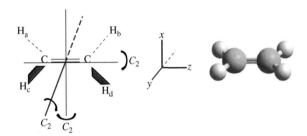

Figure 4.9 *Ethylene sketch and molecular model with the bond axis on the z axis. Model produced with the HyperChem™ program version 8.0.10*

Figure 4.10 *Benzene. The atom symbols and hydrogen atoms are removed for simplicity*

3. *Benzene (C$_6$H$_6$; Figure 4.10):* The highest order rotation axis for benzene is C_6, which is perpendicular to the molecular plane, σ_h. There are many more symmetry elements (C_3, C_2, S_6, etc.) in addition to i. Continuing with the scheme at **step 5**, the molecule has D symmetry due to the $6C_2 \perp C_6$ (only two are drawn) and the existence of a σ_h in the molecular plane takes precedence over the $6\sigma_v$s that are coincident with the C_2 symmetry elements. Thus, the benzene molecule has D_{6h} symmetry.

4. *trans-Co(NH$_3$)$_4$Cl$_2$]$^+$ (Figure 4.11):* The symmetry of this molecule is less than that of O_h due to the presence of chloride ions; the hydrogen atoms are ignored in this analysis. Following the scheme to step 3, there is a higher order C_4 axis passing through the Cl–Co–Cl axis or z axis. At step 5, there are $4C_2 \perp C_4$ (only two are drawn) observed bisecting the equatorial sides and angles, thereby confirming a D symmetry. The xy plane is a σ_h which is $\perp C_4$; thus, the point group is D_{4h}.

5. *cis-Co(NH$_3$)$_4$Cl$_2$]$^+$ (Figure 4.12):* Ignoring the hydrogen atoms again, this ion does not possess a C_4 axis as the one above. The highest C_n (step 4) is a twofold rotation axis bisecting the Cl–Co–Cl bond angle along the z axis. There are $2\sigma_v$: one coincident with the C_2 or z axis and the other is the yz plane. Thus, the point group is C_{2v}.

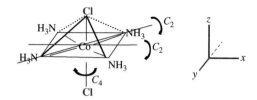

Figure 4.11   *trans-Tetraaminedichloridocobalt(III) ion*

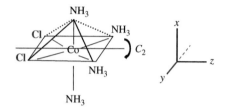

Figure 4.12   *cis-Tetraaminedichloridocobalt(III) ion. Note the difference in the Cartesian coordinate axis between this figure and Figure 4.11*

6. The hydrogen peroxide molecule, H_2O_2, has three forms to consider (Figure 4.13).
    A. The nonplanar equilibrium configuration.
        The top structure in Figure 4.13 shows the hydrogen peroxide molecule in its nonplanar form. There is a C_2 axis and no other proper axis. There are no planes of symmetry or S_n. The group is therefore C_2.
    B. The cis-planar configuration ($\theta = 0°$).
        The middle structure in Figure 4.13 is the cis-planar structure of H_2O_2. The C_2 axis, of course, remains. There are still no other proper axes. The molecule now lies in the xz plane, which is a plane of symmetry, and the yz plane, another plane of symmetry, intersects the molecular plane and contains the C_2 axis. The group is C_{2v}, which is the point group for water.
    C. The trans-planar configuration ($\theta = 90°$).
        The C_2 axis is still present, and there are no other proper axes. There is now a σ_h, which is the molecular plane (xy plane). The group is C_{2h}.
7. *1,3,5,7-Tetramethylcyclooctatetraene* ($C_{12}H_{16}$): Figure 4.14 shows the molecule top down (z axis perpendicular to the plane of the paper) and side on (z axis is vertical in the plane of the paper). At step 4, there is a C_2 axis but there is also a S_4 axis (C_4 followed by rotation through the yz plane). There are no additional independent symmetry elements. The group is therefore S_4.
8. *Allene* (C_3H_4) *Figure 4.15:* The structure on the left of Figure 4.15 shows the C=C=C bond axis and the xz plane, and the right structure shows the xy plane looking down on the bond axis, which contains both the coincident C_2 and S_4 symmetry elements. A way of looking at these model structures is that the hydrogen atoms occupy alternate positions of a cube that is elongated with the three carbon atoms at the center. There are two C_2 rotational axes perpendicular to the bond axis so the group must be a D type group; however, using the procedure at step 5, there is no σ_h. There are two vertical or dihedral planes, which lie between the C_2 axes perpendicular to the bond axis. Thus, the point group is D_{2d}. Another example of D_{nd} symmetry is cyclic crown-like S_8, which is D_{4d} as there is no σ_h. There is a C_4, $4C_2 \perp C_4$ (see Xs bisecting the S–S bonds in Figure 4.15), and $4\sigma_v(\sigma_d)$ that contain the 1,1; 2,2; 3,3; and 4,4 atoms across the S_8 ring in Figure 4.15.

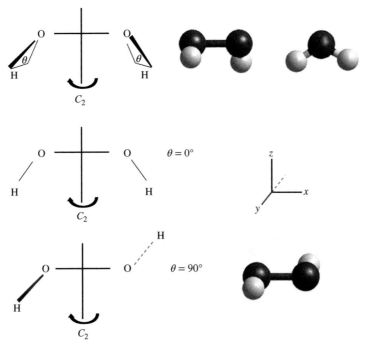

Figure 4.13 *Possible structures and their point groups for* H_2O_2. *The top structure shows a model on the left similar to the drawing (⊥ to the O–O bond axis) and the model on the right looking down the O–O bond axis. Models produced with the HyperChem™ program version 8.0.10*

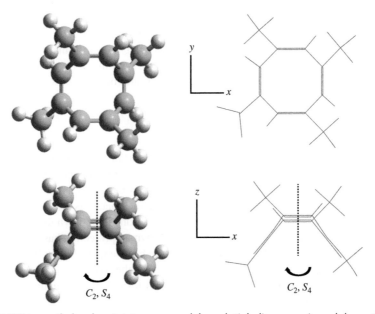

Figure 4.14 *1,3,5,7-Tetramethylcyclooctatetraene; models and stick diagrams viewed down the z axis (top) and the y axis (bottom). Models produced with the HyperChem™ program version 8.0.10*

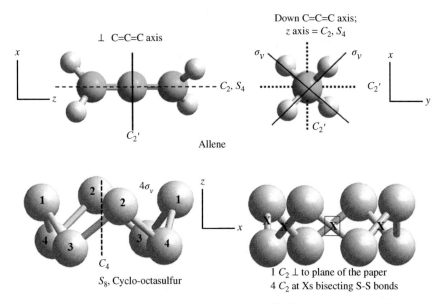

Figure 4.15 *Allene models viewed in the xz plane (left, ⊥ to the C=C=C bond axis) and in the xy plane (right) down the C=C=C bond axis. S_8 models with the xz plane in the plane of the paper and containing the C_4 axis; the X in the box shows one of the C_2⊥ to the C_4 axis and the xz plane. Models produced with the HyperChem™ program version 8.0.10*

4.4   Optical Isomerism and Symmetry

Molecules, which are **not superimposable** on their mirror images, are normally termed optically active (**chiral**), but the best term is that they are **dissymmetric**. This latter term is used rather than asymmetric because asymmetric means having no symmetry. Dissymmetric molecules are those that either have no symmetry or have **only** axes of proper rotation. Chiral molecules are those with the point groups C_1 (asymmetric), C_n (dissymmetric), and D_n (dissymmetric). The following are examples of molecules that possess symmetry but are chiral.

4.4.1   Dichloro-Allene Derivatives ($C_3H_2Cl_2$)

Substituting a chlorine atom for one hydrogen on each terminal carbon atom in allene from Figure 4.15 gives two possible structures, which are mirror images of each other but are not superimposable on each other, as shown in Figure 4.16. Notice that the point group is C_2 as there is a C_2 axis that is normal to the bond axis, but no other symmetry elements. The chlorine atoms are above the bond axis and the hydrogen atoms below the bond axis in the models of Figure 4.16. The Newman projections look down the bond axis and show the rotation about the C_2 axis. On the left of Figure 4.16, the B Newman projection has the Cl-C-H in solid lines for the left part of molecular model A whereas the <u>B</u> projection has the Cl-C-H in solid lines for the right part of molecular model B. Also, the arrows indicate the Cl atoms for the molecular models that are the same Cl atoms with the solid line to the H atom in the two dimensional form. In these dichloro-allenes, the terminal carbons are not "asymmetric" because of the double bonds.

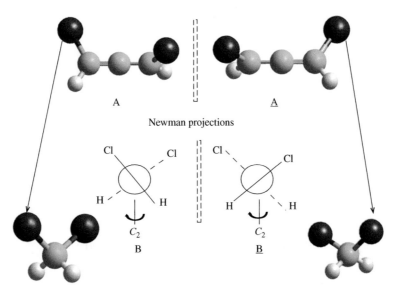

Figure 4.16 *Dichloro substituted allene optical isomers (1 Cl on each terminal C). Carbon (blue), chlorine (black), hydrogen (white). Molecular models A with A ⊥ to the C=C=C bond axis are enatiomers, and their Newman projections B with B looking down the C=C=C bond axis are enantiomers. Models produced with the HyperChem™ program version 8.0.10*

4.4.2 Tartaric Acid

Figure 4.17 shows two structures of tartaric acid that are mirror images of each other and not superimposable on each other. Notice that there is again a C_2 axis normal to the C–C bond axis, and it is perpendicular to the plane of the paper in the model structures whereas the C_2 axis is in the plane of the paper for the Newman projections (bottom part of Figure 4.17). In tartaric acid, both central carbons are termed "asymmetric" as there are four different substituents on each of the central carbon atoms supposedly indicating their chirality. However, the compound does possess symmetry. Again, the point group is C_2.

In both examples, the reader should also build models to confirm the non-superimposition of mirror images, and finally the symmetry point group.

4.4.3 Cylindrical Helix Molecules

Similar to the disubstituted allene and tartaric acid isomers above, a perfect cylindrical helix has a mirror image that possesses C_2 symmetry as the C_2 axis is perpendicular to the coils of the helix. Figure 4.18 shows this arrangement as well as linear S_{10} with a helix-like structure. A conical helix has C_1 symmetry.

4.5 Fundamentals of Group Theory

The material presented so far is mechanical in operation. Identification of symmetry elements and identification of point groups helps understand the geometry of molecules, so it is important to use **molecular models** to gain practice and become proficient in understanding symmetry elements and their operations.

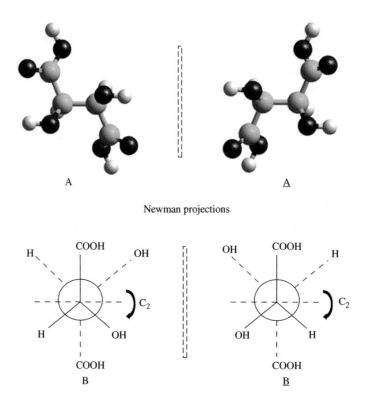

Newman projections

Figure 4.17   *Tartaric acid enantiomers. Top models with carbon (blue), oxygen (black), hydrogen (white); bottom models show two dimensional Newman projections. Models produced with the HyperChem™ program version 8.0.10*

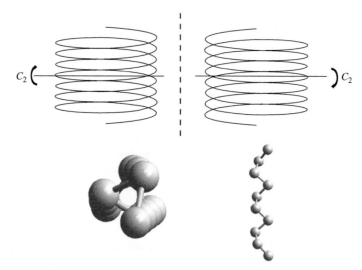

Figure 4.18   *Cylindrical helices (similar to springs) are dissymmetric. Bottom part shows linear S_{10} looking down the helix (left) and vertically (right). Models produced with the HyperChem™ program version 8.0.10*

The determination of point groups is an essential and primary step in the elucidation of more interesting molecular properties such as bond hybridization, molecular orbital wave functions, transition probabilities, and the number of infrared and Raman fundamentals. This section provides some fundamentals of group theory including how to generate and use a character table. *Although this material is not absolutely essential for an understanding of chemical bonding, it is extremely useful, and the methods shown will be referred to in subsequent chapters describing bonding.* The point groups C_{2v}, C_{3v}, and T_d are first used as examples because several molecules of environmental interest [H_2O, NH_3 and CH_4 (NH_4^+, SO_4^{2-}), respectively] are represented by these point groups. Then applications are demonstrated for determining the number of vibrations of a molecule (number of infrared and Raman fundamentals), orbitals used in bonding for the central atom molecule, and group or symmetry adapted linear combinations for terminal atoms bound to a central atom. The techniques are similar as bonds, atoms, and orbitals transform with similar symmetry properties. An understanding of matrix algebra is useful but not essential for the material presented.

4.5.1 C_{2v} Point Group

Every symmetry operation can be expressed as a 3×3 matrix, also known as a transformation matrix, which shows how the vectors of a point (e.g., position of an atom) vary in Cartesian coordinate space. The operation is represented by multiplying the original x, y, and z coordinates (a 1×3 matrix) by this 3×3 matrix to yield the new coordinates where x', y', and z' are the new coordinates. As a first example, the identity operation, E, results in no change of the vectors or coordinates ($+x$, $+y$, and $+z$ do not change) such that the matrix yields the following full equation. The values of 1 and 0 in the 3×3 matrix indicate that a vector was unchanged or changed, respectively, after performing the operation.

$$\begin{array}{ccc} x & y & z \\ \end{array}$$
$$\begin{pmatrix} x' \\ y' \\ z' \end{pmatrix} = \begin{pmatrix} 1 & 0 & 0 \\ 0 & 1 & 0 \\ 0 & 0 & 1 \end{pmatrix} \begin{pmatrix} x \\ y \\ z \end{pmatrix} = \begin{array}{ccc} x & +0y & +0z \\ 0x & +y & +0z \\ 0x & 0y & +z \end{array}$$

The sum of the on-diagonal elements for all three vectors from upper left to lower right is 3, and the value for these on-diagonal elements is given the notation $\mathbf{f(R)}$, which is the matrix value for the class (denoted by $\mathbf{R}$) and which will be used to determine molecular properties. Each class of operations has a value of $\mathbf{f(R)}$.

Figure 4.19 shows the change in the x, y, and z vectors of an atom after a C_2 operation on the z axis. Note that x becomes $-x$ and y becomes $-y$ whereas z is unchanged. *The sign of the wave function for the p_x, p_y, and p_z orbitals will also transform in a similar way as these vectors so this description will be useful for bonding discussions later.*

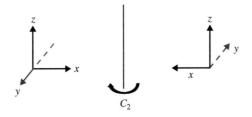

Figure 4.19 *Change of x, y and z vectors after a C_2 rotation*

Table 4.4 *Character table for C_{2v} point group*

Point group	Symmetry classes (R)					Order of group
C_{2v}	E	C_2	$\sigma_v xz$	$\sigma_v yz$		$h = 4$
A_1	1	1	1	1	z	x^2, y^2, z^2
A_2	1	1	−1	−1	R_z	xy
B_1	1	−1	1	−1	x, R_y	xz
B_2	1	−1	−1	1	y, R_x	yz

The full mathematical operation for C_2 is represented by the following matrix equation where $x' = -x$ and $y' = -y$. The right hand section is the final product of the two matrices.

$$\begin{pmatrix} -x \\ -y \\ z \end{pmatrix} = \begin{pmatrix} x' \\ y' \\ z' \end{pmatrix} = \begin{pmatrix} -1 & 0 & 0 \\ 0 & -1 & 0 \\ 0 & 0 & 1 \end{pmatrix}\begin{pmatrix} x \\ y \\ z \end{pmatrix} = \begin{matrix} -x & +0y & +0z \\ 0x & -y & +0z \\ 0x & 0y & +z \end{matrix}$$

The corresponding 3×3 matrix for the C_2 rotation is different from E as expected. The sum of the on-diagonal elements for the C_2 operation gives a $\mathbf{f(R)}$ value of −1.

The vertical xz plane, which would be the plane of the water molecule, results in no change for the x and z vectors, but y becomes $-y$. The vertical yz plane results in no change for the y and z vectors, but x becomes $-x$. The sum of the on-diagonal elements for both the σ_v xz plane and σ_v yz plane operations gives a value of 1 for $\mathbf{f(R)}$ for each plane.

$\sigma_v xz$ plane

$$\begin{pmatrix} x \\ -y \\ z \end{pmatrix} = \begin{pmatrix} x \\ y' \\ z \end{pmatrix} = \begin{pmatrix} 1 & 0 & 0 \\ 0 & -1 & 0 \\ 0 & 0 & 1 \end{pmatrix}\begin{pmatrix} x \\ y \\ z \end{pmatrix}$$

$\sigma_v yz$ plane

$$\begin{pmatrix} -x \\ y \\ z \end{pmatrix} = \begin{pmatrix} x' \\ y \\ z \end{pmatrix} = \begin{pmatrix} -1 & 0 & 0 \\ 0 & 1 & 0 \\ 0 & 0 & 1 \end{pmatrix}\begin{pmatrix} x \\ y \\ z \end{pmatrix}$$

Every symmetry operation for every point group has a transformation matrix, and these can now be complied for the C_{2v} point group from the above matrices. The complete set of matrices in every point group is called a mathematical group. In describing the mathematics of symmetry operations, it is important to note that the new arrangement is not identical to the previous or initial arrangement after performing the operation (as shown in Figure 4.2 where both hydrogen atoms are interchanged). Every point group is represented by a **character table**, which contains these matrices, and Table 4.4 shows the character table for the C_{2v} point group.

4.5.2 Explanation of the Character Table

1. **At the top left**, the Schoenflies symbol for the point group and all the symmetry elements in a class R of the point group are arranged from left to right by the number of symmetry operations in each class (E, C_n, etc.). The sum of all of the coefficients for each operation is **the order of the group, h, *at the top right***.

2. Numbers or characters (χ) ***in the center*** are in a square matrix, and the values of a character from left to right correspond to a particular **irreducible representation** (A, B, E, T) of the point group ***on the far left***. Note that the irreducible representations are not italicized whereas the classes of the symmetry operations are. The number of irreducible representations equals the number of classes as the matrix is a square. The vertical characters under each operation are called **dimensions**.

3. ***On the right*** are functions (transformations of the Cartesian coordinate x, y, z axes that indicate the p orbitals and dipole moment vectors for the x, y, z axes for infra-red spectroscopy; rotations about x, y, z

axes or R_x, R_y, R_z); note that rotation about the z axis (R_z) leaves the xy plane unaffected but changes the xz and yz planes from their original position (a 180° rotation inverts each of these two planes). The other rotations show similar behavior.

4. **On the far right**, quadratic functions (representing d and f orbitals, and polarizability tensors for Raman spectroscopy activity). Polarizability tensors have functions that transform as x^2, y^2, z^2, xy, yz, xz.

4.5.3   Generation of the Irreducible Representations (C_{2v} Case)

The irreducible representations (A_1, A_2, B_1, B_2) are generated from the transformation matrices above where the only positions in the square block diagonalized matrix that are important are the 1,1 or x,x; 2,2 or y,y; and 3,3 or z,z positions. For the operations above,

the z vector is +1 for E, C_2, σ_v xz, and σ_v yz;
the x vector is +1 for E and σ_v xz whereas it is −1 for C_2 and σ_v yz; and
the y vector is +1 for E and σ_v yz whereas it is −1 for C_2 and σ_v xz.

At this point for the C_{2v} point group, there are three irreducible representations (A_1, B_1, B_2), and there must be a fourth, which is not identical to the other three, as the order of the group is 4. One of the several criteria of irreducible representations is that different irreducible representations are **orthogonal** to each other as defined by the mathematical Equation 4.1.

$$\sum_R g(C)\chi_i(R)\chi_j(R) = 0 \tag{4.1}$$

χ_i and χ_j are the characters for each symmetry operation or class (R) for each irreducible representation, and $g(C)$ is the number of operations in each class. These are multiplied together and then summed to equal 0.
For A_1 and B_1, the following result is obtained.

$$E \qquad C_2 \qquad \sigma_v xz \qquad \sigma_v yz$$
$$(1)(1)(1) + (1)(1)(-1) + (1)(1)(1) + (1)(1)(-1) = 1 + -1 + 1 + -1 = 0$$
$$g_E A_1 B_1 \qquad g_{C_2} A_1 B_1 \qquad g_\sigma A_1 B_1 \qquad g_\sigma A_1 B_1$$

The last irreducible representation must be orthogonal to the other three (A_1, B_1, B_2), and it is readily seen by inspection that A_2 should have a value of +1 for E and C_2, and −1 for σ_v xz and σ_v yz. Note that multiplying A_1 by B_1 returns the values of the characters for B_1 because A_1 is a totally symmetric representation where all the characters are one for all the operations.

4.5.4   Notation for Irreducible Representations

The **shorthand notation or labels** for irreducible representations is defined according to their change in symmetry properties once the following symmetry operations are performed:
  Letters (uppercase is used for representations and lower case is used for orbital assignments)
  A, B (or a,b) indicate an one-dimensional representation (e.g., one orbital) where

A (or a) is symmetric to the principal axis C_n.
B (or b) is antisymmetric to the principal axis C_n.
E (or e) indicates a two-dimensional representation (e.g., two orbitals of similar energy).
T (or t) indicates a three-dimensional representation (e.g., three orbitals of similar energy).

Subscripts are defined as

1 indicates symmetric to C_2 perpendicular to C_n, or to σ_v.
2 indicates antisymmetric to C_2 perpendicular to C_n, or to σ_v.
g indicates symmetric to i.
u indicates antisymmetric to i.

Superscripts are defined as

$'$ indicates symmetric to σ_h.
$''$ indicates antisymmetric to σ_h.

4.5.5   Some Important Properties of the Characters and their Irreducible Representations

1. All groups include a **totally symmetric representation** where all the characters are one for all the operations (e.g., A_1 in C_{2v}). The s orbitals are spherically symmetric and transform as this irreducible representation.
2. The characters of the E operation must always be positive.
3. The number of irreducible representations of a group is equal to the number of classes in the group as they generate a square matrix.
4. The sum of the squares of the characters or dimensions for each irreducible representation directly under E, the identity operation, equals the order of the group, h. For the other symmetry classes the squares must be multiplied by $g(C)$.

$$h = \sum_i [\chi_i(E)]^2 \tag{4.2}$$

5. For any irreducible representation, the sum of the squares of the characters multiplied by the number of operations, $g(C)$, in the class equals the order the group, h.

$$h = \sum_R [g(C)\, \chi_i(R)]^2 \tag{4.3}$$

For B_1 in C_{2v} symmetry: $1(1)^2 + 1(-1)^2 + 1(1)^2 + 1(-1)^2 = 4.$
For A_1 in C_{3v} symmetry (see below): $1(1)^2 + 2(1)^2 + 3(1)^2 = 6.$

6. The product of two nondegenerate (A, B) irreducible representations gives one of the other irreducible representations. Multiplying B_1 by B_2 in C_{2v} symmetry gives A_2 as a result.

$$
\begin{array}{cccccccc}
E & C_2 & \sigma_v xz & \sigma_v yz & E & C_2 & \sigma_v xz & \sigma_v yz \\
(1)(1)+ & (-1)(-1)+ & (1)(-1)+ & (-1)(1) = +1 & +1 & -1 & -1 = 0 \\
B_1B_2 & B_1B_2 & B_1B_2 & B_1B_2 & = \text{the characters for } A_2
\end{array}
$$

Because B_1 and B_2 transform as x and y, respectively, then $B_1 \times B_2 = A_2$ or $x(y) = xy$. Similarly

$$B_1 \times B_1 = A_1, \text{also} x(x) = x^2; A_2 \times B_2 = B_1; A_2 \times B_1 = B_2$$

4.5.6   Nonindependence of x and y Transformations (Higher Order Rotations)

4.5.6.1 C_{3v} Point Group

The character table for this point group is in Table 4.5.

Table 4.5 *Character table for C$_{3v}$ point group*

C$_{3v}$	E	2C$_3$	3σ$_v$		h = 6
A$_1$	1	1	1	z	x² + y², z²
A$_2$	1	1	−1	R$_z$	
E	2	−1	0	(x, y)(R$_x$, R$_y$)	(x² − y², xy)(xz, yz)

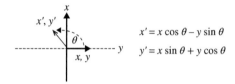

$$x' = x \cos\theta - y \sin\theta$$
$$y' = x \sin\theta + y \cos\theta$$

Figure 4.20 *Transformation of x, y into x', y' after a proper rotation of 120°(C$_3$)*

C_{3v} has some significant differences from the C_{2v} point group above. The coefficients in front of each symmetry element at the top of the C_{3v} character table indicate the number of symmetry elements in the class that behave identically in space, and this number is important in describing the character table of a group as discussed below. There are now numbers different from 1 and −1. The value under E of 2 indicates a degeneracy and the functions show (x, y) rather than x, y.

The transformation matrices do not always follow the simple x,x; y,y; and z,z transformations noted above for C_{2v} symmetry. For example, a C_3 rotational axis in C_{3v} symmetry (e.g., NH_3 or CH_3Cl) results in new x and y that are not independent of each other; that is, **they are degenerate**. Thus, the characters for the symmetry operation are the vector sums of x and y so need to be calculated using trigonometric functions according to Figure 4.20. Here the arrow does not change its length, but its vector is now a function of x and y on a 120° rotation.

Substituting the trigonometric functions into a 3 × 3 matrix gives the matrix known as the **general transformation matrix**, which is used for these types of calculations. For C_{3v} symmetry, the z component does not change and does not mix with x and y so z transforms differently from x and y.

$$\begin{pmatrix} \cos\theta & -\sin\theta & 0 \\ \sin\theta & \cos\theta & 0 \\ 0 & 0 & 1 \end{pmatrix}$$

To calculate the new x [or x'] and y [or y'] for a C_3 rotational axis, both the x and y values in a row must be calculated and substituted into the general transformation matrix as follows. The angle is 120° [or $2\pi/3$].

$$x' = x \cos(120) - y \sin(120) = -0.5x - 0.866y$$

$$y' = x \sin(120) + y \cos(120) = 0.866x - 0.5y$$

$$C_3 = \begin{pmatrix} \begin{pmatrix} -0.5 & -0.866 \\ 0.866 & -0.5 \end{pmatrix} & 0 \\ 0 & 0 \\ 0 & 0 & [1] \end{pmatrix}$$

The 3×3 matrix is now evaluated as a 2×2 and a 1×1 (lower right in brackets above) matrix where the 2×2 matrix becomes

$$\begin{pmatrix} -0.5 & -0.866 \\ 0.866 & -0.5 \end{pmatrix} = -1$$

The value of the character of the 2×2 matrix is calculated only with the on-diagonal elements from upper left to lower right; thus, the character is $-0.5 + (-0.5) = -1$.

There are two C_3 operations as rotation can occur clockwise twice (or clockwise and counterclockwise) without giving the E operation. The sum of all of the on-diagonal elements gives an **f(R)** value of 0. A general equation to determine the sum of the on-diagonal characters, **f(R)**, for a proper axis of rotation, C_n, is Equation 4.4.

$$f(R) = 2 \cos(2\pi/n) + 1 \quad \text{for } C_n \tag{4.4}$$

The other operations must be calculated in a similar way for consistency. The identity operation is below and the character value is 2 for the x, y system. The sum of all of the on-diagonal elements for the E operation again gives an **f(R)** value of 3.

$$E = \begin{pmatrix} \begin{pmatrix} 1 & 0 \\ 0 & 1 \end{pmatrix} & 0 \\ 0 & 0 & (1) \end{pmatrix} . \text{ Lastly, the } \sigma_v(xz) \text{ operation is } \sigma_v = \begin{pmatrix} \begin{pmatrix} 1 & 0 \\ 0 & -1 \end{pmatrix} & 0 \\ 0 & 0 & (1) \end{pmatrix}$$

The sum of the 2×2 matrix is $1 + (-1) = 0$. The sum of all the on-diagonal elements for σ_v gives an **f(R)** value of 1. There are $3\sigma_v$ or vertical planes in C_{3v} symmetry that are identical as two of the atoms remain unchanged whereas the other two atoms are changed in space (see CH_3Cl in Figure 4.8). This is in contrast to water in C_{2v} symmetry where the two vertical planes are distinctly different as the hydrogen atoms vary differently in space when performing the σ operation. The xz plane shows no change for any of the atoms whereas the yz plane shows an interchange of the hydrogen atoms.

The three 2×2 matrices defining the x and y vectors give the following characters that result in the E irreducible representation. Because x and y are not independent of each other, E is called **doubly degenerate**.

E	C_3	σ_{xz}
2	-1	0

The three 1×1 matrices defining the z vector give the following characters that result in the A_1 irreducible representation, which satisfies point 1 above.

E	C_3	σ_{xz}
1	1	1

Thus, there are two irreducible representations defined by the calculation using the general transformation matrix, and a third is needed. Using rule 4 in section 4.5.5, the identity operation, E, must be 1 for this third irreducible representation so that the order of the group is 6. The C_{3v} character table above shows that the irreducible representation A_1 defines the z vector whereas the irreducible representation E defines the (x, y) vector as well as the rotations about the x and y axes (R_x, R_y). The only component of rotations and translations in Cartesian coordinate space not identified is rotation about the z or C_3 axis (R_z), which has a character value of 1 for C_3 as there is no change. Thus A_2 must be the third irreducible representation as it must be orthogonal to A_1 and can only be so if the vertical planes $(3\sigma_v)$ have a character of -1. These conditions for A_2 also satisfy rule 5 and the character table given above for C_{3v} is complete.

4.6 Selected Applications of Group Theory

The use of group theory and character tables will be used when necessary to explain bonding and other characteristics of a molecule. The C_{2v} and C_{3v} point groups demonstrated what irreducible representations the p and d orbitals transform as. Table 4.6 provides information about the values of characters *pertinent to orbitals* and how they change after performing a symmetry operation.

Before using these to describe bonding, one of the best ways to demonstrate the wealth of information that is contained in character tables is to predict how free molecules exhibit translations (x, y, and z), rotations, and vibrations. It is also one of the more important chemical uses of group theory, and the approach as shown here leads to a description of bonding. The vibrations can be detected by infra-red and Raman spectroscopy. In infrared spectroscopy, vibrations corresponding to dipole moment changes result in the absorption of IR light from the ground to an excited state (Figure 1.2). In Raman spectroscopy, incident visible light undergoes scattering, and most of the scattered light has the same energy or frequency as the incident light. However, a small fraction of the incident light is shifted in energy that corresponds to energy or frequency differences in the vibrational states of the molecule (similar to Figures 5.8 and 5.9).

All motions can be represented by the $3N$ **Cartesian coordinate vectors of the N atoms** in a given molecule, and these vectors serve to form a **reducible** representation within the molecule's point group that can be reduced into a number of irreducible representations of the point group. Before demonstrating this, it is easy to calculate the number of vibrations for a molecule.

The number of vibrations for **linear** molecules can be calculated as $3N - 5$ where five represents three translations and two rotations (the z axis contains the atoms and any rotation about the z axis shows no change in the molecule). Thus for a molecule with two atoms, $3(2) - 5 = 1$ vibration and for three atoms $[3(3) - 5 = 4]$ vibrations and so on.

The number of vibrations for **nonlinear** molecules can be calculated as $3N - 6$ where six represents three translations and three rotations. For a nonlinear molecule such as H_2O with three atoms, it is $3(3) - 6 = 3$ vibrations.

Because the irreducible representations of translations and rotations are given in the character table they can be separated from the vibrations, which are identified by the process of elimination. To assign the translations, vibrations, and rotations, it is necessary to first generate the reducible representation for the molecule and then mathematically reduce the reducible representation into the irreducible representations.

4.6.1 Generation of a Reducible Representation to Describe a Molecule

The procedure is not difficult, but there are six principal steps. In this process, the reducible representation for all the vectors of the atoms in the molecule, Γ_{all}, are first obtained. Γ_{all} is then reduced to give all of the irreducible representations that make up the translations, rotations, and vibrations of the molecule. Then

Table 4.6 Summary of select character (operation change) values found after performing a symmetry operation on an atom, bond, or orbital

Character	Meaning
1	No change in vector, orbital
−1	Vector or orbital changes sign
0	Complex change

the irreducible representations for the translations and rotations are obtained from the character table, which leaves the vibrations, Γ_{vib}. The vibrations are the internal coordinates of the molecule and are composed of stretching and bending modes, which are used to produce another reducible representation(s) (e.g., Γ_{A-B} and Γ_{A-B-A} where A–B indicates a bond and A–B–A indicates a bond angle), which is then reduced to irreducible representations for proper spectroscopic assignment. Nakamoto's book [6] is recommended for those who seek an in-depth understanding of the vibrational analysis of inorganic compounds. The steps using this procedure are demonstrated for the water molecule with C_{2v} symmetry. Then other simple molecules of importance are described.

1. Determine the molecule's point group.
2. For each class of symmetry operation, R, determine the number of atoms that are **unchanged** after the symmetry operation is performed. This result is given the notation $n_u(R)$.
3. The factor $f(R)$ is determined from all of the **on-diagonal elements of a symmetry operation (all x, y, z vectors)** as above and several values are given in Table 4.7. Values of $f(R)$ can be calculated for C_n as in Equation 4.4, and for S_n with Equation 4.5.

$$f(R) = 2\cos(2\pi/n) - 1 \quad \text{for } S_n \tag{4.5}$$

Multiply $n_u(R)$ and $f(R)$ for each class of operations. The products obtained are the characters of the reducible representation for the $3N$ Cartesian coordinate vectors.

To demonstrate these three steps, the water molecule has C_{2v} symmetry and three atoms. Thus, for all of the symmetry operations, the products of $n_u(R)$ and $f(R)$ give the following characters. Note that all three atoms are unchanged after performing the operations E and σ_v (xz) whereas only the oxygen atom is unchanged after performing the operations C_2 and σ_v (yz).

$$
\begin{array}{ccccccc}
 & n_u(R) & \times & f(R) & = & \chi_{\text{vectors}}(R) \\
E & 3 & \times & (3) & = & 9 \\
C_2 & 1 & \times & (-1) & = & -1 \\
\sigma_v xz & 3 & \times & (1) & = & 3 \\
\sigma_v yz & 1 & \times & (1) & = & 1
\end{array}
$$

In summary, the reducible representation is given the **symbol Γ_{all}** for all vector changes of the atoms and is written out horizontally with its characters as the irreducible representations are.

$$
\begin{array}{ccccc}
 & E & C_2 & \sigma_v xz & \sigma_v yz \\
\Gamma_{\text{all}} & 9 & -1 & 3 & 1
\end{array}
$$

4. Reduce the reducible representation using the following **reduction formula** (Equation 4.6).

$$c(i) = \frac{1}{h} \sum_R g(C)\,\chi_i(R)\,\chi(R) \tag{4.6}$$

Table 4.7 *Some values for* f(R). *Note that* $E = C_1$; $\sigma = S_1$; *and* $i = S_2$

R	E	σ	i	C_2	C_3^1, C_3^2	C_4^1, C_4^3	C_6^1, C_6^5	S_3^1, S_3^2	S_4^1, S_4^3	S_6^1, S_6^5
$f(R)$	3	1	−3	−1	0	1	2	−2	−1	0

where $c(i)$ is the number of times an ith irreducible representation occurs in the reducible representation; h is the order of the group, $g(C)$ is the number of operations in the class, $\chi_i(R)$ is the character of the irreducible representation, and $\chi(R)$ is the character of the reducible representation.

$$c(A_1) = \tfrac{1}{4}[(1)(9) + 1(+1)(-1) + 1(1)(3) + 1(1)(1)] \qquad = \tfrac{1}{4}(12) = 3$$
$$c(A_2) = \tfrac{1}{4}[1(1)(9) + 1(+1)(-1) + 1(-1)(3) + 1(-1)(1)] = \tfrac{1}{4}(4) \ = 1$$
$$c(B_1) = \tfrac{1}{4}[1(1)(9) + 1(-1)(-1) + 1(1)(3) + 1(-1)(1)] \ = \tfrac{1}{4}(12) = 3$$
$$c(B_2) = \tfrac{1}{4}[1(1)(9) + 1(+1)(-1) + 1(1)(3) + 1(1)(1)] \qquad = \tfrac{1}{4}(8) \ = 2$$

Thus, $\Gamma_{all} = 3A_1 + A_2 + 3B_1 + 2B_2$.

5. Using the right side of the character table (C_{2v} in this case), determine the rotations and the translations that make up the irreducible representation from Γ_{all}. Then the remainder of the calculated irreducible representations from Γ_{all} are the vibrations. For C_{2v} symmetry and water,

Translations	Rotations
A_1, z	A_2, R_z
B_1, x	B_2, R_x
B_2, y	B_1, R_y

Now subtract those irreducible representations from Γ_{all} to give the remaining irreducible representations that are the three molecular vibrations, Γ_{vib}.

$$\Gamma_{vib} = \Gamma_{all} - [A_1 + A_2 + 2B_1 + 2B_2] = 3A_1 + A_2 + 3B_1 + 2B_2 - [A_1 + A_2 + 2B_1 + 2B_2]$$
$$\Gamma_{vib} = 2A_1 + B_1$$

6. The three vibrations are **internal coordinates** of a molecule. The dipole moment vectors are represented by x, y, and z as in the character table, and they are all infrared active. For H_2O, there are two O–H bond distances and one H–O–H angle and each of these vibrations represent a single vector or internal coordinate (unlike an atom which has three vectors of x, y, and z). To specify the irreducible representations for the specific vibrations, another reducible representation can be made by using the following operation for the **two O–H bonds** (once they are determined then the remaining H–O–H angle can be). From the data in Table 4.6, multiply the number of bonds unchanged, $n_b(R)$, when performing a symmetry operation by 1 as the operation did not cause a vector change after performing the operation.

	$n_b(R)$	$\times$	operation change	$=$	$\chi_{bond}(R)$
E	2	$\times$	1	$=$	2
C_2	0	$\times$	0	$=$	0
$\sigma_V(xz)$	2	$\times$	1	$=$	2
$\sigma_V(yz)$	0	$\times$	0	$=$	0

These characters now represent the change in the vector of the **bond or vibration** after the operation is performed, and the symbol for the reducible representation $\Gamma_{O\text{-}H}$ represents the bond. [This procedure can also be done for the angle being studied and the notation would be Γ_{H-O-H}.]

	E	C_2	$\sigma_v xz$	$\sigma_v yz$
$\Gamma_{O\text{-}H}$	2	0	2	0

Γ_{O-H} can be reduced using Equation 4.6 with the operation change character, $\chi_{bond}(R)$, replacing $\chi(R)$. The equations below show that only A_1 and B_1 are possible O–H bond stretches from Γ_{vib} above.

$$c(A_1) = \tfrac{1}{4}[1(1)(2) + 1(+1)(0) + 1(1)(2) + 1(1)(0)] \ \ = \tfrac{1}{4}(4) = 1$$

$$c(B_1) = \tfrac{1}{4}[1(1)(2) + 1(-1)(0) + 1(1)(2) + 1(-1)(0)] = \tfrac{1}{4}(4) = 1$$

Thus one bond stretch transforms as A_1 (symmetric to C_2) and the other as B_1 (antisymmetric to C_2) whereas by the process of elimination, the H–O–H bond angle is an in-plane bending mode that transforms as A_1 (symmetric bending to C_2 axis). Figure 4.21 shows the vibrations and their respective irreducible representations for H_2O, which have similar symmetry to the two symmetry adapted linear combination of orbitals for the O–H bonds in water (see Section 4.7.1.1).

4.6.2   Determining the IR and Raman Activity of Vibrations in Molecules

The vibrations of a molecule can be determined by infrared and Raman spectroscopy; however, not all vibrations are infrared and Raman active. Polar molecules have a permanent electric dipole moment, which indicates that the charge distributions on opposite sides of the molecule are different. Absorption of infrared radiation can occur when vibration results in a dipole moment change, which is a function of the x, y, z, Cartesian coordinates, whereas a Raman transition can occur when a vibration causes a change in the polarizability of a molecule, which is a binary product of two Cartesian coordinates; for example, xz or x^2.

For molecules that have an inversion center, each of the x, y, z, Cartesian coordinates transform into $-x, -y, -z$ whereas the binary product of these will transform into a positive function. Thus, if a molecule has a center of inversion (centrosymmetric), then none of its modes of vibration can be both IR and Raman active; this is known as the **exclusion rule**. However, a particular vibrational mode may be both IR and Raman inactive in non-centrosymmetric molecules. Molecules that have permanent dipole moments are those with the point groups C_1, C_s, C_n, and C_{nv}; these are also termed symmetry allowed with regard to a dipole moment. All other point groups have symmetry elements that prohibit a dipole, and they are termed symmetry forbidden.

For example, in a simple homonuclear diatomic molecule such as H_2 or N_2, there is only one internal coordinate, which is the H–H or N≡N bond, and this is the symmetric stretching vibration (A_{1g} or Σ_g^+ in $D_{\infty h}$) as the bond does not change after performing any symmetry operation. There is no permanent electric dipole moment so the vibration is infrared inactive; however, the vibration is Raman active as the polarizability (z^2) of the molecule changes on the bond axis.

For a heteronuclear diatomic molecule such as H–Cl, the bond is again the only internal coordinate and does not change after any symmetry operation. Both the dipole moment and polarizability transform as A_1 or Σ^+ (see $C_{\infty v}$ character table in Appendix 4.1). Thus, this vibrational mode (symmetric stretching vibration) is both infrared and Raman active.

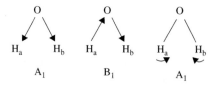

$$A_1 \qquad\qquad B_1 \qquad\qquad A_1$$

Figure 4.21   *The schematic shows the normal vibrations of the water molecule: symmetric stretch A_1, asymmetric stretch B_1, and symmetric bend (A_1 in-plane symmetric or scissors-like bending vibration)*

For the water molecule, all three vibrational modes create a change in the dipole moment. Thus all three vibrational modes are IR active. Inspection of the C_{2v} character table shows that these three vibrational modes are also Raman active as the polarizability tensors transform as A_1, B_1, and B_2. In the next section, the methane molecule and its vibrations are described.

4.6.3 Determining the Vibrational Modes of Methane, CH_4

Methane is a tetrahedral molecule with five atoms, and the character table for the T_d point group, which has five symmetry classes (R), is given below in Table 4.8. Figure 4.22 shows tetrahedral models with four single or sigma bonds and six H–C–H bond angles.

The five atoms give 15 Cartesian coordinate displacement vectors, and we can use Table 4.7 to help generate the reducible representation for these vectors. For E, all five atoms show no displacement of the vectors so that the character equals 15. For the C_3 operation, two atoms do not change but the vectors change so the character equals 0. For both the C_2 and S_4 operations, the character equals -1 in analogy to the water molecule (one atom is unchanged) whereas for the vertical planes, σ_d, three atoms in one plane do not change so generate a character of $+3$.

$$n_u(R) \times f(R) = \chi_{\text{vectors}}(R)$$

E	5	$\times$	(3)	$=$	15
C_3	2	$\times$	(0)	$=$	0
C_2	1	$\times$	(-1)	$=$	-1
S_4	1	$\times$	(-1)	$=$	-1
σ_d	3	$\times$	(1)	$=$	3

Table 4.8 Character table for T_d point group

T_d	E	$8C_3$	$3C_2$	$6S_4$	$6\sigma_d$		$h = 24$
A_1	1	1	1	1	1		$x^2 + y^2 + z^2$
A_2	1	1	1	-1	-1		
E	2	-1	2	0	0		$(2z^2 - x^2 - y^2, x^2 - y^2)$
T_1	3	0	-1	1	-1	(R_x, R_y, R_z)	
T_2	3	0	-1	-1	1	(x, y, z)	(xy, xz, yz)

Figure 4.22 *The left and center structures show the atoms in methane in relation to a cube, and the right structure shows one of the C_3 axes vertically in the plane of the paper. Models produced with the HyperChem™ program version 8.0.10*

The reducible representation for all the vectors of the atoms is below.

$$
\begin{array}{cccccc}
 & E & 8C_3 & 3C_2 & 6S_4 & 6\sigma_d \\
\Gamma_{all} & 15 & 0 & -1 & -1 & 3
\end{array}
$$

This representation reduces to $\Gamma_{all} = A_1 + E + T_1 + 3T_2$, and the calculation is reproduced here only for these irreducible representations.

$$c(A_1) = (1/24)[1(1)(15) + 8(1)(0) + 3(1)(-1) + 6(1)(-1) + 6(1)(3)] \quad = (1/24)(24) = 1$$

$$c(E) = (1/24)[1(2)(15) + 8(-1)(0) + 3(2)(-1) + 6(0)(-1) + 6(0)(3)] \quad = (1/24)(24) = 1$$

$$c(T_1) = (1/24)[1(3)(15) + 8(0)(0) + 3(-1)(-1) + 6(1)(-1) + 6(-1)(3)] = (1/24)(24) = 1$$

$$c(T_2) = (1/24)[1(3)(15) + 8(0)(0) + 3(-1)(-1) + 6(-1)(-1) + 6(1)(3)] = (1/24)(72) = 3$$

The character table indicates that the three rotations transform as T_1 and that the three translations transform as T_2; therefore the nine vibrations [$3(5) - 6 = 9$] transform as $A_1, E, 2T_2$. The C–H bonds and the H–C–H bond angles are the internal coordinates for methane and contribute to these nine vibrations.

The reducible representation for **the 4 C–H bonds** is obtained as follows. Each of the vectors for the four bonds does not change after the E operation has been performed so the character equals 4. Only one bond along the C_3 axis is unchanged so the character equals 1. On performing a C_2 operation, all bonds change so the character equals 0; the same is true for the S_4 operation. The vertical planes contain two of the bonds, which are unchanged after the operation, so the value of the character equals 2. These operations generate the following reducible representation, Γ_{C-H}, or Γ_{tetra}, for tetrahedral structures with five atoms.

$$
\begin{array}{cccccc}
 & n_b(R) & \times & \text{operation change} & = & \chi_{bond}(R) \\
E & 4 & \times & 1 & = & 4 \\
C_3 & 1 & \times & 1 & = & 1 \\
C_2 & 0 & \times & 0 & = & 0 \\
S_4 & 0 & \times & 0 & = & 0 \\
\sigma_d & 2 & \times & 1 & = & 2
\end{array}
$$

$$
\begin{array}{cccccc}
 & E & 8C_3 & 3C_2 & 6S_4 & 6\sigma_d \\
\Gamma_{C-H} & 4 & 1 & 0 & 0 & 2
\end{array}
$$

Γ_{C-H} reduces to A_1 and T_2 (see Section 5.6.1) and the calculations for only these are presented.

$$c(A_1) = (1/24)[1(1)(4) + 8(1)(1) + 3(1)(0) + 6(1)(0) + 6(1)(2)] \quad = (1/24)(24) = 1$$

$$c(T_2) = (1/24)[1(3)(4) + 8(0)(1) + 3(-1)(0) + 6(-1)(0) + 6(1)(2)] = (1/24)(24) = 1$$

The reducible representation for the six H-C-H bond angles is obtained as below; note that 4 C-H and 6 H-C-H bond angles give 10 vibrational modes so there is one redundancy. The notation, $N_{angles}(R)$, is the number of angles unchanged. Each of the vectors for the six bond angles does not change after the E operation has been performed so the character equals 6. On performing C_2 and S_4 operations, all bond angles change so the character equals 0. C_2 does exchange atoms but the two bond angles that C_2 bisect are unchanged in the plane. The operation, σ_d, has a redundancy for two H–C–H bond angles so the character equals 2. These operations generate the following reducible representation, Γ_{H-C-H}.

$$N_{angles}(R) \times \text{operation change} = \chi_{bond}(R)$$

	$N_{angles}(R)$		operation change		$\chi_{bond}(R)$
E	6	$\times$	1	$=$	6
C_3	0	$\times$	0	$=$	0
C_2	2	$\times$	1	$=$	2
S_4	0	$\times$	0	$=$	0
σ_d	2	$\times$	1	$=$	2

	E	$8C_3$	$3C_2$	$6S_4$	$6\sigma_d$
Γ_{H-C-H}	6	0	2	0	2

Γ_{H-C-H} reduces to A_1, E, and T_2, and the calculations for only these are presented.

$$c(A_1) = (1/24)[1(1)(6) + 8(1)(0) + 3(1)(2) + 6(1)(0) + 6(1)(2)] = (1/24)(24) = 1$$

$$c(E) = (1/24)[1(2)(6) + 8(-1)(0) + 3(2)(2) + 6(0)(0) + 6(0)(2)] = (1/24)(24) = 1$$

$$c(T_2) = (1/24)[1(3)(6) + 8(0)(0) + 3(-1)(2) + 6(-1)(0) + 6(1)(2)] = (1/24)(24) = 1$$

As noted above, this analysis provides one extra vibrational mode, which must be A_1 as it is the only single irreducible representation. This extra mode arises from the H-C-H bond angle analysis as it is not possible for all six bond angles to change independently.

Inspection of the T_d character table with consideration of the irreducible representations composing Γ_{C-H} and Γ_{H-C-H} indicates that A_1 is only due to C-H stretching vibration; E is only due to angle bending or deformation; and T_2 consists of both bond stretching and angle bending vibrations. For the methane molecule, the **C–H bond displacement vectors** transform into two irreducible representations, A_1 and T_2. From the T_d character table, A_1 transforms like $x^2 + y^2 + z^2$, indicating it is Raman active, but IR inactive. T_2 transforms like (x, y, z) and (xy, xz, yz) indicating that these modes are both IR and Raman active. **The H-C-H bond angle** vectors transform as E and T_2. E is Raman active but IR inactive whereas T_2 is both Raman and IR active. Thus infrared spectroscopy should show both stretching and bending vibrational modes for a total of two signals whereas Raman spectroscopy shows that all three irreducible representations are active.

The spectroscopic analysis of simple gaseous molecules is of great importance not only in the laboratory but also in determining the existence of these molecules in the atmosphere on other planets and solar systems and in the cosmos or interstellar space. The scientific literature is replete with the analysis of simple molecules, which has shed light on the conditions of other planetary and exosolar systems as scientists search for the possible existence of life (for example microbial) elsewhere in the universe.

4.6.4 Determining the Irreducible Representations and Symmetry of the Central Atom's Atomic Orbitals that Form Bonds

The reducible representation, which is used to determine the number of vibrations for bonds, can also be used to determine the symmetry of the orbitals of the central atom in a molecule as well as the symmetry of the orbitals of similar terminal atoms bound to the central atom (these atoms combine to form group orbitals). *All three of these items are components of the internal coordinates of the molecule; e.g., for water, there are two O–H bonds, two hydrogen atoms and two 1s hydrogen atom orbitals.* The technique shown to determine the irreducible representations from the reducible representation for the bond vibrations of a molecule is the same for determining the symmetry of the orbitals of the central atom in a molecule as well as the orbitals binding to the central atom. In this section, the orbitals of the central atom are described. After this section, the terminal atoms bound to the central atom are described to give a full combination of atomic orbitals to form molecular orbitals (These two sections are useful for chapters 5 and 8 on bonding).

4.6.4.1   Tetrahedral Case

Many organic and inorganic molecules and ions such as NH_4^+, SO_4^{2-}, SiF_4 and $AlCl_4^-$ exhibit tetrahedral geometry like CH_4. As shown in Figure 4.22, there are four bonds that lead to the reducible representation Γ_{tetra} so that $\Gamma_{tetra} = A_1 + T_2$.

This same analysis shows that the four sigma bonds are not equivalent and are made from two sets of the four atomic orbitals of the central atom. Looking at the right of the character table, T_2 is a combination of the set (x, y, z) or of the set (xy, xz, yz) indicating that these sets are **triply degenerate**. These combinations are due to the p orbitals or the d orbitals that are in between the Cartesian coordinate axes. A_1 is a totally symmetric irreducible representation, and the s orbital from the central atom is totally symmetric. From a simplified valence bond theory perspective, the central atom in a tetrahedral molecule has either sp^3 or sd^3 hybridization with four equivalent orbitals (see Figure 5.5).

4.6.4.2   Octahedral Case

The analysis for the six bonds to the central atom for octahedral molecules (e.g., SF_6, CoF_6^{3-}) is shown in Figure 4.23. Table 4.9 is the character table for O_h symmetry.

The reducible representation for the six bonds is obtained as follows. Each of the vectors for the six bonds does not change after the E operation has been performed so the character equals 6. Only two bonds along

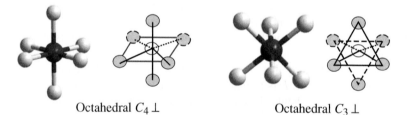

Octahedral $C_4 \perp$ Octahedral $C_3 \perp$

Figure 4.23   *The left structures show the atoms in SF_6 with a C_4 axis normal to the paper, and the right structures show a C_3 axis normal to the paper. Models produced with the HyperChem™ program version 8.0.10*

Table 4.9   *Character table for O_h point group*

O_h	E	$8C_3$	$6C_2$	$6C_4$	$3C_2(=C_4^2)$	i	$6S_4$	$8S_6$	$6\sigma_h$	$6\sigma_d$		$h = 48$
A_{1g}	1	1	1	1	1	1	1'	1	1	1		$x^2 + y^2 + z^2$
A_{2g}	1	1	−1	−1	1	1	−1	1	1	−1		
E_g	2	−1	0	0	2	2	0	−1	2	0		$(2z^2 - x^2 - y^2, x^2 - y^2)$
T_{1g}	3	0	−1	1	−1	3	1	0	−1	−1	(R_x, R_y, R_z)	
T_{2g}	3	0	1	−1	−1	3	−1	0	−1	1		(xy, xz, yz)
A_{1u}	1	1	1	1	1	−1	−1	−1	−1	−1		
A_{2u}	1	1	−1	−1	1	−1	1	−1	−1	1		
E_u	2	−1	0	0	2	−2	0	1	−2	0		
T_{1u}	3	0	−1	1	−1	−3	−1	0	1	1	(x, y, z)	
T_{2u}	3	0	1	−1	−1	−3	1	0	1	−1		

the C_4 and $C_4{}^2$ axes are unchanged so the character equals 2. On performing i, C_2, C_3, S_4, and S_6 operations, all bonds change so the character equals 0. All σ_d planes contain two of the bonds, which are unchanged after the operation, giving a character value of 2 whereas σ_h planes contain four bonds that are unchanged after performing the operation. These operations generate the following reducible representation, Γ_{oct}, for octahedral structures with six bonds.

$$n_b(\mathbf{R}) \times \textbf{operation change} \;=\; \chi_{bond}(\mathbf{R})$$

E	6	$\times$	1	=	6
C_3	0	$\times$	0	=	0
C_2	0	$\times$	0	=	0
C_4	2	$\times$	1	=	2
$C_4{}^2$	2	$\times$	1	=	2
i	0	$\times$	0	=	0
S_4	0	$\times$	0	=	0
S_6	0	$\times$	0	=	0
σ_h	4	$\times$	1	=	2
σ_d	2	$\times$	1	=	2

	E	$8C_3$	$6C_2$	$6C_4$	$3C_2(=C_4{}^2)$	i	$6S_4$	$8S_6$	$6\sigma_h$	$6\sigma_d$
Γ_{oct}	6	0	0	2	2	0	0	0	4	2

Γ_{oct} reduces to A_{1g}, E_g and T_{1u}, and the calculations for only these are presented.

$$c(A_{1g}) = (1/48)[1(1)(6) + 8(1)(0) + 6(1)0 + 6(1)2 + 3(1)(2) + 1(1)0 + 6(1)(0) + 8(1)0 + 6(1)(2) + 6(1)2]$$

$$= (1/48)(48) = 1$$

$$c(E_g) = (1/48)[1(2)(6) + 8(-1)(0) + 6(0)0 + 6(0)2 + 3(2)(2)+1(2)0 + 6(0)(0) + 8(-1)0 + 6(2)(2) + 6(0)2]$$

$$= (1/48)(48) = 1$$

$$c(T_{1u}) = (1/48)[1(3)(6) + 8(0)(0) + 6(-1)0 + 6(1)2 + 3(-1)(2) + 1(-3)0 + 6(-1)(0) + 8(0)(0)$$

$$+6(1)(2) + 6(1)2] = (1/48)(48) = 1$$

This analysis shows that the six sigma bonds are not equivalent because they are made from three sets of the six atomic orbitals of the central atom. Looking at the right of the O_h character table, T_{1u} is a combination of the set (x, y, z) indicating that the three p orbitals are triply degenerate. E_g is a combination of the set $(2z^2 - x^2 - y^2; x^2 - y^2)$ indicating that two d orbitals are doubly degenerate; the $2z^2 - x^2 - y^2$ function is abbreviated as z^2. A_1 is a totally symmetric irreducible representation, and the s orbital from the central atom is totally symmetric. From the valence bond theory perspective, the central atom in an octahedral molecule has sp^3d^2 hybridization with six equivalent orbitals.

4.6.4.3 NH₃ – The Pyramidal Case

Figure 4.24 shows the ammonia molecule of C_{3v} symmetry with which the reducible representation for the 3 N–H bonds is obtained.

Each of the vectors for the three bonds does not change after the E operation is performed so the character equals 3. All bonds along the C_3 axis are changed so the character equals 0. Each vertical plane contains

Figure 4.24   *A drawing and molecular model of the ammonia molecule. Model produced with the HyperChem™ program version 8.0.10*

one bond, which is unchanged after the operation, so the value of the character equals 1. These operations generate the following reducible representation, Γ_{pyr}, for a pyramidal structure with four atoms.

$$
\begin{array}{cccccc}
 & \mathbf{n_b(R)} & \times & \textbf{operation change} & = & \chi_{\textbf{bond}}\mathbf{(R)} \\
E & 3 & \times & 1 & = & 3 \\
C_3 & 0 & \times & 0 & = & 0 \\
\sigma_v & 1 & \times & 1 & = & 1
\end{array}
$$

$$
\begin{array}{cccc}
 & E & 2C_3 & 3\sigma_v \\
\Gamma_{pyr} & 3 & 0 & 1
\end{array}
$$

Γ_{pyr} reduces to A_1 and E as shown below.

$$c(A_1) = (1/6)[1(+1)(3) + 2(+1)(0) + 3(1)(1)] \quad = (1/6)\ (6) = 1$$
$$c(B_1) = (1/6)[1(+1)(3) + 2(+1)(0) + 3(-1)(1)] = (1/6)\ (0) = 0$$
$$c(E) = (1/6)[1\,(+2)(3) + 2(-1)(0) + 3(0)(1)] \quad = (1/6)\ (4) = 1$$

In this analysis, the three bonds are not considered equivalent as they are made from two separate sets of the four atomic central atom orbitals. Inspection of the character table (Table 4.5) shows that A_1 transforms as the s and the p_z orbitals, and E is from the (x, y) set. Again, the three bonds are not equivalent as indicated in the molecular orbital theory approach (section 5.6.2).

4.6.4.4 $CH_3^+(BH_3)$ and H_3^+ – Trigonal Planar Case

The methyl cation and the H_3^+ cation have D_{3h} symmetry (Appendix 4.1) with the structures shown in Figure 4.25. The H_3^+ cation is an example of a two-electron three-atom bond without a central atom (see Section 5.5.1) and is of fundamental importance in interstellar space.

Each of the vectors for the three bonds does not change after the E and σ_h operations are performed so the character equals 3 for each. All bonds along the C_3 and S_3 axis are changed so the character equals 0. Each vertical plane and C_2 axis contains one bond, which is unchanged after the operation, so the value of

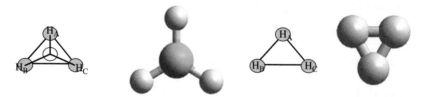

Figure 4.25   *The trigonal planar methyl and H_3^+ cations. Models produced with the HyperChem™ program version 8.0.10*

the character for each equals 1. These operations generate the following reducible representation, Γ_{trig}, for a pyramidal structure with four atoms.

$$\mathbf{n_b(R)} \times \textbf{operation change} = \chi_{bond}(\mathbf{R})$$

	$n_b(R)$	$\times$	operation change	$=$	$\chi_{bond}(R)$
E	3	$\times$	1	$=$	3
C_3	0	$\times$	0	$=$	0
C_2	1	$\times$	1	$=$	1
σ_h	3	$\times$	1	$=$	3
S_3	0	$\times$	0	$=$	0
σ_v	1	$\times$	1	$=$	1

	E	$2C_3$	$3C_2$	σ_h	$2S_3$	$3\sigma_v$
Γ_{trig}	3	0	1	3	0	1

Γ_{trig} reduces to A_1' and E' as shown below.

$$c(A_1') = (1/12)[1(+1)(3) + 2(+1)(0) + 3(1)(1) + 1(1)3 + 2(1)0 + 3(1)1] \qquad = (1/12)(12) = 1$$

$$c(A_2'') = (1/12)[1(+1)(3) + 2(+1)(0) + 3(-1)(1) + 1(-1)3 + 2(-1)0 + 3(1)1] = (1/12)(0) = 0$$

$$c(E') = (1/12)[1(+2)(3) + 2(-1)(0) + 3(0)(1) + 1(2)3 + 2(-1)0 + 3(0)1] \qquad = (1/12)(12) = 1$$

In this analysis, the three bonds are not considered equivalent as they are made from two separate sets of the four atomic central atom orbitals. Inspection of the character table (Table 4.5) shows that A_1' transforms as the s and the p_z orbitals, and E' is from the (x, y) set. Again, the three bonds are not equivalent as indicated in the molecular orbital theory approach.

4.7 Symmetry Adapted Linear Combination (SALC) of Orbitals

Although the method to obtain wave functions is described in detail later, some aspects of molecular orbitals naturally follow from this discussion of group theory. *In this section, which can be postponed and referred to later for the discussion of chemical bonding*, the process is described for H_2O, NH_3, CH_3^+, NO_2^-, and CO_3^{2-}. In the previous section, the types of atomic orbitals that are involved in bonding for the central atom were described. Now, these central atom orbitals of a given symmetry must be combined with the terminal atoms orbitals of similar symmetry in order to form bonds with proper orbital overlap. When there are two or more similar atoms bound to the central atom, then those atoms can be combined into a group or ligand group orbital of a particular symmetry. The resulting group orbitals are called Symmetry Adapted Linear Combinations or SALCs. Because the sign of the wave function for all of the atomic orbitals is known, the terminal atoms' atomic orbitals can be arranged by inspection with the central atom's designated orbital for overlap or non-overlap. However, this procedure does not give symmetry information readily for molecules with many atoms. Thus, the **projection operator** method is used. It is now described for water and ammonia.

4.7.1 Sigma Bonding with Hydrogen as Terminal Atom

4.7.1.1 *H₂O*

For H_2O, refer to Figure 4.2, which shows the two hydrogen atoms with separate labels, A and B. Because there are only two ways to linearly combine the two 1s orbitals for these hydrogen atoms, this system results in the following two combinations where the 1s orbitals are added together or subtracted from each other.

$$1s_A + 1s_B = H_A + H_B \quad \text{or} \quad 1s_A - 1s_B = H_A - H_B$$

These two expressions, which are not normalized, can be shown to transform as two irreducible representations *if the signs of the wave functions are specified and the operations performed.* Because water has C_{2v} symmetry, the two SALC orbital combinations $[(1s_A + 1s_B) \text{ and } (1s_A - 1s_B)]$ transform as $A_1(a_1)$ and $B_1(b_1)$ irreducible representations. The $(1s_A + 1s_B)$ combination does not change the sign of the wave function after any symmetry operations so all operations have a character of 1. However, the $(1s_A - 1s_B)$ does invert sign after C_2 and $\sigma_v yz$ operations so their characters are -1. Compare these values with the C_{2v} character table as well as the reducible representation described above, Γ_{O-H}, which reduces to A_1 and B_1.

	E	C_2	$\sigma_v xz$	$\sigma_v yz$
$(1s_A + 1s_B)$ or A_1	1	1	1	1
$(1s_A - 1s_B)$ or B_1	1	-1	1	-1
Γ_{O-H}	2	0	2	0

The projection operator method is used for molecules with more atoms to consider as the atom movement is more complex. It is used here to describe water, and provides the same information as just shown. The method specifically notes how one particular atom, e.g., H_A, changes into itself or to another atom, e.g., H_B, once a specific symmetry operation is performed. Equation 4.7 gives the mathematical formula for the procedure where ϕ is the resulting group or SALC orbital, and $R\psi$ is the "wave function" describing each individual orbital of the terminal atoms that will comprise ϕ. Once the function ϕ is determined, it is combined with the central atom's orbitals to give the full wave function of the molecule.

$$\phi = \sum_R \chi_i(R)R\psi \tag{4.7}$$

Step 1) To generate $R\psi$, assign the point group for the molecule first. Then use the following table to keep track of how H_A changes after each symmetry operation from the character table. For E and $\sigma_v\ xz$ operations, H_A does not change; after the C_2 and $\sigma_v\ yz$ operations, H_A becomes H_B.

	E	C_2	$\sigma_v xz$	$\sigma_v yz$	
H_A transforms to	H_A	H_B	H_A	H_B	$R\psi$

Step 2) Multiply $R\psi$ for each operation by the characters for each irreducible representation generated from the reducible representation as in Equation 4.6; for water, these are A_1 and B_1. Then sum these.

$$\mathbf{A_1}\ \mathbf{R\Psi} \quad (1)H_A + (1)H_B + (1)H_A + (1)H_B$$

$$\mathbf{B_1}\ \mathbf{R\Psi} \quad (1)H_A + (-1)H_B + (1)H_A + (-1)H_B$$

Adding the products together gives the SALCs, ϕ, for each combination or irreducible representation.

$$\Sigma\mathbf{A_1}\ \mathbf{R\Psi} \quad \text{becomes} \quad \mathbf{SALC\ for}\ \phi_{A1} = H_A + H_B + H_A + H_B = 2(H_A + H_B) = \mathbf{H_A + H_B}$$

$$\Sigma\mathbf{B_1}\ \mathbf{R\Psi} \quad \text{becomes} \quad \mathbf{SALC\ for}\ \phi_{B1} = H_A - H_B + H_A - H_B = 2(H_A - H_B) = \mathbf{H_A - H_B}$$

These two linear combinations of the two hydrogen atomic orbitals are simplified as there are no coefficients and are thus not normalized. The non-normalized wave functions are still useful to show combinations and describe the general bonding process. The function, $1s_A + 1s_B$ or $H_A + H_B$, transforms as A_1 after performing all the symmetry operations; the function, $1s_A - 1s_B$ or $H_A - H_B$, transforms as B_1 after performing all the symmetry operations.

These two SALC functions of A_1 and B_1 symmetry are then paired with the irreducible representations representing the atomic orbitals of the central atom with the same symmetry as calculated above. A_1 for oxygen transforms as a 2s and a $2p_z$ orbital; the 2s orbital combines with the linear combination, $H_A + H_B$. B_1 for oxygen transforms as a $2p_x$ orbital, which combines with the linear combination, $H_A - H_B$. Thus, two bonding and two antibonding orbitals are produced.

In the molecular orbital theory approach, the electrons are delocalized across all three atoms for each bond that is represented as a_1 and b_1 (orbitals are lowercase whereas irreducible representations are uppercase). As will be shown later with use of photoelectron spectroscopy data, the $2p_z$ orbital of oxygen interacts strongly with the two hydrogen atoms having the A_1 symmetry so is not a lone pair of electrons or non-bonding. The oxygen $2p_y$ orbital does not interact with the hydrogen atoms, and is thus a lone pair of electrons that is non-bonding.

4.7.1.2   *NH₃*

The process is now shown for the ammonia molecule with C_{3v} symmetry. It is important to note one major feature in using the projection operator method; that is, all symmetry operations must be explicitly calculated when using equation 4.7. For example, there are two C_3 operations indicated as one class in the character table; however, each C_3 operation results in a different orientation in space for H_A as shown in step 1. The vertical planes also have to be specified separately as σ_A contains the N-H_A bond; σ_B contains the N-H_B bond and σ_C contains the N-H_C bond.

Step 1:

$$
\begin{array}{ccccccc}
 & E & C_3 & C_3{}^2 & \sigma_A & \sigma_C & \sigma_B \\
H_A \text{ transforms to} & H_A & H_B & H_C & H_A & H_B & H_C & \mathbf{R\psi}
\end{array}
$$

Step 2:   Multiply Rψ by the character of the irreducible representations; then sum the products.

$$\mathbf{A_1}\ \mathbf{R\Psi}\quad (1)H_A + (1)H_B + (1)H_C + (1)H_A + (1)H_B + (1)H_C$$

$$\mathbf{A_2}\ \mathbf{R\Psi}\quad (1)H_A + (1)H_B + (1)H_C + (-1)H_A + (-1)H_B + (-1)H_C$$

$$\mathbf{E}\ \mathbf{R\Psi}\quad (2)H_A + (-1)H_B + (-1)H_C + (0)H_A + (0)H_B + (0)H_C$$

which gives SALCs ϕ.

$$\mathbf{A_1}\ \mathbf{R\Psi}\quad \phi_{A1} = H_A + H_B + H_C + H_A + H_B + H_C = 2(H_A + H_B + H_C) = \mathbf{H_A + H_B + H_C}$$

$$\mathbf{A_2}\ \mathbf{R\Psi}\quad \phi_{A2} = H_A + H_B + H_C - H_A - H_B - H_C = \mathbf{0}$$

$$\mathbf{E}\ \mathbf{R\Psi}\quad \phi_E = 2H_A - H_B - H_C = \mathbf{H_A - H_B - H_C}$$

Only two linear combinations of the three hydrogen atomic orbitals are found with one being degenerate with E symmetry; thus there are two orbitals of similar symmetry and energy. The resulting combinations are given without coefficients and are thus not normalized. The function, $1s_A + 1s_B + 1s_C$ or $H_A + H_B + H_C$, transforms as A_1 after performing all the symmetry operations; the function, $1s_A - 1s_B - 1s_C$ or $H_A - H_B - H_C$, transforms as E after performing all the symmetry operations.

These two functions with their irreducible representations are then paired with the appropriate irreducible representations representing the atomic orbitals of the central atom. A_1 for nitrogen transforms as a 2s and a

$2p_z$ orbital; the 2s orbital combines with the linear combination, $H_A + H_B + H_C$. E for nitrogen transforms as a $2p_x$, $2p_y$ degenerate set, which combines with the linear combination, $H_A - H_B - H_C$. Again these bonding orbitals represent a pair of electrons distributed across three atoms. The nitrogen $2p_z$ orbital is a filled non-bonding orbital, which can donate a pair of electrons, and transforms as an A_1 irreducible representation.

4.7.1.3 $CH_3{}^+$

The process is now shown for $CH_3{}^+$, which also has a C_3 major axis with D_{3h} symmetry (Appendix 4.1).

Step 1:

	E	C_3	$C_3{}^2$	C_2	C_2'	C_2''	σ_h	S_3	$S_3{}^2$	σ_A	σ_C	σ_B	
H_A transforms to	H_A	H_B	H_C	H_A	H_B	H_C	H_A	H_B	H_C	H_A	H_B	H_C	**Rψ**

Step 2:   Multiply Rψ by the character of the irreducible representations determined from the reducible representation; then sum the products.

A_1' **RΨ** $(1)H_A + (1)H_B + (1)H_C + (1)H_A + (1)H_B + (1)H_C + (1)H_A + (1)H_B + (1)H_C$
$\qquad + (1)H_A + (1)H_B + (1)H_C$

E' **RΨ** $(2)H_A + (-1)H_B + (-1)H_C + (0)H_A + (0)H_B + (0)H_C + (2)H_A + (-1)H_B + (-1)H_C$
$\qquad + (0)H_A + (0)H_B + (0)H_C$

which gives SALCs ϕ.

A_1' **RΨ** $\phi_{A1} = H_A + H_B + H_C + H_A + H_B + H_C = 4(H_A + H_B + H_C) = \mathbf{H_A + H_B + H_C}$

E' **RΨ** $\phi_E = 4H_A - 2H_B - 2H_C = \mathbf{H_A - H_B - H_C}$

Notice the similarity in the final irreducible representations (A and E) and SALCs between NH_3 and $CH_3{}^+$ even though they have different point groups.

4.7.2   Sigma and Pi Bonding with Atoms Other than Hydrogen as Terminal Atom

4.7.2.1 $NO_2{}^-$ σ and π Bonding (O_3, SO_2)

These three species have C_{2v} symmetry as does water; however, all three atoms in each species now have p orbitals to consider for a total of 12 orbitals rather than the 6 for water. The six additional orbitals are involved with π type bonding molecular orbitals. For molecules that have terminal atoms with p orbitals, a reducible representation is made based on how each p orbital set changes with each symmetry operation and the sign of the wave function is also considered. These are reduced to irreducible representations that are then combined with the same irreducible representations of the central atom. Figure 4.26 shows the bonding arrangement for the nitrite ion as Lewis structures (see Sections 5.1.1 and 5.7.3) and the spatial representation for the p_x, p_y, and p_z orbitals for each atom with which reducible representations can be determined.

The s orbitals of the two oxygen atoms will transform as $A_1(a_1)$ and $B_1(b_1)$ irreducible representations similar to the two hydrogen atoms in water. The s and p_z orbitals of the central nitrogen atom also transform as A_1 whereas the p_x orbital transforms as B_1. To complete the group theory analysis of the molecule, the p orbitals of the terminal oxygen atoms need to be determined.

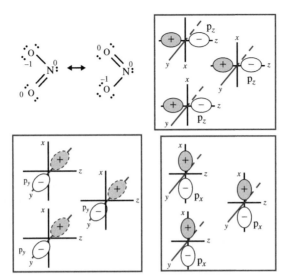

Figure 4.26 *Nitrite ion structure with all of the p orbitals including the sign of the wave function. The xz plane is in the plane of the paper with the y axis normal to the paper*

The p_x orbitals for the two O atoms in nitrite are in the plane of the paper and produce the following reducible representation after performing the C_{2v} symmetry operations where the operation change indicates whether the orbital and the sign of its wave function are unchanged, changed or inverted in sign. Here $n_u(R)$ indicates the number of orbitals affected by the operation.

$$
\begin{array}{ccccccc}
 & n_u(\mathbf{R}) & \times & \text{operation change} & = & \chi(\mathbf{R}) \\
E & 2 & \times & (1) & = & 2 \\
C_2 & 0 & \times & (0) & = & 0 \\
\sigma_v xz & 2 & \times & (1) & = & 2 \\
\sigma_v yz & 0 & \times & (0) & = & 0 \\
\end{array}
$$

$$
\begin{array}{ccccc}
 & E & C_3 & \sigma_v xz & \sigma_v yz \\
\chi(\mathbf{R})p_x & 2 & 0 & 2 & 0 \\
\end{array}
$$

The p_x orbital set reduces to A_1 and B_1 as shown above for the s orbital set of H_2O in C_{2v} symmetry. Thus, the SALC wave functions will be similar to those for the water s orbital set.

$$A_1 = (O_A p_x + O_B p_x)$$
$$B_1 = (O_A p_x - O_B p_x)$$

The p_z orbitals for the two O atoms in nitrite are in the plane of the paper and produce the following reducible representation after performing the C_{2v} symmetry operations where the operation change indicates whether the orbital and the sign of its wave function are unchanged, changed, or inverted in sign.

$$\mathbf{n_u(R)} \times \text{ operation change } = \chi\mathbf{(R)}$$

	$n_u(R)$		operation change		$\chi(R)$
E	2	$\times$	(1)	$=$	2
C_2	0	$\times$	(0)	$=$	0
$\sigma_v xz$	2	$\times$	(1)	$=$	2
$\sigma_v yz$	0	$\times$	(0)	$=$	0

	E	C_3	$\sigma_v xz$	$\sigma_v yz$
$\chi(R)p_z$	2	0	2	0

The p_z orbital for nitrogen transforms as A_1 and the oxygen atom p_z set also reduces to A_1 and B_1 as shown above for the H_2O s orbital set in C_{2v} symmetry. Thus, the SALC wave functions will be similar to those for the H_2O s orbital set.

$$A_1 = (O_A p_z + O_B p_z)$$
$$B_1 = (O_A p_z - O_B p_z)$$

The p_y orbitals for the two O atoms in nitrite are normal to the plane of the paper and produce the following reducible representation after performing the C_{2v} symmetry operations. Note that the sign of the wave function for these orbitals is inverted after performing the $\sigma_v xz$ operation.

$$\mathbf{n_u(R)} \times \text{ operation change } = \chi\mathbf{(R)}$$

	$n_u(R)$		operation change		$\chi(R)$
E	2	$\times$	(1)	$=$	2
C_2	0	$\times$	(0)	$=$	0
$\sigma_v xz$	2	$\times$	(-1)	$=$	-2
$\sigma_v yz$	0	$\times$	(0)	$=$	0

	E	C_3	$\sigma_v xz$	$\sigma_v yz$
$\chi(R)p_y$	2	0	-2	0

By inspection, the p_y orbital set reduces to A_2 and B_2, and generates one of the nitrite ion's π bonding set of orbitals. These SALC combinations give a different sign to the a and b orbitals as the y axis is normal to the xz plane of the molecule.

$$A_2 = (O_A p_y - O_B p_y)$$
$$B_2 = (O_A p_y + O_B p_y)$$

Combination of Atomic Orbitals from N with O Atoms to Form Molecular Orbitals. Coupling the p orbital analysis for the two oxygen atoms with the s and p orbitals from the nitrogen atom gives the orbitals indicated next and that are shown in Figures 5.49 and 5.50.

The nitrogen 2s orbital transforms as A_1 and B_1. The oxygen s and p_x SALC orbitals transform as A_1 to give the $1a_1$ and $2a_1$ bonding orbitals with the nitrogen 2s orbital whereas the nitrogen p_x and oxygen SALC s orbitals combine to form $1b_1$; there is also a $5a_1$ antibonding combination between the oxygen p_x and nitrogen s orbitals. The oxygen p_x SALC orbitals transform as B_1 and interact with the nitrogen p_x orbital to form $2b_1$; there is also a $4b_1$ antibonding combination between the oxygen p_x SALCs and the nitrogen p_x orbital. This analysis yields 6 of the 12 molecular orbitals.

$$1a_1 = \text{Ns} + (O_A\text{s} + O_B\text{s})$$

$$2a_1 = \text{Ns} + (O_A\text{p}_x + O_B\text{p}_x)$$

$$1b_1 = \text{Np}_x + (O_A\text{s} + O_B\text{s})$$

$$2b_1 = \text{Np}_x + (O_A\text{p}_x + O_B\text{p}_x)$$

$$5a_1{}^* = \text{Ns} - (O_A\text{p}_x + O_B\text{p}_x)$$

$$4b_1{}^* = \text{Np}_x - (O_A\text{p}_x + O_B\text{p}_x)$$

The nitrogen 2p_z transforms as A_1 and the oxygen 2p_z SALC orbitals transform as A_1 and B_1; thus there can be only a total of three molecular orbitals. These combinations lead to $3a_1$, $3b_1$, and $4a_1$, which are bonding, nonbonding, and antibonding type orbitals, respectively. The nitrogen 2p_z orbital interacts more strongly with the O 2p_z orbitals (not the O s orbital) because the p_z orbitals are closer in energy.

$$3a_1 = \text{Np}_z + (O_A\text{p}_z + O_B\text{p}_z)$$

$$3b_1 = \text{Np}_z + (O_A\text{p}_z - O_B\text{p}_z) \quad \text{or} \quad 3b_1 = (O_A\text{p}_z - O_B\text{p}_z)$$

$$4a_1{}^* = \text{Np}_z - (O_A\text{p}_z + O_B\text{p}_z)$$

The nitrogen 2p_y transforms as B_2 and the oxygen 2p_y SALC orbitals transform as A_2 and B_2. Again only three combinations are possible: bonding $1b_2$, nonbonding $1a_2$, and antibonding $2b_2$ orbitals.

$$1b_2 = \text{Np}_y + (O_A\text{p}_y + O_B\text{p}_y)$$

$$1a_2 = \text{Np}_y + (O_A\text{p}_y - O_B\text{p}_y) \quad \text{or} \quad 1a_2 = (O_A\text{p}_y - O_B\text{p}_y)$$

$$2b_2{}^* = \text{Np}_y - (O_A\text{p}_y + O_B\text{p}_y)$$

4.7.2.2 $CO_3{}^{2-}$ σ and π Bonding

As for nitrite, reducible representations need to be obtained for the p orbital system for the three oxygen atoms. Figure 4.27 shows the 2p orbital sets for the terminal oxygen atoms that bind to carbon in the carbonate ion, $CO_3{}^{2-}$.

   The p_x orbitals for the three O atoms in carbonate are in the plane of the paper and produce the following reducible representation after performing the D_{3h} symmetry operations where the operation change indicates whether the orbital and the sign of its wave function are unchanged, changed, or inverted in sign.

	$n_u(R)$	$\times$	operation change	$=$	$\chi(R)$
E	3	$\times$	(1)	$=$	3
C_3	1	$\times$	(0)	$=$	0
C_2	1	$\times$	(−1)	$=$	−1
σ_h	3	$\times$	(1)	$=$	3
S_3	1	$\times$	(0)	$=$	0
σ_v	1	$\times$	(−1)	$=$	−1

	E	C_3	$3C_2$	σ_h	$2S_3$	$3\sigma_v$
$\chi(R)\text{p}_x$	3	0	−1	3	0	−1

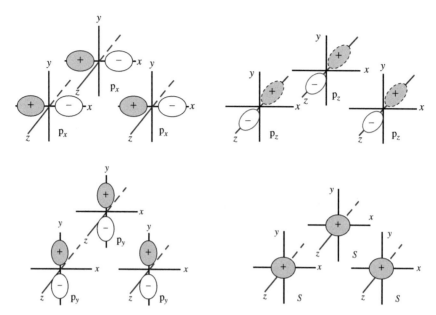

Figure 4.27   *The 2s and three 2p orbital sets for the oxygen atoms in CO_3^{2-}. The p_z orbital set is $\perp$ to the plane of the paper, which contains σ_h*

The p_x orbital set reduces to A_2' and E' as shown below for all symmetry operations, D_{3h}. Thus, the SALC wave functions will be similar to those for the CH_3^+ s orbital set.

$$c(A_1') = (1/12)[1(+1)(3) + 2(+1)(0) + 3(1)(-1) + 1(1)3 + 2(1)0 + 3(1) - 1] \quad = (1/12)(0) = 0$$

$$c(A_2') = (1/12)[1(+1)(3) + 2(+1)(0) + 3(-1)(-1) + 1(1)3 + 2(1)0 + 3(-1) - 1] = (1/12)(12) = 1$$

$$c(E') = (1/12)[1(+2)(3) + 2(-1)(0) + 3(0)(-1) + 1(2)3 + 2(-1)0 + 3(0) - 1] \quad = (1/12)(12) = 1$$

$$c(A_1'') = (1/12)[1(1)(3) + 2(1)(0) + 3(1)(-1) + 1(-1)3 + 2(-1)0 + 3(-1) - 1] \quad = (1/12)(0) = 0$$

$$c(A_2'') = (1/12)[1(1)(3) + 2(1)(0) + 3(-1)(-1) + 1(-1)3 + 2(-1)0 + 3(1) - 1] \quad = (1/12)(0) = 0$$

$$c(E'') = (1/12)[1(+2)(3) + 2(-1)(0) + 3(0)(-1) + 1(-2)3 + 2(1)0 + 3(0) - 1] \quad = (1/12)(0) = 0$$

The p_y orbitals for the O atoms in carbonate are also in the plane of the paper and produce the following reducible representation after performing the D_{3h} symmetry operations. The p_y orbital is in the plane of the paper, and the $3C_2$ and $3\sigma_v$ symmetry elements contain one of the orbitals.

	$n_u(R)$	×	operation change	=	$\chi(R)$
E	3	×	(1)	=	3
C_3	1	×	(0)	=	0
C_2	1	×	(1)	=	1
σ_h	3	×	(1)	=	3
S_3	1	×	(0)	=	0
σ_v	1	×	(1)	=	1

$$
\begin{array}{ccccccc}
 & E & C_3 & 3C_2 & \sigma_h & 2S_3 & 3\sigma_v \\
\chi(\mathbf{R})p_y & 3 & 0 & 1 & 3 & 0 & 1
\end{array}
$$

The p_y orbital set reduces to A_1' and E' as shown below. The SALC wave functions for E' will be similar to those for the CH_3^+ s orbital set.

$$c(A_1') = (1/12)[1(+1)(3) + 2(+1)(0) + 3(1)(1) + 1(1)3 + 2(1)0 + 3(1)1] = (1/12)(12) = 1$$

$$c(E') = (1/12)[1(+2)(3) + 2(-1)(0) + 3(0)(1) + 1(2)3 + 2(-1)0 + 3(0)1] = (1/12)(12) = 1$$

The p_z orbitals for the O atoms in carbonate are perpendicular to the plane of the paper and produce the following reducible representation after performing the D_{3h} symmetry operations.

$$
\begin{array}{ccccc}
 & n_u(\mathbf{R}) & \times & \text{operation change} & = \chi(\mathbf{R}) \\
E & 3 & \times & (1) & = \quad 3 \\
C_3 & 1 & \times & (0) & = \quad 0 \\
C_2 & 1 & \times & (-1) & = \quad -1 \\
\sigma_h & 3 & \times & (-1) & = \quad -3 \\
S_3 & 1 & \times & (0) & = \quad 0 \\
\sigma_v & 1 & \times & (1) & = \quad 1
\end{array}
$$

$$
\begin{array}{ccccccc}
 & E & C_3 & 3C_2 & \sigma_h & 2S_3 & 3\sigma_v \\
\chi(\mathbf{R})p_z & 3 & 0 & -1 & -3 & 0 & 1
\end{array}
$$

The p_z orbital set reduces to A_2'' and E'' as shown below. The SALC wave functions for E'' will be similar to those for the CH_3^+ s orbital set.

$$c(A_2'') = (1/12)[1(1)(3) + 2(1)(0) + 3(-1)(-1) + 1(-1) - 3 + 2(-1)0 + 3(1)1] = (1/12)(12) = 1$$

$$c(E'') = (1/12)[1(+2)(3) + 2(-1)(0) + 3(0)(-1) + 1(-2) - 3 + 2(1)0 + 3(0)1] = (1/12)(12) = 1$$

Coupling this analysis of the three sets of oxygen p orbitals with the oxygen s orbitals as in CH_3^+ (the oxygen s orbitals are similar to the hydrogen s orbitals and transform as A_1' and E') and the carbon orbitals gives the carbonate molecular orbitals that are summarized below and in Figure 4.28.

The oxygen s orbital set (A_1' and E') combines with the carbon 2s orbital (transforms as A_1') to give the $1a_1'$ σ bonding orbital; there is also a $3a_1'$ antibonding combination. The carbon p_x and p_y orbitals (E') combine with the E' oxygen set to give the $1e'$ set of weakly bonding s orbitals.

The oxygen p_y orbital set (E') interacts with the carbon p_x and p_y orbitals (transform as E') to give the $2e'$ set of σ bonding orbitals with carbon. INVERT the carbon p_x and p_y orbitals to obtain the $4e'$ σ antibonding set, which is not shown. The oxygen p_y orbital set also transforms as A_1' and combines with the carbon 2s orbital (A_1') to give the $2a_1'$ bonding orbital.

The oxygen p_x orbital set (E') interacts with the carbon p_x and p_y orbitals (transform as E') to give the $3e'$ set of σ nonbonding orbitals with carbon. The oxygen p_x orbital set also transforms as A_2', which is another σ nonbonding orbital.

The carbon $2p_z$ transforms as A_2'', and the three oxygen p_z orbitals transform as A_2'' and E''; thus, only the $1a_2''$ is a π bonding combination; there is also a $2a_2''$ π antibonding combination. The $1e''$ set from the oxygen p_z orbitals is a π nonbonding set of two orbitals.

The description for carbonate is similar for NO_3^-, SO_3, and BF_3. Compare these results with Figure 5.51 and its discussion in Chapter 5 (Section 5.8).

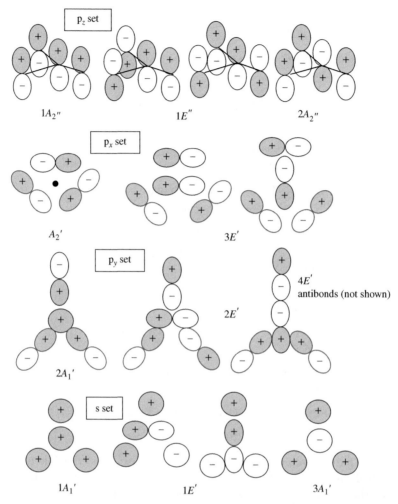

Figure 4.28 *Carbonate ion molecular orbitals. Oxygen $2p_y$ orbitals point to the carbon atom; thus, the p_x are $\perp$ to the p_y orbitals and point away from the carbon atom. No s–p mixing of the central atom is considered. The xy plane is the plane of the paper for the p_x and p_y orbital sets. The p_z orbital set is redrawn so that the z axis is in or parallel to the plane of the paper to better observe the π orbital overlap*

APPENDIX 4.1

Some additional useful character tables

C_{4v}	E	$2C_4$	C_2	$2\sigma_v$	$2\sigma_d$		$h = 8$
A_1	1	1	1	1	1	z	$x^2 + y^2, z^2$
A_2	1	1	1	−1	−1	R_z	
B_1	1	−1	1	1	−1		$x^2 - y^2$
B_2	1	−1	1	−1	1		xy
E	2	0	−2	0	0	$(x, y)\,(R_x, R_y)$	(xz, yz)

C_{5v}	E	$2C_5$	$2C_5^2$	$5\sigma_v$		$h = 10, \alpha = 72$
A_1	1	1	1	1	z	$x^2 + y^2, z^2$
A_2	1	1	1	-1	R_z	
E_1	2	$2\cos\alpha$	$2\cos2\alpha$	0	$(x, y)\,(R_x, R_y)$	(xz, yz)
E_2	2	$2\cos2\alpha$	$2\cos\alpha$	0		$(x^2 - y^2, xy)$

$C_{\infty v}$	E	$2C_\infty^\phi$	$\ldots$	$\infty\sigma_v$		$h = \infty$
$A_1 \equiv \Sigma^+$	1	1	$\ldots$	1	z	$x^2 + y^2, z^2$
$A_2 \equiv \Sigma^-$	1	1	$\ldots$	-1	R_z	
$E_1 \equiv \Pi$	2	$2\cos\phi$	$\ldots$	0	$(x, y)\,(R_x, R_y)$	(xz, yz)
$E_2 \equiv \Delta$	2	$2\cos2\phi$	$\ldots$	0		$(x^2 - y^2, xy)$
$E_3 \equiv \Phi$	2	$2\cos3\phi$	$\ldots$	0		
$\ldots$	$\ldots$	$\ldots$	$\ldots$	$\ldots$		

D_{3h}	E	$2C_3$	$3C_2$	σ_h	$2S_3$	$2\sigma_v$		$h = 12$
A_1'	1	1	1	1	1	1		$x^2 + y^2, z^2$
A_2'	1	1	-1	1	1	-1	R_z	
E'	2	-1	0	2	-1	0	(x, y)	$(x^2 - y^2, xy)$
A_1''	1	1	1	-1	-1	-1		
A_2''	1	1	-1	-1	-1	1	z	
E''	2	-1	0	-2	1	0	(R_x, R_y)	(xz, yz)

D_{4h}	E	$2C_4$	C_2	$2C_2'$	$2C_2''$	i	$2S_4$	σ_h	$2\sigma_v$	$2\sigma_d$		$h = 16$
A_{1g}	1	1	1	1	1	1	1	1	1	1		$x^2 + y^2, z^2$
A_{2g}	1	1	1	-1	-1	1	1	1	-1	-1	R_z	
B_{1g}	1	-1	1	1	-1	1	-1	1	1	-1		$(x^2 - y^2)$
B_{2g}	1	-1	1	-1	1	1	-1	1	-1	1		xy
E_g	2	0	-2	0	0	2	0	-2	0	0	(R_x, R_y)	(xz, yz)
A_{1u}	1	1	1	1	1	-1	-1	-1	-1	-1		
A_{2u}	1	1	1	-1	-1	-1	-1	-1	1	1	z	
B_{1u}	1	-1	1	1	-1	-1	1	-1	-1	1		
B_{2u}	1	-1	1	-1	1	-1	1	-1	1	-1		
E_u	2	0	-2	0	0	-2	0	2	0	0	(x, y)	

$D_{\infty h}$	E	$2C_\infty^\phi$	...	$\infty\sigma_v$	i	$2S_\infty^\phi$	...	∞C_2		$h = \infty$
Σ_g^+	1	1	...	1	1	1	...	1		$x^2 + y^2, z^2$
Σ_g^-	1	1	...	−1	1	1	...	−1	R_z	
Π_g	2	$2\cos\phi$	...	0	2	$-2\cos\phi$	...	0	(R_x, R_y)	(xz, yz)
Δ_g	2	$2\cos2\phi$	...	0	2	$2\cos2\phi$	...	0		$(x^2 - y^2, xy)$
...	...	...	...	...	...	...	...	...		
Σ_u^+	1	1	...	1	−1	−1	...	−1	z	
Σ_u^-	1	1	...	−1	−1	−1	...	1		
Π_u	2	$2\cos\phi$	...	0	−2	$2\cos\phi$	...	0	(x, y)	
Δ_u	2	$2\cos2\phi$	...	0	−2	$-2\cos2\phi$	...	0		
...	...	...	...	...	...	...	...	...		

References

1. Gartman, A. and Luther, III, G. W. (2013) Comparison of pyrite (FeS_2) synthesis mechanisms to reproduce natural FeS_2 nanoparticles found at hydrothermal vents. *Geochimica Cosmochimica Acta* **120**, 447–458.
2. Cotton, F. A. (1990) *Chemical Applications of Group Theory* (3rd ed.), Wiley, NY.
3. Jaffé, H. H. and Orchin, M. (1977) *Symmetry in Chemistry*, R. E. Krieger Pub. Co., Huntington, NY.
4. Bunker, P. R. and Jensen, P. (2005) *Fundamentals of Molecular Symmetry*, CRC Press, Boca Raton, FL, pp. 358.
5. Kettle, S. F. A. (2007) *Symmetry and Structure: Readable Group Theory for Chemists* (3rd ed.), John Wiley & Sons, Inc. pp. 436.
6. Nakamato, K. (1997) *Infrared and Raman Spectra of Inorganic and Coordination Compounds* (5th ed.), Wiley-Interscience, NY.

5

Covalent Bonding

5.1 Introduction

In 1916, after Bohr's development of the hydrogen atom and before Schrodinger's development of quantum mechanics, Lewis [1] proposed that atoms share electrons to form covalent bonds so that each atom has an octet or eight electrons around it. This sharing of eight electrons is known as the **octet rule**. For the main group elements, an octet of electrons results in a noble gas configuration. A single bond shares two electrons, which are represented by two dots or a single dashed line; a double bond shares four electrons and a triple bond shares six electrons. Any other electrons that are not in bonds are called lone pairs or nonbonding pairs. The placement of the electrons around each atom is called a **Lewis structure**, which is a two-dimensional arrangement. A three-dimensional spatial arrangement comes from other bonding theories including valence shell electron pair repulsion theory, valence bond theory (VBT), and molecular orbital theory (MOT).

5.1.1 Lewis Structures and the Octet Rule

A convenient method to write Lewis structures with each atom having eight electrons is to follow these four rules, which work in myriad instances.

1. Count the total number of valence electrons of all atoms in the formula. If the formula has a charge, add electrons to the valence electrons if the formula is negative. Subtract electrons from the valence electrons, if the formula is positive.
2. Count the total number of electrons needed for each atom to have an octet except for hydrogen, which can have only two electrons.
3. Subtract rule #1 from rule #2. This gives the number of electrons that are not available for an octet and which must be shared to form bonds for each atom to have an octet. Divide this number by 2 giving the number of bonds for the formula.

Inorganic Chemistry for Geochemistry and Environmental Sciences: Fundamentals and Applications, First Edition. George W. Luther, III.
© 2016 John Wiley & Sons, Ltd. Published 2016 by John Wiley & Sons, Ltd.
Companion Website: www.wiley.com/go/luther/inorganic

For example, $SO_4{}^{2-}$

1.   S has 6 electrons in its valence shell (1 S-atom) = 6
     O has 6 electrons in its valence shell (4 O-atoms) = 24
     $\qquad\qquad$ 2 electrons for − 2 charge = 2
     $\qquad\qquad\qquad$ Total electrons available = 32

2.   S needs 8 electrons × 1 S-atom = 8
     O needs 8 electrons × 4 O-atoms = 32
     $\qquad\qquad$ Total electrons needed = 40

3.   40 − 32 = 8 electrons; therefore there are 4 bonds (8/2); 8 of the 32 electrons are in bonds and the remaining 24 of the 32 electrons are in lone pairs or nonbonding pairs that will be distributed around the oxygen atoms to obtain octets.

Once this calculation is complete, it is possible to assemble the electrons to form a proper Lewis structure for the sulfate ion. Here, the sulfur atom is the central atom and the four oxygen atoms are terminal atoms bound to the sulfur. The Lewis structure results in four single bonds between each oxygen atom and the central sulfur atom. Each bond is normally drawn as a line or dash and the lone pairs as two dots. All the nonbonding pairs of electrons are then distributed around each oxygen atom so all atoms have an octet. Figure 5.1 shows the correct structure for $SO_4{}^{2-}$.

4.   To fully account for all charges on each atom, it is necessary to calculate the formal charge (F.C.) so that the sum of all formal charges on each atom equals the charge of the formula and the Lewis structure. The formal charges are calculated from the Lewis structure and are defined mathematically as

F.C. = # valence shell electrons − (# electrons in bonds/2) − (# electrons not in bonds).

For the sulfur atom, $S = 6 − (8/2) − 0 = +2$
For each oxygen atom, $O = 6 − (2/2) − 6 = −1$
As there are four oxygen atoms with a −1 one charge and one sulfur atom with a +2 charge, the total charge for sulfate is −2.

Owing to repulsive effects, a Lewis structure is not correct (or a stable species) when two adjacent atoms have the same charge. The most electronegative atom will normally be the central atom and the more electropositive atoms will be bound to that. However, the formal charges calculated may not reflect a negative charge on the most electronegative atom. For example, carbon monoxide (CO) has a total of 10 valence electrons when 16 are needed; thus, CO has a triple bond between the carbon and the oxygen atoms. To complete an octet, CO has the Lewis structure (:C≡O:). The formal charge on the carbon is −1 and the formal charge on the oxygen atom is +1 so that the molecule is neutral. The cyanide ion (CN^-) has the same number of valence electrons as carbon monoxide so we can expect the same Lewis structure with the N atom replacing the O atom. Molecules and ions, which have the same number of valence electrons, are called **isoelectronic**. The isoelectronic principle or analogy is useful to understand bonding. Again, the formal charge on carbon will be −1 and the formal charge on the nitrogen atom will be zero. For both the cyanide ion and carbon monoxide, the negative charge on the carbon atoms indicates that the carbon will donate a pair of electrons to an electron acceptor or Lewis base. For these two cases, the carbon will actually bond with iron in hemoglobin and cause asphyxiation when CO (HCN) enters the lungs.

The octet rule suggests that all electrons are paired so that all structures are diamagnetic. For example, the ozone molecule, O_3, has two possible Lewis structures as shown in Figure 5.1. Because the double bond could be on either side of the central oxygen atom, both structures are considered correct and are called **resonance structures**; the two headed arrow indicates that both structures combined represent the ozone molecule. It is common to sum the double bond and single bond, and then divide by two indicating that there are one

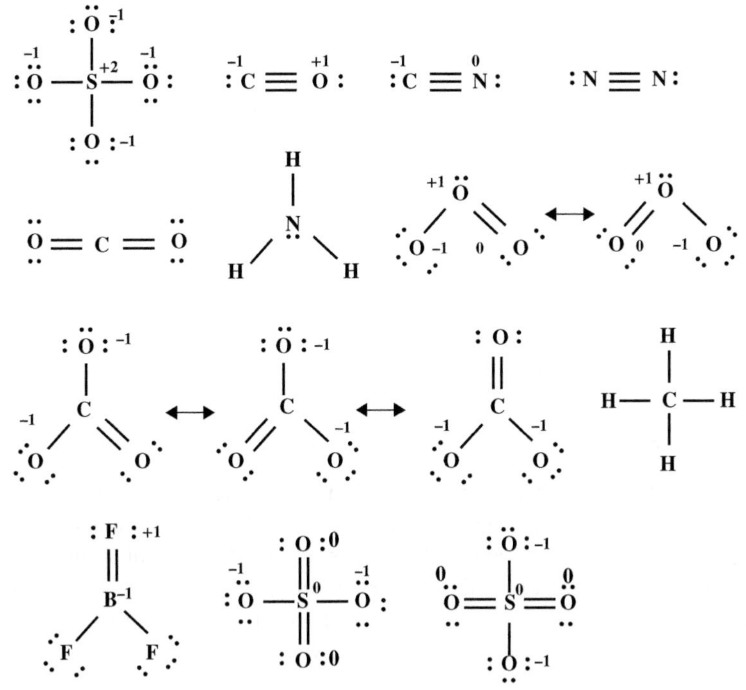

Figure 5.1 *Lewis structures for sulfate ion, carbon monoxide, cyanide ion, ozone, carbonate, and boron trifluoride with the formal charges for each atom in each species. The formal charges are 0 for all the atoms of ammonia, carbon dioxide, dinitrogen, and methane*

and a half bonds between the central oxygen atom and each of the terminal oxygen atoms. In fact, the bond lengths are 1.5 bonds. Sulfur dioxide, SO_2, and the nitrite ion, NO_2^-, also have 18 valence electrons and are isoelectronic with ozone; thus, they would have an identical Lewis structure to ozone with the sulfur and nitrogen being the central atoms in each of these chemical species. For the nitrite ion, there would be a -1 charge. The carbonate ion (CO_3^{2-}) has three resonance structures and the number of bonds between each C and O atom is 1.33. For boron trifluoride, the Lewis structure could be written as one B–F bond for a total of three bonds, and this representation would be an electron-deficient molecule. However, Figure 5.1 shows the possible double bond character in BF_3 that results in a total of three resonance structures and four bonds for the molecule (note the differences in formal charge between the isoelectronic BF_3 and the carbonate ion). Likewise, the sulfur–oxygen bond in sulfate is not one for all sulfur–oxygen bonds; it is 1.5 based on experimental evidence. The bottom portion of Figure 5.1 shows two more of the possible six forms for SO_4^{2-}, which can have two sulfur–oxygen double bonds. Note the difference of the formal charges for these two structures versus the one where sulfate only has sulfur oxygen single bonds.

Another way to represent the ozone molecule is to use "o" and "x" to represent electrons with different spins. In this way, it is possible to have eight electrons around each oxygen atom in ozone and 1.5 bonds from the central O atom to each of the terminal O atoms. Figure 5.2 shows this particular Lewis structure. In addition, there are molecules with an odd number of electrons that can be represented with an octet around each atom. The molecule NO is nitric oxide, which is produced as a by-product of combustion of fossil fuels with air. NO is a free radical and there are a total of 11 valence electrons so that we can predict 2.5 bonds

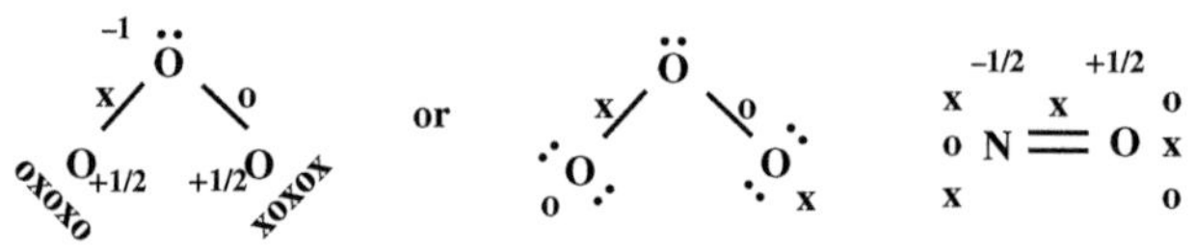

Figure 5.2 *Lewis structures for ozone and nitric oxide with the formal charges for each atom in each species. Here fractional bonds and radicals are possible*

(five electrons are shared). The remaining six electrons are divided equally between the N atom and the O atom. The formal charge is $-1/2$ for the N atom and $+1/2$ for the O atom so that the molecule is neutral. Because of the negative charge on the N atom, the N atom is expected to donate electrons. In fact, the NO molecule can donate 1, 2, or 3 electrons to a metal ion.

5.1.2   Valence Shell Electron Pair Repulsion Theory (VSEPR)

This theory was developed by Gillespie and Nyholm [2], and it is a method to draw correct Lewis structures in order to give three-dimensional geometries. This theory also allows for arrangements of less than eight electrons and greater than eight electrons (**expanded octets**) around the central atom in a given molecule or ion. This theory takes into account the optimal geometry to minimize repulsions by atoms and lone pairs of electrons in molecules; i.e., to place all electron pairs in positions of a given geometry as far apart as possible. There are seven relatively simple rules that can be used to give correct Lewis structures and geometries.

1.   Count all the electrons (e^-s) on the **central atom**. Include charges of ions in this count recalling to add electrons for negative charges and to subtract electrons for positive charges on the molecule/ion.
2.   Count each **halogen as donating** one e^- to the central atom. Hydrogen atoms are also considered to donate one electron to the central atom.
3.   Count each **oxygen as accepting** two e^-s from the central atom; thus, no electrons are added to the formal count of valence electrons.
4.   Add the total of rules 1 and 2 and then divide by 2 to obtain the number of chemical bonds and lone pairs of electrons to the central atom.
5.   Place all the pairs of electrons around the central atom according to the following geometries.
     NOTE: In considering the geometry of a molecule or ion, double and triple bonds count as one pair of electrons only; that is, they take only one position in the geometrical structure. For example, the carbon atoms in ethylene would be considered to have three pairs of electrons because each C atom contributes to the double bond; the carbon atoms in acetylene would be considered to have only two pairs of electrons as each C atom is involved in the triple bond.
     Examples of the total number of pairs of electrons and the energetically most stable geometry for each are provided in Table 5.1. A square planar geometry is also possible for a total of four pairs of electrons around the central atom, and a square pyramid geometry is possible for a total of five pairs of electrons around the central atom (one pair removed from the octahedral geometry in Table 5.1). These two geometries are important in transition metal chemistry (Chapters 8 and 9).
6.   For the **final molecular geometry**, consider the **atoms only** (not the lone pairs of electrons); however, it is necessary to consider how the lone pairs of electrons would interact with other electron pairs to make subtle changes in geometry to the bond angle. The order of decreasing electron pair repulsive interactions is as follows:
     lone pair–lone pair > lone pair–bonding pair > bonding pair–bonding pair.

Table 5.1   *Typical numbers of electron pairs, geometry, and bond angles when using the valence shell electron pair repulsion theory*

Number of pairs	Geometry	Bond angles	Examples
2	Linear	180°	BeH_2, $TlCl$, $HC{\equiv}CH$, $O{=}C{=}O$
3	Trigonal planar	120°	BF_3, $H_2C{=}CH_2$
4	Tetrahedral	109.47°	CF_4, SiF_4, NH_3, H_2O
5	Trigonal bipyramid	120°; 90°; 180°	PF_5, XeF_2
6	Octahedral	90°; 180°	SF_6, $SiF_6{}^{2-}$
7	Pentagonal bipyramid	72°; 90°; 180°	IF_7

For methane, ammonia, and water, there are a total of four pairs of electrons; methane has four bonding pairs, ammonia has three bonding pairs, and water has two bonding pairs. Initially, these three compounds would be considered to have tetrahedral geometry, but the lone pair–lone pair interactions for H_2O would be substantial and result in a smaller H–O–H bond angle as below.

$$CH_4 \quad 109.47° \text{ H–C–H bond angles (no lone pairs)}$$

$$NH_3 \quad 106.8° \text{ H–N–H bond angles (one lone pair)}$$

$$H_2O \quad 104.5° \text{ H–O–H bond angle (two lone pairs)}$$

Therefore, CH_4 is a tetrahedral molecule; NH_3 is a pyramidal molecule and H_2O is a bent molecule. Figure 5.3 shows the number of arrangements for molecules and ions for two through six pairs of electrons around an atom.

7.   When the octet rule is obeyed and the central atom is from the third row or below in the periodic table, the lone pair of "s" electrons are inactive in forming bonds for AX_2 and AX_3 compounds (also called the inert "s" pair effect). Thus, bonds are formed primarily from the "p" orbitals. For H_2S, H_2Se, and H_2Te, the bond angles are 92°, 91°, and 89.5°, respectively. For PH_3, AsH_3, and SbH_3 the bond angles are 93.8°, 91.8°, and 91.3°, respectively.

The trigonal bipyramid structure is an important case because the five geometric positions around the central atom are not equivalent based upon bond angles. The trigonal planar portion has three bonds to the central atom (the **equatorial** positions), which are shorter than the two bonds to the central atom that are above and below the trigonal plane (these are considered **axial** positions). Typically, these axial positions have atoms that are highly electronegative such as fluorine. For example, the molecule ClF_3O would have one fluorine atom in each of the axial positions whereas the trigonal planar positions are occupied by one fluorine atom, one oxygen atom, and one lone pair of electrons.

The Lewis structure and valence shell electron pair repulsion theory models do not give us any information on the atomic orbitals that are used to form electron pair bonds. Two major theories have been used to formally describe bonding, and they are the VBT and MOT, which are described in the following sections.

5.2   Valence Bond Theory (VBT)

Wave functions can be used to describe the electrons that interact between atoms to form chemical bonds, and the solution of the wave functions will give information on the bond energy and the bond distance, which are

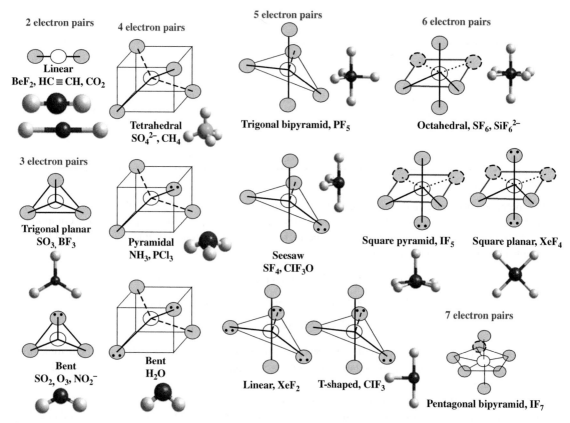

Figure 5.3   *Possible geometrical arrangements for two through seven coordination to the central atom are given for the total number of electron pairs (bonding and lone pairs) with examples for each. In drawn structures, white circles represent the central atoms and the blue circles represent the terminal atoms bound to the central atom. Lone pairs are given as two black dots in the blue circles. Model structures (CO_2, BeF_2, BF_3, NO_2^-, CH_4, NH_3, H_2O, PF_5, SF_4, ClF_3, SF_6, IF_5, and XeF_4) produced using the version 8.0.10 HyperChem™ program from Hypercube, Inc.*

well known from experimental observations. Figure 5.4 shows the Morse curve for H_2, where the potential energy (V) of the interaction of the two hydrogen atoms is plotted versus the internuclear distance of the two approaching atoms using Equation 5.1. D_{eq} is the dissociation energy measured from the equilibrium position, and α is a constant for each molecule (in this case H_2) that includes D_{eq}.

$$V = D_{eq}\left(1 - e^{\alpha(r - r_{eq})}\right)^2 \tag{5.1}$$

At distances where the atoms are far apart there is no interaction, and as the atoms approach each other, they begin to interact to form a bond, which is where the potential energy reaches a minimum value at the bond length (or equilibrium distance) between the two hydrogen atoms. The bond dissociation energy or bond strength is the difference in energy between the separation of atoms at infinity and the potential energy minimum. For H_2, the bond length is 74.2 pm and the bond energy is 436 kJ mol^{-1}. As the internuclear separation decreases toward zero between atoms A and B, the potential energy increases dramatically due to repulsive forces.

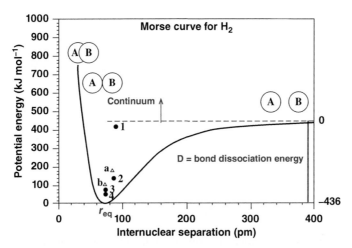

Figure 5.4   *Morse curve for the H$_2$ molecule. The energies on the y-axis on the left are from Equation 5.1. The energies on the right y-axis indicate the zero point (dashed line) where the atoms do not form a bond and the potential energy minimum, which is the bond energy for H$_2$. As noted in the text, the circles are energies calculated for various valence bond wave function modifications; the triangles are energies calculated for various molecular orbital wave function modifications*

5.2.1   H$_2$ and Valence Bond Theory

The simplest molecule is H$_2$ and it is possible to describe the various ways that the electrons will be around each atom in the molecule using VBT [3]. First, each electron remains with its original atom so is attracted by only one nucleus as in the following wave function with its corresponding detailed Lewis structure.

$$\Psi_{(b)} = \Psi_{(A)(1)}\Psi_{(B)(2)}$$

$$H_{(A)(1)}{}^{\bullet}{}^{\bullet}H_{(B)(2)}$$

Here (A)(1) refers to atom A having electron 1 and (B)(2) refers to atom B having electron 2 with no sharing of electrons between the atoms. This bonding representation is not very effective as it results in a stabilization of only 24 kJ mol^{-1} and a bond distance of 90 pm (#1 on Figure 5.4).

Heitler and London [3] added a covalent interaction where each electron is shared between the two atoms so the nuclei cannot distinguish the two electrons, which must have opposite spin to occupy the same region of space between the two atoms in a chemical bond. This interaction is known as the **Exchange Energy**, and the corresponding wave function and Lewis structure are given as

$$\Psi_{(b)} = \Psi_{(A)(1)}\Psi_{(B)(2)} + [\Psi_{(B)(1)}\Psi_{(A)(2)}] \text{ or } H_{(A)(1)}H_{(B)(2)} + H_{(A)(2)}H_{(B)(1)}$$

$$H_A - H_B$$

The bond energy resulting from this wave function is 303 kJ mol^{-1} and the bond distance is 86.9 pm (#2 on Figure 5.4). It is obvious that the sharing of electrons enhances the stability of the bond. The wave function for the exchange energy can be improved by the addition of shielding where electron 1 shields nucleus A from electron 2 and electron 2 shields nucleus B from electron 1. This improvement to the wave function gives a bond energy of 365 kJ mol^{-1} and a bond distance of 74.3 pm (#3 on Figure 5.4).

Addition of ionic interactions where both electrons reside on one atom but at different periods of time yields the following wave function.

$$\Psi_{(b)} = \Psi_{(A)(1)}\Psi_{(B)(2)} + \Psi_{(B)(1)}\Psi_{(A)(2)} + \lambda[\Psi_{(A)(1)}\Psi_{(A)(2)} + \Psi_{(B)(1)}\Psi_{(B)(2)}]$$

Or

$$H_{(A)(1)}H_{(B)(2)} + H_{(A)(2)}H_{(B)(1)} + \lambda[H_{(A)(1)}H_{(A)(2)} + H_{(B)(1)}H_{(B)(2)}]$$

where λ is a weighting factor to allow for more or less ionic bonding depending on the molecule under consideration. Ionic contributions improve the bond energy to 388 kJ mol^{-1} (#4 on Figure 5.4), which is a modest increase as two electrons around one H atom would experience some repulsive effects. The second term in Lewis structures is equivalent to the two resonance structures below:

$$: H_A^- \; H_B^+ \leftrightarrow H_A^+ \; H_B : ^-$$

with the major covalent and ionic bonding contributions for the hydrogen molecule given as

$$H - H \leftrightarrow : H_A^- \; H_B^+ \leftrightarrow H_A^+ + H_B : ^-$$

5.2.2   Ionic Contributions to Covalent Bonding

For molecules composed of atoms from the same element, it is clear that ionic contributions will not be very significant compared to the covalent contributions. Thus, the H–H bond has no dipole moment or charge separation and is called a nonpolar molecule. However, for molecules composed of two atoms from different elements such as hydrogen and fluorine to form HF, the ionic contributions will become more important. This is readily seen by comparing the bond energies for the homonuclear diatomic molecules H_2 (432.0 kJ mol^{-1}) and F_2 (154.8 kJ mol^{-1}) with the heteronuclear molecule HF (565 kJ mol^{-1}; see Appendix 5.1 for a table of bond energies). Pauling [3] noticed that the bond energy for heteronuclear molecules such as HF was larger than the arithmetic mean of the bond energies for the two homonuclear diatomic molecules. This energy difference is called **the ionic resonance energy** and can be used to calculate the **electronegativity** difference between two atoms in a chemical bond. As shown in the calculation below, the bond energy (BE) for HF is almost twice that of the arithmetic mean of H_2 and F_2 so there should be significant ionic bonding in the molecule HF.

$$\Delta E = BE_{HF} - 0.5(BE_{HH} + BE_{FF}) = 565 - 0.5(432.0 + 154.8) = 293.4$$

Pauling [3] defined the electronegativity difference in electron volts (eV) between two atoms in a bond with Equation 5.2.

$$\chi_A - \chi_B = 0.102 \, (\Delta E)^{0.5} \tag{5.2}$$

Substituting the value for ΔE above for HF gives an electronegativity difference of 1.68. Assigning the electronegativity of fluorine as 3.98, the electronegativity of hydrogen can be calculated to be 2.30. This is close to the value in Table 3.7. The slight difference to the hydrogen atom value of 2.20 in Table 3.7 is due to the consideration of other experimental data for these atoms. For example, performing the same calculation for H_2 with Cl_2 (239.7 kJ mol^{-1}) to form HCl (428.0 kJ mol^{-1}) gives an electronegativity difference of 0.98. The electronegativity difference for the H and Cl atoms given in Table 3.7 is 0.96.

Because of the significant difference in electronegativity values between H and F, the covalent H–F bond has a dipole moment and thus partial ionic or polar character, which will permit HF to align to the respective electric poles when placed in an electric field. The H–F bond is described as $H^{\delta+} - F^{\delta-}$ to indicate the polar nature of the bond. The partial charge can be calculated from the electric dipole moment, μ, which is equal to the product of the charge at one end or pole of the molecule and the distance between the poles of the molecule, d.

$$\mu = qd \tag{5.3}$$

The dipole moment is given in Debye units where one Debye unit is 3.336×10^{-30} Coulomb meter (C m) and the charge of an electron is 1.6022×10^{-19} C. The observed dipole moment and bond distance for HF are 1.98 Debye units (6.60×10^{-30} C m) and 91.7 pm, respectively, so the calculated charge, q, is 0.72×10^{-19} C. Thus, the ionic charge as a fraction of an electron is 0.72×10^{-19} C/(1.6022×10^{-19} C electron^{-1}) or 0.45 electron, and the percent ionic character is 45%. HF could now be represented as $H^{+0.45} - F^{-0.45}$. A similar calculation for H–Cl (1.03 Debye and 127.5 pm bond distance) shows that the percent ionic character is 16.8%. The measurement of dipole moments is an excellent way to describe the polarity of a bond in diatomic molecules (as well as overall polarity in polyatomic molecules). When the dipole moment is larger, it is due to the greater ionic character in the bond and the charge separation in the bond.

5.2.3 Polyatomic Molecules and Valence Bond Theory

To describe polyatomic molecules, VBT developed the concept of hybridization to explain bonding. To describe the bonding in the molecule BeH_2 requires an unpairing of the two electrons in the 2s atomic orbital of Be so that there are two unpaired electrons to combine with the individual electrons from each hydrogen atom. Here one of the 2s electrons is promoted (which costs some energy but is offset by removing electron–electron repulsions and the lowering of the 2p orbital energies on mixing) to a vacant 2p orbital so that there are two unpaired electrons, which can now combine with the electrons from the two hydrogen atoms. Two sp hybrid orbitals, which are equivalent in energy, form from the mixing.

$$\frac{\uparrow\downarrow}{2s^2} \rightarrow \frac{\uparrow}{2s^1} \; \frac{\uparrow}{2p_z{}^1} \; \frac{}{2p_x{}^0} \; \frac{}{2p_y{}^0} \rightarrow \frac{\uparrow}{sp} \; \frac{\uparrow}{sp} \; \frac{}{2p_x{}^0} \; \frac{}{2p_y{}^0}$$

The energetics for mixing is qualitatively described in Figure 5.5 for the carbon atom. Because BeH_2 is linear and the z axis is the C_∞ axis, the Be hybrid orbitals are termed sp hybrids and the bond angle is 180°. The same process describes the bonding in BeF_2, but the electrons from F also come from the 2 p_z atomic orbitals of the two F atoms.

The promotion of s electrons into vacant p orbitals in VBT is consistent with the VSEPR model approach, but provides further information on the atomic orbitals that are used in bonding. The promotion of s electrons into vacant d orbitals for elements in rows 3 and higher also permits expanded octets. Table 5.2 shows the

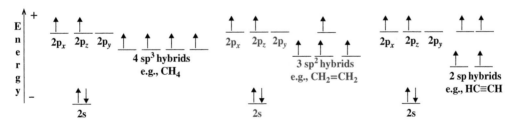

Figure 5.5 *Simplified mixing of carbon atomic orbitals to give hybrid orbitals of equivalent energy for tetrahedral* $CH_4(sp^3)$, *planar* $H_2C{=}CH_2(sp^2)$, *and linear* $HC{\equiv}CH$ (sp)

Table 5.2 *Some of the possible geometries from hybrid orbitals composed from atomic orbitals*

Number of pairs	Geometry	Hybrid orbitals	Atomic orbitals combined
2	Linear	sp; sd	s, p_z ; s, d_z^2
3	Trigonal planar	sp^2	s, p_x, p_y
4	Tetrahedral	sp^3; sd^3	s, p_x, p_y, p_z; $s, d_{xz}, d_{yz}, d_{xy}$
5	Trigonal bipyramid	sp^3d	s, p_x, p_y, p_z, d_z^2
5	Square pyramid	sp^3d	$s, p_x, p_y, p_z, d_{x^2-y^2}$
6	Octahedral	sp^3d^2	$s, p_x, p_y, p_z, d_{x^2-y^2}, d_z^2$

hybrid orbitals used for some important geometries where only s, p, and d orbitals are combined; combining these with f orbitals permits more bonds. The atomic orbitals, which comprise these sets of hybrid orbitals, are consistent with those found from a group theory analysis of bonding for the central atom (Section 4.6.4).

5.3 Molecular Orbital Theory (MOT)

In the MOT [4, 5], atomic orbitals from different atoms will be combined to form molecular orbitals, which will be filled according to Hund's rule and the Pauli exclusion principle as in atomic theory. The total number of molecular orbitals that are formed equals the sum of all the atomic orbitals available from each atom. Every atomic orbital that is available from each atom is used to form molecular orbitals so that the total energy of the molecule can be calculated. For bonding purposes, the principal atomic orbitals that are used to describe a molecule come from the valence or outer shell of each atom. These outer orbitals are called **frontier orbitals** and the combination of frontier molecular orbitals from one molecule with another molecule permits prediction of a symmetry allowed or symmetry forbidden (kinetically hindered) reaction when a reaction is thermodynamically favorable (see Section 5.4).

5.3.1 H$_2$

In VBT, the first molecule described was H_2. However, the molecule ion H_2^+, which has one-electron shared between two H atoms, each with one 1s orbital, is described first. The simplest mathematical function that can describe this behavior in the form of an electronic wave function is the linear combination of atomic orbitals (LCAO) to form molecular orbitals. There are two possible linear combinations. For H_2^+, electron 1 can reside in the linear positive combination of both 1s orbitals (near both atoms) to form a molecular orbital, which has the following wave function.

$$\psi_{(b1)} = \psi_{(A)(1)} + \psi_{(B)(1)}$$

Here "b" indicates the orbital designated for bonding, A and B indicate the atoms, and 1 indicates the first electron. The other linear negative combination of these two orbitals is written as

$$\psi_{(a1)} = \psi_{(A)(1)} - \psi_{(B)(1)}$$

For H_2^+, electron 1 can reside in the negative linear combination of both orbitals where "a" indicates the orbital designated for antibonding, which is the absence or cancellation of bonding.

The general method for writing wave functions for all electrons using the linear combination of atomic orbitals is given in Equation 5.4.

$$\psi = \sum_i c_i \varphi_i \tag{5.4}$$

where φ equals each atomic orbital (e.g., s, p, d orbital) and c is the coefficient for the original atomic orbital used in forming a molecular orbital. The coefficient is not normally specified when atoms from the same element are combined to form a molecule. A shorthand notation typically used is

$$\psi_{(b)} = 1s_A + 1s_B \qquad\qquad \psi_{(a)} = 1s_A - 1s_B$$

where the actual atomic orbitals used in combination are given. Here for H_2^+, the 1s orbitals from each atom are specified. As in atomic theory, the square of each wave function gives the probability of finding the electron around each atom in the molecule.

For H_2, a wave function is necessary for each electron; thus the mathematics becomes more complicated.

$$\psi_{(b1)} = \psi_{(A)(1)} + \psi_{(B)(1)}, \text{where electron 1 can reside near both atoms}$$

$$\psi_{(b2)} = \psi_{(A)(2)} + \psi_{(B)(2)}, \text{where electron 2 can reside near both atoms}$$

where b equals the molecular orbital designated for each bonding electron. The complete wave function is then the product of the one-electron wave functions. Thus the result for H_2 is,

$$\psi_{(b)} = \psi_{(b1)}\psi_{(b2)} = [\psi_{(A)(1)}\psi_{(B)(2)} + \psi_{(B)(1)}\psi_{(A)(2)}] + [\psi_{(A)(1)}\psi_{(A)(2)} + \psi_{(B)(1)}\psi_{(B)(2)}]$$

Covalent contribution   Ionic contribution

Note that this wave function explicitly gives both covalent and ionic interactions and only differs from the valence bond approach by the coefficient λ. Thus the ionic and covalent contributions appear to be equivalent. Solution of this wave function gives an energy of 260 kJ mol^{-1} and a bond distance of 85 pm (see "a" in Figure 5.4). Addition of shielding effects as discussed in the VBT approach provides a potential energy of 337 kJ mol^{-1} and a bond distance of 73 pm (see "b" in Figure 5.4).

To understand chemical bonding, it is important to provide the electron distribution in each of these orbitals. For simplicity, H_2^+ is used and squaring the wave functions for the bonding and antibonding orbitals gives the following equations.

bonding: $\quad \psi(b) = \sigma = 1s_A + 1s_B; \quad \psi^2(b) = 1s_A{}^2 + 2(1s_A 1s_B) + 1s_B{}^2$

antibonding: $\psi(a) = \sigma^* = 1s_A - 1s_B; \quad \psi^2(a) = 1s_A{}^2 - 2(1s_A 1s_B) + 1s_B{}^2$

5.3.1.1   Overlap Integral

The difference between these two equations is the **cross term** S ($s_A \times s_B$), which is known as the **overlap integral** and the sign of the cross term indicates the bonding type. Figure 5.6 shows the plots of these wave functions with their probability functions. These plots are from the radial distribution function for the hydrogen atom where $n = 1$ and $l = 0$. Figure 5.6a shows the radial distribution function versus distance for hydrogen atoms A and B; here there is no linear combination of the two atomic orbitals. Figure 5.6b shows the positive linear combination of the two 1s orbitals from hydrogen atoms A and B to form a sigma (σ) bond. Figure 5.6c shows a plot of the probability distribution function for Figure 5.6b where the electron density is directly between the two nuclei. Figure 5.6b and c shows that the electron density between the nuclei of the two hydrogen atoms is increased from that found in Figure 5.6a. Thus the overlap integral S is greater than zero ($S > 0$).

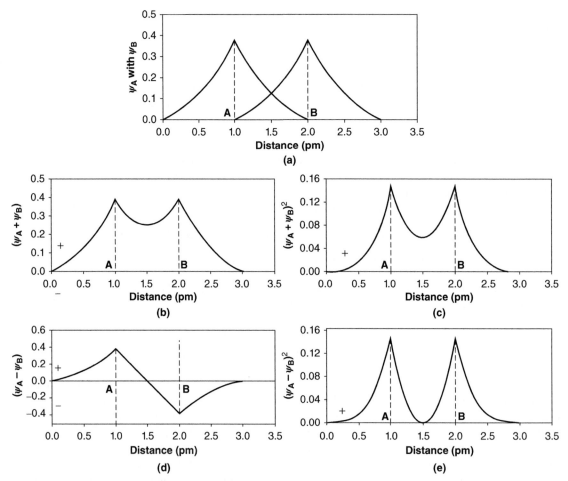

Figure 5.6 *(a) Plots of the individual radial distribution functions for each hydrogen atom in the absence of bonding. (b) Positive linear combination of the individual radial distribution functions to form a bond. (c) Probability plot for the positive linear combination. (d) Negative linear combination of the individual radial distribution functions to form an antibond. (e) Probability plot for the negative linear combination*

Figure 5.6d shows the negative linear combination of the two 1s orbitals from the hydrogen atoms A and B to form a sigma antibond, which is represented as σ^* where the * indicates an antibond. The wave function goes through zero so there is a node. Figure 5.6e shows the probability distribution plot for Figure 5.6d indicating that there is no electron density between the two; thus, the overlap integral S is less than zero ($S < 0$). Figure 5.6d and e shows that the electron density between the nuclei of the two hydrogen atoms is decreased compared to the individual atoms in Figure 5.6. The antibonding orbital is a repulsive situation. As will be shown later, the filling of electrons in antibonding orbitals during a chemical reaction can result in the breaking of a chemical bond. Later $S = 0$ will be shown to be a nonbonding orbital.

Figure 5.7 shows a simplified molecular orbital description for the hydrogen molecule; there are three major items to describe. On the lower left is the shape of the orbital and the electron density contour for the bonding situation where electron density can be found between and around the two hydrogen atoms. In the

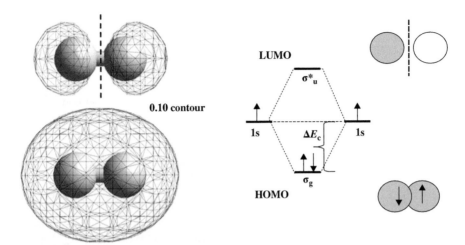

Figure 5.7   *The molecular orbitals for the hydrogen molecule. On the left are the electron density contours (3-D isosurface maps) representing 90% probability (0.1 contour) of finding the electrons calculated at **the AM1 level** with HyperChem™ version 8.0.10; the blue color indicates the positive part of the wave function and the black color indicates the negative part of the wave function (this convention will be used throughout the book). In the center is the molecular orbital energy diagram indicating the atomic orbitals used to form the molecular orbitals. On the right are 2-D sketches of the original orbitals from the hydrogen atoms to form the molecular orbitals*

upper left is a representation for the antibonding situation where electron density cannot be found between the two hydrogen atoms; there is a nodal surface or plane indicated by the dashed line. On the lower right is a simplified sketch describing the positive linear combination of the two 1s orbitals to overlap to form the bonding orbital. The upper right sketch is a description for the negative linear combination of the two 1s orbitals to form the antibonding orbital. In the center is the normal representation for the energy level diagram, and the dashed lines between the atomic orbitals and the molecular orbitals indicate which atomic orbitals combine to form the molecular orbitals. The hydrogen 1s orbitals are of equivalent energy, but when they combine, the bonding orbital is more stable and of lower energy than the two atomic orbitals used to combine it. The difference between the energy for the hydrogen atomic orbitals and the bonding orbital for the hydrogen molecule is known as the **exchange energy** (ΔE_c) and the two electrons pair up and occupy the molecular orbital. Likewise, the antibonding orbital is less stable than the two atomic orbitals as its energy is more positive than the hydrogen 1s orbitals. Because there are only two electrons total between the two hydrogen atoms, the antibonding orbital is not filled. The total number of bonds is called the bond order (B.O.) and can be calculated with the following equation.

$$\text{Bond order} = (\#\ \text{bonding electrons} - \#\ \text{antibonding electrons})/2$$

The bonding orbital is fully occupied and is called the highest occupied molecular orbital (HOMO) and the antibonding orbital is unoccupied and called the lowest unoccupied molecular orbital (LUMO). Because the sign of the wave function for the HOMO is positive everywhere, it is symmetric to the center of inversion and given the subscript g (σ_g). Because the sign of the wave function for the LUMO inverts to the center of inversion, it is antisymmetric to the center of inversion and is given the subscript u (σ_u^*).

The energy level diagram in Figure 5.7 is commonly used to show the energetics of H_2. Unfortunately the energies of the atomic orbitals and of the molecular orbitals are not in this description. Figure 5.8 shows

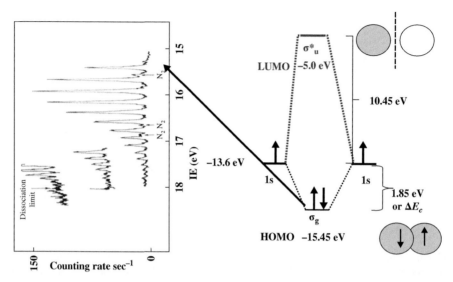

Figure 5.8   *On the left is the UV photoelectron spectrum for the hydrogen molecule [all photoelectron spectra are excited with the He emission of 21.22 eV]. On the right are the energies of the HOMO and the LUMO orbitals for the hydrogen molecule; note the significant energy gap of 10.45 eV between the hydrogen atomic orbitals and the LUMO energy for the hydrogen molecule (Source: Turner 1970 [6], Figure 3.5. Reproduced with permission from Wiley, Inc.)*

a more detailed representation of the energies of the orbitals as well as the UV photoelectron spectrum of the hydrogen molecule, which gives one electronic transition with several lines due to molecular vibrations (discussed in the next section) compared to the single sharp peaks in the spectrum of the hydrogen atom (not shown) and the helium atom in Figure 3.15.

5.3.1.2   HOMO and LUMO Energies

Koopmans' theorem [4, 5] states that the first ionization potential energy (IP) of a molecule is equal to the negative of the orbital energy of the highest occupied molecular orbital; likewise, the electron affinity (EA) of a molecule is equal to the negative of the orbital energy of the lowest unoccupied molecular orbital. A table of HOMO and LUMO energies for many important small molecules and ions is given at the end of the chapter (Appendix 5.2).

$$E_{HOMO} = -IP \qquad E_{LUMO} = -EA$$

Here, the energy of the HOMO for H_2 is determined from the UV photoelectron spectrum and has a value centered at -15.45 eV due to the loss of the electron from the σ_g bonding orbital.

$$\sigma_g{}^2 \rightarrow \sigma_g{}^1 + e^- \quad \text{or} \quad H_2 \rightarrow H_2{}^+ + e^-$$

The several sharp peaks composing this band are the result of the vibrations of the molecule ion $H_2{}^+$, which is formed on ionization of the hydrogen molecule. This feature is different from single atoms as shown in Figure 3.15 for the noble gases and as will be shown below for nonbonding orbitals in molecules that are not involved with bonding. Figure 5.9 shows a Morse curve with all the vibrational levels for the hydrogen molecule. Several peaks, which are represented by the four blue arrows of different height and thus energy, in a band or energy level indicate a bonding orbital whereas only one peak for an energy level indicates

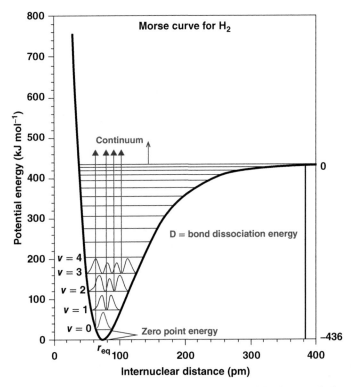

Figure 5.9   *The Morse curve for H_2 showing the vibrational levels possible along with the plot of the probability functions for the vibrational levels from 0 to 3. These levels are partially populated and give rise to several sharp vibrational energy peaks (only four are represented with blue arrows) for the hydrogen molecule ion (H_2^+) UV photoelectron spectrum in Figure 5.8*

no or negligible bonding. The energy difference between each of these peaks is in the IR region of the electromagnetic spectrum (Figure 1.2). The exchange energy (ΔE_c) in Figure 5.8 is found to be 1.85 or 3.70 eV when accounting for both electrons occupying the orbital, and this total energy is 358 kJ mol^{-1}, which is close to that found in Figure 5.4. The LUMO is determined from the electron affinity of the hydrogen molecule and has an energy of −5.0 eV [7]. The difference in energy between the hydrogen atomic orbitals and the LUMO is 10.45 eV so the energy gap is significantly higher than the gap of 3.70 eV between the hydrogen atomic orbitals and the HOMO. These data indicate that putting an electron into the antibonding orbital would result in the cancellation of bonding and the breaking of the H–H bond (see Sections 9.12–13). In effect, these energy differences show the repulsive effects between the nuclei when electrons occupy antibonding orbitals.

5.3.2   Types of Orbital Overlap

So far, only the formation of single bonds from the s orbitals has been considered. Figure 5.10 shows many possible orbital overlaps for the formation of sigma, pi (π), and delta (δ) bonds where $S > 0$. As noted above, σ bonds have electron density on the internuclear axis between the two nuclei. The π bonds have electron density above and below the internuclear axis such that two lobes of one orbital interact with two lobes of another orbital. The δ bonds also have electron density above and below the internuclear axis and are formed when two d orbitals combine such that all four lobes of each orbital interact with each other.

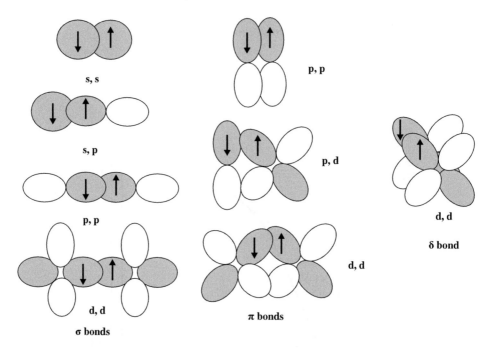

Figure 5.10 *Common types of atomic orbital overlap (S > 0) to form bonds. The blue color indicates the positive sign of the atomic wave function whereas white indicates the negative side of the atomic weight function*

5.3.3 Writing Generalized Wave Functions

For many molecules, it is possible to find atomic orbital combinations and relative energy level diagrams in the literature [8]. This applies to many small inorganic and organic molecules in particular. The following is a brief description of a **useful qualitative** approach for writing wave functions for a molecule using the linear combination of atomic orbitals (LCAO) approach.

1. Choose a geometry for the molecule under study. VSEPR or valence bond descriptions can be useful in the choice.
2. Choose valence (outer shell) atomic orbitals (AOs) of similar energy and symmetry for overlap to form the combinations (e.g., bonds). The energies of the orbitals to be combined should be within about 6 eV (579 kJ mol^{-1}) for different atoms. This is a "rule of thumb." Here it is important to have information about the energies of the atomic orbitals to be used (knowing the experimental energies of the molecular orbitals that are formed is most useful).
3. Combine the AOs according to the signs of the wave functions for the orbitals. **$S > 0$ for bonding; $S < 0$ for antibonding; $S = 0$ for nonbonding combinations.** Sketch a diagram of the orbitals with their signs for the linear combinations to aid selections. Remember the number of molecular orbitals formed must equal the number of atomic orbitals at the start.
4. Write an MO energy level diagram with bonding choices at highest negative energy and antibonding at the lowest negative energy.

5. Fill the electrons into the MO energy level according to Hund's rule and the Pauli exclusion principle.
6. Calculate the bond order (the number of bonds) from

$$\text{Bond order} = (\text{\# bonding electrons} - \text{\# anti-bonding electrons})/2$$

As the bond order increases, the bond length decreases and the bond energy increases.

5.3.4 Brief Comments on Computational Methods and Computer Modeling

Although the discussion has involved the qualitative molecular orbital approach, there are powerful computational techniques to calculate the properties for molecules including solids. There are two main approaches to solving the Schrodinger equation numerically for polyatomic molecules using the linear combination of atomic orbital approach [4]. **Ab initio** methods calculate structures and molecular properties from first principles using the atomic orbitals from each of the atoms (Equation 5.4); the atomic orbitals chosen are referred to as a **basis set**. A correct molecular Hamiltonian is used and a solution is attempted (see Equation 3.10) without the use of experimental data to aid the calculation. The Hartree–Fock or self-consistent field (SCF) method is the primary approximation and is applied to all electron–electron repulsions. The wave functions for each electron are solved in an iterative manner so that the molecular orbitals are refined until an energy minimum, which no longer changes in value, is achieved; i.e., becomes self-consistent.

Semi-empirical methods solve the Schrodinger equation with a simpler Hamiltonian that uses approximations and adjustable parameters including experimental data to provide best fits. An important approach is the neglect or partial **neglect of differential overlap** terms; for example, many of the integrals from the complete wave function are not used in order to simplify the calculation. This makes the Schrodinger equation easier to solve and requires less computational time; however, these methods are not as accurate as the **ab initio** method. To overcome the use of the neglect of differential overlap approximation, empirical and adjustable parameters such as valence orbital ionization energies for the atoms (Table 5.3) are introduced into the Schrodinger equation to help achieve the best fit. In the book and especially the following discussion, results from the HyperChem™ program (version 8.0.10) are presented as above in Figure 5.7. The primary semi-empirical method used is the AM1 method, which is recommended for most of the molecules discussed below.

Table 5.3 *Valence Orbital Ionization energies in electron volts (eV); many determined by ESCA. These ionization energies are not the successive potential energies given in Table 3.6 and are used as estimates in semi-empirical MO calculations*

Atom	1s	2s	2p	Atom	3s	3p	Atom	4s	4p
H	13.6			Na	5.1		K	4.3	
He	24.6			Mg	7.6		Ca	6.1	
Li		5.5		Al	11.3	5.9	Zn	9.4	
Be		9.3		Si	14.9	7.7	Ga	12.6	6.0
B		14.0	8.3	P	18.8	10.1	Ge	15.6	7.6
C		19.4	10.6	S	20.7	11.6	As	17.6	9.1
N		25.6	13.2	Cl	25.3	13.7	Se	20.8	10.8
O		32.2	15.8	Ar	29.2	15.8	Br	24.1	12.5
F		40.2	18.6				Kr	27.5	14.3

Density functional theory (DFT) is an alternate third method to determining molecular structure and properties and is as powerful as the **ab initio** method. Here, the total energy depends on the total electron density rather than solving the energy of the wave function for each electron. DFT methods express the Schrodinger equation as a set of equations called the **Kohn–Sham** equations; these are also solved in an iterative manner until they are self-consistent in finding an energy minimum. The DFT method does not require as much computer time and total calculations as the **ab initio** method. Reference [9] gives a 50-year perspective of this method.

5.3.5   Homonuclear Diatomic Molecules (A_2)

Using the generalized wave function approach, it is now possible to look at the possible formation of homonuclear diatomic molecules (A_2) starting with a combination of helium atoms through the elements in the second row of the periodic table.

He_2 has the same linear combination of atomic orbitals as the hydrogen molecule as given below. Note that many notations can be used for the wave function.

$$\psi_{(b)} = \sigma_g = \sigma_{1s} = 1s_A + 1s_B$$

$$\psi_{(a)} = \sigma_u{}^* = \sigma_{1s}{}^* = 1s_A - 1s_B$$

For H_2, only the bonding orbital is filled and it can be represented in shorthand notation as $\sigma_{1s}{}^2$. For He_2 both the bonding and antibonding orbitals are completely filled and can be given in shorthand notation as $\sigma_{1s}{}^2\sigma_{1s}{}^{*2} = \sigma_g{}^2\sigma_u{}^{*2} = KK$ where KK indicates the filling of the first energy level of the periodic table. Thus, He_2 is predicted to not exist because the antibonding electrons cancel out the bonding electrons for a bond order of zero.

Li_2 is similar to H_2 and Be_2 is similar to He_2 as lithium and beryllium combine their outer 2s orbitals to form sigma bonding and antibonding orbitals. The linear combination of their atomic orbitals is given in the following wave functions.

$$\psi_{(b)} = \sigma_g = \sigma_{2s} = 2s_A + 2s_B; \quad \psi_{(a)} = \sigma_u{}^* = \sigma_{2s}{}^* = 2s_A - 2s_B$$

The shorthand notation for Li_2 is $KK\sigma_{2s}{}^2$ or $KK\sigma_g{}^2$ and the shorthand notation for Be_2 is $KK\sigma_{2s}{}^2\sigma_{2s}{}^{*2}$ or $KK\sigma_g{}^2\sigma_u{}^{*2}$. The energy level diagram for Li_2 is similar to that in Figures 5.7 and 5.8. Be_2 is also predicted not to exist because the antibonding electrons cancel out the bonding electrons for a bond order of zero.

The linear combination of atomic orbitals for the molecules B_2, C_2, N_2, O_2, and F_2 requires combination of the three 2p orbitals for these elements. In the *absence of any s–p mixing* or hybridization, these can be given as

$$\sigma_{2p} = 2p_{zA} - 2p_{zB} = \sigma_g \qquad \sigma^*{}_{2p} = 2p_{zA} + 2p_{zB} = \sigma_u{}^*$$

$$\pi_{2px} = 2p_{xA} + 2p_{xB} = \pi_u \qquad \pi^*{}_{2px} = 2p_{xA} - 2p_{xB} = \pi_g{}^*$$

$$\pi_{2py} = 2p_{yA} + 2p_{yB} = \pi_u \qquad \pi^*{}_{2py} = 2p_{yA} - 2p_{yB} = \pi_g{}^*$$

Figure 5.11 demonstrates the linear combination of the p_x, the p_y, and the p_z orbitals. As the z axis is the bond axis, the $2p_z$ orbitals are linearly combined to form a σ bond and a σ^*. Because the p orbitals have ungerade symmetry, the sigma 2p orbital (σ_{2p}) is formed by the negative combination of the two p_z orbitals. The sigma 2p antibonding orbital ($\sigma_{2p}{}^*$) is formed by the positive combination of the two p_z orbitals. The p_x (and p_y) orbitals on each atom are parallel to each other and combine to form a π bonding orbital (π_{px}) and a π antibonding ($\pi_{px}{}^*$) orbital. Here, the electron density is above and below the internuclear axis between the

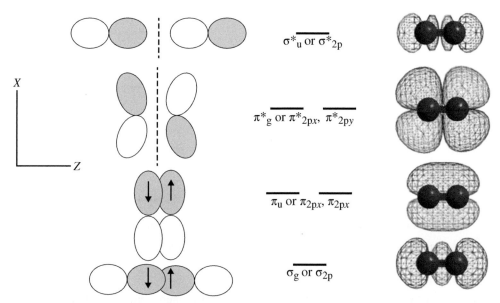

Figure 5.11 *Linear combination of the 2p orbitals to give σ and π bonds and antibonds; only the π bond and antibond on the x-axis are shown. The z axis is the bond axis, which contains the C_∞ axis of rotation. The predicted energy level diagram is in the center, and the molecular orbitals with 0.10 electron density contours are on the right (calculated for N_2 with the HyperChem™ program version 8.0.10 from Hypercube, Inc.)*

two atoms. A positive linear combination of the p_x (and p_y) orbital results in overlap so that $S > 0$; a negative linear combination of the p_x (and p_y) orbital does not result in overlap of the orbitals as $S < 0$; note the node between the atoms in Figure 5.11. Because both the p_x and p_y orbitals are perpendicular to the bond axis, the linear combinations result in orbitals of the same energy; that is, they are called degenerate. Note that the π^* orbitals have electron density pointed away from each atom. The final result of these linear combinations is the formation of six molecular orbitals from six p atomic orbitals; there are three bonding orbitals and three antibonding orbitals. Figure 5.11 also shows the notation of the molecular orbitals based on the sign of the wave function to the center of inversion. The σ_{2p}, π_{px}^*, and π_y^* orbitals are called gerade whereas the σ_{2p}^*, π_{px}, and π_{py} orbitals are called ungerade.

The arrangement of the eight molecular orbitals formed from the 2s and 2p atomic orbitals into energy levels requires data from electronic absorption spectroscopy and photoelectron spectroscopy as well as from quantum mechanical calculations. Figure 5.12 shows two energy arrangements for the eight molecular orbitals that explain the bonding of the diatomic molecules in the second row of the periodic table.

For the diatomic molecules B_2, C_2, and N_2, the arrangement of the molecular orbitals according to their energies is on the left of Figure 5.12. For the diatomic molecules O_2 and F_2, the arrangement is on the right of Figure 5.12. The major noticeable difference between these arrangements is the ordering of the energies for the π_u and σ_g bonding molecular orbitals. B_2, C_2, and N_2 are formed from elements that have low electronegativity or effective nuclear charge to radius ratio (Z_{eff}/r). Inspection of Figure 3.17 and Table 5.3 indicates that the energy difference between the 2s and 2p orbitals for B, C, and N is significantly smaller than the energy difference between the 2s and 2p orbitals for O and F. For B, C, and N, the 2s and 2p orbitals on each atom are able to mix via hybridization prior to forming molecular orbitals. Because the energy difference between the 2s and 2p orbitals is so great for O and F, hybridization is not possible and the 2s orbitals and 2p orbitals for

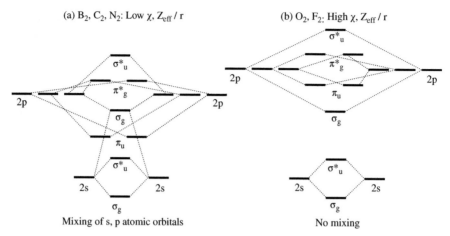

(a) B_2, C_2, N_2: Low χ, Z_{eff} / r   (b) O_2, F_2: High χ, Z_{eff} / r

Mixing of s, p atomic orbitals   No mixing

Figure 5.12   *Qualitative ordering of the energy levels for the diatomic molecules of the elements in the second row of the periodic table*

each atom only interact with their counterpart on the other atom. The energy order and filling of the molecular orbitals for the diatomic molecules from Li_2 to F_2 is in Figure 5.13.The blue solid lines and arrows for the electrons in the $3\sigma_g$ orbital show the change in ordering of the energy levels with the $1\pi_u$ on increasing atomic number across the second row.

The results from Figure 5.13 indicate the following for each diatomic molecule of the second row of the periodic table. The filling of electrons follows Hund's rule and the Pauli Exclusion Principle as was done in the filling of atomic orbitals.

Li_2 has one bond and is diamagnetic. The actual bond energy is 110 kJ mol^{-1} with a bond distance of 267 pm. The shorthand notation for filling of the molecular orbitals is $KK\sigma_{2s}^{2}$ or $KK\sigma_g^{2}$.

Be_2 is predicted not to exist as there are an equivalent number of bonding and antibonding electrons. The shorthand notation for filling of the molecular orbitals is $KK\ \sigma_{2s}^{2}\sigma_{2s}^{*2}$ or $KK\sigma_g^{2}\sigma_u^{*2}$.

B_2 has one bond and is paramagnetic (two electrons for a spin multiplicity of 3) as the $1\pi_u$ bonding orbitals are filled before the $3\sigma_g$ orbital. The actual bond energy is 272 kJ mol^{-1} with a bond distance of 159 pm. The shorthand notation for filling of the molecular orbitals is $KK\ \sigma_{2s}^{2}\sigma_{2s}^{*2}\pi_x^{1}\pi_y^{1}$ or $KK\sigma_g^{2}\sigma_u^{*2}\pi_{ux}^{1}\pi_{uy}^{1}$.

C_2 has two bonds and is diamagnetic as the $1\pi_u$ bonding orbitals are again filled before the $3\sigma_g$ orbital. The actual bond energy is 602 kJ mol^{-1} with a bond distance of 124 pm. The shorthand notation for filling of the molecular orbitals is $KK\ \sigma_{2s}^{2}\sigma_{2s}^{*2}\pi_x^{2}\pi_y^{2}$ or $KK\sigma_g^{2}\sigma_u^{*2}\pi_{ux}^{2}\pi_{uy}^{2}$.

N_2 has three bonds and is diamagnetic as the $1\pi_u$ bonding orbitals and the $3\sigma_g$ orbital are now all filled. The actual bond energy is 941 kJ mol^{-1} with a bond distance of 110 pm. The shorthand notation for filling of the molecular orbitals is $KK\ \sigma_{2s}^{2}\sigma_{2s}^{*2}\pi_x^{2}\pi_y^{2}\sigma_{2pz}^{2}$ or $KK\sigma_g^{2}\sigma_u^{*2}\pi_{ux}^{2}\pi_{uy}^{2}\sigma_{gpz}^{2}$.

O_2 has two bonds and is paramagnetic (two electrons for a spin multiplicity of 3) as the $1\pi_g^{*}$ antibonding orbitals are now partially filled. The actual bond energy is 493 kJ mol^{-1} with a bond distance of 121 pm. Note that the HOMO and LUMO orbitals are the same for O_2 due to the partial filling of the $1\pi_g^{*}$ orbitals. These are called **singly occupied molecular orbitals (SOMO)** and have an impact on the reactivity of O_2. The shorthand notation for filling of the molecular orbitals is $KK\ \sigma_{2s}^{2}\sigma_{2s}^{*2}\pi_x^{2}\pi_y^{2}\sigma_{2pz}^{2}\pi_{px}^{*1}\pi_{py}^{*1}$ or $KK\sigma_g^{2}\sigma_u^{*2}\pi_{upx}^{2}\pi_{upy}^{2}\sigma_{gpz}^{2}\pi_{gx}^{*1}\pi_{gy}^{*1}$.

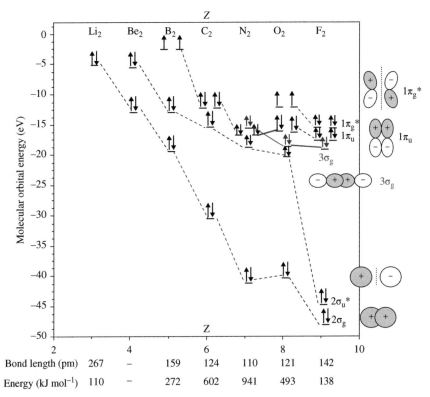

Figure 5.13 *Orbital diagrams for the diatomic molecules of the second row of the periodic table (Li_2 through to F_2). Energies from ESCA data [6] for N_2, O_2 and from IP data for Li_2, C_2, and F_2 (Appendix 5.2). AM1 level calculations are used for σ_g and σ_u^* orbital energies for the other molecules, and Be_2 orbital energies are calculated without a bonding interaction*

F_2 has one bond and is diamagnetic. The actual bond energy is 138 kJ mol^{-1} with a bond distance of 142 pm. The shorthand notation for filling of the molecular orbitals is KK $\sigma_{2s}{}^2\sigma_{2s}{}^{*2}\pi_x{}^2\pi_y{}^2\sigma_{2pz}{}^2\pi_{px}{}^{*2}\pi_{py}{}^{*2}$ or KK$\sigma_g{}^2\sigma_u{}^{*2}\pi_{upx}{}^2\pi_{upy}{}^2\sigma_{gpz}{}^2\pi_{gx}{}^{*2}\pi_{gy}{}^{*2}$.

Ne$_2$ is not shown but is predicted not to exist as there is an equivalent number of bonding and antibonding electrons, and all orbitals are completely filled.

The prediction of paramagnetism for O_2 in accordance with experimental data was a major reason for the acceptance of the molecular orbital approach. The X-ray and UV photoelectron spectra of these diatomic molecules provided energies for the molecular orbitals and confirmed the ordering of the molecular orbitals in Figures 5.12 and 5.13. The experimental results above also show the effect of increased bond energy and decreased bond distance with increased bond order or number of bonds. Comparing the bond energy and bond distance of the diatomic molecules with single bonds [Li_2, B_2, and F_2] shows that B_2 has the highest bond energy, which is likely due to both electrons being in different orbitals and having the same electron spin so that repulsive effects are minimized. The bond distance in molecules with only one bond decreases for Li_2, B_2, and F_2 in accordance with the increased effective nuclear charge on going across a row.

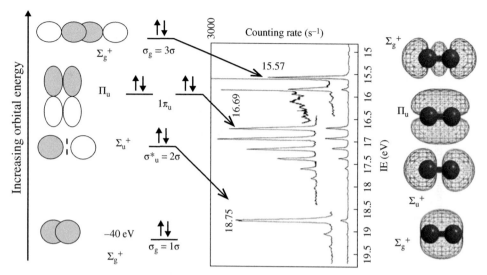

Figure 5.14   *Atomic orbitals (left) combined to form the molecular orbitals of N_2 on the right (0.10 contour plots produced with the HyperChem™ program version 8.0.10). The UV photoelectron spectrum correlates with the ordering in Figure 5.12A [10]. The shorthand notation (1σ, 2σ, etc.) is described in the next section (5.3.6) (Source: Turner 1970 [6], Figure 3.5. Reproduced with permission from Wiley, Inc.)*

Figure 5.14 shows the UV photoelectron spectrum for N_2 to N_2^+. The spectrum for the π bond system shows that these bonds are slightly lower in energy than the σ_g orbital formed from the combination of the nitrogen atom $2p_z$ orbitals. In addition, it also shows the expected fine structure for strong bonding as observed for H_2^+ above. The spectrum confirms that the HOMO orbital is the σ_g orbital, which is weakly bonding due to the lack of detailed fine structure. In support of weak bonding, the bond distance for N_2 is 109.76 pm and for N_2^+ is 111.6 pm and the loss of bonding is 102 kJ mol^{-1} from the N_2 value of 941 kJ mol^{-1} (11% loss in bonding). The low reactivity of N_2 toward donating electrons is related to this weakly bonding orbital and its very stable energy. The σ_u^* orbital is weakly antibonding and the σ_g is bonding [5, 6]. The group theory notation for the molecular orbitals is also given in Figure 5.14 (confirm using Table 4.6 and the $D_{\infty h}$ character table in Chapter 4, Appendix 4.1 and compare with Figure 5.48 for CO_2).

5.3.6   Heteronuclear Diatomic Molecules and Ions (AB; HX) – Sigma Bonds Only

A description of heteronuclear diatomic molecules in the absence of π bonding is first considered. Here one of the atoms has a higher effective nuclear charge or electronegativity than the other atom. In this case, the generalized wave functions require the use of coefficients for each atomic orbital from each atom to adequately describe the molecular orbitals that are formed. For example, the combination of the wave functions of atoms A and B to form a bonding orbital, ψ_b, would have the following form.

$$\psi_b = c_A \psi_A + c_B \psi_B$$

When A is the more electronegative atom, the bonding orbital would have more contribution from atom A. Thus, ψ_A will have assigned to it the coefficient of higher value, which is c_A so that $c_A > c_B$. The wave functions of atoms A and B to form an antibonding orbital would have a similar form. Here the coefficient of

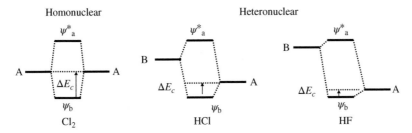

Figure 5.15 *Relative molecular orbital energy diagrams for a diatomic molecule, Cl₂, compared to a polar molecule HCl and a highly polar molecule HF*

higher value is assigned to the wave function for atom B, as the antibonding orbital will have more contribution from atom B.

$$\psi^*_a = c_B \Psi_A - c_A \Psi_B$$

Figure 5.15 shows a schematic diagram comparing a homonuclear diatomic molecule such as Cl_2 with two heteronuclear diatomic molecules with different electronegativity. Br–Cl would have a smaller electronegativity difference than H–F or H--Cl. The exchange energy, ΔE_c, decreases with increasing electronegativity difference between atoms A and B because ionic bonding or polar character in the bond increases as shown above for H–F and H–Cl.

The first molecule described is H–Cl, and only the valence atomic orbitals are needed. The hydrogen atom has one 1s orbital and the chlorine atom has one 3s orbital and three 3p orbitals. Figure 5.16 shows the energies of all the atomic orbitals as well as the bonding and nonbonding molecular orbitals formed from their combination. On the right of Figure 5.16 are the 3-D isosurface electron density plots of the five orbitals at the 0.05 contour level. The bonding and antibonding combination from the atomic orbitals are also drawn, and the dashed vertical line indicates the node of the antibonding orbital. The hydrogen atom 1s orbital combines with the $3p_z$ orbital of the chlorine atom (note that these two orbitals are very close in energy) to form a sigma bond (the 2σ MO that exhibits a broad peak with vibrational structure). Likewise, these two orbitals also combine to form an antibonding sigma orbital (3σ), which is the LUMO orbital. Although the s, p_x, and p_y orbitals from the chlorine atom are not involved with bonding to a significant extent, the 1σ MO shows that there is some electron density near the hydrogen atom's 1s orbital. This minor amount of bonding character likely results from hybridization or s–p mixing of the Cl atom's orbitals. The HOMO orbitals are the π orbitals (p_x and p_y orbitals from the chlorine atom) and are thus nonbonding orbitals.

The generalized wave functions for the bonding and antibonding orbitals in HCl would have the following form where the coefficients for the H and Cl orbitals would follow $C_{Cl} > C_H$ as chlorine is more electronegative than hydrogen. The sketches of the molecular orbitals are drawn to show the relative contribution (by size) of the atomic orbitals; i.e., a larger contribution from the atomic orbital with the largest coefficient.

$$\psi_{(b)} = \sigma = 2\sigma = C_H(1s_H) + C_{Cl}(3p_{zCl})$$
$$\psi_{(a)} = \sigma^* = 3\sigma = C_{Cl}(1s_H) - C_H(3p_{zCl})$$

Because HCl is composed of two different atoms, the use of gerade or ungerade in shorthand naming of orbitals is not needed so the following is a convention for the numbering and naming of the molecular orbitals. Here the first σ-like orbital is given the number 1 for the first valence orbital. Because this orbital comes from the chlorine 3s orbital, it is a nonbonding orbital. The other σ orbitals are increased accordingly. The first π

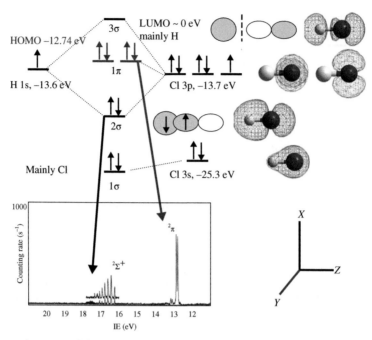

Figure 5.16   *The combination of the hydrogen and chlorine atomic orbitals to form HCl. The C_∞ axis is the z axis. The atomic orbital energies are from ESCA, and the UV photoelectron spectrum of HCl is shown; arrows indicate the orbitals of HCl. To the right of the chlorine atomic orbitals are line drawings for both the bonding and antibonding orbitals. To the far right are orbitals produced with HyperChem™ version 8.0.10 (Source: Turner 1970 [6], Figure 3.20. Reproduced with permission from Wiley, Inc.)*

orbitals are given the number 1 and any subsequent π orbitals of higher energy are given the number 2. (In some texts, the 1σ orbital will be given the symbol 5σ when the chlorine 1s and 2s orbitals are given the 1σ and 3σ designations, respectively.)

There are several molecules and ions that are isoelectronic with the HCl molecule and their molecular orbitals will have similar characteristics as shown in Figure 5.16. These chemical species include HF, HBr, HI, OH^-, and HS^-. Any differences for these isoelectronic species are attributed to the energies of the originating atomic orbitals that form the resulting molecular orbitals. Another important chemical species that would have this MO arrangement is the OH• (hydroxyl radical) where one electron is removed from one of the two 1π orbitals, thus making this a very reactive species. Figure 5.17a shows the energies of the HOMO orbitals for the hydrogen halides and the HOMO orbitals for the anions, X^-. The gap between the energies for the hydrogen halide HOMO orbitals and the HOMO of their respective anions is greater than 6 eV; the large differences are an indication of the **hardness** of the halogens. The energy gap between the halogen species decreases in the order fluorine > chlorine > bromine > iodine and is consistent with the hardness for the halogens. Figure 5.17b shows a similar pattern for water, hydrogen peroxide, oxygen, and their anions. Note that removing hydrogen atoms from a molecule results in a new HOMO that is unstable relative to the parent compound. In the case of a neutral molecule such as O_2, the addition of an electron(s) results in the HOMO becoming more unstable relative to the parent compound. The more unstable or higher in energy the HOMO is, the better it will be as an electron pair donor or nucleophile (Figure 5.17b). Note that OH^- has a higher HOMO energy than all of the halide ions so is a better electron donor.

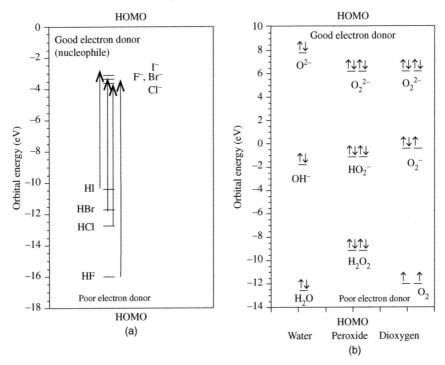

Figure 5.17   *(a) Comparison of the HOMO energies for HX compounds with the orbital energies of their anions; although not drawn all orbitals are filled with a pair of electrons. (b) Comparison of the HOMO energies for water, hydrogen peroxide, and oxygen and their anions*

5.3.7   Heteronuclear Diatomic Molecules and Ions (AB) – Sigma and Pi Bonds

There are several heteronuclear diatomic molecules and ions that are important in the environment. Among these are carbon monoxide (CO), cyanide ion (CN⁻) and nitric oxide (NO). Carbon monoxide occurs naturally and is produced during incomplete hydrocarbon combustion. The cyanide ion is a degradation product during the decomposition of the peptide bond in natural organic matter. Nitric oxide is also produced during incomplete hydrocarbon combustion (N_2 reacts with O_2) and is an intermediate in many biological processes (section 12.7). In this section, CO, which is isoelectronic with CN⁻, is described in detail.

Carbon monoxide is formed from the combination of the valence carbon and oxygen atomic orbitals, and for the moment, ***no s–p mixing*** or hybridization of the oxygen or carbon atoms is considered. Figure 5.18 shows the energies of both the carbon and oxygen 2s and 2p orbitals. The form of the wave functions and distribution of their energies would be close to what is shown in Figure 5.12a, but with some differences. Because the oxygen 2s orbital is extremely stable and about 13 eV lower in energy compared to the carbon 2s orbital, negligible combination of the oxygen 2s orbital with any of the carbon atomic orbitals is predicted, and this 2s orbital is assigned as a nonbonding molecular orbital in carbon monoxide. There are now seven atomic orbitals remaining for combination to form seven molecular orbitals. The oxygen 2p orbitals are of similar energy to all the other carbon atomic orbitals. To form a σ bonding molecular orbital, the 2s orbital of carbon will combine with the $2p_z$ orbital of oxygen as shown in Figure 5.18. A corresponding σ* also results. This leaves the p_x and the p_y orbitals from oxygen to react with the p_x and the p_y orbitals from carbon

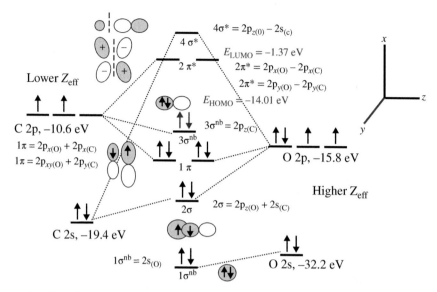

Figure 5.18 *Molecular orbital energy diagram for CO with sketches of the atomic orbitals used to form the molecular orbitals. No s–p mixing or hybridization for the oxygen atom is considered. The size of the atomic orbitals indicates their relative contribution to the molecular orbital.*

to form two π bonding orbitals and two π antibonding orbitals. The remaining carbon $2p_z$ orbital becomes a nonbonding sigma orbital (σ^{nb}), and is the HOMO for carbon monoxide. This HOMO is used in bonding as an electron donor. Because there are two σ nonbonding orbitals and three σ bonding orbitals, the bond order is 3. This molecular orbital analysis indicates that in the HOMO, the carbon atom is negative and the oxygen atom positive in carbon monoxide, which is consistent with the Lewis octet rule and the formal charges of each atom. The LUMO orbitals have significant contribution from the carbon atom and are π^* orbitals, which are able to accept electrons (π acceptors; Section 8.6.3) when bound to transition metals in low oxidation states; e.g., Fe(CO)$_5$ (Section 9.9) and for abiotic organic compound formation (Section 9.13.4).

The shorthand notation for numbering and naming of the molecular orbitals for CO is similar to that described above for HCl. The first σ-like orbital is given the number 1 and all higher energy σ and σ^* orbitals increase accordingly. Because the 1σ orbital comes from the oxygen 2s orbital, it is a nonbonding orbital. The first π orbitals are given the number 1 and the π^* orbitals are given the number 2 (In some texts, the $1\sigma^{nb}$ orbital will be given the symbol $3\sigma^{nb}$ when the 1s orbitals of carbon and oxygen are explicitly considered).

This orbital analysis is not entirely consistent with the UV photoelectron spectrum of CO, which is given in Figure 5.19 along with the actual molecular orbital energy level diagram, the sketch of the atomic orbitals that are combined to form the molecular orbitals, and the shape of the molecular orbitals. The resulting molecular orbitals also show the relative contribution of the carbon and oxygen atomic orbitals on combination. The vibrational fine structure or extra peaks for the 1π orbitals in the UV photoelectron spectrum indicate that they are strongly bonding (similar to H$_2^+$ above). The 2σ orbital is slightly bonding rather than strongly bonding as shown in Figure 5.18, but the 3σ orbital has mainly one sharp peak and is nonbonding as expected. The 1σ orbital is primarily a bonding orbital as it shows a broad peak with an energy centered at -40 eV, and is much more stable than the 2s orbital of the oxygen atom at -32.2 eV. Thus, *s–p mixing* or hybridization of the oxygen atom 2s and $2p_z$ orbitals occurred [an equal mixing of the oxygen 2s and $2p_z$ orbitals would result in an energy of 24 eV for the two hybridized orbitals, and the energies of these hybridized oxygen orbitals will

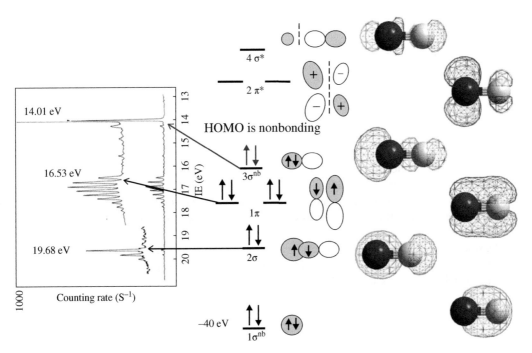

Figure 5.19 *UV photoelectron spectrum for CO with the energy level diagram along with the atomic orbital combinations from Figure 5.18 used to form the molecular orbitals, which are on the right (0.1 contour level; carbon is blue and hydrogen white; produced with HyperChem™ version 8.0.10). The assignment of the peaks and the molecular orbitals demonstrate s–p mixing or hybridization of the oxygen (Source: Turner 1970 [6], Figure 3.10. Reproduced with permission from Wiley, Inc.)*

be comparable to the carbon orbitals]. These hybridized orbitals will form two group orbitals (Sections 4.7 and 5.5.2) that will combine with the carbon atom 2s and $2p_z$ orbitals; thus, all the σ orbitals will have some contribution from the carbon and oxygen 2s and $2p_z$ atomic orbitals. The π bonds are more oxygen like so oxygen orbitals are drawn larger whereas the π* are more carbon like so carbon orbitals are drawn larger. The HOMO $3\sigma^{nb}$ orbital (-14.1 eV) is largely *nonbonding* and more carbon like; thus, the carbon atom donates the pair of electrons in a reaction. The analysis indicates that the bond order for CO is still three.

The cyanide ion and nitric oxide will have similar molecular orbital energy diagrams and molecular orbital shapes. CN^- is isoelectronic with CO and has the same Lewis octet structure with the carbon atom having a negative charge and the nitrogen atom being neutral. For both CN^- and CO, it is the carbon atom that will bind with iron in hemoglobin, resulting in asphyxiation. Comparing the molecular orbital diagrams of CO (Figure 5.19) with N_2 (Figure 5.14) gives an indication why N_2 is unreactive; the HOMO for N_2 is a weak σ bonding orbital whereas the HOMO for CO is a σ nonbonding orbital that has negative charge on the C atom. Thus, for N_2 to donate a pair of electrons requires loss of bond energy between the two nitrogen atoms, which is not the case for CO.

It is also useful to compare the ionization energies for CO and N_2 with the diatomic molecules O_2 and NO to show the importance of the energies of the HOMOs. For NO, there is one extra electron in one of the π* orbitals so the molecule is paramagnetic. Likewise for O_2, there are two extra electrons, one in each of the π* orbitals so the molecule is paramagnetic.

The relevant information for CO, N_2, O_2, and NO are

$$CO = Be_2\pi^4\sigma^2{}_{2p}{}^{nb} \qquad \text{112.8 pm; 3 bonds; IE} = -14.10 \text{ eV}$$

$$N_2 = Be_2\pi^4\sigma^2{}_{2pz} \qquad \text{110 pm; 3 bonds; IE} = -15.58 \text{ eV}$$

$$O_2 = Be_2\sigma^2{}_{2p}\pi^4\pi^2{}* \qquad \text{121 pm; 2 bonds; IE} = -12.07 \text{ eV}$$

$$\text{vs NO} = Be_2\sigma^2{}_{2p}\pi^4\pi^1{}* \quad \text{115 pm; 2.5 bonds; IE} = -9.26 \text{ eV}$$

Loss of an electron from each of these molecules shows significant differences in the ionization energy of the HOMO orbitals. For N_2, an electron is lost from a weak σ bonding orbital and results in loss of bond energy; in CO, the electron is lost from a nonbonding orbital so there is no loss of bond energy. For O_2 and NO, an electron is lost from a π^* orbital and this electron loss results in increasing the bond order (with a decrease in the bond distance) and a partial gain of bond energy that helps to compensate for the energy required to remove the electron.

Similarly, when oxygen is reduced, it accepts an electron into its π^* orbital to form a superoxide ion, and there is a decrease in bond order to 1.5. On accepting a second electron, the peroxide ion forms and the bond order becomes 1.0. In both cases, the bond distance increases.

$$O_2 + e^- \rightarrow O_2{}^-\text{(superoxide)} \quad \text{126 pm; 1.5 bonds}$$

$$O_2 + 2e^- \rightarrow O_2{}^{2-} \text{ (peroxide)} \quad \text{149 pm; 1.0 bond}$$

5.4   Understanding Reactions and Electron Transfer (Frontier Molecular Orbital Theory)

It is possible to predict reactions between molecules or electron transfer between molecules once the highest occupied molecular orbitals and the lowest unoccupied molecular orbitals are available for these molecules. The following rules were developed by Pearson [5, 11].

1. The highest occupied molecular orbitals (HOMO) donate electrons and the molecule acts as a reductant. The lowest unoccupied molecular orbitals (LUMO) accept electrons and the molecule acts as an oxidant. Electrons flow from the HOMO to the LUMO. For two-electron transfer reactions, a Lewis base donates to a Lewis acid (also section 7.7).
2. The symmetry of the HOMO and the LUMO must be similar for good overlap or a bonding interaction to occur; that is, the molecular orbitals must be positioned for good overlap.
3. The energies of the HOMO and LUMO must be comparable for net positive overlap, and the energy of the LUMO should be lower than that of the HOMO or no more than 6 eV above the HOMO. See Appendix 5.2 for a table of HOMO and LUMO energies.
4. For electron transfer to occur from the HOMO to LUMO, the net effect of electron transfer must correspond to bonds being made and broken. The bonds being made or broken must be consistent with the expected end products of the reaction.

If all these criteria are met, then a reaction is termed **symmetry allowed**. If not, the reaction is termed **symmetry forbidden**.

For electron transfer processes (see Marcus theory, Section 9.7.1), items **1–3** above must occur to be ***symmetry allowed***. For electron transfer processes, bonds can also be broken, but many electron transfer reactions only involve electron transfer from one molecule or ion to another.

5.4.1 Angular Overlap

To illustrate whether there is proper symmetry or overlap and the combination of orbitals, Figure 5.19 demonstrates three possible interactions of two p orbitals from two different atoms, M and L. Figure 5.19a shows two p orbitals on the z axis interacting head on to form a linear σ bond (S_σ). The p_x orbitals have an angular distribution function based on the cosine function. The overlap integral for M bonding with L is defined as S_{ML} and is related to S_σ by the equation

$$S_{ML} = S_\sigma \cos \theta$$

In Figure 5.20a, θ is 0° and $\cos \theta$ equals 1, which is the maximum overlap possible; that is, $S_{ML} = S_\sigma$. In Figure 5.20b, an angular σ bond is possible when θ is 45° and $S_{ML} = 0.707 S_\sigma$, indicating that the bond is only 70% as strong as a linear σ bond. When the two orbitals (p_z versus p_x or p_z versus p_y) are at an angle of 90° (Figure 5.20c), $\cos \theta$ equals 0 and there is no possible bonding interaction ($S = 0$). This latter case is known as the **orthogonality** of the p orbitals and the interaction is symmetry forbidden.

There are numerous possibilities for molecular orbital overlap, and Figure 5.21 provides a number of possible interactions. Figure 5.21c shows another nonbonding interaction that might occur when an orbital on the z axis from S_2^{2-} would interact with a metal, M, d_{xz} orbital.

5.4.2 H⁺ + OH⁻

A symmetry-allowed reaction will have no activation energy when the HOMO is of higher energy than the LUMO; that is, the electrons will flow to a more stable orbital. An example of this type of reaction is the acid–base neutralization reaction of the H^+ from a strong acid with OH^- from a strong base, which is an instantaneous reaction.

$$H^+ + OH^- \rightarrow H_2O$$

This reaction also describes the **proton affinity** (PA, Section 7.6) for the hydroxide ion, which is isoelectronic with HCl (Figure 5.16). Figure 5.22a shows the H^+ 1s atomic orbital, which has an energy of -13.6 eV, and the OH^- nonbonding 1π or p_x orbital, which has an energy of -1.825 eV. Here, the flow of the electrons is from the

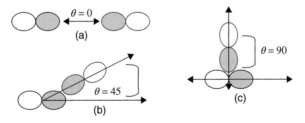

Figure 5.20 (a) linear sigma bond between 2 p_z orbitals; (b) angular sigma bond between 2 p_z orbitals; and (c) nonbonding sigma interaction between 2 p_z orbitals

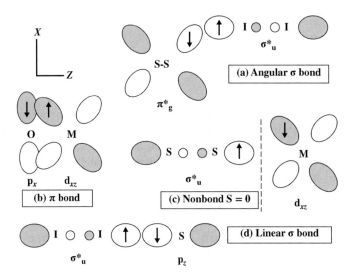

Figure 5.21   *Some examples of (a) an angular σ bond between two diatomic molecules, (b) the π bond between a p_x orbital and a d_{xz} orbital, (c) a nonbond between a σ* orbital of the diatomic molecule with a d orbital from a metal, (d) and a linear σ bond between a σ* orbital of a diatomic molecule and a p_z orbital of an atom (results in a redox reaction)*

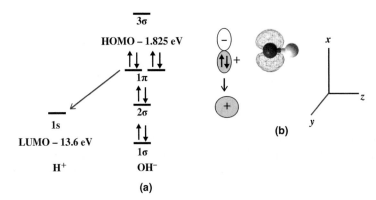

Figure 5.22   *(a) The orbitals and their energies from the H+ and OH− involved in bonding to form water. (b) The respective orbitals that are used to form water including the hydroxide ion 1π HOMO at the 0.1 contour level produced with HyperChem™ version 8.0.10. For this figure and the discussion of all linear molecules, the Cartesian coordinate system will be the same*

OH− to the H+ so the reaction is a downhill reaction with no **activation energy** (E_a) to overcome. Figure 5.22b shows the overlap of the 1π HOMO from the hydroxide ion with the 1s atomic orbital for the proton; the signs of the wave functions for the two orbitals are positive so there is overlap and the reaction is symmetry allowed. The product, H_2O, will be angular and not linear based on the orbital overlap in Figure 5.22b.

5.4.3 $H_2 + D_2$

To gain a better understanding of possible reactivity for thermodynamically favorable reactions, it is important to consider possible reactions that are either symmetry forbidden or have significant activation energy to

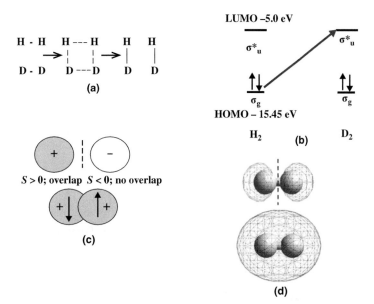

Figure 5.23 *(a) An idealized reaction sequence for* $H_2 + D_2$*; (b) molecular orbital diagrams* $H_2 + D_2$*; (c) line drawing of the frontier molecular orbitals* $H_2 + D_2$*; (d)* $H_2 + D_2$ *molecular orbitals (0.1 contour level; produced with HyperChem™ version 8.0.10)*

overcome for reaction to occur. The reactions between H_2 and several diatomic molecules are now considered. The first reaction is the isotopic exchange reaction between H_2 and D_2 to form HD.

$$H_2 + D_2 \rightarrow 2\,HD$$

Figure 5.23a shows a possible exchange mechanism to form the product HD. Figure 5.23b gives the molecular orbital diagrams and the energies of the MOs for both H_2 and D_2. Here electron transfer would have to occur from the σ HOMO of H_2 to the σ^* LUMO of D_2 followed by the formation of H–D bonds. This process has significant uphill activation energy of over 11 eV so violates rule 3 above. In addition, Figure 5.23c shows that this reaction is a symmetry-forbidden process (rule 2) as there is positive overlap between one H atom in H_2 with one D atom in D_2, but there is negative overlap between the other H and D atoms. Thus, the overall contribution to forming an H–D bond is zero. Figure 5.23d shows the molecular orbitals for the two reactants. From this analysis, the reaction is predicted to not occur under normal pressures and temperatures. However, the reaction could occur by breaking the H–H bond and/or the D–D bond so that free radicals would be created and react to form the product H–D.

5.4.4 $H_2 + F_2$

The reaction between H_2 and F_2 to form 2 HF molecules has two possibilities that can be considered, and these are electron transfer from HOMO H_2 to LUMO F_2 and HOMO F_2 to LUMO H_2. Figure 5.24a shows the possible reaction for H_2 with F_2, and Figure 5.24b shows the molecular orbital diagrams. Process 1 in Figure 5.24b shows electron transfer from HOMO H_2 to LUMO F_2; this process also has a significant uphill activation energy of over 12 eV. In addition, Figure 5.24c demonstrates that there is positive overlap between one H atom and one F atom but negative overlap for the other H atom and F atom in the H_2 and F_2 molecules. Thus, as in the isotopic exchange of hydrogen molecules above, this is a symmetry-forbidden reaction. Figure 5.24b

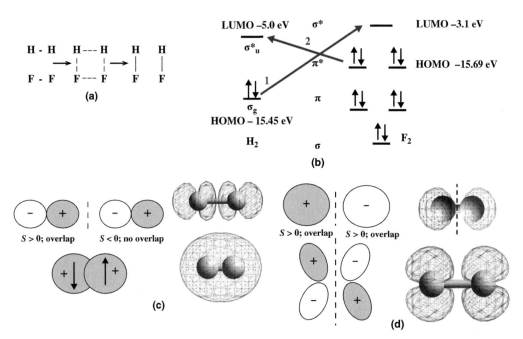

Figure 5.24 *(a) An idealized reaction sequence for $H_2 + F_2$; (b) molecular orbital diagrams $H_2 + F_2$; (c) line drawing and 0.1 contour plots of the frontier molecular orbitals $H_2(\sigma_g) + F_2(\sigma_p{}^*)$; (d) line drawing and molecular orbital (0.1 contour level; produced with HyperChem™ version 8.0.10) plots for $H_2(\sigma_u{}^*) + F_2(\pi^*)$*

and d shows process 2 where electron transfer would occur from HOMO F_2 to LUMO H_2. Although there is complete positive overlap between the π^* orbital of F_2 with the σ^* orbital of H_2, this process would result in the strengthening of the F–F bond rather than bond breaking, which is required to form H–F. This process also has significant uphill activation energy of over 11 eV as shown in Figure 5.24b (blue arrow). Although process 2 is a symmetry-allowed reaction, the energetics of the process do not permit formation of the end products as both rules 3 and 4 are violated. To form HF from these reactants would require free radical formation by homolytic cleavage of the H–H bond and or the F–F bond.

5.4.5 $H_2 + C_2$

The reaction of H_2 and C_2 to form acetylene, $H–C{\equiv}C–H$ is shown in Figure 5.25a. Figure 5.25b shows the molecular orbitals for both H_2 and C_2. Here, there is significant uphill activation energy for the donation of electrons from HOMO H_2 to LUMO C_2 in violation of rule 3. Figures 5.25c and d show that the process is symmetry allowed as a result of complete positive overlap from HOMO H_2 to LUMO C_2.

5.4.6 $H_2 + N_2$ (also $CO + H_2$)

The possible reaction between H_2 and N_2 to form H–N=N-H is given in Figure 5.26a, as H–N=N–H is a possible intermediate in the formation of ammonia but has not been detected. Here, one possible electron transfer would be from HOMO H_2 to LUMO N_2 (**1** in Figure 5.26b; Figure 5.26c). The molecular orbital energy level diagrams for the reactants are given in Figure 5.26b, and their energies indicate that there is a very large activation energy of over 17 eV to overcome, partially because of the unstable positive energy for

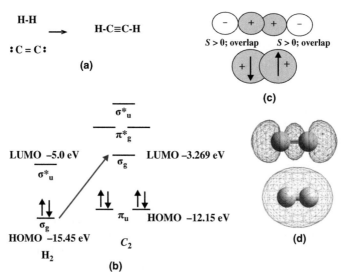

Figure 5.25   *(a) An idealized reaction sequence for $H_2 + C_2$; (b) molecular orbital diagrams $H_2 + C_2$; (c) line drawing of the frontier molecular orbitals for $H_2(\sigma_g) + C_2(\sigma_g)$; (d) molecular orbital (0.1 contour level; produced with HyperChem™ version 8.0.10) plots for $H_2(\sigma_g) + C_2(\sigma_g)$*

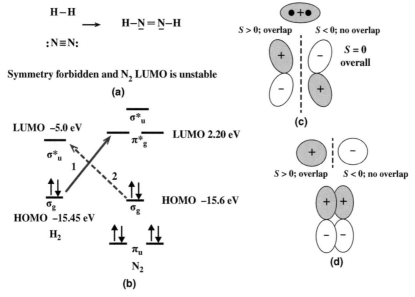

Figure 5.26   *(a) An idealized reaction sequence for $H_2 + N_2$; (b) molecular orbital diagrams $H_2 + N_2$; (c) line drawing of the frontier molecular orbitals HOMO $H_2(\sigma_g)$ + LUMO N_2 (π^*); (d) line drawing of the frontier molecular orbitals LUMO $H_2(\sigma_u^*)$ + HOMO N_2 (σ_g)*

the LUMO of N_2 [10]. In addition, Figure 5.26c shows that the net overlap between the σ_g HOMO orbital of H_2 and the π^* LUMO orbital of N_2 is zero. Another possibility for electron transfer is HOMO N_2 to LUMO H_2 (**2** in Figure 5.26b; Figure 5.26d). Again this possibility has a large activation energy of over 11 eV to overcome and is symmetry forbidden. The frontier molecular orbital analysis for this reaction shows that the reaction is energetically and symmetry forbidden, and that N_2 is inert and one of the most difficult small molecules to **activate**. The direct reaction of H_2 and N_2 to form NH_3 occurs at high temperature ($> 400°C$) and high pressure (>100 atm) over an iron catalyst; this is known as the **Haber process (Section 9.13.5)**. The reaction of CO (isoelectronic with N_2) with H_2 has similar problems (see Figure 5.19).

5.4.7   Dihalogens as Oxidants

From the above analyses, it is tempting to conclude that many diatomic molecules may not be very reactive chemical species. However, F_2 and the other dihalogen (X_2) molecules are reactive because halogen atoms have high electronegativity and Z_{eff}. These properties result in the energies of their LUMO orbitals being negative and stable as shown in Figure 5.27a. Thus, the LUMO is able to accept electrons so that the halogen–halogen bond breaks with formation of the anion (X^-), and all X_2 molecules are excellent oxidizing agents as evidenced by iodometric titrations with reductants such as thiosulfate, bisulfide, and polysulfide anions (S_x^{2-}).

$$I_2 + 2S_2O_3^{2-} \rightarrow 2I^- + S_4O_6^{2-}$$

$$I_2 + HS^- \rightarrow 2I^- + S(0) + H^+$$

$$I_2 + S_x^{2-} \rightarrow 2I^- + xS(0)$$

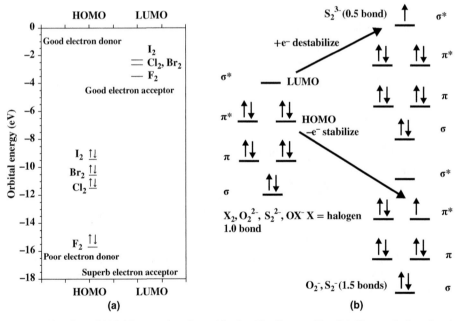

Figure 5.27 *(a) HOMO and LUMO energies plotted for F_2, Cl_2, Br_2, and I_2. (b) Change in bond order when an electron is removed or accepted by a dihalogen*

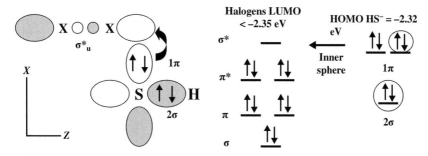

Figure 5.28 *Molecular orbitals used during the reaction of HS⁻ donating an electron pair to dihalogens. MO energy level diagrams for the reactants*

Figure 5.27a also notes the areas in the plot for good electron donors and acceptors. Good electron acceptors have stable orbitals characterized with negative energy values. Poor electron donors are those with very stable orbitals characterized with negative energy values; conversely, good electron donors have unstable orbitals characterized with positive energy values.

Isoelectronic with the dihalogen molecules are the peroxide anion (O_2^{2-}), the persulfide anion (S_2^{2-}), and the hypohalous acid anions (OX^-). Figure 5.27b shows the effect of adding one electron to the LUMO in decreasing the bond order as well as the effect of removing one electron from the HOMO of these species in increasing the bond order (also Figure 5.24d; the reaction of H_2 and F_2).

Figure 5.28 shows a schematic representation for the donation of a pair of electrons from the bisulfide ion (isoelectronic with HF and OH^-) to a dihalogen – HS^- HOMO $1\pi_x$ to X_2 LUMO $\sigma_{pz}*$ or σ_u*. The direct donation would be an inner sphere process with formation of the intermediate below prior to electron transfer. There is no significant activation energy to overcome.

$$[\text{H-}\overline{\underline{\text{S}}}: \rightarrow :\overline{\underline{\text{X}}}\text{-}\overline{\underline{\text{X}}}:]^-$$

This process would be similar for other two electron reductants reacting with X_2 and peroxide.

5.4.8 O_2 as an Oxidant and its Reaction with H_2S and HS^-

O_2 is normally perceived as a strong oxidant in nature. However, the two electron transfer reactions of H_2S and HS^- with F_2, Cl_2, Br_2, and I_2 are instantaneous whereas the HS^- reaction with H_2O_2 has a half-life in freshwater of 15 minutes [12] and with O_2 a half-life of 2 days [13]. Because O_2 has two partially occupied (SOMO) orbitals (Figure 5.29), it cannot accept two electrons into one orbital from two electron donor reductants such as sulfide. O_2 would have to react with a two-electron donor by accepting one electron at a time or the reductant would have to donate two electrons into each of the SOMOs of O_2 as below.

$$O_2 + H_2S \rightarrow HS\bullet + O_2^- + H^+ \qquad \text{(one-electron transfer)}$$
$$O_2 + H_2S \rightarrow S^0 + H_2O_2 \qquad \text{(two-electron transfer)}$$

Figure 5.29a shows the thermodynamics of the reactions for the dihalogens, O_2 and H_2O_2 with H_2S and HS^-. The dihalogens and H_2O_2 have no thermodynamic barrier for the reaction with H_2S and HS^- as these oxidants can accept two electrons into a vacant $\sigma*$ LUMO. There is no thermodynamic barrier for the two-electron reduction of O_2 to H_2O_2, but there is a significant kinetic barrier. The two-electron transfer

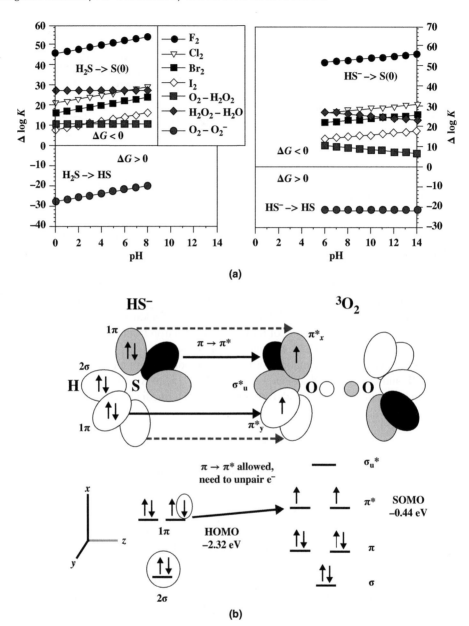

Figure 5.29 *(a) Thermodynamics from the redox half-reactions in Table 2.2 [Reactions F1, CL1, Br1, IO1, O2, O3, and O6 minus S1 through S4]. The pK$_a$ of H$_2$S is ~7 so sulfide speciation changes with pH. OH$^-$ also reacts with the dihalogens to form HOX, which reacts with sulfide species such as X$_2$. (b) Overlap of the frontier MOs for HS$^-$ and O$_2$ along with their energy level diagrams*

reactivity pattern generally follows the energy of the lowest unoccupied molecular orbital for these oxidants [$F_2(-3.1 \text{ eV}) > I_2(-2.55 \text{ eV}) > Br_2(-2.53 \text{ eV}) > Cl_2(-2.35 \text{ eV}) > O_2H(-1.19 \text{ eV}) > O_2(-0.44 \text{ eV})$] with O_2 being the least reactive. However, the one-electron reduction of O_2 to O_2^- (superoxide) is thermodynamically unfavorable at all pH values because the unstable free radicals, O_2^- and HS•, would be the products. Thus, O_2 should not accept one electron at a time from these electron donors.

The reaction of HS$^-$ and $^3O_2$ is used near hydrothermal vents where the production of organic matter by **chemosynthesis** occurs, and a two-electron transfer is thermodynamically favorable but kinetically hindered. Figure 5.29b shows the HS$^-$ and O_2 energy level diagrams and frontier orbitals, which give insight into the slow reactivity of O_2. The energies of the degenerate HOMOs and LUMOs are within 6 eV so are reasonable for electron transfer. However, an electron from each 1π HS$^-$ HOMO would have to enter a corresponding π^* O_2 SOMO simultaneously. This would result in a significant *kinetic barrier* to electron transfer. The HOMO–LUMO energy gap of 1.88 eV is accessible at 660 nm so the reaction may occur via photolysis. The kinetic barriers to reaction are overcome by organisms at hydrothermal vents as the rates of this reaction are enhanced by > 40,000 by microbial mediation [14]. It is not clear how the organisms deal with the transfer of electrons, but one way may be to convert $^3O_2$ to $^1O_2$ shown in Figure 5.30. The kinetic barrier would be removed as two electrons could transfer from HS$^-$ to $^1O_2$ readily.

Figure 5.30 shows the MO energy level diagram and orbital occupation for the ground and excited states of O_2, as well as the energy needed to produce the excited states from the ground state. The singlet oxygen state, $^1\Delta_g$, where two electrons are paired in one π^* orbital from the original SOMOs, is very reactive as two electrons can be accepted from a reductant in the other empty π^* orbital. The other singlet state, $^1\Sigma^+_g$, where the electrons are of opposite spin and in different π^* SOMOs, is also not reactive. The singlet O_2 state, $^1\Delta_g$, is one of the reactive oxygen species (ROS), which are formed as transients in nature. The superoxide ion MO energy level diagram and orbital occupation are in the lower left of Figure 5.30.

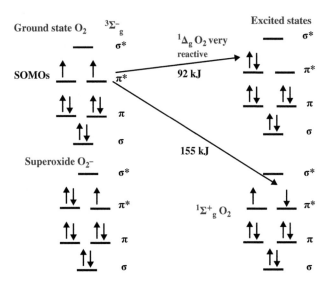

Figure 5.30   *MO energy level diagrams and orbital populations for the ground and excited states of O_2 and for the superoxide ion*

5.4.8.1   Term Symbols for Linear Molecules

The shorthand notation for symbols to describe the energy states for linear molecules such as O_2 above is now described and is similar to the Russell–Saunders discussion for atoms in Chapter 3. In this discussion, the term symbols for the different O_2 molecules ($D_{\infty h}$ symmetry) will be provided as an example. The full term symbol indicating the energy and state of the molecular orbital with respect to the spin and angular momentum about the z axis is described as $^{2S+1}\Lambda_{\Omega}{}^{\pm}$.

S (spin multiplicity) has the same significance as already described for atoms. For ground state O_2, the sum of the spins is 1 and $2S + 1 = 3$ for a triplet state. The angular momentum of the molecular orbitals is described based on the angular momentum, m_l, of the atomic orbitals, which are combined to form them as below.

Molecular orbitals	m_l	Λ	Atomic orbitals
σ	0	0	$s, p_z, d_z{}^2$
π	± 1	1	p_x, p_y, d_{xz}, d_{yz}
δ	± 2	2	$d_{xy}, d_{x^2-y^2}$

Because there are π_x and π_y bonding and antibonding orbitals, $\Lambda = 1$ for π_x and -1 for π_y. M_L (for the molecular orbital) $= m_{l1} + m_{l2} + \ldots + m_{ln}$. Thus, $\Lambda = |M_L|$

$$\text{where } \Lambda = \quad 0 \quad 1 \quad 2 \quad 3 \quad \ldots$$
$$\Sigma \quad \Pi \quad \Delta \quad \Phi \quad \ldots$$

For the O_2 ground state, the sum of the M_L for all electrons will be zero so Λ is Σ. At this point, the term symbol is $^3\Sigma$.

If spin–orbit coupling is important, then the value of the total angular momentum (Ω) can be indicated as a right subscript. For Σ states, Ω equals S, and for singlet states, Ω equals Λ so Ω is rarely given. Otherwise, Ω is determined as follows.

$$\Omega = \Lambda + S, \Lambda + S - 1, \ldots, |\Lambda - S|$$

The symbol g is given for a state symmetric to i (inversion) and u for a state antisymmetric to i. Here, the product of the wave functions of the original atomic orbitals combining to make the molecular orbital is used to determine g and u. The direct products of orbitals is

$$g \times g = g; \quad u \times u = g; \quad g \times u = u$$

For ground state $^3O_2$, the π orbitals are $g \times g = g$ and the term symbol is now $^3\Sigma_g$.

The $\pm$ symbol is needed for Σ states and is "+" when the state is symmetric to σ (reflection in any plane containing the molecular axis) and is "−" when the state is antisymmetric to σ. Typically a "−" state occurs when two electrons of parallel spin are in different π (or δ) orbitals. For ground state $^3O_2$, the full term symbol is now $^3\Sigma_g{}^-$.

For the singlet oxygen state, $^1\Delta_g$, $2S + 1 = 1$ and $\Lambda = 2$ as both electrons are paired in the same orbital ($\pi_x{}^*$). As two p_x orbitals on two different O atoms are combined to form the $\pi_x{}^*$ orbital, the orbital is symmetric to inversion and is gerade.

5.4.8.2 *More on Sulfide and Halide Oxidation*

The oxidation of H_2S and HS^- by a two-electron transfer process results in the formation of a sulfur atom (S^0) in aqueous solution. S^0 reacts with HS^- to form a linear polysulfide anion, HS_2^- or S_2^{2-}. Continued reaction of additional S^0 atoms (from sulfide oxidation) with the polysulfide in solution results in the formation and precipitation of cyclic S_8 as below.

$$HS^- + H_2O_2 \rightarrow 1/8 \; S_8 + H_2O + OH^-$$

$$1/8 \; S_8 + SH^- \rightarrow S_2H^-$$

$$1/8 \; S_8 + S_2H^- \rightarrow S_3H^- \; \ldots\ldots$$

$$1/8 \; S_8 + S_8H^- \rightarrow S_8 + HS^-$$

The terminal S atom of polysulfides can oxidize further to form linear S_x–S–SO_3. Cleavage of the S–S bonds can lead to S_x and $S_2O_3^{2-}$ (thiosulfate) or S_{x+1} and SO_3^{2-} (sulfite). SO_3^{2-} and $S_2O_3^{2-}$ have been identified as products or intermediates in sulfide oxidation back to SO_4^{2-}.

The halides are isoelectronic with HS^- and have the same thermodynamic barrier to one-electron oxidation by $^3O_2$ [15]. The two-electron reactions of halides with H_2O_2 and O_3 to form X_2 and HOX are thermodynamically favorable. The reaction kinetics with H_2O_2 are kinetically hindered, and a variety of marine eukaryotic organisms as well as marine and terrestrial microorganisms have developed haloperoxidases, which contain metal centers that activate the oxidant, to increase the halide oxidation rates [16].

5.5 Polyatomic Molecules and Ions

The discussion has been limited to diatomic molecules and their reactivity. The rest of the chapter describes the formation of the molecular orbitals for molecules and ions containing three or more atoms. The description starts with H_3^+, AH_2, AH_3, and AH_4 species containing only sigma bonds and ends with species having sigma and pi bonds (AB_2 and AB_3). The Group Theory section (Chapter 4.5 through 4.7) can be used in conjunction with these examples, but is not totally necessary. The reactivity of these polyatomic molecules and ions will be briefly discussed in this chapter and in the later chapters that describe transition metal bonding and reactivity.

5.5.1 H_3^+ Molecular Cation

The simplest polyatomic molecule is the triatomic molecular hydrogen cation, H_3^+, which was originally found in electrical discharges by J. J. Thompson in 1911 using mass spectrometry. Although this may seem to be an unimportant or trivial chemical species, it is one of the most abundant ions in the universe and is stable in the interstellar medium [17–19]. H_3^+ is produced by the following reaction as cosmic rays are capable of easily ionizing H_2 to H_2^+.

$$H_2 + H_2^+ \rightarrow H_3^+ + H$$

H_3^+ is an important reactant for interstellar chemistry as it is a H^+ donor or acid (Section 7.6.1) that reacts with other elements or simple molecules according to the following general equation.

$$H_3^+ + X \rightarrow H_2 + HX^+$$

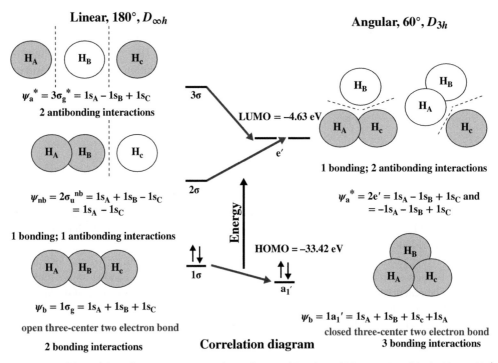

Figure 5.31 *Possible H_3^+ bonding arrangements from the combination of H atom 1s orbitals. Linear H_3^+ has σ, σ^{nb}, and σ^* orbitals. Angular H_3^+ has a σ bonding a_1' orbital and two σ antibonding orbitals e'. The $1a_1'$ orbital wave function contains the $1s_A$ orbital twice to denote that the system is delocalized*

Thus, knowing the structure and molecular orbitals of H_3^+ is critical to assessing its reactivity in the universe and how it protonates other atoms and small molecules to form organic compounds. There are two possible ways to combine the three hydrogen atoms: first as a linear molecule with only the center hydrogen atom binding the other two hydrogen atoms (an **open three-center two-electron bond**) and second as a triangular molecule with all three hydrogen atoms bonding to each other (a **closed three-center two-electron bond**). Figure 5.31 shows both geometric arrangements for H_3^+ along with the generalized wave functions and the relative MO energy level diagram as three atomic orbitals combine to form three molecular orbitals (see Sections 4.6.4.4 and 4.7 for the formation of group orbitals or **symmetry-adapted linear combinations**, SALC). The change in geometry and orbital energies from linear to triangular is known as a **correlation diagram**. The shorthand notation for orbitals or irreducible representations is given in **Section 4.5.4**.

5.5.1.1   *Open Three-Center Two-Electron (3c-2e) Bond*

For the linear open three-center two-electron bond, the combining of atomic orbitals is straightforward and gives σ bonding, nonbonding, and antibonding orbitals. The bonding orbital has the central atom combining with both terminal atoms to give two bonding interactions; the nonbonding orbital has the central atom combining with one terminal atom to give one bonding interaction and the other terminal atom to give one antibonding interaction; the antibonding arrangement has the central atom combining with both terminal atoms to give two antibonding interactions. Only the 1σ bonding orbital would be occupied, and the electrons would be delocalized across the three hydrogen atoms. However, the angular arrangement is more stable.

5.5.1.2   *Closed three-center two-electron (3c-2e) bond*

For the closed three-center two electron bond, the combining of atomic orbitals is more complex as all three atoms combine with each other to form a sigma bonding orbital with three bonding interactions, $1a_1'$, which is more stable than the 1σ bonding MO above. This bonding wave function explicitly shows atom H_C combining with atom H_A unlike the linear arrangement so that each atom has two bonding interactions for a total of three bonding interactions for H_3^+. In this case, cyclic delocalization of the two electrons across the three hydrogen atoms is superior to the linear arrangement above. For the other possible combinations, two atoms and their atomic orbitals (e.g., H_A and H_C) can have one bonding interaction with each other, and the third atom and its atomic orbital (H_B) would have two antibonding interactions with the first two (H_A and H_C). Another identical set is shown in Figure 5.31 indicating that these two orbitals are degenerate and of similar energy. These are termed e' antibonding orbitals and are predicted to be intermediate in energy to the nonbonding and antibonding MOs for the linear geometry. The triangular arrangement has been confirmed experimentally [18]. The HOMO is the bonding orbital $1a_1'$, which is much more stable than the H_2 molecule due to the electron delocalization between three atoms. The bond distance between the atoms in H_3^+ is 88 pm.

5.5.2   BeH$_2$ – Linear Molecule with Sigma Bonds Only

Beryllium dihydride is the next molecule of increasing complexity to describe using MOT; it only exists as a monomer in the gas phase but exists as a polymeric solid phase at ambient conditions [20]. It is a linear molecule as discussed above using the valence shell electron pair repulsion theory. Here, beryllium has a filled 2s orbital and these electrons will bind with the electrons from each hydrogen atom to form two bonds. In addition, beryllium has three p orbitals, which could potentially be involved with bonding; however, only one of these p orbitals will be used. Figure 5.32 shows the combinations for the four atomic orbitals of beryllium with the two atomic orbitals from the two hydrogen atoms to give the six molecular orbitals for beryllium dihydride.

The generalized wave functions for the bonding and the antibonding orbitals are given in Figure 5.32 and take into account the different effective nuclear charges for beryllium and hydrogen. In addition, the two hydrogen orbitals can be combined to give two "group" or symmetry-adapted linear combination (SALC, Section 4.7) orbitals of the following form.

$$(1s_A + 1s_B) \text{ and } (1s_A - 1s_B)$$

The parentheses indicate a group or SALC orbital. The SALC is generated using the principles of group theory (Section 4.7). The $1s_A + 1s_B$ combination has the same sign as the wave function for each orbital. On performing the operations for the $D_{\infty h}$ point group, the atoms and the sign of each wave function remain unchanged so the two s orbitals transform as the A_{1g} (also Σ_g^+) irreducible representation. For the $1s_A - 1s_B$ combination, the signs of the wave function for each orbital change. On performing the operations for the $D_{\infty h}$ point group, the atoms and the sign of each wave function change for the C_2, i, and other operations so the two s orbitals transform as the A_{1u} (also Σ_u^+) irreducible representation.

These then combine with the beryllium 2s and 2p atomic orbitals. For example, the σ_g bonding molecular orbital and the $\sigma_g{}^*$ antibonding orbital from the combination of all the beryllium and hydrogen atom s orbitals are given as

$$\psi_b = \sigma_g(\Sigma_g^+) = a2s + b(1s_A + 1s_B) \tag{5.5}$$

$$\psi_a = \sigma_g{}^* = b2s - a(1s_A + 1s_B) \tag{5.6}$$

Note that the $(1s_A + 1s_B)$ group orbital is used for both of these combinations. The coefficient "a" before the 2s orbital of beryllium is smaller than the coefficient "b" before the $1s_A + 1s_B$ orbital combination for the two

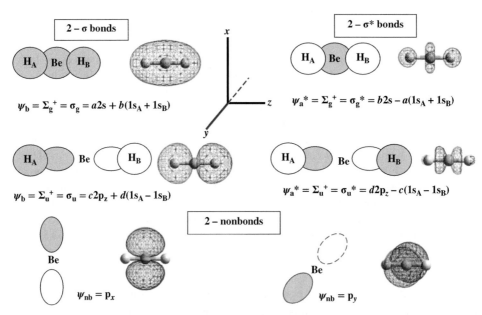

Figure 5.32   *The wave functions derived from the combination of the six atomic orbitals from beryllium and the two hydrogen atoms to form six molecular orbitals. The z axis is the bond axis (C_∞ axis). Contour plots (0.1 level) of the molecular orbitals produced using the HyperChem^{TM} program version 8.0.10*

hydrogen atoms based on the energies in Figure 5.33. For the corresponding σ_g* antibonding orbital when all the atomic s orbitals are combined, the negative sign is placed before the group orbital and the coefficients are reversed as described above for HCl. At this point, only two molecular orbitals have been formed (one bond and one antibond).

The second bonding and antibonding combination comes from the beryllium $2p_z$ atomic orbital and the same hydrogen atomic orbitals, but with the $(1s_A - 1s_B)$ group orbital. Because the $2p_z$ orbital is ungerade, the group orbitals for the hydrogen atoms must now be subtracted from each other to have the proper symmetry match for the p_z orbital. Coefficients "c" and "d" are used to discriminate this bonding combination of the beryllium $2p_z$ atomic orbital from that with the beryllium 2s atomic orbital. Thus the bonding orbital, σ_u, and the antibonding orbital, σ_u*, become Equations 5.7 and 5.8.

$$\psi_b = \sigma_u(\Sigma_u^+) = c2p_z + d(1s_A - 1s_B) \tag{5.7}$$

$$\psi_a = \sigma_u* = d2p_z - c(1s_A - 1s_B) \tag{5.8}$$

The corresponding antibonding orbital has a negative sign before the group orbital and the coefficients before the $2p_z$ atomic orbital beryllium and the group orbital from the hydrogen atoms are reversed. At this point, two bonding and two antibonding molecular orbitals have been created from the four atomic orbitals (the beryllium 2s and $2p_z$ atomic orbitals and the two hydrogen atom 1s atomic orbitals). The remaining $2p_x$ and $2p_y$ atomic orbitals from beryllium are nonbonding orbitals and do not contribute anything to the bond order of the molecule; these are also degenerate in energy (e_{1u}, Π_u). Both σ bonding orbitals will have two electrons in each and the electrons are delocalized over all three atoms. Figure 5.33 shows the relative molecular orbital energy level diagram with the electron filling of the orbitals for beryllium dihydride.

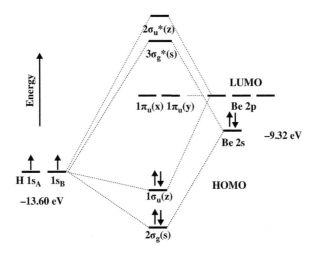

Figure 5.33 *Idealized molecular orbital energy diagram for* BeH$_2$

5.5.3 H$_2$O – Angular Molecule with Sigma Bonds Only

For the water molecule, there are the same total number of atomic orbitals, which can combine to form six molecular orbitals. However, the oxygen atom has six electrons in its outermost state compared to the two electrons in beryllium so there are a total of eight electrons, which results in an angular molecule. Figure 5.34 shows how the atomic orbitals for the oxygen atom and the two hydrogen atoms combine to form the six molecular orbitals; note that the z axis is the C_2 axis of H$_2$O. There is a σ bonding (2a$_1$) and antibonding (4a$_1$) combination of the three s atomic orbitals as well as a σ bonding (1b$_1$) and antibonding (2b$_1$) combination of the two s hydrogen atomic orbitals with the p$_x$ orbital from oxygen. The generalized wave functions for these bonding and antibonding orbitals (5.9–5.12) have some differences from those described for the beryllium hydride molecule above with the coefficients including appropriate consideration for the differences in Z_{eff} for the oxygen and hydrogen atoms and the angular bonding.

$$\psi_b = \phi_{2A1} = a2s + b(1s_A + 1s_B) \tag{5.9}$$

$$\psi_a = \phi_{4A1}{}^* = b2s - b(1s_A + 1s_B) \tag{5.10}$$

$$\psi_b = \phi_{1B1} = c2p_x + d(1s_A - 1s_B) \tag{5.11}$$

$$\psi_a = \phi_{2B1}{}^* = d2p_x - c(1s_A - 1s_B) \tag{5.12}$$

Group theory details for water are provided in Section 4.7.1.1. Because water has C_{2v} symmetry, the two SALC orbital combinations [(1s$_A$ + 1s$_B$) and (1s$_A$ − 1s$_B$)] transform as the A$_1$(a$_1$) and B$_1$(b$_1$) irreducible representations.

The nonbonding orbitals show a major difference between BeH$_2$ and water. The p$_z$ orbital (which also transforms as an A$_1$ irreducible representation) of the oxygen atom actually interacts in a bonding manner with both of the hydrogen atoms so has the following form for its wave function.

$$\psi_b = \phi_{3A1} = a2s + b(1s_A + 1s_B) \tag{5.13}$$

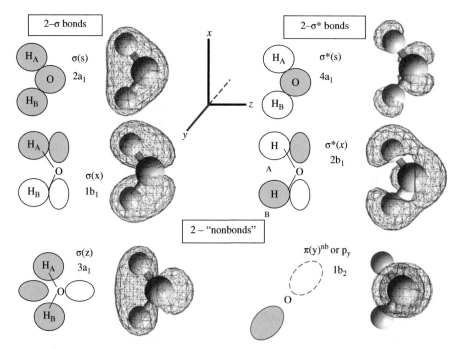

Figure 5.34   *The atomic orbitals combined to form the molecular orbitals for H_2O along with their respective electron contour plots (0.1 contour level) produced using the HyperChem™ program version 8.0.10*

The p_y orbital of the oxygen atom (transforms as $1b_2$) is a nonbonding HOMO orbital, which is perpendicular to the plane of the water molecule. Although an individual H_2O molecule does not form a stable H_2O^- species, bulk water does solvate electrons, and water clusters do bind excess electrons [21]. The LUMO energy for the $4a_1$ LUMO in Figure 5.35 is for the smallest cluster of six water molecules that accepts an electron; higher clusters are more stable [21].

Figure 5.35 shows the molecular orbital energy diagram for water. The $2a_1$ or σ (s) molecular orbital is primarily composed by the oxygen 2s orbital; the diagram does not show hybridization or s–p mixing for the oxygen atom. Although the interaction of the p_z orbital of the oxygen atom with the two hydrogen atom s orbitals, $\sigma(z)$, or $3a_1$ might be considered a nonbonding orbital, there is significant bonding, which is shown in the UV photoelectron spectrum of water. The $\pi(y)^{nb}$ HOMO orbital is primarily composed by the oxygen $2p_y$ orbital and shows some fine structure in the UV photoelectron spectrum. The energy for the HOMO indicates a very stable orbital that will not be a strong electron donor, thus leading to the dissociation of water from M^{x+} ions (lability, Section 9.2.1). Figure 5.17 shows a HOMO energy comparison of H_2O with other oxygen species.

H_2S, H_2Se, and H_2Te are isoelectronic with H_2O when considering the outermost electrons for each molecule, and these compounds will have similar molecular orbitals. The major difference between H_2S and H_2O is that H_2S has a bond angle of 92° compared to 104.5° for H_2O. To account for the bond angle differences, the oxygen atom in H_2O exhibits hybridization or s–p mixing, which is not that important for the sulfur atom.

The change in bond angle for triatomic molecules from 180° to 90° results in a change in the energies and their ordering for the four molecular orbitals that are fully occupied. Figure 5.36 shows a Walsh

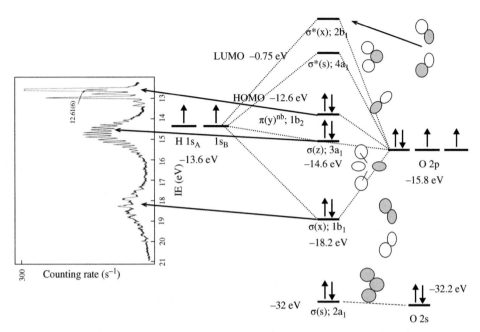

Figure 5.35 *The molecular orbital energy level diagram correlated with the UV photoelectron spectrum for water. Two different shorthand annotations for the labeling of the molecular orbitals are given (Source: Turner 1970 [6], Figure 4.26. Reproduced with permission from Wiley, Inc.)*

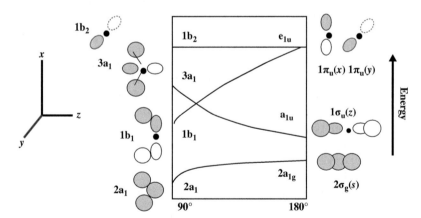

Figure 5.36 *Idealized Walsh correlation energy diagram for AH$_2$ bonding orbitals on changing the bond angle. The z-axis is coincident with the major rotational axis of the molecules; e.g., H$_2$O (C$_{2v}$) and BeH$_2$ (D$_{\infty h}$). If the x and z coordinates are exchanged for C$_{2v}$ symmetry, then the 1b$_1$ and 3a$_1$ orbitals exchange; i.e., the 1b$_1$ increases and the 3a$_1$ decreases in energy with no energy crossover occurring on changing the bond angle*

correlation diagram [4, 5, 8] for the change in energy for AH_2 compounds as the bond angle varies from 180° to 90°. The bonding $2a_1$ orbital becomes more stable than the bonding $2a_{1g}(2\sigma_g)$ orbital as discussed for the H_3^+ molecule. The $3a_1$ orbital in angular geometry becomes less stable relative to the bonding $a_{1u}[\sigma_u(z)]$ orbital in linear geometry as overlap between the H atom 1s orbitals with the central atom p_z orbital is less efficient in angular geometry (as the H atom orbitals are no longer on the z axis). The $1b_1$ orbital in angular geometry becomes more stable than the $e_{1u}(p_x, p_y)$ orbitals in linear geometry as the p_x orbital now interacts with the H atom 1s orbitals, which are no longer on the z axis but are in the xz plane. The $1b_2$ nonbonding orbital, which is not involved in bonding, does not experience any energy change with geometry change from 180° to 90° as it is still perpendicular to the xz plane. In summary for the angular H_2X molecules, the $2a_1$, $1b_1$, and $3a_1$ orbitals are bonding orbitals as seen in the X-ray photoelectron spectrum for water; the $1b_2$ orbital is more nonbonding.

5.6   Tetrahedral and Pyramidal Species with Sigma Bonds only (CH_4, NH_4^+, SO_4^{2-})

5.6.1 CH_4

Before discussing NH_3 and other pyramidal chemical species, methane and other tetrahedral species including the ammonium ion are discussed. The total number of atomic orbitals available for combination to form molecular orbitals is eight (carbon 2s and three 2p atomic orbitals plus a 1s orbital for each of the four hydrogen atoms). Obviously, there will be four bonding orbitals and four antibonding orbitals. A convenient way to look at the bonding interactions is to put the carbon atom at the center of the cube and the four hydrogen atoms at alternating vertices of the cube. Figure 5.37 shows this arrangement with all of the s atomic orbitals from all five atoms in the lower portion of figure. The bonding interaction, $\sigma(s)$, shows a group orbital, which is the positive combination of all the four hydrogen atom orbitals. The antibonding interaction, $\sigma(s)^*$, shows a negative sign for this group orbital. The top portion of Figure 5.37 shows the interaction of a p_x orbital with another set of group orbitals for the four hydrogen atoms, σp_x. Because of the ungerade symmetry for the carbon p_x orbital, the sign of the wave function for hydrogen atoms 3 and 4 has a different sign from the wave functions for hydrogen atoms 1 and 2 in order to have proper symmetry overlap with the carbon orbital. For simplicity, coefficients are not provided in the wave functions shown in Figure 5.37. Likewise the antibonding configuration, σp_x^*, has the same group orbital with a negative sign. The p_y and p_z orbitals will combine to give a similar set of molecular orbitals as the p_x orbitals, but the group orbitals will be different to reflect their ungerade symmetry on their respective axes. Thus, four bonding and four antibonding orbitals result for methane, and the molecular orbitals from the carbon atomic p orbitals produce a triply degenerate set, t_2, as shown in Section 4.6.3. The resulting orbital contour plots for the σ_s, σ_s^*, σp_s, and σp_x^* molecular orbitals are also given at the 0.05 contour level.

The four 1s orbitals make one SALC combination that is bonding and another that is antibonding.

$$\psi_b = \phi_{A1} = a2s + b(1s_1 + 1s_2 + 1s_3 + 1s_4) \qquad \psi_b^* = \phi_{A1}^* = a2s - b(1s_1 + 1s_2 + 1s_3 + 1s_4)$$

From the group theory analysis of the T_d bonds for methane (Section 4.6.3), the $(1s_1 + 1s_2 + 1s_3 + 1s_4)$ SALC transforms as A_1. The other three SALC combinations $(1s_1 + 1s_2 - 1s_3 - 1s_4)$, $(1s_1 - 1s_2 + 1s_3 - 1s_4)$, and $(1s_1 - 1s_2 - 1s_3 + 1s_4)$ transform as the triply degenerate t_2 set (x, y, z). Thus, the interaction of the carbon 2p orbitals with the three hydrogen 1s orbitals makes three bonding and antibonding combinations

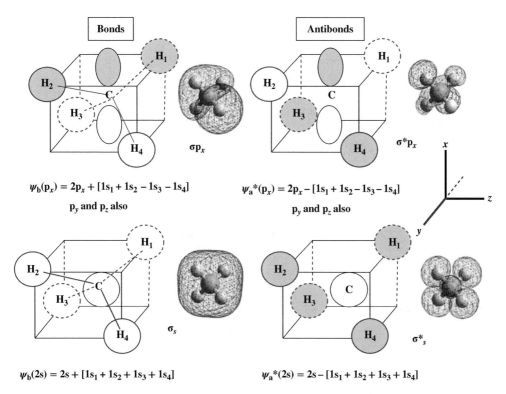

$\psi_b(p_x) = 2p_x + [1s_1 + 1s_2 - 1s_3 - 1s_4]$

p_y and p_z also

$\psi_a{}^*(p_x) = 2p_x - [1s_1 + 1s_2 - 1s_3 - 1s_4]$

p_y and p_z also

$\psi_b(2s) = 2s + [1s_1 + 1s_2 + 1s_3 + 1s_4]$

$\psi_a{}^*(2s) = 2s - [1s_1 + 1s_2 + 1s_3 + 1s_4]$

Figure 5.37 *The bonding interactions for all the s orbitals for carbon and hydrogen atoms and the p_x orbital of carbon with four hydrogen atoms. The p_y and p_z orbitals for carbon will combine with the hydrogen atoms in a similar manner as the p_x orbital of carbon. Orbital contour plots (produced using the HyperChem^{TM} program version 8.0.10) are drawn to represent the cubic arrangement of the carbon and hydrogen atoms*

of similar energy with the following form.

$$\psi_b = \phi_{T2} = a2p_x + (1s_1 + 1s_2 - 1s_3 - 1s_4) \quad \psi_a{}^* = \phi_{T2}{}^* = a2p_x - (1s_1 + 1s_2 - 1s_3 - 1s_4)$$

$$\psi_b = \phi_{T2} = a2p_y + (1s_1 - 1s_2 + 1s_3 - 1s_4) \quad \psi_a{}^* = \phi_{T2}{}^* = a2p_y - (1s_1 - 1s_2 + 1s_3 - 1s_4)$$

$$\psi_b = \phi_{T2} = a2p_z + (1s_1 - 1s_2 - 1s_3 + 1s_4) \quad \psi_a{}^* = \phi_{T2}{}^* = a2p_z - (1s_1 - 1s_2 - 1s_3 + 1s_4)$$

Figure 5.38 shows the molecular orbital energy level diagram and the UV photoelectron spectrum for CH_4. The three HOMOs of methane show vibrational fine structure, indicating that these are bonding orbitals. The LUMO is the antibonding orbital, $a_1{}^*$, whose energy is taken from [22, 23]. Although the oxidation of CH_4 by O_2 ($E_{SOMO} = -0.44$ eV) is thermodynamically favorable, the stable HOMO energy and the unstable LUMO energy (large HOMO–LUMO gap) indicate that methane is a stable molecule. Thus, activation of both CH_4 and O_2 (see Section 10.6.1) are required in biochemical reactions. The same is also true for the ammonium ion. The oxidation of both methane and ammonium ion by chemical and/or microbial mediation in the environment

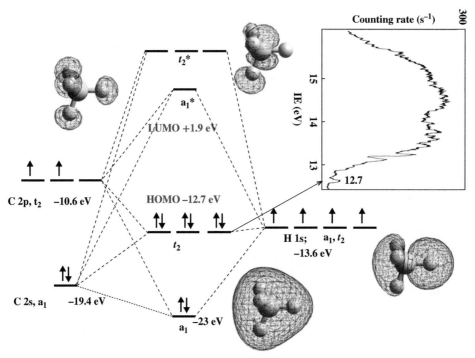

Figure 5.38 *UV photoelectron spectrum of the HOMO and molecular orbital energy level diagram for methane. Orbital contour plots (produced with the HyperChem program version 8.0.10) are drawn along one of the C_3 axes; – in this case the z-axis from Figure 5.37 (Source: Turner 1970 [6], Figure 6.1. Reproduced with permission from Wiley, Inc.)*

is an area of intense study. SO_4^{2-} would have a similar molecular orbital energy level diagram, but the wave functions would differ as the oxygen atoms would use p orbitals rather than s orbitals for bonding with sulfur.

5.6.2 NH_3 (C_{3v})

The total number of atomic orbitals available for combination to form molecular orbitals is seven (carbon 2s and three 2p atomic orbitals plus a 1s orbital for each of the three hydrogen atoms). Three sigma bonding orbitals, three sigma antibonding , and a nonbonding orbital are predicted. Again a convenient way to look at all bonding and antibonding interactions is to put the nitrogen atom at the center of the cube and the three hydrogen atoms with the lone pair of electrons at alternating vertices of the cube. Figure 5.39 shows the generalized wave functions for the seven possible atomic orbital combinations to form the molecular orbitals for NH_3, and these agree with the group theory analysis in Section 4.6.4.3. As in methane, the s orbitals from the three hydrogen atoms and the one nitrogen atom combine to form one sigma bond, $1a_1$, and one sigma antibond, $3a_1$. The p_x and p_y orbitals interact with the three s orbitals from the hydrogen atoms to give a degenerate set of bonding orbitals, 1e, and a degenerate set of antibonding orbitals, 2e.

The nitrogen p_z orbital (transforms as A_1) is on the C_3 major axis of the molecule and can interact with the three hydrogen atoms to produce the HOMO in a similar manner as the $3a_1$ molecular orbital produced for the water molecule (the p_z orbital for the O atom with the two hydrogen atom s orbitals in Figure 5.35). Although

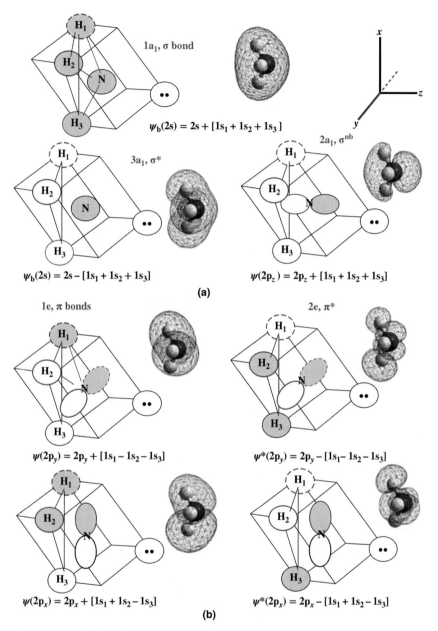

Figure 5.39 *The atomic orbital combinations giving the seven molecular orbitals for* NH_3 *and their orbital contours (0.05 level) produced using the HyperChem program version 8.0.10). (a) The "a" orbital set. (b) The "e" orbital set. Note the z or C_3 axis coincides with the p_z orbital of ammonia (C_{3v}) and is similar to the z-axis for methane in Figure 5.37. Coefficients for the wave functions are omitted for simplicity*

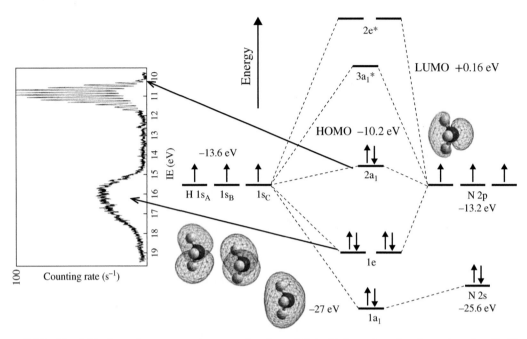

Figure 5.40 *The UV photoelectron spectrum, molecular orbitals (0.1 contour level produced with Hyper-Chem™ program version 8.0.10), and the molecular orbital energy diagram for* NH_3 *(Source: Turner 1970 [6], Figure 14.1. Reproduced with permission from Wiley, Inc.)*

the HOMO is anticipated to be primarily a nonbonding orbital that can donate a pair of electrons, the UV photoelectron spectrum in Figure 5.40 shows significant fine structure, which indicates that the orbital has bonding properties. The HOMO $2a_1$ molecular orbital has an energy of -10.2 eV, which is less stable than the HOMO of water. Thus, NH_3 is a pyramidal molecule and a better electron donor than H_2O. The LUMO, $3a_1$, is from the antibonding combination of all the s orbitals and is unstable with an energy estimated as $+0.16$ eV based on the solvated electron in ammonia [24].

5.6.3 BH_3 and the Methyl Cation, CH_3^+ (D_{3h})

5.6.3.1   *The Hypothetical* BH_3

The molecule BH_3 does not exist as a discrete entity, but exists as a dimer known as diborane B_2H_6, which has the two B atoms in BH_2 groups bridged by two hydrogen atoms in two open three-center two-e bonds (Section 5.5.1.1). Nevertheless, BH_3 is more convenient to describe in molecular orbital terms relative to the boron halides, BX_3, as BH_3 has only hydrogen s orbitals that bind with boron whereas the halide p orbitals can also bind with boron. BH_3 is also an electron-deficient compound as the total number of valence electrons available for bonding total only six. Thus, the theoretical BH_3 is a planar molecule in comparison to NH_3, which has a pyramidal shape. Figure 5.41 shows the combinations of the seven atomic orbitals to produce the seven molecular orbitals for BH_3, which agree with the group theory analysis in Section 4.6.4.4. In this case, the z axis is perpendicular to the plane of the paper and the p_z orbital is not involved in σ bonding with the H atoms. It is a vacant nonbonding orbital, $1a_2''$, that can readily accept a pair of electrons from a Lewis base or electron donor. As in NH_3, the s orbitals from the three hydrogen atoms and the one nitrogen atom

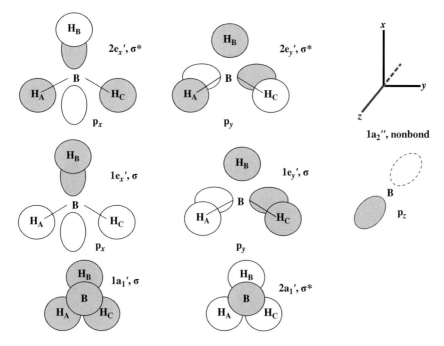

Figure 5.41   *The atomic orbital combinations for the seven molecular orbitals of* BH_3 *and* CH_3^+

combine to form one sigma bond, $1a_1'$, and one sigma antibond, $2a_1'$. The p_x and p_y orbitals interact with the three s orbitals from the hydrogen atoms to give a degenerate set of σ bonding orbitals, $1e'$, and a degenerate set of σ antibonding orbitals, $2e'$. The generalized wave functions are similar to those for NH_3 except for the coefficients and the fact that the overlap is more directional.

Figure 5.42 shows the molecular orbital energy diagram for the hypothetical BH_3 as well as the orbital contour plots of the molecular orbitals at the 0.05 contour level. For the boron trihalides, the halogen atoms can also use p orbitals for bonding with boron. In the earlier discussion on the use of the VSEPR and the Lewis octet rule (Figure 5.1), the possibility of π bonding between the p_z orbital from the boron atom and the p orbitals from the fluorine atoms in BF_3 was noted. This type of (p–p) π bonding is similar to that described for the homonuclear and heteronuclear diatomic molecules, and results in a stronger bond between the boron atom and the halide atoms.

As in the linear case for BeH_2 and the angular case for water, a Walsh **correlation diagram** can be derived for the planar BH_3 and a pyramidal NH_3 case. The main feature is the way that the p_z orbital is a purely non-bonding orbital for the planar case as it is not occupied with electrons whereas the p_z orbital in the pyramidal case is occupied with electrons and interacts with the three hydrogen atomic orbitals to give a more stable molecular orbital.

5.6.3.2 CH_3^+

The methyl cation (CH_3^+) is isoelectronic with BH_3 and has a similar molecular orbital energy level diagram and orbital contour plots for the seven molecular orbitals. Figure 5.43 shows the major tetrahedral and planar species for compounds and ions that have the central atom from the second row of the periodic table binding with hydrogen atoms. The tetrahedral species are stable and unreactive when compared with the pyramidal

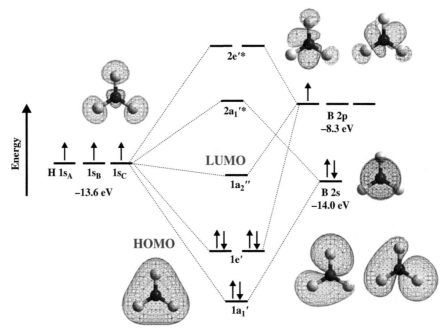

Figure 5.42 *Molecular orbital energy diagram for* BH_3 *and orbital contour plots (0.10 contour level produced with the HyperChem^{TM} program version 8.0.10) for the seven molecular orbitals*

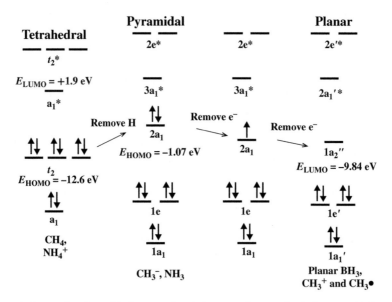

Figure 5.43 *The relative molecular orbital energy level diagrams for tetrahedral, pyramidal, and planar species. The energies for the HOMO and LUMO orbitals are given for* CH_3^- *and* CH_3^+ *with the appropriate number of electrons filling the molecular orbitals;* CH_3^+ *will only have six electrons whereas the* $CH_3\bullet$ *will have seven electrons*

species that can donate a pair of electrons and the planar species which can accept a pair of electrons. Note that for the tetrahedral molecules the HOMO is very stable and the LUMO is unstable. For the pyramidal CH_3^- species, the HOMO is slightly stable but more able to donate a pair of electrons than even the halide anions (see Figure 5.17 for HOMO energies). For the planar CH_3^+ species, the LUMO is very stable and will accept electrons readily. The methyl radical ($CH_3\bullet$), which has one more electron than CH_3^+, has been experimentally determined to be a planar species whereas other seven electron chemical species can have a pyramidal shape [25].

5.6.3.3 *Methylation Reactions*

Methyl lithium is a versatile reagent used in the laboratory, but in the environment, methyl groups enter into several different kinds of reactions as shown below. The second reaction shows how acetate under appropriate light conditions can release carbon dioxide and form a methyl anion that can be transferred to a metal cation. Methyl halides can transfer CH_3^+ to a metal. The last reaction shows the unprotonated and water soluble zwitterion dimethylsulfoniopropionate (DMSP), which can decompose to dimethyl sulfide and acrylic acid. DMSP is produced by salt marsh plants such as *Spartina Alterniflora* and by phytoplankton as **alleochemicals** (reagents that are released as toxins so other organism do not eat it) and as reagents to regulate osmotic pressure. DMSP is also an excellent reagent to transfer CH_3^+ to metals and nonmetals alike. Dimethyl sulfide (DMS) and other methylated compounds are formed by natural processes involving DMSP, and the transport of methylated compounds from the ocean surface to the atmosphere is considered important in forming clouds; e.g., DMS oxidizes to sulfate nanoparticles that create **cloud condensation nuclei** (CCN) or albedo [26].

$$Li^+CH_3^- + Hg^{2+} \rightarrow CH_3Hg^+ + Li^+$$

$$CH_3COO^- + Hg^{2+} + h\upsilon \rightarrow CH_3Hg^+ + CO_2 \text{ (formally methyl anion transfer)}$$

$$CH_3^+I^- + Hg \rightarrow CH_3Hg^+ + I^- \text{ (formally methyl cation transfer)}$$

This is an example of an **oxidative addition reaction** (Section 9.12); the E_{LUMO} is −0.2 eV for CH_3I.

$$(CH_3)_2S^+ - CH_2CH_2COO^-(DMSP) \rightarrow (CH_3)_2S + H_2C = COOH$$

5.7 Triatomic Compounds and Ions Involving π Bonds (A_3, AB_2, and ABC)

5.7.1 A_3 Linear Species

For triatomic compounds and ions involving main group elements of row 2 and below in the periodic table, there are a total of four atomic orbitals (one s and three p outermost orbitals) from each atom that are used to produce molecular orbitals. The best way to demonstrate the combination of these atomic orbitals is to use the linear azide ion, N_3^-, without any mixing or hybridization of the nitrogen 2s and 2p orbitals prior to forming molecular orbitals. The azide ion has a total of 16 electrons including the negative charge; it is an extremely strong base and is used often as a metabolic inhibitor to kill microbes and thus prevent bacterial mediation of a reaction of environmental significance [27]. Figure 5.44 is split into three sections to show the linear combination of all these atomic orbitals to form molecular orbitals; the wave functions are provided, but coefficients are not necessary in this case as all the atoms in this anion are from nitrogen. The lower section is the combination of the s orbitals from three nitrogen atoms to form a sigma bonding, antibonding, and nonbonding set of molecular orbitals. This arrangement is similar to the linear or open three-center combination of atomic

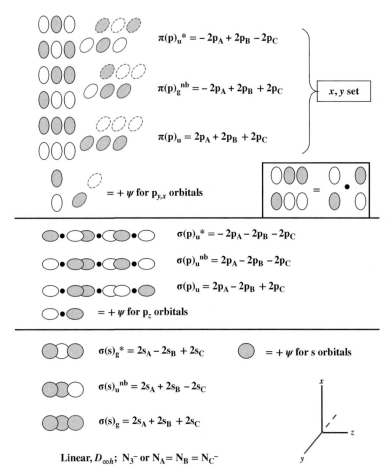

Figure 5.44 *Combination of the 12 atomic orbitals from three nitrogen atoms to form the four bonding, four nonbonding, and four antibonding molecular orbitals of the azide ion ($D_{\infty h}$) with no s–p mixing on the central N atom. A positive wave function corresponding to a particular orientation of an atomic orbital is defined in each of the three sections*

orbitals for the molecular cation, H_3^+. Another sigma combination of orbitals is given in the central section of Figure 5.44 where p_z orbitals, which are on the major rotational axis (C_∞), are used to form a bonding, antibonding and nonbonding set of molecular orbitals. This arrangement is similar to BeH_2 except that the hydrogen s atomic orbitals are now replaced by nitrogen p_z orbitals. *In writing generalized wave functions, it is important to specify the direction of any p orbital so that the sign of the wave function is known.* The black dots in the central portion of Figure 5.44 indicate the nitrogen atoms. These two σ sets account for six of the twelve molecular orbitals to be formed, and are on the internuclear axis. The p_x (and p_y) orbitals, which are perpendicular to the bond axis, are now combined to form a π bonding, antibonding and nonbonding set of molecular orbitals. For the π case, the bonding interaction has all three orbitals overlapping so that the electrons are delocalized across all three atoms; for the nonbonding interaction, two adjacent orbitals overlap

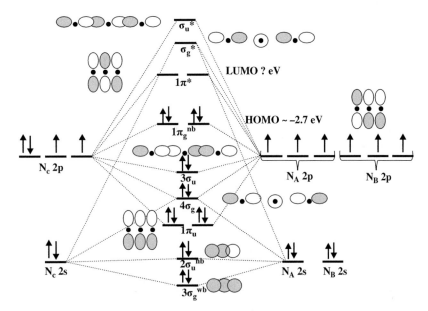

Figure 5.45 *The relative energy level diagram for the azide ion showing the atomic orbitals with s–p mixing on the central N atom (see next section) used to create the molecular orbitals*

and one of the terminal orbitals gives an antibonding interaction so that $S = 0$ overall; for the antibonding interaction, the orbital for the central atom does not overlap with either of the terminal atoms so that $S < 0$. The upper portion of Figure 5.44 shows another convention for the drawing of the three atomic orbitals in a nonbonding interaction (boxed set); here, only the terminal atomic orbitals are drawn and the central atomic orbital is not drawn and is represented by a black dot. In this book, the three atomic orbitals will always be drawn when a reaction with another species occurs so that it will be easier to ascertain chemical reactivity.

Figure 5.45 shows the energy level diagram for N_3^- with the order of the molecular orbitals, which are consistent with the X-ray photoelectron spectrum [28]. The azide ion is a good electron donor as the HOMO π nonbonding orbitals have an energy between the energies of the halide anions and the hydroxide ion (see Appendix 5.2). The energy of the LUMO has not been determined experimentally, but putting an electron into the antibonding π LUMO would result in a -2 charge and a loss of bonding.

Another important anion is the well-known tri-iodide ion, I_3^-. The same atomic orbitals are used to form the same 12 molecular orbitals shown above for the azide ion, but I_3^- has a total of 22 electrons. I_3^- has a bond order of one, indicating that two bonding electrons are being shared across the three iodine atoms (open three-centered two-electron bond similar to H_3^+). I_3^- is also a linear molecule as predicted by VSEPR theory, where the central iodine atom would have 7 electrons plus the 1 electron for the negative charge with each of the other two iodine atoms donating 1 electron to the central atom for a total of 10 electrons around the central iodine atom. To minimize the repulsive forces between the lone pairs of electrons, the three lone pairs would occupy the equatorial positions of a trigonal bipyramidal structure and the two terminal iodine atoms would occupy the axial positions. Figure 5.46 shows the idealized energy level diagram for I_3^-. The energies for both the HOMO and the LUMO are positive, indicating that these are not particularly stable orbitals and that I_3^- should be an electron donor.

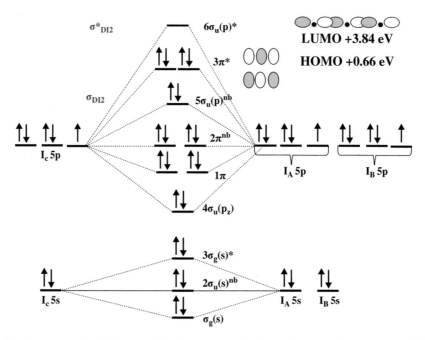

Figure 5.46   *The* I_3^- *energy level diagram with the atomic orbitals used to create the HOMO and the LUMO (no s–p mixing). The* $5\sigma_u(\sigma_{DI2})$ *and* $6\sigma_u(\sigma^*_{DI2})$ *orbitals are labeled to compare with Figure 7.3*

5.7.2   AB$_2$ Linear Species CO$_2$ (COS and N$_2$O)

Before considering other homonuclear triatomic species such as ozone that are bent molecules, one of the most important molecules of environmental interest, carbon dioxide, is now discussed. CO_2 also has 16 electrons and is isoelectronic with the azide ion; thus it is a linear molecule. However, the energies of the atomic orbitals for oxygen are more stable than those for carbon as shown in Figure 5.47 and as noted above for CO. The combination of the 12 atomic orbitals is thus slightly different from what was discussed for the azide ion. In the absence of any s–p mixing or hybridization, the two 2s orbitals of oxygen are expected to be nonbonding orbitals. Thus for the central carbon atom, the 2s and $2p_z$ orbitals are expected to interact with the $2p_z$ orbitals from the two oxygen atoms; these four atomic orbitals result in two bonding molecular orbitals [$\sigma_g(s)$ and $\sigma_u(z)$] and two antibonding molecular orbitals [$\sigma_g(s)^*$ and $\sigma_u(z)^*$]. The generalized wave functions for these molecular orbitals are given below, and the coefficients are omitted for simplicity (a and b refer to the two different oxygen atoms). These atomic orbital combinations would be similar to the molecule BeF$_2$ and other beryllium dihalides.

$$\sigma_g(s) = 2s_C + (2p_{za} - 2p_{zb}) \quad \sigma_g(s)^* = 2s_C - (2p_{za} - 2p_{zb})$$
$$\sigma_u(z) = 2p_{zC} - (2p_{za} + 2p_{zb}) \quad \sigma_u(z)^* = 2p_{zC} + (2p_{za} + 2p_{zb})$$

Six of the 12 molecular orbitals have now been described, and the π bonding, nonbonding, and antibonding sets for the p_x and the p_y orbitals remain. These are essentially identical to that described above for the azide ion. The major difference for these orbitals is that the HOMO π nonbonding orbitals have more contribution from the oxygen atomic p orbitals than from the carbon atomic p orbital as a result of the electronegativity difference between carbon and oxygen. The energy of the HOMO is -13.77 eV and indicates that it is also

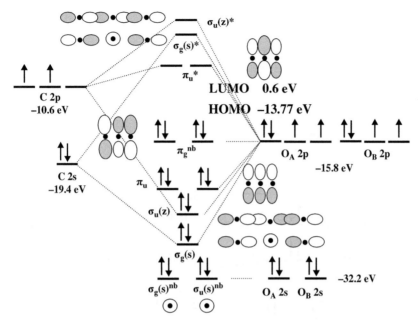

Figure 5.47 *Molecular orbital energy level diagram and possible atomic orbital combinations for* CO_2. *The diagram and the atomic orbitals used for combination show s–p mixing. Figure 5.48 shows the molecular orbitals and their assignment based on ESCA data*

a very stable molecular orbital. The LUMO π^* orbitals have more contribution from the carbon atomic p orbital than from the oxygen atomic p orbitals, and are not very good electron acceptors because the LUMO energy is positive at 0.6 eV [29]. Thus, CO_2 is a poor electron acceptor, which has consequences for its solubility in water and its fixation into organic carbon via photosynthesis and chemosynthesis.

The UV photoelectron spectrum for CO_2 is given in Figure 5.48. The HOMO π_g orbitals are assigned at -13.77 eV, and the peak has very little fine structure so these are nonbonding orbitals (π_g^{nb}). The π_u bonding orbitals are centered near -17.5 eV and have extended vibrational fine structure so are bonding orbitals. The four σ orbitals are assigned as shown, but require further comment. The σ_g(s) (-19.4 eV) and σ_u(z) (-18.1 eV) orbitals are single peaks indicating that they are nonbonding orbitals. The σ_g(s) and the σ_u(z) orbitals give a continuous band centered near -38 eV (not shown), which is significantly lower in energy than the 2s atomic orbitals for the oxygen atoms at -32.2 eV. Because the 2s orbitals of the oxygen atoms are much more stable than the 2s orbital of the carbon atom, some s–p mixing or hybridization of the oxygen atoms must occur prior to their combination with the 2s and the $2p_z$ orbitals from the carbon atom. As noted above for carbon monoxide, an equal mixing of the oxygen 2s and $2p_z$ orbitals would result in an energy of -24 eV for the hybridized orbitals, which should permit combination of these carbon and oxygen orbitals. Thus, the σ_g(s) and σ_u(s) orbitals at -38 eV are assigned as σ bonding orbitals from the combination of the carbon 2s and $2p_z$ orbitals with the hybridized 2s and $2p_z$ oxygen atomic orbitals rather than nonbonding oxygen 2s orbitals. Thus, carbon dioxide has a total bond order of four.

Carbonyl sulfide, COS, and nitrous oxide, N_2O, are isoelectronic with CO_2 and have similar UV photoelectron spectra to CO_2 indicating that the molecular orbital assignments are also similar. Both COS and N_2O are important environmental gases and N_2O is formed during denitrification. All of these compounds are stable. CO_2 and N_2O along with CH_4 are among the most important **greenhouse gases** affecting climate on the earth.

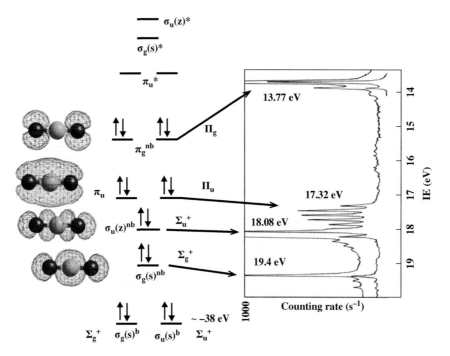

Figure 5.48 CO_2 *UV photoelectron spectrum with the assignment of the molecular orbitals (including group theory notation; produced with the HyperChem™ program version 8.0.10) and the filling of the 16 electrons into the orbitals (Source: Turner 1970 [6], Figure 4.16. Reproduced with permission from Wiley, Inc.)*

5.7.3 O_3, NO_2^-, and SO_2 (Angular Molecules)

Ozone, sulfur dioxide, and the nitrite ion are important in environmental processes. All of these chemical species are diamagnetic and have a total of 18 valence electrons. As a result they are angular molecules as noted previously using the Lewis octet rule and the valence shell electron pair repulsion theory. The contribution of the atomic orbitals to the molecular orbitals and the molecular orbital diagram of the homonuclear ozone molecule will be similar to that of the azide ion; however, as in the sigma bonding triatomic linear case for BeH_2 changing to the angular case for water or H_2S, the π type orbitals are no longer degenerate on a change from linear geometry. Figure 5.49 shows the Walsh correlation diagram for the molecular orbitals on changing from linear to angular geometry with s–p mixing or hybridization. The NO_2^- is given on the right and is consistent with the group theory analysis in Section 4.7.2.1.

Figure 5.50 gives the relative molecular orbital energy level diagrams for O_3, NO_2^-, and SO_2. The contribution of the nitrogen (sulfur) atomic orbital with the terminal oxygen atomic orbitals to form the molecular orbitals for the heteronuclear NO_2^- (SO_2) will be similar to that of carbon in carbon dioxide as the nitrogen atom has lower electronegativity than the oxygen atoms. If the x and z coordinates are exchanged for C_{2v} symmetry as in water, then the b_1 and b_2 orbital labels exchange also.

Because the degeneracy of the π^* orbitals is removed as shown in Figure 5.50, O_3, NO_2^-, and SO_2 are diamagnetic. The $1a_2$ and $3b_1$ orbitals are similar to the π^{nb} orbitals in carbon dioxide whereas the $4a_1$ and $2b_2$ are similar to the π^* orbitals in carbon dioxide. For NO_2^-, the HOMO has an energy of -2.5 eV, and because the two electrons are delocalized across the three atoms in the nitrite ion, it is possible to have the

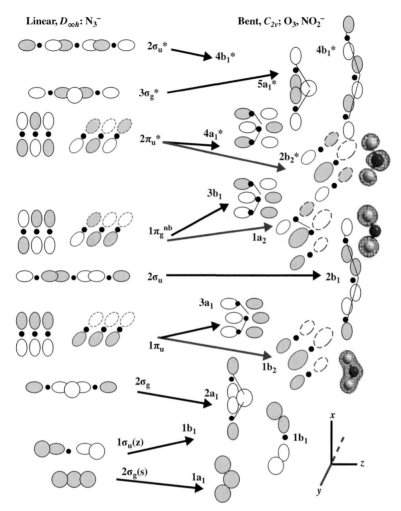

Linear, $D_{\infty h}$: N_3^-

Bent, C_{2v}; O_3, NO_2^-

$2\sigma_u^*$ → $4b_1^*$ $4b_1^*$

$3\sigma_g^*$ $5a_1^*$

$2\pi_u^*$ $4a_1^*$ $2b_2^*$

$3b_1$

$1\pi_g^{nb}$ $1a_2$

$2\sigma_u$ → $2b_1$

$3a_1$

$1\pi_u$ $1b_2$

$2\sigma_g$ $2a_1$

$1b_1$ $1b_1$

$1\sigma_u(z)$

$2\sigma_g(s)$ $1a_1$

x
z
y

Figure 5.49 *Walsh-type correlation diagram for linear to angular triatomic molecules with π bonding (hybridization or s–p mixing is included). The direction of the arrows indicates the relative stability of the orbitals for the bent molecules relative to the orbitals for the linear molecules. The orbitals derived from p_y orbitals are perpendicular to the plane of the paper*

nitrogen atom or one of the oxygen atoms donate a pair of electrons and form a bond with an electron acceptor such as a metal cation (**linkage isomers;** Section 8.4.4). It is also possible to have another atom bridge across an N–O bond (Figure 8.4). For SO_2, the energies of the HOMO and the LUMO are -12.5 eV and -1.1 eV, respectively; SO_2 can also bind metals as described above for NO_2^-. For O_3, the energies of the HOMO and the LUMO are -12.43 eV and -2.3 eV, respectively. The stable LUMO energies make O_3 and SO_2 excellent electron acceptors or oxidants. All three are potential oxygen atom transfer reagents. For example, in Chapter 2 (Section 2.7.2.1), UV dissociation of O_3 was shown to lead to $^3O_2$ and O atoms, ($|\underline{O}|$ in Equation 2.10). Over 90 % of the O atoms exist in the diamagnetic electronic configuration $2s^2 2p_x^2 2p_z^2 2p_y^0$ with the atomic term symbol ^1D. The empty orbital makes the O atom a powerful oxidant that can accept one or two electrons readily.

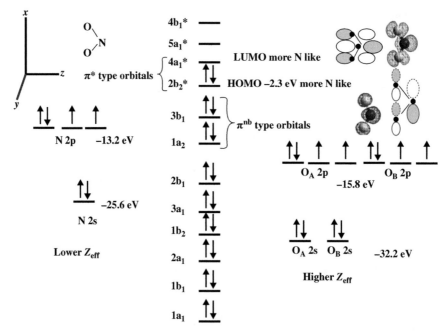

Figure 5.50 *The molecular orbital diagram and possible atomic orbital combinations for the HOMO and LUMO of NO_2^- (O_3 and SO_2 are similar); the z-axis is the C_2 rotational axis. Note that the $3b_1$ and $4a_1$ orbitals are in the yz plane (the $2b_2^*$ orbital is perpendicular to the plane of the paper)*

NO_2^- forms on addition of an electron to the **nitrogen dioxide** radical, NO_2, which is a key atmospheric pollutant produced during the combustion of fossil fuels.

$$NO_2 + e^- \rightarrow NO_2^-$$

NO_2 has a SOMO $2b_2^*$ orbital (-2.273 eV), which allows NO_2 to be an effective oxidant in the atmosphere (Section 2.7.2). The buildup of NO_2 in a localized area due to low winds, which would remove NO_2 to another location, gives that atmosphere an orange red (smog) color. The coloration of concentrated **nitric acid** on exposure to light is also due to the formation of NO_2, which enhances the oxidizing properties of nitric acid. In the atmosphere, NO_2 also reacts readily with OH• to form HNO_3. Equations 2.31–2.33 show that SO_2 also reacts in three steps with OH•, O_2, and H_2O to form sulfuric acid, H_2SO_4. The dissolution of NO_2^-, NO_2, and SO_2 during rain events leads to acid rain that falls on land and waterways. The NO_x species also add fixed nitrogen as fertilizer to those systems, which can lead to phytoplankton growth including **algal blooms**.

5.8   Planar Species (BF_3, NO_3^-, CO_3^{2-}, SO_3)

These compounds and ions are 24 electron systems with 16 orbitals. BF_3 and SO_3 are electron acceptors whereas NO_3^- and CO_3^{2-} are electron donors. The combination of atomic orbitals to give the molecular orbitals for CO_3^{2-} is described in the group theory section (chapter 4.7.2.2). The molecular orbital energy level diagram along with the shapes of the MOs is in Figure 5.51. For CO_3^{2-}, the HOMO is the π_{nb} or

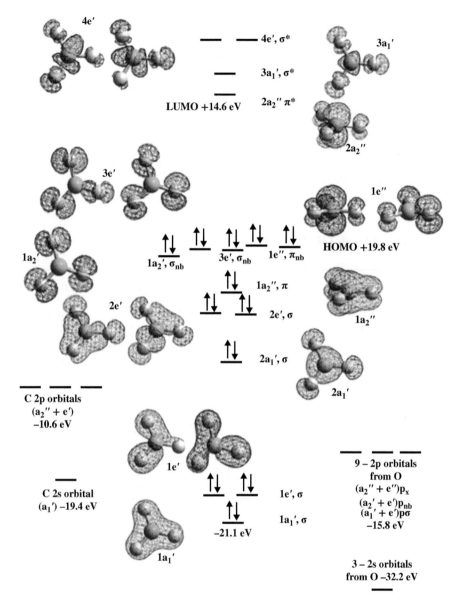

Figure 5.51 *Carbonate ion energy level diagram with atomic and molecular orbitals. The group theory nota-tion indicates the atomic orbitals used for bonding. Molecular orbitals and their energies calculated with the Hyperchem™ program version 8.0.10 (AM1 level; 0.05 contour level)*

$1e''$ degenerate orbitals, which bind with H^+ in an acid–base titration as they are positive in energy, and the electron density is maximized on one of the oxygen atoms. The unstable LUMO is the $2a_2''^*$ or π^* orbital. This analysis is what would be predicted using Lewis structures and the VBT.

The orbital arrangement is similar for NO_3^-, but the energies are different due to the higher Z_{eff}/r for N over C. The NO_3^- HOMO (π_{nb} or $1e''$) is more stable at -3.11 eV, and the $2a_2''^*$ or π^* is the LUMO

(+5.5 eV). SO_3 has the same HOMO and LUMO orbitals, but the LUMO is now calculated to be stable with an energy of -3.0 eV (the experimental value is -1.7 eV; Appendix 5.2) indicating that it will accept electrons. This is consistent with the reaction of SO_3 with OH^- to form HSO_4^-, and its two-electron reduction to form sulfite, SO_3^{2-}. This MO diagram can explain the formation of a fourth bond between BF_3 and other BX_3 compounds with amines (X_3B-NR_3). However, BF_3 is less reactive with amines than the other BX_3 compounds. The LUMO energy for BF_3 is near 0 eV, likely due to the strong π bonding between B and F, which is consistent with the HOMO energy for BF_3 being very stable at -15.95 eV [30]. The HOMO energies for the other boron halides range from -9.36 to -11.62 eV. AM1 level MO calculations of BF_3 also suggest that it may be less reactive than the other BX_3 compounds because the $3a_1'^*$ (σ^*) orbital is lower in energy than the $2a_2''^*$ (π^*) orbital. The $3a_1'^*$ (σ^*) orbital has significant B 2s character but provides less overlap for an attacking nucleophile or Lewis base.

APPENDIX 5.1

Bond energies for selected bonds [31–33]

Bond	kJ mol^{-1}	Bond	kJ mol^{-1}	Bond	kJ mol^{-1}	Bond	kJ mol^{-1}
H–H	432.0	C–C	345.6	N–N (N_2H_4)	247	Si–O	452
H–O	458.8	C=C	602	N=N	418	Si=O	642
H–C	411	C≡C	835.1	N≡N	941.69	Ge–O	350
H–N	386	C–N	304.6	P–P (P_4)	201	N–O	201
H–P	322	C=N	615	As–As (As_4)	146.4	N=O	607
H–S	363	C≡N	887	O–O (H_2O_2)	207.1	P–O	335
H–Se	276	C–P	264	O=O	493.59	P=O	544
H–Te	238	C–O	357.7	S–S (S_2)	421.58	As–O	301
H–F	565	C=O	798.9	S–S (S_8)	226	S–O	468.3
H–Cl	428.02	C≡O	1072	S–S (H_2S_2)	268	S=O	532.2
H–Br	362.3	C–S	272	S=S	424.7	O–F	189.5
H–I	294.6	C=S	573	Se–Se (Se_6)	172	O–Cl	218
		C–Si	318	Se=Se	272	O–Br	201
Li–F	573	Si–Si	222			O–I	201
Li–Cl	464	Ge–Ge	188	O–CH_3	280		
Li–Br	418	C–CH_3	293	N–CH_3	272	F–F	154.8
Li–I	347	N–CH_3	272	S–CH_3	293	Cl–Cl	239.7
Na–F	477	O–CH_3	280	Zn–CH_3	188	Br–Br	190.16
Na–Cl	408	F–CH_3	452	Cd–CH_3	226	I–I	148.95
Na–Br	363	Cl–CH_3	334	Hg–CH_3	247		
Na–I	304	Br–CH_3	293	Pb–CH_3	209	Hg–F	268
		I–CH_3	234			Hg–Cl	225
						Hg–Br	185
K–F	490	Rb–F	490	Cs–F	481	Hg–I	145
K–Cl	423	Rb–Cl	444	Cs–Cl	416	Be–F	632
K–Br	379	Rb–Br	385	Cs–Br	481	Be–Cl	461
K–I	326	Rb–I	331	Cs–I	315	Be–Br	372
						Be–I	289

APPENDIX 5.2

Energies of LUMOs (of oxidants) and HOMOs (of reductants) for important small molecules from several compilations [34–43]. A value for M^{2+} is estimated based on water exchange for labile complexes [44]. Negative values indicate stable orbitals. For more discussion on the 2nd EA for O and S atoms (HOMO energy O^{2-}, S^{2-}), see Refs. [43, 45, 46]

LUMO (eV)	Species	LUMO (eV)	Species	HOMO (eV)	Species	HOMO (eV)	Species
7.62	OI^-	−2.093	S_3	7.74	O^{2-}	−9.1	H_2S_2
2.3	HCN	−2.10	O_3	6.18	O_2^{2-}	−9.19	H_2O_2
2.2	N_2	−2.2	$SCN\bullet$	4.73	S^{2-}	−9.26	NO
1.9	CH_4	−2.273	$NO_2\bullet$	1.36	S_4^{2-}	−9.3	HOI
0.5	CO_2	−2.276	$ClO\bullet$	0.5	FeS	−9.39	I_2
0.0	BF_3	−2.35	Cl_2	−0.44	O_2^-	−9.538	$CH_3 - I\bullet$
−0.03	$NO\bullet$	−2.32	$HS\bullet$	−0.516	S_5^{2-}	−9.75	NO_2
−0.080	$CH_3\bullet$	−2.353	$BrO\bullet$	−1.078	HO_2^-	−10.07	CS_2
−0.16	NH_3	−2.378	$IO\bullet$	−1.825	OH^-	−10.16	NH_3
−0.2	$CH_3 - I\bullet$	−2.4	NO	−1.907	H_2S_2	−10.39	HI
−0.313	HCO	−2.50	HOI	−2.10	O_3	−10.47	H_2S
−0.47	O_2	−2.53	Br_2	−2.276	ClO^-	−10.51	Br_2
−0.5	OCS	−2.55	I_2	−2.3	NO_2^-	−11.2	OCS
−0.6	HNO_3	−2.6	$OCN\bullet$	−2.32	HS^-	−11.49	Cl_2
−0.75	H_2O	−2.7	$N_3\bullet$	−2.353	BrO^-	−11.67	HBr
−0.8	CS_2	−3.1	F_2	−2.378	IO^-	−12.07	O_2
−1.078	$HO_2\bullet$	−3.269	C_2	−2.7	N_3^-	−12.15	C_2
−1.1	H_2S	−3.6	$NCO\bullet$	−3.06	I^-	−12.43	O_3
−1.107	SO_2	−3.7	$NO_3\bullet$	−3.11	NO_3^-	−12.5	SO_2
−1.263	PH_3	−3.82	$CN\bullet$	−3.36	Br^-	−12.7	CH_4
−1.37	CO	−4.226	$I_3\bullet$	−3.4	F^-	−12.61	H_2O
−1.47	N_2O	−5.0	H_2	−3.5	FeS_2	−12.74	HCl
−1.59	$CH_3 - I\bullet$	−15	M^{2+}	−3.61	Cl^-	−12.89	N_2O
−1.67	S_2			−3.82	CN^-	−13.59	HCN
−1.7	SO_3			−4.67	S_3^{2-}	−13.77	CO_2
−1.825	$OH\bullet$			−5.145	Li_2	−14.01	CO
−1.861	$S-CH_3$			−8.5	HOI	−14.1	CN
−1.867	$CH_3 - S\bullet$			−9.04	S_8	−15.43	H_2
−1.907	H_2S_2			−9.1	H_2S_2	−15.58	N_2
−2.00	$CNS\bullet$			−9.19	H_2O_2	−15.69	F_2
−2.077	S_8 as S			−9.26	NO	−16.01	HF

References

1. Lewis, G. N. (1916) The atom and the molecule. *Journal of the American Chemical Society*, **38**, 762–785.
2. Gillespie, R. J. (2008) Fifty years of the VSEPR model. *Coordination Chemistry Reviews*, **252**, 1315–1327.
3. Pauling, L. (1960) *The Nature of the Chemical Bond* (3rd ed.), Cornell University Press, Ithaca, NY, pp. 644.
4. Engel, T. and Reid, P. (2013) *Quantum Chemistry and Spectroscopy* (3rd ed.), Pearson Education, Boston, MA, pp. 507.
5. DeKock, R. L. and Gray, H. B. (1980) *Chemical Structure and Bonding*, The Benjamin/Cummings Publishing Co., Menlo Park, CA, pp. 491.
6. Turner, D. W., Baker, C., Baker, A. D. and Brundle, C. R. (1970) *Modern Photoelectron Spectroscopy*, Wiley-Interscience, London, pp. 386.
7. Jordan-Thaden, B and others. (2011) Structure and stability of the negative hydrogen molecular ion. *Physical Review Letters*, **107**, 193003.
8. Gimarc, B. M. (1979) *Molecular Structure and Bonding: The Qualitative Molecular Orbital Approach*, Academic Press, New York.
9. Becke, A. D. (2014) Perspective: fifty years of density-functional theory in chemical physics. *Journal of Chemical Physics*, **140**, 18A301: 1–18.
10. Schulz, G. L. (1973) Resonance in electron impact on diatomic molecules. *Reviews of Modern Physics*, **45**, 423–486.
11. Pearson R. G. (1976) *Symmetry Rules for Chemical Reactions: Orbital Topology and Elementary Processes*, Wiley-Interscience, NY, pp. 548.
12. Hoffmann, M. R. (1977) Kinetics and mechanism of oxidation of hydrogen sulfide by hydrogen peroxide in acidic solution. *Environmental Science and Technology*, **11**, 61–66.
13. Millero, F. J., Hubinger, S., Fernandez, M. and Garnett, S. (1987). Oxidation of H_2S in seawater as a function of temperature, pH, and ionic strength. *Environmental Science and Technology*, **21**, 439–443.
14. Luther, III G. W., Findlay, A. J., MacDonald, D. J., Owings, S. M., Hanson, T. E., Beinart, R. A. and Girguis, P. R. (2011) Thermodynamics and kinetics of sulfide oxidation by oxygen: a look at inorganically controlled reactions and biologically mediated processes in the environment. *Frontiers in Microbiology/Microbial Physiology and Metabolism*, **2**, 1–9, Article 62, *Front. Microbiol.* doi: 10.3389/fmicb.2011.00062, April 9.
15. Luther III, G. W. (2011) Thermodynamic redox calculations for one and two electron transfer steps: Implications for halide oxidation and halogen environmental cycling, In *Aquatic Redox Chemistry* (eds. P. G. Tratnyek, T. J. Grundl and S. B. Haderlein). American Chemical Society (ACS) books, pp. 15–35.
16. Butler, A. and Sandy M. (2009) Mechanistic considerations of halogenating enzymes. *Nature*, **460**, 848–854.
17. Oka T. (2006) Interstellar H_3^+. *Proceedings of the National Academy of Sciences*, **103**, 12235–12242.
18. Oka, T. (2012) Chemistry, astronomy and physics of H_3^+. *Philosophical Transactions of the Royal Society A*, **370**, 4991–5000. (See the entire issue for more details on H_3^+).
19. Oka T. (2013) Interstellar H_3^+. *Chemical Reviews*, **113**, 8738–8761. (See the entire issue for more details on Astrochemistry).
20. Shayesteh, A., Tereszchuk, K., Bernath, P. F. and Colin, R. (2003) Infrared emission spectra of BeH_2 and BeD_2. *Journal of Chemical Physics*, **118**, 3622–3627.

21. Coe, J. V., Lee, G. H., Eaton, J. G., Arnold, S. T., Sarkas, H. W., Bowen, K. H., Ludewigt, C., Haberland, H. and Worsnop, D. R. (1990) Photoelectron spectroscopy of hydrated electron cluster anions, $(H_2O)^-_{n=2-69}$. *Journal of Chemical Physics*, **92**, 3980–3982.

22. Yang, G., Liu, C., Han, X. and Bao, X. (2010) Charge effects on alkanes and the potential applications in selective catalysis: insights from theoretical studies. *Molecular Simulation*, **36**, 204–211.

23. Zhan, C.-G., Nichols, J. A. and Dixon, D. A. (2003) Ionization potential, electron affinity, electronegativity, hardness, and electron excitation energy: molecular properties from density functional theory orbital energies. *Journal of Physical Chemistry A*, **107**, 4184–4195.

24. Almeida, T. S., Coutinho, K., Costa Cabral, B. J. and Canuto, S. (2008) Electronic properties of liquid ammonia: a sequential molecular dynamics/quantum mechanics approach. *The Journal of Chemical Physics*, **128**, 014506.

25. Pauling, L. (1969) Structure of the methyl radical and other radicals. *Journal of Chemical Physics*, **51**, 2767–2769.

26. Vallina, S. M. and Simó, R. (2007) Strong relationship between DMS and the solar radiation dose over the global surface ocean. *Nature*, **315**, 506–508.

27. Emerson S., Kalhorn S., Jacobs L., Tebo B. M., Nealson, K. H. and Rosson, R. A. (1982) Environmental oxidation rate of manganese (II): bacterial catalysis. *Geochimica et Cosmochimica Acta*, **46**, 1073–1079.

28. Lee, T.H., Colton, R. J., White, M. G. and J. W. Rabalais (1975) Electronic structure of hydrazoic acid and the azide ion from X-ray and ultraviolet electron spectroscopy. *The Journal of the American Chemical Society*, **97**, 4845–4851.

29. Compton R. N., Reinhardt P. W. and Cooper C. D. (1975) Collisional ionization of Na, K, and Cs by CO_2, COS, and CS_2: molecular electron affinities. *Journal of Chemical Physics*, **63**, 3821–3827.

30. King, G. H., Krishnamurthy, S. S., Lappert, M. F. and Pedley, J. B. (1972) Bonding studies of compounds of boron and the group 4 elements. Part 9. Photoelectron spectra and bonding studies of halogeno-, dimethylamino-, and methyl-boranes, BX_3 and BX_2Y. *Faraday Discussions Chemical Society*, **54**, 70–83.

31. Darwent, B. deB. (1970) Bond dissociation energies in simple molecules. NSRDS-NBS 31, National Bureau of Standards.

32. Cottrell, T. L. (1958) *The strength of Chemical Bonds* (2nd ed.), Butterworths, London, pp. 317.

33. Williams, R. J. P. (1988) The transfer of methyl groups: a general introduction, In *The Biological Alkylation of Heavy Elements* (eds P. J. Craig and F. Glocking), The Royal Society of Chemistry, pp. 5–19. Special Publ. 66.

34. Hotop, H. and Lineberger, W. C. (1975) Binding energies in atomic negative ions. *Journal of Physical and Chemical Reference Data*, **4**, 539.

35. Hotop, H. and Lineberger, W. C. (1985) Binding energies in atomic negative ions:II. *Journal of Physical and Chemical Reference Data*, **14**, 731–750.

36. Andersen, T., Haugen, H. K. and Hotop, H. (1999) Binding energies in atomic negative ions:III. *Journal of Physical and Chemical Reference Data*, **29**, 1511–1533.

37. Drzaic, P. S., Marks, J. and Brauman, J. I. (1984) Electron photodetachment from gas phase molecular anions, In *Gas Phase Ion Chemistry* (ed. M. T. Bowers) Vol. 3, Academic Press, Inc., New York, pp. 167–211.

38. Radzic, A. A. and Smirnov, B. M. (1985) Reference data on atoms, molecules, and ions, In *Springer Series in Chemical Physics* Vol. 31, Springer-Verlag, Berlin, pp. 375–438.

39. Lowe, J. P. (1977) Qualitative molecular orbital theory of molecular electron affinities. *Journal of the American Chemical Society*, **99**, 5557–5570.

40. Lias, S. G., Bartmess, J. E., Liebman, J. F., Holmes, J. L., Levin R. D. and Mallard, W. G. (1988) Gas-phase ion and neutral thermochemistry. *Journal of Physical and Chemical Reference Data*, **17**, Supplement No. 1. American Chemical Society, American Institute of Physics, U.S. National Bureau of Standards.

41. Rienstra-Kiracofe, J. C., Tschumper, G. S., Schaeffer, H. F., Nandi, S. and Ellison, G. B. (2002) Atomic and molecular electron affinities: photoelectron experiments and theoretical computations, *Chemical Reviews*, **102**, 231–282.

42. Refaey, K. M. and Franklin, J. L. (1976) Endergonic ion-molecule-collision processes of negative ions. III. Collisions of I^- on O_2, CO, and CO_2. *International Journal of Mass Spectrometry and Ion Physics*, **20**, 19–32.

43. Pearson, R. G. (1991) Negative electron affinities of nonmetallic elements. *Inorganic Chemistry*, **30**, 2856–2858.

44. Pearson, R. G. (1988) Absolute electronegativity and hardness: application to Inorganic Chemistry. *Inorganic Chemistry*, **27**, 734–740.

45. Holbrook, J. B., Sabry-Grant, R., Smith, B. C. and Tandel, T. V. (1990) Lattice enthalpies of ionic halides, hydrides, oxides, and sulfides: second-electron affinities of atomic oxygen and sulfur. *Journal of Chemical Education*, **67**, 304–307.

46. Harding, J. H. and Pyper, N. C. (1995) The meaning of the oxygen second-electron affinity and oxide potential models. *Philosophical Magazine Letters*, **71**, 113–121.

6

Bonding in Solids

6.1 Introduction

At the beginning of the 20th century, inorganic chemists and geochemists began to realize that the external symmetry found in solids was related to the arrangement of atoms in the solid. Soon they centered much of their attention on common minerals and ionic solids, ultimately leading to a comprehensive description of ionic bonding and how cations and anions packed in a crystal. With the advent of x-ray diffraction techniques, the seven crystal classes and the fourteen unit cells (**Bravais** lattices) for solids were described. Because many simple ionic solids (M^+X^-) assume one of the three cubic unit cells, which gave the length of the unit cell and the internuclear distance, it was possible for Victor Goldschmidt and Linus Pauling to produce the first summary of ionic radii for simple cations and anions. The first part of this chapter discusses band theory and the bonding of atoms in metals and semiconductors, which is a natural extension of covalent bonding in the previous chapter. This is followed by how atoms pack in a crystal and a description of the crystal classes and their unit cells. Ionic bonding models are then discussed.

6.2 Covalent Bonding in Metals: Band Theory

6.2.1 Atomic Orbital Combinations for Metals

The bonding in pure metals can be simply described using the molecular orbital approach. In the previous chapter, the discussion was limited to simple molecules with less than six or seven atoms, but this is now extended to describe **band theory**. Figure 6.1 shows the linear arrangement for "s" orbitals when two, three, and four atoms combine to produce bonding (ψ_b), nonbonding (ψ_n), and antibonding (ψ_a) orbitals. When the number of atoms is odd numbered, nonbonding orbitals are produced.

Figure 6.2 shows the relative energies for the combinations of "s" orbitals up to 20 atoms. In a solid with an infinite number of atoms, all the orbitals will overlap to form a **band** of orbitals (right of Figure 6.2). Half of the orbitals within the band are bonding orbitals and the other half are antibonding orbitals. If the band is only partially filled, the metal will conduct electricity when an applied electric field is imposed upon it. The smooth parabola-like curve describes the relative number of orbitals at a given energy (see density of states (DOS) Section 6.2.5). The larger number of orbitals at the center of the band is due to the larger number of orbital combinations near the center.

Inorganic Chemistry for Geochemistry and Environmental Sciences: Fundamentals and Applications, First Edition. George W. Luther, III.
© 2016 John Wiley & Sons, Ltd. Published 2016 by John Wiley & Sons, Ltd.
Companion Website: www.wiley.com/go/luther/inorganic

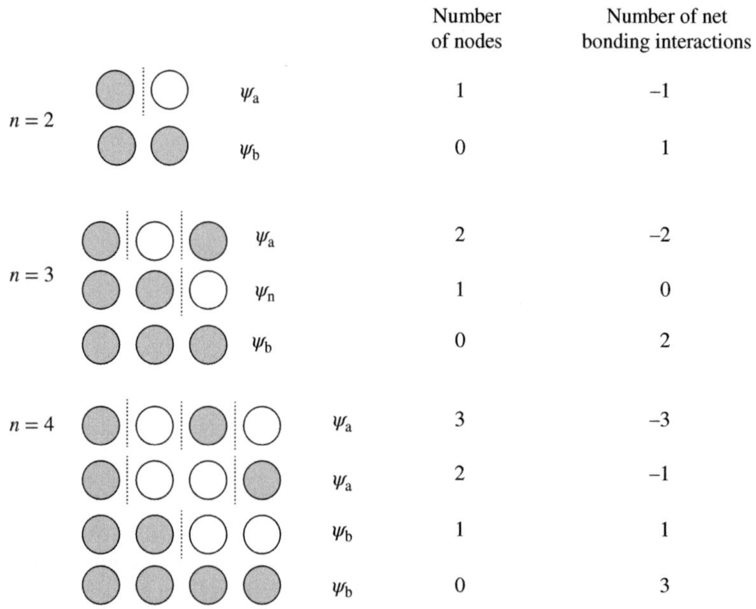

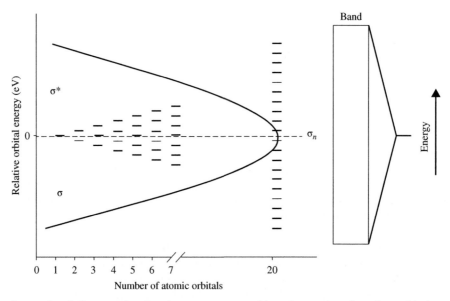

			Number of nodes	Number of net bonding interactions
$n = 2$		ψ_a	1	−1
		ψ_b	0	1
$n = 3$		ψ_a	2	−2
		ψ_n	1	0
		ψ_b	0	2
$n = 4$		ψ_a	3	−3
		ψ_a	2	−1
		ψ_b	1	1
		ψ_b	0	3

Figure 6.1 *The arrangement of "s" orbitals in a one-dimensional linear array. The dotted lines indicate a node and a net antibonding interaction*

Figure 6.2 *Energy level diagram showing the arrangement of bonding and antibonding orbitals to produce a band of orbitals*

6.2.2 Metal Conductors

For metals and elements that primarily have covalent bonding, Figure 6.3a shows the general arrangement of the bands for all the s and p orbitals for metals in the third row of the periodic table. The 1s, 2s, and 2p bands are core bands that are closer to the nucleus. The energies of the 2s and 2p bands are significantly different (e.g., see Figure 3.19) so that there is no overlap of these bands; thus, there is a band gap. In this example, the 3s band is completely full (Mg), but it overlaps with the 3p band resulting in mixing and combining of the orbitals so that there is no band gap and the conduction of electricity is possible. Figure 6.3b shows the upper band for the partially filled 2s band for lithium, and Figure 6.3c shows the overlap of the 3s and 3p bands for aluminum. In all cases, there is no energy gap between the higher energy bands so these are called **conduction** bands.

6.2.3 Semiconductors and Insulators

Figure 6.4 shows schematic band theory energy level diagrams for an insulator and a semiconductor, and Table 6.1 gives **band gap** energy data between the valence and conductor bands for common materials. For NaCl, Figure 6.4a shows a large energy or band gap between the 3p orbitals of the chloride ion (the valence band) and the 3 s orbitals (conduction band) for the sodium ion; thus, sodium chloride cannot conduct electricity. The reason for the large energy gap between the Na^+ and the Cl^- is due to the transfer of an electron from sodium to chlorine to give localized electron clouds or density exhibiting mainly ionic bonding (Section 6.3).

Figure 6.4b is an example of an **intrinsic** semiconductor such as silicon where each Si atom has four tetrahedral covalent bonds. For Si, there is a very small band gap between the σ and σ^* orbitals, which are the HOMO and the LUMO orbitals or bands, respectively. For semiconductors, the HOMO orbitals are called the **valence** band, and the LUMO orbitals are called the **conduction** band. Excitation of electrons from the valence band to the conduction band is required before the semiconductor can conduct electricity; however, the number of electrons excited is very dilute compared to a true metal. The excitation of electrons creates a negative conduction band and a positive valence band, which contains positive holes. Table 6.2 gives electrical

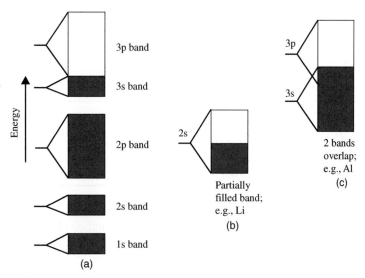

Figure 6.3 *Energy level diagrams for the bands of (a) Mg, (b) Li, and (c) Al. Blue coloration indicates that the bands are filled with electrons, and white coloration indicates that the bands are empty*

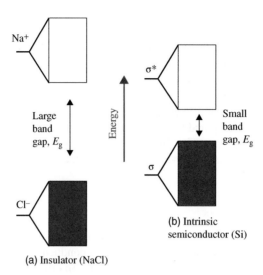

Figure 6.4   *Schematic diagram showing the relative energy gaps, E_g, for insulators, conductors, and intrinsic semiconductors*

Table 6.1   *The energies of the band gaps for some common materials and minerals [1, 2]*

Material	E_g (eV)	Material	E_g (eV)	Material	E_g (eV)
Germanium	0.66	Magnetite, Fe_3O_4	0.10	FeS	0.10
Silicon	1.11	Hematite, Fe_2O_3	2.20	FeS_2	0.95
Gallium Arsenide	1.35	Birnessite, MnO_2	0.25	CdS	2.53
Silicon Carbide	3.00	Anatase, TiO_2	3.20	ZnS	3.20
Diamond (Carbon)	5.47	Ilmenite, $FeTiO_3$	2.80	NaCl, SiO_2	~8.5, 9.00

Table 6.2   *Electrical conductivity (σ) at 20 °C for some common materials [3]. One siemens (S) is the reciprocal of one ohm*

Nonmetal	$\sigma(S\ m^{-1})$	Metal	$\sigma(S\ m^{-1})$
Germanium	2.17	Li	1.17×10^7
Silicon	1.56×10^{-3}	Fe	1.04×10^7
Gallium Arsenide	1×10^{-8} to 1×10^3	Au	4.52×10^7
Graphite (Carbon)	$\sim 2.0 \times 10^5$	Cu	5.98×10^7
Diamond (Carbon)	10^{-13}	Ag	6.30×10^7

conductivity data for common materials, with silver metal having the highest **electrical conductivity** (σ). [σ is the inverse of the electrical resistivity ($\rho = EAl^{-1}$) where E is the electrical resistance in ohms, A is the cross-sectional area of the material in m^2, and l is the material's length in m].

Tables 6.1 and 6.2 indicate that the best intrinsic semiconductors are Ge and Si, which are in the carbon family of the periodic table, and which have the outer electron configuration of s^2p^2. A general rule for semiconductors is that their valence electron configuration adds up to eight with the general formula $M^{IV}M^{IV}$. This electron configuration can be achieved by $A^{III}B^{V}$ (e.g., GaAs, Group 13 with 15 elements) and $A^{II}B^{VI}$ (CdS, Group 12 with 16 elements).

6.2.4   Fermi Level

At 0 °K, the electrons fill or occupy the molecular orbitals, which make up their band, according to the Aufbau principle. Thus, for lithium metal (Figure 6.5a), all of the lowest energy orbitals are completely filled, and the highest occupied orbital, which is at the center of the band in Figure 6.5a, is called the **Fermi level** (μ) [4]. Only the electrons nearest the Fermi level move within the metal lattice to conduct electricity. On increasing temperature, the electrons collide with the atoms so that the electrons scatter and lose their energy causing electrical resistance and a decrease in conductivity. Electrical resistance also increases as the cross-sectional area of the atom decreases and the conductor becomes longer. Impurities in the metal also lead to more electron – atom collisions and a decrease in electrical conductivity. Figure 6.5b shows that the Fermi level increases in energy on an increase in temperature for a conductor as the electrons from the lower energy states are excited to higher energy states.

For small band gaps (Figure 6.5c), the electrical conductivity for semiconductors will increase with increasing absolute temperature in contrast to the behavior found for metals. At thermodynamic equilibrium, the electron population (P) for the orbitals in a band of energy (E) is given by the **Fermi–Dirac distribution** in Equation 6.1 [4] where μ is the total chemical potential of electrons or **Fermi level** and k is the Boltzmann constant.

$$P = \frac{1}{1 + e^{(E-\mu)/(kT)}}$$ (6.1)

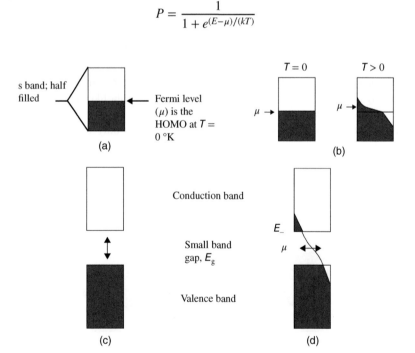

Figure 6.5   *(a) The Fermi level for lithium lies at the center of the band for the 2s orbitals at 0 °K. (b) The change in Fermi level on an increase in temperature. (c) Semiconductor at 0 °K. (d) The change in Fermi level for a semiconductor on an increase in temperature*

The ratio of probabilities that two states across the band gap (Figure 6.5d) will be occupied by an electron is given by the Arrhenius-like temperature dependence expression in Equation 6.2 where σ represents the conduction band and σ_0 represents the valence band.

$$\sigma = \sigma_0 e^{\left(-\frac{E_g}{2kT}\right)} \tag{6.2}$$

On increasing temperature, the Fermi level is found at the center of the band gap energy, E_g, and E_- is the lowest energy orbital for the conduction band. The energy of activation for electron transfer is equal to $\frac{1}{2}E_g = E_- - \mu$. Figure 1.1 (Chapter 1) shows that thermal excitation at 25 °C only provides 0.03 eV of energy, which is insufficient energy for electrons to move from the valence to the conduction band. The visible region of the electromagnetic spectrum is from 400 to 800 nm (Figure 1.2), which corresponds to a range from 3.10 to 1.55 eV, respectively. Because most semiconductors have E_g values typically below 2 eV, absorption of visible light is enough to affect electron transfer between the valence and conduction bands. Table 6.1 shows that many oxide and sulfide minerals found in the environment have band gap energies accessible in the visible region.

For an insulator such as NaCl, the band gap is about 8.5 eV. The ultraviolet region in the electromagnetic spectrum is from 200 to 400 nm, which corresponds to a range from 6.20 to 3.10 eV, respectively. Thus, for insulators, thermal, visible, and ultraviolet activation do not provide enough energy to permit conduction of electricity. Only at much higher temperatures where the solid can melt is electrical conductivity possible via the mobility of ions in the melt. ZnS and TiO_2 have band gap energies accessible in the UV region (Table 6.1).

6.2.5 Density of States (DOS)

The number of energy levels in a band is not an even distribution as shown in Figures 6.3 through 6.5. The energy levels of the orbitals are actually packed together more closely or densely at some energies over other energies within the band (e.g., Figure 6.2). The term DOS is defined as the number of energy levels in a given energy range divided by the width of that energy range and has a parabola-like shape. Figure 6.6a shows that the greatest density of the energy states occurs near the center of the band and the lowest density occurs above and below the energy of that center. As shown in Figure 6.2, there is only one possible molecular orbital

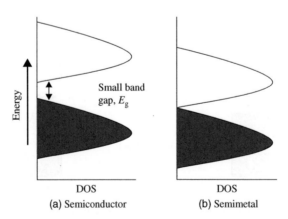

Figure 6.6   *(a) The density of states for a semiconductor with a small band gap. (b) The density of states for a semi-metal such as graphite*

combination at the low and high energy extremes of the band. For the lattice of a metal with an infinite number of atomic orbitals (referred to as an **extended lattice**), there are more bonding and antibonding molecular orbital arrangements toward the center of the band. Thus, there are numerous orbitals at the center of the band and if one draws a parabolic shape around these orbitals, a smooth energy band (as in Figures 6.2 and 6.6) is produced. Figure 6.6a shows a representation for the DOS of the valence and conduction bands in a semiconductor. For a semi-metal such as graphite (Figure 6.6b), the high-energy region of the fully occupied DOS and the lower energy region of the completely unfilled DOS meet so that electrical conduction occurs. The planar nature of the carbon atoms in graphite (similar in planar structure to benzene) accounts for graphite's high electrical conductivity (Table 6.2) compared to the tetrahedral structure of the carbon atoms in diamond.

6.2.6 Doping of Semiconductors

The data in Table 6.1 indicate that intrinsic semiconductors have approximately a million times lower electrical conductivity than pure metals. The electrical conductivity of intrinsic semiconductors can be improved by adding impurities or dopants uniformly throughout the lattice; thus, they become **extrinsic semiconductors**. There are two ways to dope intrinsic semiconductors as shown in Figure 6.7. First, for an intrinsic semiconductor such as Si or Ge with an electron configuration of s^2p^2, adding an impurity atom with one more electron such as As (s^2p^3) provides a donor band of slightly lower energy (Figure 6.7b) than the conduction band of the pure intrinsic semiconductor. This is termed an **n-type semiconductor** as the extrinsic semiconductor has negative charges due to the added electrons that carry the electrical charge. Overall, the dopant lowers the band gap energy and increases the electrical conductivity by about 10^4. Second, adding an impurity atom with one less electron such as boron or gallium (s^2p^1) provides an acceptor band of slightly higher energy (Figure 6.7c) than the valence band of the pure intrinsic semiconductor. This is known as a **p-type semiconductor** as the extrinsic semiconductor has positive holes or charges that carry the electrical charge. In both n-type and p-type semiconductors, the energy gap may be so small that electrons can be promoted from the donor band to the conduction band (valence band to the acceptor band) via thermal activation. In some sensor

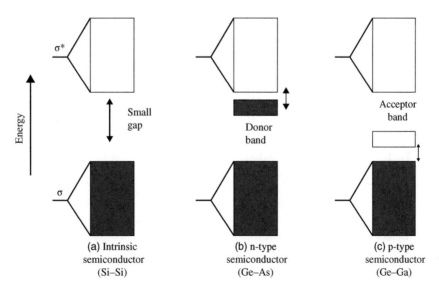

Figure 6.7 *Representative energy level diagrams for (a) an intrinsic semiconductor (Si); (b) an intrinsic semiconductor doped with a donor atom (As); (c) and intrinsic semiconductor doped with an acceptor atom (Ga or B)*

applications where thermal activation would provide background electrical noise, the semiconductor sensor is cooled with liquid nitrogen.

6.2.7   Structures of Solids

6.2.7.1   *Packing of Metal Atoms in a Crystal*

The covalent analysis of solids does not describe how the atoms of the metal pack into an **extended lattice**, which is a three-dimensional infinite and regular array of points or spheres that represent atoms, ions, or molecules. As a first approximation, each atom and its neighboring atom are considered perfect spheres and are bound identically to each other within the lattice; i.e., each atom has the same number of nearest neighbors within the lattice. For pure metals, the atoms pack in the lattice with one of the following high symmetries: **hexagonal close packed** (hcp), **cubic close-packed** (ccp), **body centered cube** (bcc), and **primitive cube** (P).

Figure 6.8 displays the possible ways to closely pack the same atoms as perfect spheres in three dimensions. Figure 6.8a shows the top view of four rows of atoms (open circles; this row is referred to as row A) in the same plane with each atom touching its neighbor in the same row and the rows adjacent to it. Each atom has six nearest neighbors in the same plane. To place atoms on a plane above that, there are two mutually exclusive possibilities labeled T_1 and T_2 because placing another atom on top of one of those positions results in a tetrahedral arrangement of four atoms (there is a tetrahedral (T) hole between these four atoms in which a smaller atom could be placed). Figure 6.8b shows another plane of atoms that overlay the T_1 sites in Figure 6.6a (the circles outlined in blue; this plane is referred to as plane B); the blue colored atom now has six nearest neighbors in row A and three nearest neighbors from plane B. Figure 6.8c shows a third plane of atoms that overlay the T_2 sites (the dashed circles outlined in blue; this plane is referred to as plane C and creates a second series of tetrahedral holes). When all three (or more) planes are placed on top of each other (e.g., *ABCABC* or *ABABAB*), the blue colored atom now has a total of 12 nearest neighbors: 6 within plane *A*, 3 arranged in a trigonal plane from plane *B*, and 3 more from plane *C* (or *B*). Figure 6.8d shows that there are octahedral sites represented by the opposing triangles between planes *A* and *B*. These are in addition to the tetrahedral sites between two planes *A* and *B*. The number of octahedral sites between any two planes is one half of the number of the tetrahedral sites. These are important in describing ionic lattices below (Section 6.3).

Figure 6.9 shows the two primary ways of packing atoms in a lattice. In the sequence from left to right in Figure 6.9a, the first three schematics are a top down view showing (1) atoms from plane *A* have atoms from plane *B* placed above T_1 sites, (2) atoms from plane *A* have atoms from plane *B* placed underneath T_1 sites, (3) the resulting combination of planes as *B,A,B*, and (4) the side view of this combination. The *ABAB*

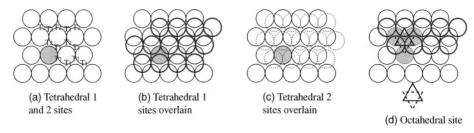

| (a) Tetrahedral 1 and 2 sites | (b) Tetrahedral 1 sites overlain | (c) Tetrahedral 2 sites overlain | (d) Octahedral site |

Figure 6.8   *(a) Atoms of a metal as circles in a plane or row A. (b) Atoms from row B overlaying the* T_1 *sites from plane A. (c) Atoms from a third plane C overlaying the* T_2 *sites from plane A. (d) Formation of octahedral sites between plane A atoms and atoms from plane B that overlay the* T_1 *sites from plane A*

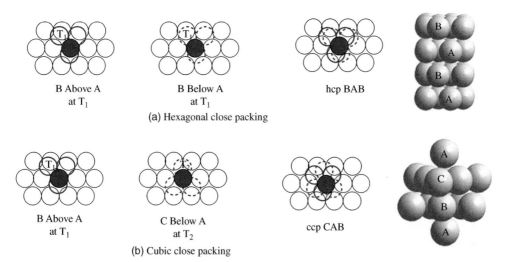

Figure 6.9 *Schematic diagrams of the stacking of planes or sheets of atoms in (a) hexagonal close packing and (b) cubic close packing lattice arrays. Bold spheres are out of the plane of the paper and dashed spheres are below the plane of the paper. Space filled models created using the Hyperchem™ program version 8.0.10*

layering of rows is called hcp. In the sequence from left to right in Figure 6.9b, the first three schematics are a top down view showing that (1) atoms from plane *A* have atoms from plane *B* placed above T_1 sites, (2) atoms from plane *A* have atoms from plane *C* placed underneath T_2 sites, (3) the resulting combination of planes as *C,A,B* (or *ABC*), and (4) the side view of this combination. The *ABCABC* layering of planes is called ccp. Both the hcp and ccp schemes result in 12 nearest neighbors for each individual atom.

6.2.7.2 Crystal Systems

A crystal is an infinite repetition in three dimensions of a parallelepiped called the **unit cell**, which is the smallest volume of the crystal containing all the structural and symmetry information. Thus, the crystal structure is known if all the unit cell parameters are known, and unit cells are constructed to conform to the outer symmetry of the crystal while containing the smallest integer number of atoms (ions, molecules). The unit cell edges have lengths *a*, *b*, and *c* and are parallel to the crystal axes, *x*, *y*, and *z*. The system to which a crystal belongs can be obtained from the relationship between the unit cell lengths (a, b, c) and the angles (α, β, γ) formed between the cell edges [*ab* for α; *bc* for β; *ac* for γ]. There are seven lattice or crystal systems that are defined by the lengths (a, b, c) and the angles (α, β, γ) that the lattice points occupy in three-dimensional space (Table 6.3). Within the seven crystal systems, there are 14 **unit cells** or Bravais lattices. Figure 6.10 shows the three unit cells in the cubic system: the primitive or simple cube (P), the body-centered cube (bcc) and the face-centered cube (fcc). There is only one unit cell in the hexagonal system. Only examples from these systems will be discussed in some detail. To better show the unit cells, the spheres of the atoms are drawn so they do not touch as discussed in Figures 6.8 and 6.9.

The total number of atoms in a unit cell can be readily calculated as follows for the cubic structures.

First, each corner of a unit cell (Figure 6.10a) is shared with 8 other unit cells and counts as $1/8$ of an atom to each unit cell; thus, for a primitive cube, the total number of atoms in the unit cell equals 1 from $(1/8 (8)) = 1$.

Second, an atom in the center of the unit cell counts as one atom; thus, for a bcc (Figure 6.10b), the total number of atoms is $(1 + \frac{1}{8}(8)) = 2$.

Third, an atom occupying a face is shared between two unit cells and counts as $\frac{1}{2}$ of an atom to each. For a face-centered cube (Figure 6.10d), the total number of atoms is $(\frac{1}{2}(6) + \frac{1}{8}(8)) = 4$.

Fourth, in the event that an edge is occupied as in Figure 6.10f, an atom on the edge is shared with four other unit cells so that it contributes $\frac{1}{4}$ of an atom to each unit cell.

The primitive cube (Figure 6.10a) and the bcc (Figure 6.10b) are not close packed structures as the hexagonal and cubic closed packing schemes are. Figure 6.10c shows two representations for the face-centered cube. The structure on the right shows black spheres that correspond to plane *A* (in the ABC close packing scheme); the white spheres correspond to plane *B* and the blue spheres correspond to plane *C*. The distinct dashed triangles drawn through the blue and white spheres help describe the C_3 axis, which passes through

Table 6.3   *The seven crystal systems, their lattice parameters, and unit cell types. The ≠ sign indicates that two or more items are not equivalent. P = primitive; BC = body-centered; FC = face-centered*

Crystal system	Lattice parameters		Unit cells (Bravais lattice) types
Cubic	$a = b = c$	$\alpha = \beta = \gamma = 90°$	P, BC, FC
Hexagonal	$a = b \neq c$	$\alpha = \beta = 90°; \gamma = 120°$	P
Tetragonal	$a = b \neq c$	$\alpha = \beta = \gamma = 90°$	P, BC
Rhombohedral	$a = b = c$	$\alpha = \beta = \gamma \neq 90°$	P
Orthorhombic	$a \neq b \neq c$	$\alpha = \beta = \gamma = 90°$	P, BC, FC, base centered
Monoclinic	$a \neq b \neq c$	$\alpha \neq \beta \neq 90°; \beta = 90°$	P, base centered
Triclinic	$a = b \neq c$	$\alpha \neq \beta \neq \gamma \neq 90°$	P

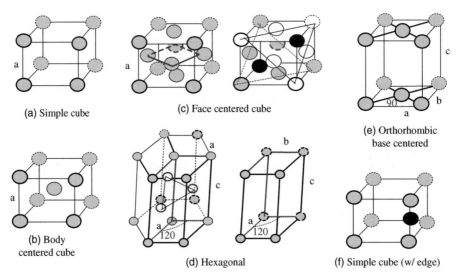

Figure 6.10   *Examples of unit cells for (a) simple or primitive cube; (b) body-centered cube; (c) face-centered cube; (d) primitive hexagon; (e) base-centered orthorhombic; (f) primitive cube with an edge occupied. Dashed spheres are behind the plane of the paper and bold spheres are in front of the plane of the paper*

the black spheres, of an octahedron. Figure 6.10d shows the position of atoms occupying a primitive hexagon on the left. The unit cell is a rectangular box as shown on the right. The three atoms in the trigonal plane are in the faces of the interior planes (they could also occupy the centers of each of three rectangular boxes), and there are two tetrahedral holes on each edge of each plane above and below the trigonal plane. Figure 6.10e demonstrates the base-centered system that is found in the orthorhombic and monoclinic crystal classes, and Figure 6.10f shows one edge occupied in a primitive cube that is shared by four other unit cells.

The coordination number of one metal atom to the others in hexagonal and cubic close packing is 12; in a primitive cube it is 6 and in a bcc it is 8. At normal temperature and pressure, cubic close packing structures are found for Al and members of the Ni and Cu families, which have some of the highest values for electrical conductivity. The alkali metals, the members of the Vanadium and chromium families, and Fe are found in the body-centered structure. Most of the other metals are found in the hexagonal structure. The increased coordination number leads to an increase in metallic or covalent bonding as all the d orbitals can be used to form σ, π, and δ bonds (as in Figure 6.11) in addition to bonds formed from s and p orbitals. The result is higher melting points for the metals in the center of the transition series, which have partially filled d orbitals that overlap to form more bonds. For example, the melting points (in °C) for elements in row 4 are 64 (K), 839 (Ca), 1539 (Sc), 1660 (Ti), 1890 (V), 1857 (Cr), 1245 (Mn), 1535 (Fe), 1495 (Co), 1452 (Ni), 1083 (Cu), 420 (Zn), and 30 (Ga).

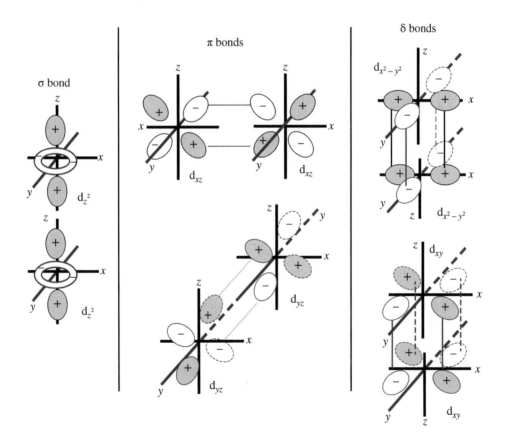

Figure 6.11   *Overlap of d orbitals on adjacent atoms to form σ, π, and δ bonds*

On increasing the temperature, the highly ordered or close packed structures undergo disorder resulting in structures with less efficient packing and lower coordination number (**entropy effect**). Increasing pressure tends to favor the highly ordered close packed structures.

6.3   Ionic Solids

Ionic solids are crystals composed of an extended lattice of positive and negative ions, which are stabilized by maximizing the attractive forces and minimizing the repulsive forces in the lattice. In a purely ionic solid, a metal gives up its electron(s) to a nonmetal, and the ions formed have electron clouds that cannot be distorted and are considered hard spheres (see Sections 3.8.7 and 7.8.3 for Na^+). The forces or bonds between the positive and negative ions are very strong and they are omnidirectional; thus, they have high melting points. They are characterized by low electrical conductivity in the solid (see Figure 6.4 for the band gap of an insulator), but on melting, they have high electrical conductivity. Under compression or external pressure, they tend to break apart due to their brittle nature. Ionic solids are soluble in polar solvents of high dielectric constant ($c\varepsilon_0$) as the attractive forces (E_c) between the cation and anion decrease according to Coulomb's law (Equation 6.3) where Z^+ and Z^- are the ionic charges for the cation and the anion, respectively; ε_0 is the vacuum permittivity [8.854×10^{-12} C^2 m^{-1} J^{-1}] and c is a constant for the solvent ($c_{H_2O} = 82$; $c_{NH_3} = 25$; $c_{vacuo} = 1$)]; and r (or r_0) is the bond distance between the cation and the anion. In water, E_c (the Coulombic attraction of ions) is 1.2% of that in vacuum.

$$E_c = \frac{Z^+ Z^-}{4\pi(c\varepsilon_0)r} \tag{6.3}$$

The breakup of the lattice and the solvation of both the cation and the anion lead to an increase in entropy and a more favorable heat of solvation. Ionic solids that readily dissolve in water are those of the group 1 and group 2 metals bound to halogens, hydroxide, and oxide. Ionic solids with polyatomic cations and anions such as ammonium, nitrate and carbonate also dissolve in water to an appreciable extent. Although aluminum, silicon, and transition metal oxides (and sulfides) are frequently thought of as ionic solids, they have significant covalent bonding between the metal and the oxygen (sulfur) so are sparingly soluble in water at best.

6.3.1   Solids AX Stoichiometry

6.3.1.1   *Sodium Chloride*

Ionic solids crystallize in the seven crystal systems shown in Table 6.3 and Figure 6.10, but only ionic solids with high symmetry will be discussed. One of the simplest to describe is the NaCl or **rock salt** lattice, which crystallizes in a fcc structure. Both the chloride ion and the sodium ion have a coordination number of six as shown in Figure 6.12. Inspection of the figure shows that the blue sphere marked as Cl^- has six sodium ions as nearest neighbors; there are 12 chloride ions as next nearest neighbors (as expected for ccp of anions); then there are eight sodium ions after that and so on to maximize the attractive forces and minimize the repulsive forces. Figure 6.9b shows the packing of perfect spheres, which can represent the chloride ions in Figure 6.12 even though the chloride ions should repel each other if they touch each other (using space-filled models the larger ions touch or are very close to touching each other). The sodium ions fill every one of the octahedral holes between the layers of the chloride ions. Figure 6.12 shows that the bond distance, r, for the nearest neighbors in NaCl is the sum of the sodium ion radius (r_+) and the chloride ion radius (r_-). The next nearest neighbors have the distance h (the diagonal of the square planar face) and the next nearest neighbors have

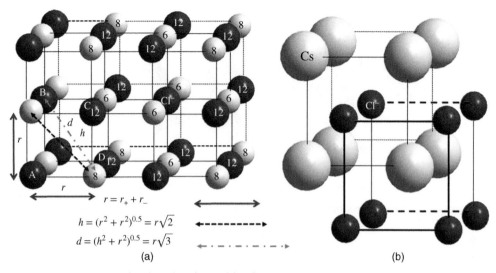

$$r = r_+ + r_-$$

$$h = (r^2 + r^2)^{0.5} = r\sqrt{2}$$

$$d = (h^2 + r^2)^{0.5} = r\sqrt{3}$$

(a)   (b)

Figure 6.12 (a) Face-centered cube of sodium chloride or rock salt. The equations indicate the geometrical relationships for the FCC structure; the arrows are not drawn to 3D scale. (b) Body-centered cube of cesium chloride. Both unit cells produced using Hyperchem™ version 8.0.10

the distance d (the diagonal across the cube). The Cl^- ions marked as A, B, C, D are discussed in the **spinel** section (6.3.2.2.2).

6.3.1.2 Cesium Chloride

Cesium chloride (CsCl; Figure 6.12b) crystallizes in the body-centered cubic structure, which has the primitive cubic lattice of cesium ions interpenetrating with the primitive lattice of chloride ions. Thus, the coordination number for both the cesium and chloride ions is eight.

6.3.1.3 Zinc Sulfide

Figure 6.13 describes the two distinct crystal lattices for zinc sulfide (ZnS). **Sphalerite** or **zinc blende** is cubic ZnS (Figure 6.13a), and **wurtzite** is a hcp structure (Figure 6.13b). Sphalerite is the favored structure at low temperatures, and wurtzite is the favored structure at higher temperatures. In Figure 6.13a, the ccp sulfide ions for layers A, B, and C are clearly marked, and one half of the tetrahedral sites between these layers is occupied by the zinc ions. The white solid and dashed lines show a six-membered ring of alternating zinc and sulfide ions in chair conformation. Superimposed upon that ring is a Zn_3S moiety where each Zn ion bridges a sulfide ion from the ring with the sulfur anion at the top; this is similar to the **adamantane** structure of carbon atoms in organic chemistry. Figure 6.13b shows the A and B layers of the hcp sulfide ions, and again one half of the tetrahedral sites are occupied by the zinc ions. The dark solid and dashed lines show a six-membered ring of alternating zinc and sulfide ions in the boat conformation. In these structures as well as in the cubic structures in Figure 6.12, there is a significant amount of volume that is not occupied by ions (or spheres). Note that in Figure 6.13 the sulfur ions are very close to touching each other. Carbon as **diamond** also exhibits the zinc blende structure.

Figure 6.14 displays the unit cells of the sphalerite and wurtzite structures. In Figure 6.14a, the black spheres indicate the A layer, the white spheres indicate the B layer, and the blue spheres indicate the C layer of sulfide

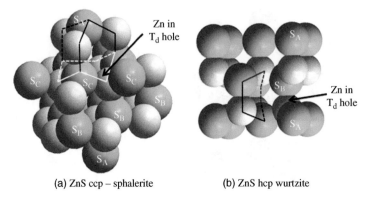

(a) ZnS ccp – sphalerite   (b) ZnS hcp wurtzite

Figure 6.13   *Space-filled models produced using Hyperchem™ version 8.0.10 for (a) cubic close packing structure for sphalerite. (b) Hexagonal close packing structure for wurtzite*

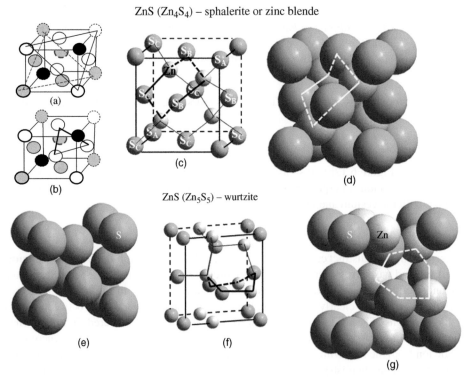

Figure 6.14   *(a and b) Two different ways to view the cubic close packing arrangement of sulfide ions. (c) The unit cell for sphalerite. (d) Space-filled model for the unit cell of sphalerite. (e) Hexagonal close packing arrangement of sulfide ions. (f) The unit cell for wurtzite. (g) Space-filled model for the unit cell of wurtzite. C, D, E, F, and G produced using Hyperchem™ version 8.0.10*

ions, respectively. In Figure 6.14b, the triangle indicates the C_3 axis; a zinc ion slips in the tetrahedral space between the triangle and the black sphere. Figure 6.14c and d shows the cubic close packed unit cell for sphalerite, and the A, B, and C layers for the sulfide anions are labeled. The solid and dashed black lines in Figure 6.14c (white lines in Figure 6.14d) show the six-membered ring structure with alternating zinc and sulfide ions in a chair conformation. Eight sulfur ions occupy the corners of the cube and six sulfur ions occupy the faces of the cube for a total of four sulfur ions in the unit cell. All four zinc ions are within the unit cell for a total of four zinc ions in the unit cell. The overall unit cell stoichiometry is Zn_4S_4.

Figure 6.14e shows the sulfur ions in a hcp structure. Figure 6.14f and g shows how the zinc ions pack into the tetrahedral holes of the sulfur ions shown in Figure 6.14e. Figure 6.14f shows black solid and dashed lines (white lines in Figure 6.14g) outlining the six-membered ring structure with alternating zinc and sulfide ions in a boat confirmation. Eight sulfur ions occupy the corners of the unit cell, two sulfur ions occupy a face, and three sulfur ions are in the center of the unit cell for a total of five sulfur ions in the unit cell. Four zinc ions are within the unit cell and four more occupy edges of the unit cell for a total of five zinc ions within the unit cell. The overall unit cell stoichiometry is Zn_5S_5, and comparison of Figure 6.14d with Figure 6.14g shows more empty space between the ions in the unit cell of wurtzite.

6.3.2 Solids with Stoichiometry of AX_2, AO_2, A_2O_3, ABO_3 (Perovskite), AB_2O_4 (Spinel)

6.3.2.1 *AX_2, AO_2, A_2O_3*

Many solids have a stoichiometry that is not one to one. Thus, when these solids adopt a particular geometry such as 1:2 (or 2:1), only one half of the possible cation or anion sites are occupied. In Figure 6.15, every other calcium site is occupied in the BCC lattice. Thus, in **fluorite** or **calcium difluoride** (CaF_2), the coordination of the calcium ion is 8 whereas each fluoride ion has a coordination number of 4. K_2O also adopts the fluorite structure but the potassium and oxide ions have a coordination number of 4 and 8, respectively; this is referred to as an antifluorite structure because the cations and anions are reversed to that of CaF_2.

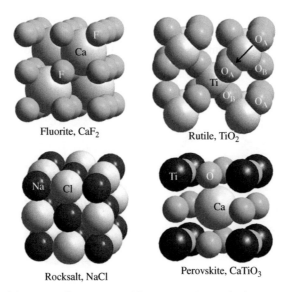

Figure 6.15 *Comparison of the unit cell structures of fluorite (BCC), rutile (hcp), rock salt (ccp), and perovskite (ccp). For rutile the unit cell is extended. Structures produced using Hyperchem™ version 8.0.10*

Rutile (TiO_2) crystallizes in a tetragonal or hexagonal closest packing structure and not the ccp or fcc structure of NaCl. Figure 6.15 shows an extended TiO_2 unit cell; the arrow indicates the direction of the packing of O anions in an AB or O_AO_B arrangement. Every other octahedral site in the lattice is occupied by Ti^{4+} ions; thus, each Ti is six coordinate and each O is three coordinate.

Corundum ($\alpha\text{-}Al_2O_3$, not shown) crystallizes in a cubic close packing of Al atoms between O atoms with $2/3$ of the octahedral sites occupied. Each Al is six coordinate and each O is four coordinate.

6.3.2.2   Ternary Compounds

ABO_3 – Perovskites.   Figure 6.15 also shows the ideal ABO_3 perovskite structure (**perovskite** is the mineral name for $CaTiO_3$). Coordination numbers of 12 and 6 are found for Ca^{2+} (A ions) and Ti^{4+} (B ions), respectively, and the structure is an example of cubic close packing of Ti atoms between O atoms (note the similarity to NaCl, Figure 6.12). The Ca^{2+} sits above and below two planes (sheets) with alternating Ti and O atoms of Ti_4O_4 stoichiometry; these planes or extended sheets are bound through four Ti–O bonds along the edge of the unit cell. There is significant interest in this structure as certain materials having it show **superconducting** properties. The ABO_3 stoichiometry also displays a hcp structure, which is not discussed.

AB_2O_4 – Spinels.   The AB_2O_4 or $A(B)_2O_4$ stoichiometry is found in the **spinels** (e.g., $MgAl_2O_4$), which are common in nature. For a **normal spinel**, the O anions form in a cubic close packed arrangement, where the A cations occupy one-eighth of all of the tetrahedral holes and the B cations occupy one-half of the octahedral holes. The spinel structure has similarity to NaCl as the O anions occupy Cl^- sites and the B cations $1/2$ of the octahedral Na^+ sites. The A cations slip into tetrahedral sites as shown in Figure 6.12 with the Cl^- sites marked A, B, C, and D. Figure 6.16 shows two structures for $MgAl_2O_4$; (a) is the unit cell showing the tetrahedral Mg cations bound to four O anions and the octahedral Al cation bound to six O anions. Figure 6.16b is the unit cell tilted to show the cubic close packing of O anions in the ABC arrangement as shown in Figure 6.9b. For an **inverse spinel**, which occurs frequently, all the A cations switch positions with $1/2$ of the B cations. Magnetite $[Fe^{3+}(Fe^{2+}Fe^3)O_4$ instead of $Fe^{2+}(Fe^{3+})_2O_4]$ is an example of an inverse spinel that is stabilized as explained by crystal field energetics when an Fe^{2+} moves to an octahedral hole (see Table 8.9). **Olivine** [Mg_2SiO_4 or $(MgFe)_2SiO_4$ where Fe^{2+} substitutes for Mg^{2+}] is an orthosilicate that forms with an hcp arrangement of O anions (not shown). The coordination numbers of the Mg and Si cations are four and six, respectively.

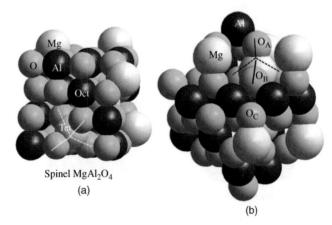

Spinel $MgAl_2O_4$

(a)

(b)

Figure 6.16   *(a) Unit cell for $MgAl_2O_4$. (b) Unit cell showing the ccp of O atoms. Structures produced using Hyperchem™ version 8.0.10*

6.3.3   Crystal Radii

In the early 20th century, X-ray diffraction techniques provided information on the interatomic distances between atoms in the unit cell of a crystal lattice. From this, the radius of the cation and the anion could be obtained as the unit cell length is related to the sum of the ionic radii for the cation and the anion. For ionic crystals, an assumption was made that the spheres of the cation and anion were not distorted; i.e., they were considered hard. For sodium chloride in the face-centered cubic structure, the unit cell length is equal to $2r_0$ where r_0 is the sum of the cation radius (r_+) and the anion radius (r_-). Using this methodology, Goldschmidt [5] began a systematic experimental program to determine the ionic radii from simple inorganic solids and minerals with halides and oxides as anions where the oxide ion radius was found to be 132 pm. Pauling [6] also calculated the ionic radii from theoretical principles and arrived at a value of 140 pm for the oxide ion. Despite this difference, there was general agreement between the Goldschmidt and Pauling tabulations. Pauling's tabulation of radii was refined by Shannon and Prewitt [7, 8] and then Shannon [9]; the latter are based on the fact that X-rays are scattered by electrons so that electron density maps can be generated. The minimum in electron density of the cation and the anion in a crystal mark the start of one ion and the end of another ion. Table 6.4 provides **crystal radii** for selected cations and anions from [9]. The radii of cations increase with an increase in coordination number as repulsions increase with the larger number of coordinating anions. The radii of the anions vary less than 5 pm with geometry changes so no specific geometry is given. The accepted anion radii are 14 pm smaller and the cation radii 14 pm larger than the traditional values of Pauling [also known as the **effective ionic radii** (IR)].

Inspection of Table 6.4 shows that there are some important trends for the crystal radii. Both cation and anion ionic radii increase on descending a group in the periodic table. The loss of electrons for cations relative to the neutral atom results in a decrease in size for the cation. For cations of the same element, an increase in positive charge results in a shorter radius for the same geometry (e.g., in tetrahedral geometry, Fe^{2+} is 77 pm and Fe^{3+} is 63 pm). On moving across a row of the periodic table, cations of the same charge will have a smaller cation radius (e.g., in tetrahedral geometry, Fe^{2+} is 77 pm and Co^{2+} is 72 pm) because of the increased nuclear charge. The effect of the poor shielding of the d electrons, which resulted in the lanthanide contraction for atomic radii, results in only a 2 pm difference between the crystal radii for six coordinate Ca^{2+} and Hg^{2+}. The gain of electrons for anions relative to the neutral atom results in an increase in size for the anion. On moving across the periodic table, anions decrease in size because of the increased nuclear charge and smaller ionic charge (e.g., O^{2-} is 126 pm and F^- is 119 pm). For ions with the same electron configuration, the ion with the highest atomic number has the smallest ionic radius (e.g., for the ions with the neon electron configuration in octahedral geometry, the ionic radii for Mg^{2+} is 86 pm; Na^+ is 116 pm; F^- is 119 pm; O^{2-} is 126 pm). Thus, cations are smaller than anions when each has the same electron configuration.

6.3.4   Radius Ratio Rule

For purely ionic solids with little or no covalent bonding, there is no distortion of their spherical electron clouds; thus, their spheres are considered to be hard. Using the close packing schemes that assume that the anions touch each other with the cation slipping into the hole between the anions, it is possible to calculate the limiting ratio of the cation radius to the anion radius to predict the coordination number of the cation and the anion. Figure 6.12 shows NaCl, a six-coordinate FCC structure, and the geometrical equations representing the distances between cations and anions. Assuming the anions touch each other, the radius of the anion is given as

$$2r_- = h = r_0 \sqrt{2}; \text{ thus } r_- = 0.707 \, r_0$$

and

$$r_+ = r_0 - 0.707 \, r_0 = 0.293 \, r_0$$

Table 6.4 *List of crystal radii in picometers (pm) for some common ions from [9]. CN = coordination number. CN = 4 is high spin T_d. CN = 6 is O_h. CN = 8 is BCC. HS = high spin; LS = low spin*

+1 Ion	CN	r (pm)	+2 Ion	CN	r (pm)	Transition metals	CN	r (pm)	Anions r (pm)
Li^+	4 6 8	73 90 106	Be^{2+}	4	41	V^{2+}	6	93	F^- 119
Na^+	4 6 8	113 116 132	Mg^{2+}	4 6 8	71 86 103	Cr^{2+}	6 LS 6 HS	87	Cl^- 167
K^+	4 6 8	151 152 165	Ca^{2+}	6 8	114 126	Mn^{2+}	4 6 LS 6 HS	80 81 97	Br^- 182
Rb^+	6 8	166 175	Sr^{2+}	6 8	132 140	Fe^{2+}	4 6 LS 6 HS	77 75 92	I^- 206
Cs^+	6 8	181 188	Ba^{2+}	6 8	149 156	Co^{2+}	4 6 LS 6 HS	72 79 88.5	O^{2-} 126 OH^- 121
Cu^+	2 4 6	60 74 91	Cd^{2+}	4 6 8	92 109 124	Ni^{2+}	4 6	69 83	S^{2-} 170
Ag^+	2 4 6	81 114 129	Hg^{2+}	2 4 6	83 110 116	Cu^{2+}	4 6	71 87	Se^{2-} 184
Au^+ Au^{3+}	6 6	151 99	Pb^{2+}	6 8	133 143	Zn^{2+}	4 6	74 88	Te^{2-} 207
Important ions with charge greater than 2									
Al^{3+}	4 6	53 67.5	Si^{4+}	4 6	40 54	Ti^{4+}	4 6	56 74.5	N^{3-} 132
Cr^{3+}	6	75.5	Mn^{3+}	4 6 LS 6 HS	72 72 78.5	Mn^{4+}	4 6	53 67	
Fe^{3+}	4 6 LS 6 HS	63 69 78.5	Co^{3+}	6 LS 6 HS	68.5 75	Pt^{2+} Pt^{4+}	4 6	74 76.5	

The minimum or limiting ratio of the cation radius to the anion radius for the cation and anion to crystallize in an octahedral geometry is given below.

$$\frac{r_+}{r_-} = \frac{0.293\, r_0}{0.707\, r_0} = 0.414$$

Table 6.5 lists minimum radius ratios for common coordination numbers, geometries, and lattice types. A value between 0.225 and 0.414 (or 4.45 to 2.42) predicts a tetrahedral geometry. Using radii from Table 6.4,

Table 6.5 *Limiting radius ratios for different geometries and coordination numbers (CN). Values in parentheses are the inverse of the smaller radius divided by the larger radius*

CN Cation:anion	Geometry	Limiting radius ratio	Lattice structure types
4:4	Tetrahedral	0.225 (4.45)	Sphalerite, wurtzite
4	Square planar	0.414 (2.42)	
6:6 and 6:3	Octahedral	0.414 (2.42)	NaCl, TiO_2
8:8 and 8:4	BCC	0.732 (1.37)	CsCl, CaF_2
12	CCP, HCP, dodecahedron	1.000 (1.000)	Metals; Ca^{2+} in perovskite

we can predict that NaCl is octahedral ($116/167 = 0.69$), CsCl is BCC ($188/167 = 1.08$), and CaF_2 is eight coordinate for Ca^{2+} ($126/119 = 1.05$). Interestingly, the sulfides of Zn^{2+}, Cd^{2+}, and Hg^{2+} are all found as four coordinate (sphalerite and wurtzite) structures and violate the radius ratio rule; e.g., for ZnS, we predict octahedral ($74/170 = 0.435$). The reason for this is that these solids are infinite covalent and not ionic lattices (see Section 12.2). The radius ratio rules assume hard anion spheres that are not distorted by the cation. Thus, the tabulation of crystal radii in Table 6.4 is best used with the oxide and fluoride ions. As will be shown in the discussion of **hard–soft** acids and bases (Section 7.8.3), these metals are soft acids and sulfide is a soft base; thus, they are highly polarizable ions. A coordination number of 12 is not found in purely ionic substances but is found in alloys and metals and in certain solids with more covalent character (e.g., Ca^{2+} in perovskite) as noted above.

6.3.5 Lattice Energy

For ionic compounds, the energy of the crystal lattice is the energy released when cations and anions (as perfect spheres or point charges) come together from infinite separation to form a perfect crystal as in Equation 6.4. The energy, E_C, in Joules when one gaseous cation and one gaseous anion combine in vacuo at 0 °K is described by the Coulombic attraction Equation 6.5, which is modified from Equation 6.3 to consider the electrostatic charge, $e = 1.602 \times 10^{-19}$ Coulomb (C), the vacuum permittivity, $\varepsilon_0 = 8.854 \times 10^{-12}$ C^2 m^{-1} J^{-1}, and r is in meters. Because of the negative charge on the anion, the Coulombic energy is negative and is representative of the Gibbs free energy of the interaction.

$$M^+_{(g)} + X^-_{(g)} \rightarrow MX(s) \tag{6.4}$$

$$E_c = \frac{Z^+ Z^- e^2}{4\pi\varepsilon_0 r_0} \tag{6.5}$$

Equation 6.5 only describes one ion pair interaction between nearest neighbors, but as shown in Figure 6.12 for sodium chloride there are significantly more interactions based on the geometry of the lattice. The first attractive interaction, r_0, has six nearest neighbors for each cation (or anion); the second interaction, $h = r_0\sqrt{2}$, is repulsive and has 12 next nearest neighbors; the third interaction, $d = r_0\sqrt{3}$, is again attractive and has eight next nearest neighbors, and so on. Continued analysis of all of the ionic interactions between cations and anions leads to a convergent series that is known as the Madelung constant (M or A), which varies with the geometry of the crystal lattice or unit cell. Thus, Equation 6.5 becomes Equation 6.6a where N equals the Avogadro's number for 1 mol of ion pairs in the fcc structure. Equation 6.6a accounts for about 70% of the forces in the lattice, and is also known as the Landé attraction expression. Equation 6.6b is the

Table 6.6   *Values of the geometrical Madelung constant for structures with different geometries and the Born exponent for different electron configurations of the ions*

Structure	Coordination number Cation:anion	Madelung Constant	Ion Configuration	n
Sphalerite, ccp	4:4	1.63806	He	5
Wurtzite, hcp	4:4	1.64132	Ne	7
Sodium chloride	6:6	1.74756	Ar, Cu^+	9
Cesium chloride	8:8	1.76267	Kr, Ag^+	10
Fluorite	8:4	2.51939	Xe, Au^+	12

general expression for a given geometry and Madelung constant, which are found in Table 6.6.

$$E_c = \frac{NZ^+Z^-e^2}{4\pi\varepsilon_0 r_0}\left(6 - \frac{12}{\sqrt{2}} + \frac{8}{\sqrt{3}} \cdots\right) \tag{6.6a}$$

$$E_c = \frac{ANZ^+Z^-e^2}{4\pi\varepsilon_0 r_0} \tag{6.6b}$$

However, ions are not perfect spheres or point charges as they have electron clouds, which repel each other at short distances and occupy volume; thus, the ions are compressible. A repulsive force must balance the Coulombic attraction between ions, and Born represented the repulsive force, E_R, in Equation 6.7 where B is a constant and the exponent n is evaluated from compressibility data. As the ions move closer to each other, the bond distance (r) decreases, and the repulsive force increases. As r increases, the value of n will increase because larger ions have higher electron density with more electron charge to be pulled toward the nucleus (see Table 6.6). The total lattice energy, U, is the sum of E_c and E_R (Equation 6.8).

$$E_R = \frac{B}{r_0^{\,n}} \tag{6.7}$$

$$U = E_c + E_R = \frac{ANZ^+Z^-e^2}{4\pi\varepsilon_0 r_0} + \frac{NB}{r_0^{\,n}} \tag{6.8}$$

Taking the derivative (Equation 6.9) of the lattice energy with respect to the bond distance, r, minimizes the lattice energy, U_O, at the equilibrium distance between ions so that the attractive and repulsive forces are balanced. Thus, the constant, B, can now be evaluated as in Equation 6.10 for a given lattice with a Madelung constant, A.

$$\frac{dU}{dr} = 0 = \frac{ANZ^+Z^-e^2}{4\pi\varepsilon_0 r_0^{\,2}} - \frac{nNB}{r_0^{\,n+1}} \tag{6.9}$$

$$B = \frac{-AZ^+Z^-e^2 r_0^{\,n-1}}{4\pi\varepsilon_0 n} \tag{6.10}$$

Substitution of B from Equation 6.10 into Equation 6.8 leads to Equation 6.11a, which is known as the Born–Landé expression. Equation 6.11b evaluates the constants in Equation 11a so that U_O is in units of kJ mol^{-1}. For ionic crystals with the same A value, increasing the bond distance r_0 results in an increase of the lattice energy (U_O becomes more positive); recall that the radius as shown in Table 6.3 also increases with an increase in the coordination number. Thus, four coordinate structures are predicted to be more stable as the lattice energy is more favorable. U_O is easily evaluated from X-ray diffraction data when A and r_0 are known. The value of U_O decreases (becomes more negative) with an increase in ionic charge and with an increase in the Madelung constant, which increases as the coordination number increases.

$$U_O = \frac{ANZ^+Z^-e^2}{4\pi\varepsilon_0 r_0} - \frac{ANZ^+Z^-e^2}{4\pi\varepsilon_0 r_0 n} = \frac{ANZ^+Z^-e^2}{4\pi\varepsilon_0 r_0}\left(1 - \frac{1}{n}\right) \tag{6.11a}$$

or

$$U_O(\text{kJ mol}^{-1}) = 1.39 \times 10^5 \frac{AZ^+Z^-}{r_0}\left(1 - \frac{1}{n}\right) \tag{6.11b}$$

Using the radii for Na^+ and Cl^-, the Madelung constant of 1.74756 (Table 6.6) and the Born exponent of 8 (Table 6.6; the average of the neon and argon electron configurations for Na^+ and Cl^-, respectively), a U_O value of -755.2 kJ mol^{-1} is obtained. The experimentally determined Born exponent is 9.1, which leads to a value of -766.2 kJ mol^{-1} for U_O. Using corrections for van der Waals forces, the zero point energy, and heat capacity effects, the generally accepted calculated value for U_O is -777.8 kJ mol^{-1}. All are in good agreement.

6.3.6 Born–Haber Cycle

According to **Hess's law**, the enthalpy of a reaction (ΔH_f) is the same whether the reaction occurs in one or several steps. Equation 6.12 shows the formation of $MX_{(s)}$ from the elements in their standard states. Equation 6.4 above describes the reaction of the gaseous M^+ and X^- ions to form the same crystal, $MX_{(s)}$. Born and Haber demonstrated a thermodynamic cycle (Equation 6.13, Figure 6.17) to calculate U_O. The cycle can also be used to estimate one of the other parameters (ΔH_{IE} or ΔH_{EA}) when U_O can be determined experimentally or calculated from Equation 6.11 or 6.14.

$$M_{(s)} + \tfrac{1}{2}X_{2(g)} \rightarrow MX_{(s)} \qquad \Delta H_f \tag{6.12}$$

$$\Delta H_f = \Delta H_{AM} + \Delta H_{AX} + \Delta H_{IE} + \Delta H_{EA} + U_O \tag{6.13}$$

There are accurate atomic experimental data for the first four terms in Equation 6.13 where ΔH_{AM} is the enthalpy of atomization of the metal, ΔH_{IP} is the enthalpy of atomization of a nonmetal, ΔH_{AX} is the enthalpy of dissociation of a diatomic molecule, and ΔH_{EA} is the electron affinity of the gaseous nonmetal. For the simple alkali halides, accurate experimental data are available for ΔH_f. Using these experimental data for sodium chloride, a U_O value of -770.3 kJ mol^{-1} is calculated. This is in excellent agreement with the calculated value of U_O using Equation 6.11.

Kapustinskii [10] realized that the internuclear distance, the stoichiometry of a compound or crystal, and the Madelung constant are interrelated. In the absence of any X-ray data that would give the crystal structure and thus the Madelung constant, the lattice energy can be calculated for compounds that are not well known using Equation 6.14 and the crystal radii from Table 6.4, where v is the number of ions per molecule based on the stoichiometry. The value of 34.5 pm indicates the repulsion of ions at short distances from each other.

$$U_O(\text{kJ mol}^{-1}) = \frac{120,200\ v\ Z^+Z^-}{r_0}\left(1 - \frac{34.5\ \text{pm}}{r_0}\right) \tag{6.14}$$

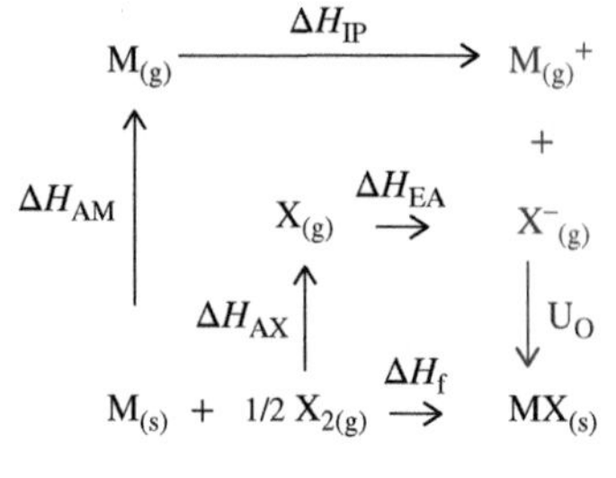

Figure 6.17   *Born–Haber cycle for the formation of a crystal, $MX_{(s)}$, from the elements in their standard states*

Table 6.7   *Thermochemical radii in picometers for environmentally relevant polyatomic ions from Refs. [10, 11] are corrected to be consistent with radii in Table 6.4*

Cations		Anions		
NH_4^+, 151	NO_3^-, 165	CO_3^{2-}, 164	PO_4^{3-}, 252	SeO_4^{2-}, 235
$(CH_3)_4N^+$, 215	NO_2^-, 178	HCO_3^-, 142	SO_4^{2-}, 244	IO_3^-, 108

There are many examples of ionic solids ($CaCO_3$ or $CaSO_4$) that have polyatomic cations (NH_4^+) and/or anions (NO_3^-, CO_3^{2-}, SO_4^{2-}). Because the ΔH_f of these ionic solids is known, use of the Born–Haber cycle permits calculation of U_O when the other ion is not polyatomic. Also, substitution of U_O into Equation 6.11 or 6.14 permits calculation of r_0 and then calculation of the r_+ or r_- of the polyatomic ion. These calculated values of the radii obtained are known as **thermochemical radii** (Table 6.7), and are reliable for spherical ions (e.g., NH_4^+, SO_4^{2-}) that will "pack" into a lattice. Many ions are planar (CO_3^{2-}) or linear (CN^-), and their calculated radii should be regarded as an estimate.

6.3.7   Thermal Stability of Ionic Solids

Large negative vaues for ΔH_f and U_O of ionic solids lead to high melting points. For ionic solids with polyatomic anions, decomposition to other products can occur. For example, the metal carbonates decompose to metal oxides on heating as represented by the following reaction.

$$MCO_{3(s)} + heat \rightarrow MO_{(s)} + CO_{2(g)}$$

Group II or alkaline earth carbonates crystallize mainly in the **calcite** ($CaCO_3$) or **aragonite** structures (Figure 6.18, Table 6.8). Only $CaCO_3$ crystallizes in both structures. In hexagonal calcite, there are two planes of alternating CO_3^{2-} ions that point in opposite directions (180° rotation to each other). The metal cation (Mg^{2+}, Ca^{2+}) is six coordinate (oxygen atoms are numbered in the bottom of Figure 6.18a) as six CO_3^{2-} ions surround (3 above and 3 below) the M^{2+} ion. There is one O atom from each CO_3^{2-} ion octahedrally arranged about one M^{2+}. In orthorhombic aragonite, there are also two planes of alternating CO_3^{2-} ions (slightly deformed from planar geometry) that also point in opposite directions. However, the M^{2+} ions (Ca^{2+}, Sr^{2+}, Ba^{2+}) are displaced laterally from the calcite structure so that the six CO_3^{2-} ions surrounding each M^{2+} ion have nine oxygen atoms binding the metal; three CO_3^{2-} ions have two O atoms (**bidentate**) binding via 1,2; 4,5; 6,7 (Figure 6.18b), and the other three CO_3^{2-} ions have only one O atom

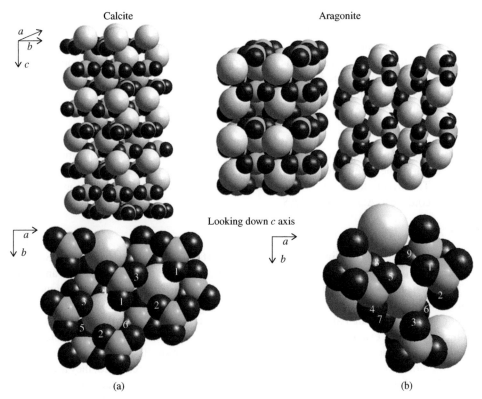

Calcite   Aragonite

Looking down *c* axis

(a)   (b)

Figure 6.18 Crystal structure representations of calcite (a; unit cell of a = 499.00 pm; b = 499.00 pm; c = 1706.15 pm; α = 90; β = 90; γ = 120) and aragonite (b; unit cell of a = 496.14 pm; b = 796.71 pm; c = 574.04 pm; α = β = γ = 90) made using Hyperchem™ version 8.0.10. Colors: White = Ca; blue = C; black = O

Table 6.8 Parameters for the alkaline earth carbonates. CN = coordination number. The MCO₃ temperature occurs at a dissociation pressure of 1 atm CO₂. Metal oxide melting points (m.p.) from Ref. [12]; MO and MCO₃ coordination numbers from Refs. [12, 13]

	$BeCO_3$	$MgCO_3$	$CaCO_3$	$SrCO_3$	$BaCO_3$
MCO_3 dissoc. T (°C)	250	540	900	~1290	1360
Cation CN in MCO_3	4	6	6 or 9	9	9
r_M/r_{CO_3}	0.25	0.524	0.695	0.804	0.908
MO m.p. (°C)	2530	2826	2613	2430	1923
Cation CN in MO	4	6	6	6	6
r_M/r_O	0.325	0.683	0.905	1.045	1.118

(**monodentate**) binding via atoms 3; 8; 9. Aragonite is denser (more stable at high pressure and lower T) and more soluble than calcite.

As shown in Table 6.8, the thermal stability of the Group II carbonates increases on going from beryllium to barium, which parallels the increase in cation radius shown in Table 6.4. The ionic radius for CO_3^{2-} is 164 pm, and the radius is 126 pm for O^{2-}. The coordination number for the cations increases on descending

Group II in the periodic table, and indicates that large cations stabilize large anions whereas small cations stabilize small anions. However, the thermal stability of the alkaline earth oxides (O^{2-} radius is 126 pm) decreases on going from beryllium to barium; this $M^{2+}O^{2-}$ pattern parallels the more stable U_O due to the decrease in the radius for metal oxides and the lower coordination number for the cations at the top of the periodic table. The smaller cations are harder (see Section 7.8.3) and can polarize the CO_3^{2-} so that one of the C–O bonds is broken to form MO solids with release of CO_2.

6.3.8 Defect Crystal Structures

All solids contain some kind of defect or imperfection in their structure and/or composition. Defects can be intrinsic, which occur in the pure material, and extrinsic, which occur when impurities substitute for part of the pure material. The driving force for defects is the increase in entropy that results from creating disorder or imperfection according to $\Delta G = \Delta H - T\Delta S$.

6.3.8.1 Point Defects (Intrinsic Defects)

There are many types of defect structures and only a few **point defects** will be described. A **Schottky** defect arises when a vacancy occurs in a perfect lattice. For example, the same number of sodium and chloride ions are missing from their normal sites (Figure 6.19a) so that the solid retains its 1:1 stoichiometry. For ionic solids with the stoichiometry of 1:2 as in CaF_2, for every calcium ion that is missing, two fluoride ions are also missing; thus, the charges are balanced and the stoichiometry is not changed. Some solids have apparent non-stoichiometry as in the examples of $Fe_{0.95}O$ and $Cu_{1.77}S$; however, these solids are stoichiometric with respect to charge. For example, in $Fe_{0.95}O$, for every 3 Fe^{2+} ions removed, they are replaced by two Fe^{3+} to achieve the atomic stoichiometry. Similarly in $Cu_{1.77}S$, for every $2Cu^+$ ions removed, they are replaced by one Cu^{2+} ion. In all these examples, the cations and anions are located at the appropriate site in the crystal lattice.

Another example of a point defect is the **Frenkel** defect where an atom or an ion has moved to an interstitial site from its normal site in the crystal lattice. AgCl has the rocksalt structure and a small number of Ag^+ ions move from octahedral to tetrahedral sites (Figure 6.19b), but the stoichiometry remains unchanged. Frenkel defects occur most often in structures with low coordination number where the radius of the cation is smaller than the radius of the anion. Table 6.4 shows that the radius of Ag^+ changes from 129 to 114 pm on going from octahedral to tetrahedral geometry (Na^+ only changes from 116 to 113 pm). Figures 6.13 and 6.14 show that there is significant open space in the crystal lattices described by the hexagonal and cubic close packing

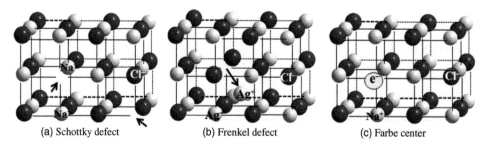

(a) Schottky defect   (b) Frenkel defect   (c) Farbe center

Figure 6.19 *(a) Schottky defect: arrows indicate missing of equal number of cation and anion sites. (b) Frenkel defect: arrow shows Ag^+ ion moved from an octahedral to a tetrahedral site. (c) Farbe center. Structures produced using Hyperchem™ version 8.0.10*

schemes so movement of the cation (or the anion) to interstitial sites can occur. Although the ionic lattice has a balanced charge, there is a slight imbalance in charge in the area of the lattice where the cation moved, and this may affect reactivity.

Another example of point defects is the **Farbe** or colored center (Figure 6.19c), which occurs when an electron in the crystal lattice is excited by heat or by radiation from visible, UV, and X-ray energy. The electron occupies a vacant anion site and a characteristic color specific to the ionic compound is observed when it is excited. For example, sodium iodide produces a yellow color when ionized by an X-ray beam during an X-ray diffraction experiment. The color can be enhanced by doping a small amount of sodium metal into the crystal of sodium iodide.

In some instances, two different atoms or ions can interchange in the crystal lattice. For example, in β-brass with alternating Cu and Zn atoms, two Cu or Zn atoms can become next nearest neighbors on heating. The **atom interchange** defect occurs more frequently in alloys. The repulsive forces of two cations or two anions together in a crystal lattice prevents ions of the same charge from interchanging to be nearest neighbors. In **ternary compounds** like the spinels, Fe^{2+} and Fe^{3+} ions can interchange between octahedral and tetrahedral sites. If one Mg and one Al ion interchange as in Figure 6.16, which shows the spinel $MgAl_2O_4$, it would be similar to interchange of a tetrahedral Fe^{2+} with an octahedral Fe^{3+} to produce the inverse spinel. Octahedral Fe^{3+} has a radius of 78.5 pm similar to the radius of tetrahedral Fe^{2+} (77 pm) whereas tetrahedral Fe^{3+} and octahedral Fe^{2+} have radii of 63 and 92 pm, respectively.

6.3.8.2 Impurities (Extrinsic Defects)

Gemstones. The incorporation in nature of low levels of one atom for another atom in a crystal structure leads to different chemical and physical properties. **Corundum** (α-Al_2O_3) crystallizes in a cubic close packing of Al atoms between O atoms with two-thirds of the octahedral sites occupied. Al is six coordinate and each O is four coordinate. In nature, Cr^{3+} substitutes stoichiometrically for Al^{3+} (0.2–1% atomic) leading to the gemstone ruby, which has a red color from the absorption of green light due to the excitation of its d electrons (see Section 8.8.5). Other gemstones also have stoichiometric replacement of a major ion with impurities.

In Figure 6.7b and c, the doping of small amounts (1 ppm or less) of material into intrinsic semiconductors leads to enhanced electrical properties.

Substitution in Biogenic Carbonates. Many organisms make shells or skeleton material from calcium carbonate. When organisms form minerals as a part of their life process, the process is called **biomineralization**. Macrofauna (clams, mussels, some corals, etc.) typically produce calcite shells or skeletons, but other corals as well as otoliths (a structure in the inner ear of fish) form aragonite. Phytoplankton such as coccolithophores and foraminifera also produce calcite skeletons whereas pteropods produce aragonite skeletons. Figure 6.18 shows that there is significant space between Ca^{2+} and CO_3^{2-} ions in calcite and aragonite, but the unit cell is largest for calcite. Thus, substitution of other metal and/or nonmetal ions for Ca^{2+} and CO_3^{2-} ions in $CaCO_3$ minerals occurs during their formation. It is easier for larger ions (Sr^{2+}, Ba^{2+}) to replace Ca^{2+} in aragonite as these larger ions have the aragonite structure.

In many types of corals, carbonate shells or skeletons that have the aragonite structure, metals such as Sr^{2+} substitute more for Ca^{2+} when waters become colder [14]. Thus, the measurement of Sr concentration over the shell's length (a measure of its life span) reflects the temperature of the waters latitudinally or seasonally (if the shell is studied over its entire length using microscopic techniques or laser ablation with elemental analysis by inductively coupled plasma -mass spectrometry, ICP-MS). These types of measurements have also been used on older material in museum and other collections to obtain data on sea surface temperature over time. This type of metal substitution is an inorganic or abiotic process that can be characterized by the temperature-dependent "distribution coefficient" (D_M; Equation 6.15).

$$D_M = \frac{\left\{\frac{M}{Ca}\right\}_{shell}}{\left\{\frac{M}{Ca}\right\}_{water}} \tag{6.15}$$

$\{M/Ca\}_{shell}$ and $\{M/Ca\}_{water}$ are the molar ratios of a given metal (M) in the shell (coral, otolith, etc.) and ambient water, respectively. Equation 6.15 ignores kinetic and metabolic controls on $CaCO_3$ formation [15]. Organisms that grow in polluted environments also easily incorporate other divalent metal ions (Zn, Cd, Hg, Pb, etc.) in their carbonate structures along with Ca^{2+}, and can be used to give a record of pollution history [16]. Likewise, because CO_3^{2-} ions are planar, spherical ions such as SO_4^{2-} can substitute for them alone or in combination with Ba as $BaSO_4$. As arsenic exists normally as AsO_3^{3-} and AsO_4^{3-} incorporation of arsenic in carbonate biominerals indicates possible pollution. The study of cations and anions in carbonate materials is a very active research area.

6.4   Nanoparticles and Molecular Clusters

Emphasis in this chapter has been on the discussion of the extended or infinite lattice of solids. However, the reaction of solution materials that form solids frequently leads to the formation of nanoparticles, which are defined as materials ranging between 1 and 100 nm in size (see Chapter 12). Nanoparticles are frequently found in the environment [17], but they are being actively synthesized for semiconductor, energy, medical, and other applications. Nanoparticle size can be calculated from spectroscopic data, and the calculations are in reasonable agreement with size information from electron microscopy. Table 6.1 gives information on the band gap for several important minerals including some that have semiconductor behavior (e.g., the metal sulfides), and these band gaps lie in the UV or visible region. The excitation of an electron from the valence band to the conductance band of a semiconductor (with formation of an electron–hole pair) is analogous to the electronic transition from the HOMO to the LUMO in a molecule. The UV–Vis spectra of nanoparticles are blue shifted from that of the bulk material; i.e., the absorption peak shifts to lower wavelength or higher energy. The difference between the energy of the peak for the bulk material and the peak of the nanoparticulate material (ΔE) is related to the radius (R) of the nanoparticle (Equation 6.16; [18]).

$$\Delta E = E_{np} - E_{bulk} = \frac{hc}{\lambda} - E_{bulk} = \frac{\pi^2 \hbar^2}{2R^2}\left(\frac{1}{m_e} + \frac{1}{m_h}\right) - \frac{1.8e^2}{\varepsilon R} \tag{6.16}$$

Here, m_e and m_h are the effective masses of the electron and the hole, respectively, e is the charge of an electron, $\hbar = (h/2\pi)$, and ε is the dielectric constant for the semiconductor ($\varepsilon = 4\pi c\varepsilon_o$, where c is the constant for the particle as in Equation 6.3). The highest wavelength, λ, of the absorption peak is related to the first excitation of an electron from the valence band to the conduction band of the nanoparticle and is used to find its band gap energy (Figure 6.20). The first term in Equation 6.16, which dominates when R is small, corresponds to the confinement energies for the electron–hole pair (i.e., particle-in-a-box), and the second term accounts for the Coulombic interaction between the electron and the hole. This process is also known as the **quantum confinement** effect. These model calculations are upper estimates because coupling of electronic states to vibrational contributions and the structure of the surface are ignored. Still, there is good agreement between these size calculations and the size determined by electron microscopy.

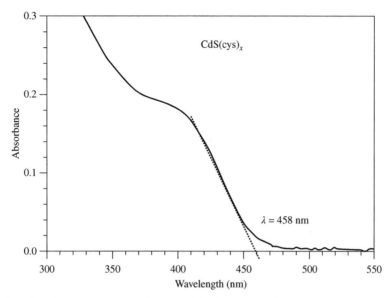

Figure 6.20 *UV/vis absorbance spectrum of a typical thiolate capped CdS nanoparticle solution. The sample had 600 μM cysteine, 150 μM Cd(II), and 150 μM S(-II) in 2.2 μM NaHCO$_3$ (pH 7.8)*

Nanoparticles do not exhibit identical chemical behavior with the bulk solid or mineral for several reasons. First, they have increased surface area. Figure 6.21 demonstrates the decreasing number and fraction (F) of surface atoms to total atoms in a cube as the number of atoms increases. For a cube with a length of 30 atoms on a side, the fraction of surface atoms is <19%. Although the individual atoms composing the nanoparticle are more exposed to the medium in which they reside, they may not be more reactive. For example, they can be effectively capped by organic material, which may prevent their dissolution or their growth to larger particles. The CdS nanoparticles giving the spectrum in Figure 6.20 were capped by the thiol-containing amino acid, cysteine, which binds Cd(II) preventing further reaction with sulfide and growth of CdS. Second, the nanoparticles in Figure 6.21 are ideal or "perfect" as they have no defects; thus, the lattice energy is maximized. As defects increase, lattice energy becomes more positive and nanoparticle reactivity, such as dissolution, can increase. Third, the band gap increases compared to the bulk solid as in Equation 6.16, which can make the LUMO band more positive and less ready to accept electrons during nanoparticle reduction reactions. Likewise, the HOMO will become more stable and less ready to donate electrons during nanoparticle oxidation reactions.

Molecular cluster is a term used to describe a very small group of atoms about the size of a unit cell or more. In Figure 6.21, the cube with eight atoms is an example of a cluster. Ferredoxins are Fe_4S_4 units with Fe–S–Fe bonds (Figure 12.8); the Fe atoms are also bound to organic substrates and are an example of a bioinorganic molecular cluster. In Figures 6.13 and 6.14, the cyclic Zn_3S_3 and Zn_4S_6 subunits of sphalerite and wurtzite are examples of possible molecular clusters that may be stabilized with capping agents (see Sections 12.2 and 12.4).

Figure 4.6 shows **cage structures** that are found in tetrahedral P_4, icosahedral $B_{12}H_{12}^{2-}$ as well as various allotropes of B, and in C_{60} fullerene. Cyclic S_8 (Figure 4.15) is another cluster example. The neutral compounds P_4O_6 and P_4O_{10} are cage structures where O atoms insert between the P–P bonds of the P_4 tetrahedron

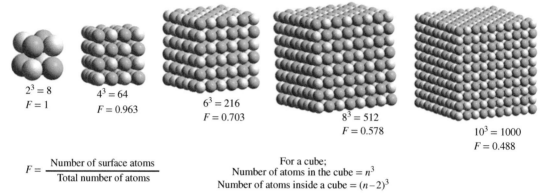

$$2^3 = 8$$
$$F = 1$$

$$4^3 = 64$$
$$F = 0.963$$

$$6^3 = 216$$
$$F = 0.703$$

$$8^3 = 512$$
$$F = 0.578$$

$$10^3 = 1000$$
$$F = 0.488$$

$$F = \frac{\text{Number of surface atoms}}{\text{Total number of atoms}}$$

For a cube;
Number of atoms in the cube $= n^3$
Number of atoms inside a cube $= (n-2)^3$

Figure 6.21   *Cubic nanoparticles (e.g., NaCl or PbS) show the decrease in the number of surface atoms with increasing size of the cube. Models produced with Hyperchem™ version 8.0.10*

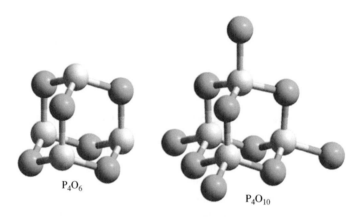

P_4O_6

P_4O_{10}

Figure 6.22   *Structures of the cage compounds P_4O_6 and P_4O_{10}. White = P; Blue = O*

(Figure 6.22). The P_4O_6 structure is similar to adamantane and the Zn_4S_6 structure in Figure 6.13a. In P_4O_6, P is +3 and in P_4O_{10}, P is +5.

The reaction of molybdate with phosphate in acidic solution forms the **phosphomolybdate anion** $[PMo_{12}O_{40}]^{3-}$, which is used to determine phosphate in natural waters.

$$PO_4^{3-} + 12MoO_4^{2-} + 27H^+ \rightarrow H_3PMo_{12}O_{40} + 12H_2O$$

Its ammonium salt $\{(NH_4)_3[PMo_{12}O_{40}]\}$ was first synthesized by Berzelius in 1826 and exhibits the α-**Keggin** cage-like structure shown in Figure 6.23. The P atom in PO_4^{3-} is tetrahedral and PO_4^{3-} is surrounded by 12 neutral MoO_3 units with the 12 Mo(VI) ions in a **cuboctahedral** arrangement (also see Figure 4.6); each O in PO_4^{3-} coordinates to three Mo. Thus, each Mo is six coordinate (MoO_6 octahedra) as there are five bridging O atoms (four Mo–O–Mo bridges and one Mo–O–P bridge for each Mo) plus one Mo=O for each Mo. There are a number of anions with the stoichiometry of $[XM_{12}O_{40}]^{n-}$ or $[(XO_4)(M_{12}O_{36})]^{n-}$ where M = Mo(VI) or W(VI) and X = P^{5+}, Si^{4+} and B^{3+}. These are known as **heteropoly** compounds or anions, which are important

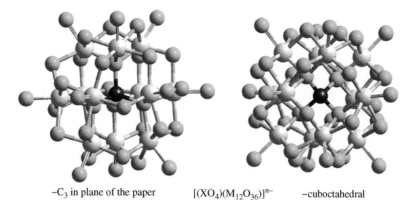

~C_3 in plane of the paper $[(XO_4)(M_{12}O_{36})]^{n-}$ ~cuboctahedral

Figure 6.23 *Two views of the same α-Keggin structure for the phosphomolybdate anion* $[PMo_{12}O_{40}]^{3-}$ *produced with Hyperchem*[TM] *version 8.0.10. White = Mo; Black = P; Blue = O*

catalysts in organic synthesis [19] and water splitting (oxidation) to form O_2 [20]. For $[PMo_{12}O_{40}]^{3-}$, mild reducing agents such as ascorbic acid can reduce Mo(VI) to Mo(V), resulting in a blue color with absorption peaks at 660 and 882 nm for PO_4^{3-} analysis.

References

1. Strehlow, W. H. and Cook, E. L. (1973) Compilation of energy band gaps in elemental and binary compound semiconductors and insulators. *Journal of Physical Chemistry Reference Data*, **2** (1), 163–198.
2. Xu, Y. and Schoonen, M.A.A. (2000) The absolute energy positions of conduction and valence bands of selected semiconducting minerals. *American Mineralogist*, **85**, 543–556.
3. Haynes, W. M. (2014) *Handbook of Chemistry and Physics* (95th ed.), CRC press, Boca Raton, FL.
4. Burdett, J. K. (1995) *Chemical Bonding in Solids*, Oxford University Press, Oxford, pp. 319.
5. Goldschmidt, V. M. (1954) *Geochemistry* (ed. A. Muir). Clarendon Press, Oxford. pp. 730.
6. Pauling, L. (1960) *The Nature of the Chemical Bond* (3rd ed.), Cornell University Press, Ithaca, NY.
7. Shannon, R. D. and Prewitt, C. T. (1969) Effective ionic radii in oxides and fluorides. *Acta Crystallographica B*, **25**, 925–946.
8. Shannon, R. D. and Prewitt, C. T. (1970) Revised values of effective ionic radii. *Acta CrystallographicaB*, **26**, 1046–1048.
9. Shannon, R. D. (1976) Revised effective ionic radii and systematic studies of interatomic distances in halides and chalcogenides. *Acta CrystallographicaA*, **32**, 751–767.
10. Kapustinskii, A. F. (1956) Lattice energy of ionic crystals. *Quarterly Reviews of the Chemical Society*, **10**, 283–294.
11. Jenkins, H. B. D. and Thakur, K. P. (1979) Reappraisal of thermochemical radii for complex ions. *Journal of Chemical Education*, **56**, 576–577.
12. Bailar, J. C., Eméleus, H. J. and Trotman-Dickenson, A. F. (1973) *Comprehensive Inorganic Chemistry*, Vol. 1, Pergamon Press, pp. 1487.
13. Wells, A. F. (1986). *Structural Inorganic Chemistry* (5th ed.), Clarendon Press, Oxford, pp. 1382.

14. Sadler, J., Webb, G. E., Nothdurft, L. D. and Dechnik, B. (2014) Geochemistry-based coral palaeoclimate studies and the potential of 'non-traditional' (non-massive *Porites*) corals: Recent developments and future progression. *Earth-Science Reviews*, **139**, 291–316.

15. Mackenzie, F. T. and Lerman, A. (2006) *Carbon in the Geobiosphere: Earth's outer shell*, Springer Dordrecht, The Netherlands, pp. 402.

16. Gillikin, D. P., Dehairs, F., Baeyens, W., Navez, J., Lorrain, A. and Andre, L. (2005) Inter- and intra-annual variations of Pb/Ca ratios in clam shells (Mercenaria mercenaria): a record of anthropogenic lead pollution? *Marine Pollution Bulletin*, **50**, 1530–1540.

17. Hochella, M. F., Lower, S. K., Maurice, P. A., Penn, R. L., Sahai, N., Sparks, D. L. and Twining, B. S. (2008) Nanominerals, mineral nanoparticles, and Earth systems. *Science*, **319**, 1631–1635.

18. Brus, L. E. (1984) Electron-electron and electron-hole interactions in small semiconductor crystallites - the size dependence of the lowest excited electronic state. *Journal of Chemical Physics*, **80**, 4403–4409.

19. Sun, M., Zhang, Z., Putaj, P., Caps, V., Lefebvre, F., Pelletier, J. and Basset, J.-M. (2014) Catalytic Oxidation of Light Alkanes (C1–C4) by Heteropoly Compounds. *Chemical Reviews*, **114**, 981–1019.

20. Kärkäs, M. D., Verho, O., Johnston, E. V. and Åkermark, B. (2014) Artificial photosynthesis: molecular systems for catalytic water oxidation. *Chemical Reviews*, **114**, 11863–12001.

7

Acids and Bases

7.1 Introduction

Acids and bases have been known since before Roman times, for their ability to transform a chemical substance into other chemical forms. For example, sulfuric acid (H_2SO_4, oil of vitriol; vitriolic acid) could be produced by the aqueous oxidation of pyrite (FeS_2), which is a common mineral used for many purposes [1]. H_2SO_4 dehydrates sugar as well as metal hydrates with sometimes dramatic color changes, and also dissolves metals via redox reactions. It is normally the chemical that is produced and sold in greatest quantity in the world each year.

$$H_2SO_4 + C_6H_{12}O_6 \rightarrow 6C_{black} + 5H_2O + \text{sulfuric acid/water mix}$$

$$H_2SO_4 + CuSO_4 \cdot 5H_2O_{(blue)} \rightarrow CuSO_{4(white)} + 5H_2O$$

$$H_2SO_4 + Fe \rightarrow H_2 + Fe^{2+} + SO_4{}^{2-}$$

$$2H_2SO_4 + Cu \rightarrow SO_2 + 2H_2O + SO_4{}^{2-} + Cu^{2+}$$

The first base was called lye, which was obtained by leaching ashes with water producing potassium hydroxide solution. Lye is a common name for bases, which are used in a variety of purposes including wood degradation into paper or fibers and soap production.

7.2 Arrhenius and Bronsted–Lowry Definitions

Common acids such as hydrochloric acid (HCl) and nitric acid (HNO_3) were not formally prepared until about the 16th century, so formal definitions of acids and bases similar to bonding theories are relatively new. In 1884, Svante Arrhenius defined an acid as a chemical species that when dissolved in water produced the hydrogen ion, H^+. Although useful, the definition is limited as it does not encompass a large variety of reactions and only considers water as solvent.

In 1923, Brønsted and Lowry defined an acid and base reaction as one that involves a hydrogen ion transfer between two reactants. The acid is the hydrogen ion donor (the Brønsted acid) and the base is the hydrogen

Inorganic Chemistry for Geochemistry and Environmental Sciences: Fundamentals and Applications, First Edition. George W. Luther, III.
© 2016 John Wiley & Sons, Ltd. Published 2016 by John Wiley & Sons, Ltd.
Companion Website: www.wiley.com/go/luther/inorganic

ion acceptor (the Brønsted base). This definition applies to all solvents and the gas phase. For example, the reaction between HCl and H_2O can occur in water as solvent or in the gas phase (an atmospheric reaction) and results in **complete** H^+ transfer to form the hydronium ion, H_3O^+. The reactions are written as follows to express the physical state of the species: aq = aqueous or water as solvent; g = gas; l = liquid (solvent other than water).

$$HCl_{(l)} + H_2O_{(l)} \rightarrow \quad Cl^-_{(aq)} \quad + \quad H_3O^+_{(aq)}$$
$$HCl_{(g)} + H_2O_{(g)} \rightarrow \quad Cl^-_{(g)} \quad + \quad H_3O^+_{(g)}$$
$$\text{acid} \quad \text{base} \quad \text{conjugate base} \quad \text{conjugate acid}$$

HCl is termed the acid and Cl^- (a product) is termed its **conjugate base**. Likewise, H_2O is the base and H_3O^+ (a product) is the **conjugate acid** of H_2O.

The reaction between HCl and NH_3 can occur in water as solvent, in liquid ammonia as solvent, or in the gas phase (forming an aerosol particle) as in the following three reactions.

$$HCl_{(l)} + NH_{3(aq)} \rightarrow Cl^-_{(aq)} + NH_4^+_{(aq)}$$

$$HCl_{(l)} + NH_{3(l)} \rightarrow Cl^-_{(l)} + NH_4^+_{(l)}$$

$$HCl_{(g)} + NH_{3(g)} \rightarrow Cl^-_{(g)} + NH_4^+_{(g)} \rightarrow [NH_4][Cl]_{(g)} \text{ or } NH_4Cl$$

Formation of $NH_4Cl_{(g)}$ as an ion pair results in **aerosol** or gas particle formation. Aerosol formation reactions are very important in the atmosphere and are dominated by the reaction of $NH_{3(g)}$ with $H_2SO_{4(g)}$. Both reactants can come from natural (e.g., SO_2 from volcanoes followed by oxidation) and anthropogenic sources, to form a variety of (nano)particles of different stoichiometry depending on the number of gaseous acid and base molecules present.

The HCl reactions result in complete hydrogen ion transfer so HCl is defined as a strong acid. The acid strength of an aqueous solution can be determined by the pH of the solution (Equation 7.1).

$$pH = -\log\{H_3O^+\} \tag{7.1}$$

Here, $\{H_3O^+\}$ is the activity of the hydrogen ion although $[H_3O^+]$, the molar hydrogen ion concentration, is frequently used. For a 0.1 M HCl solution, the pH will be 1 from Equation 7.1 whereas a 0.01 M HCl solution has a pH of 2.

HF, acetic or ethanoic acid (CH_3COOH or HAc) and carbonic acid (H_2CO_3) are examples of weak acids as complete hydrogen ion transfer to water does not occur. The reactions below show a two-edged arrow indicating an equilibrium reaction between the reactants and the products as the reaction does not result in complete dissociation of hydrogen ion from the parent acid.

$$HF_{(l)} + H_2O_{(l)} \leftrightarrow F^-_{(aq)} + H_3O^+_{(aq)} \qquad K_a = 3.5 \times 10^{-4} \text{ at } 25\,°C$$

$$HAc_{(l)} + H_2O_{(l)} \leftrightarrow Ac^-_{(aq)} + H_3O^+_{(aq)} \qquad K_a = 1.74 \times 10^{-5} \text{ at } 25\,°C$$

$$H_2CO_3 + H_2O_{(l)} \leftrightarrow HCO_3^-_{(aq)} + H_3O^+_{(aq)} \qquad K_a = 4.3 \times 10^{-7} \text{ at } 25\,°C$$

The strength of an acid is represented by the equilibrium reaction as in Equation 7.2 for HF where K_a is the **acid ionization constant** (also dissociation constant or acidity constant). Here [ ] are used to denote concentration, but for accurate work, the **activity** denoted by { } should be used. The activity of H_2O, the solvent, is assumed to be 1 when the acid solution is dilute so is omitted as in the right of Equation 7.2.

$$K_a = \frac{[F^-][H_3O^+]}{[HF][H_2O]} = \frac{[F^-][H_3O^+]}{[HF]} \tag{7.2}$$

A value of $K_a < 1$ or $pK_a > 1$ (where $pK_a = -\log K_a$) indicates that the acid does not dissociate or ionize to a significant extent; thus, the fraction of ionization can be readily calculated from the initial concentration of the weak acid and the K_a. Physical and chemical parameters need to be specified as K_a is a function of temperature as well as pressure (P) and ionic strength (abbreviated as I or μ) of the solution. Table 7.1 provides some pK_a values for common acids. Strong acids have negative values.

Table 7.1 shows that some acids can lose two or more H^+. For H_3PO_4, there are three acid ionization constants listed as $pK_{a1} = 2.12$, $pK_{a2} = 7.21$, and $pK_{a3} = 12.67$). There is only one pK_a listed for H_2S because of the uncertainty in the value of pK_{a2}, which is estimated to be > 18 [2] compared to the value of ~ 14 that is reported in many texts.

An important way to show the fraction (α) of each chemical species to the total concentration of all species versus pH is with a **speciation** or species distribution diagram. The fraction of each acid–base species is calculated using all the acid dissociation constants for the chemical species [3, 4]. Figure 7.1 shows diagrams for the H_2S and CO_2 (7.7.3.1) systems. On increasing pH, the conversion of equilibrium species from one to the other normally occurs within 3 pH units.

The dissolution of a base, such as ammonia, in water can be represented in a similar way (Equation 7.3) where K_b is the **basicity** constant. The activity of H_2O, the solvent, is assumed to be 1. Here, the extra pair

Table 7.1   *Common acids and their pK_a values at 25°C*

Acid	pK_a	Acid	pK_a	Acid	pK_a	Acid	pK_a	Acid	pK_a
HI	−11	$HClO_4$	−10	H_2SO_4	−9	HNO_3	−2	H_3PO_4	2.12
HBr	−9	$HClO_3$	−1	HSO_4^-	1.92	HNO_2	3.25	$H_2PO_4^-$	7.21
HCl	−7	$HClO_2$	2	H_2SO_3	1.81	H_3AsO_4	2.23	HPO_4^{2-}	12.67
HF	3.45	HClO	7.2	H_2S	7.04	$H_2AsO_4^-$	6.95		
						$HAsO_4^{2}$	11.5		
CH_3COOH	4.75	$CClH_2COOH$	2.85	HCOOH	3.75	H_2O_2	11.6	H_2O	14

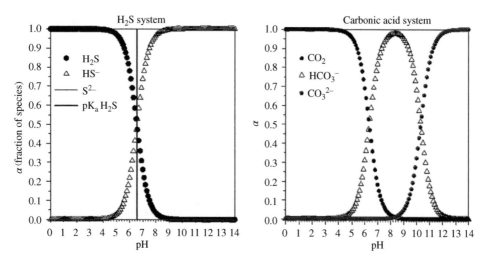

Figure 7.1   *Speciation diagrams for the H_2S (seawater) and CO_2 (freshwater) systems*

of electrons on N bond to H^+ from H_2O.

$$NH_{3(aq)} + H_2O_{(l)} \leftrightarrow NH_4^+{}_{(aq)} + OH^-{}_{(aq)} \qquad K_b = 3.5 \times 10^{-4} \text{ at } 25\,°C$$
$$\text{base} \qquad \text{acid} \qquad \text{conjugate acid} \quad \text{conjugate base}$$

$$K_b = \frac{[NH_4^+][OH^-]}{[NH_3][H_2O]} = \frac{[NH_4^+][OH^-]}{[NH_3]} \tag{7.3}$$

Replacing a H atom on NH_3 with electron-withdrawing groups such as OH produces weaker bases; e.g., NH_2OH (hydroxylamine) and H_2NNH_2 (hydrazine) withdraw electrons from the central N atom and their K_b values are 1.07×10^{-8} and 1.70×10^{-6}, respectively. Alkyl amines are stronger bases so have a higher K_b than NH_3; the CH_3 group in CH_3NH_2 ($K_b = 4.37 \times 10^{-4}$) donates electron density relative to H atoms to the central N atom.

7.3   Hydrolysis of Metal–Water Complexes

On dissolution of a salt in water, metals form complexes with water such as $Fe(H_2O)_6^{3+}$. Here, the metal is fully coordinated or hydrated with water molecules, and is called a **free metal ion**. Free metal cations, $[M(H_2O)_6]^{x+}$, **hydrolyze** in water releasing hydrogen ions and forming hydroxo species according to Equation 7.4 for $Fe(H_2O)_6^{3+}$. This reaction can be written as an equilibrium reaction that gives the hydrolysis constant K_H (Equation 7.5), which is another way to express K_a. The activity of H_2O, the solvent, is assumed to be 1.

$$Fe(H_2O)_6^{3+} + H_2O \leftrightarrow [Fe(H_2O)_5(OH)]^{2+} + H_3O^+$$
$$\text{acid} \qquad \text{base} \qquad \text{conjugate base} \qquad \text{conjugate acid} \tag{7.4}$$

$$K_H = \frac{[Fe(H_2O)_5(OH)^{2+}][H_3O^+]}{[Fe(H_2O)_6]^{3+}[H_2O]} = \frac{[Fe(H_2O)_5(OH)^{2+}][H_3O^+]}{[Fe(H_2O)_6]^{3+}} \tag{7.5}$$

The pK_H value ($-\log K_H$ or the tendency for free metal ions to hydrolyze) is similar to pK_a and generally decreases with an increase in the square of the ionic charge to radius of the cation (Z^2/r) as shown in Table 7.2. Z^2/r and Z/r are similar to electronegativity and are a good measure of the molecule's or ion's tendency to draw electrons to itself in bonds and stabilize higher oxidation states. The increasing tendency to hydrolysis can be high for elements in the same family of the periodic table (e.g., it decreases for the ions from Li to K; and Be to Ba) or for elements with two or more oxidation states (e.g., $Tl^{3+} > Tl^+$; $Fe^{3+} > Fe^{2+}$). However, the values of the radii, which are used for the Z^2/r calculation, are Shannon [7] crystal radii and not solution radii. Thus, there is not as smooth an increase of Z^2/r with decreasing pK_H; e.g., notice the slight differences going across the first transition series (Mn to Zn) for the +2 cations. There is an irregular pattern for Al and the elements below it probably due to the first filling of the d and f orbitals (Scandide and Lanthanide contractions).

The elements with the smallest pK_H will hydrolyze further, resulting in precipitation of metal oxides or metal (oxy)hydroxides. In the case of $Fe(H_2O)_6^{3+}$, the result will be precipitates such as FeOOH (goethite) and Fe_2O_3 (hematite). Precipitation can be prevented by binding with organic compounds or chelates with functional groups containing two or more O, N, and S atoms (e.g., Table 8.1).

The addition of base (OH^- or NH_3) to free metal cations is an acid–base titration and results in the precipitation of metal hydroxides, (oxy)hydroxides, or oxides. These are **condensation** reactions that lose H_2O and H^+ as H_3O^+ leading to polymeric materials (e.g., the dimer in Equation 7.6 has bridging hydroxide groups;

Table 7.2 *The first hydrolysis constants (pK$_H$) for selected metal ions at infinite dilution from [5, 6] arranged to decrease on descending the periodic table. Data for Z^2/r in units of (C^2 m^{-1} × 10^{28}) are calculated from the six coordinate radii given by Shannon [7] except for Be, which is four coordinate*

Main group	Z^2/r	pK$_H$	First transition series	Z^2/r	pK$_H$	Second/third/post transition series	Z^2/r	pK$_H$
K$^+$	1.69	14.46	**Mn^{2+}**	10.6	10.6	**Ag$^+$**	1.99	12.0
Na$^+$	2.21	14.18	**Ni^{2+}**	12.4	9.86			
Li$^+$	2.85	13.64	**Co^{2+}**	11.6	9.65	**Ac^{3+}**	18.3	<10.4
Tl$^+$	1.57	13.21	**Fe^{2+}**	11.2	9.58	**Cd^{2+}**	9.42	10.08
Ba^{2+}	6.89	13.47	**Zn^{2+}**	11.7	8.96	**Pb^{2+}**	7.72	7.71
Sr^{2+}	7.78	13.29	**Cu^{2+}**	11.8	<8.0			
Ca^{2+}	9.01	12.85				**Sn^{2+}**	7.55	3.40
Mg^{2+}	11.9	11.44	**Sc^{3+}**	26.1	4.3	**Hg^{2+}**	8.85	3.40
Lanthanides	19–23	7.6–8.5	**Cr^{3+}**	30.6	4.01	**Th^{4+}**	38.0	3.20
Be^{2+}	17.4	5.40	**Co^{3+}**	30.8	2.92	**Pu^{4+}**	41.1	1.60
Al^{3+}	34.2	4.97	**Fe^{3+}**	29.4	2.19	**U^{4+}**	39.9	0.65
In^{3+}	24.6	4.00	**V^{3+}**	29.6	2.26	**Zr^{4+}**	47.7	0.3
Ga^{3+}	30.4	2.6	**Ti^{3+}**	22.5	2.2	**Hf^{4+}**	48.3	0.25
Tl^{3+}	22.5	0.62	**Mn^{3+}**	29.4	0.08			

Section 8.3) or solids with dimensions ranging from 1 nm to 1 μm. In the range from 1 to 100 nm, these are termed **nanoparticles** and in the range from 0.1 to 1 μm, they are also called **colloidal** suspensions. The earth's crust and mantle are composed of Fe and Si **polyoxo** anions that form extended three dimensional structures. Equation 7.6 is an example of an entropy favored reaction as is Equation 12.2 (Section 12.2).

$$2Fe(H_2O)_6^{3+} \leftrightarrow [Fe_2(H_2O)_8(OH)_2]^{4+} + 2H_3O^+ \tag{7.6}$$

Other solvents undergo **solvolysis** (general term for solvent molecules binding an ion); for example, NH_3 reacts with metal ions as in Equation 7.4 to form $M(NH_3)_6^+$, which can form the species $M(NH_3)_5(NH_2) + NH_4^+$.

7.4 Hydration of Anhydrous Acidic and Basic Oxides

7.4.1 Acidic Oxides

The addition of water to anhydrous nonmetal oxides (Equations 7.7 and 7.8) from Groups 14–17 of the periodic table leads to the generation of oxoacids, which generate hydrogen ions; nonmetal oxides are thus acidic oxides. For sulfuric (H_2SO_4, $pK_a = -9$) and perchloric ($HClO_4$, $pK_a = -10$) acids, the H^+ is fully dissociated so that hydrolysis is virtually complete.

$$SO_3 + H_2O \rightarrow H_2SO_4 \rightarrow 2H^+ + SO_4^{2-} \tag{7.7}$$

$$ClO_3 + H_2O \rightarrow HClO_4 \rightarrow H^+ + ClO_4^- \tag{7.8}$$

The strength of an oxoacid is related to the electronegativity or formal charge on the central atom. Thus, nitric acid (HNO_3), perchloric acid and sulfuric acid are among the strongest acids known. Carbonic acid (H_2CO_3)

is weak and boric acid (H_3BO_3) is even weaker. There are four oxoacids of chlorine: HOCl, HOClO, HOClO$_2$, and HOClO$_3$ with pK_a values of 7.2, 2.0, −1, and −10, respectively. Increasing the number of O atoms not bound to H atoms also increases the acidity of these acids as well as the formal charge on the central atom (0, 1, 2, and 3 for these acids). A useful rule or equation to determine the pK_a for these acids is Equation 7.9 where m is the formal charge on the central atom and n is the number of O atoms not bound to H atoms [8].

$$pK_a = 8.0 - m(9.0) + n(4.0) \tag{7.9}$$

The same is true for the oxoacids of nitrogen (nitric acid, HONO$_2$, and nitrous acid, HONO) and sulfur [sulfuric acid $(HO)_2SO_2$ and sulfurous acid $(HO)_2SO$]. As shown in Table 7.1, replacing a H atom on CH_3COOH with a Cl atom results in a stronger acid, as the Cl atoms withdraw electrons from the carboxyl group, thereby releasing hydrogen ions to solution more readily.

Acidic oxides also react with base to form the conjugate base as in Equation 7.10.

$$CO_{2(aq)} + OH^- \leftrightarrow HCO_3^- \tag{7.10}$$

7.4.2 Basic Oxides

Addition of water to anhydrous metal oxides results in the production of hydroxide ions (Equations 7.11 and 7.12); metal oxides are thus basic oxides. Except for BeO, Group I and II metal oxides are basic oxides, and on descending each group, the basicity of the metal oxide increases.

$$Li_2O + 2H_2O \rightarrow 2LiOH \rightarrow Li^+ + OH^- \tag{7.11}$$

$$BaO + H_2O \rightarrow Ba(OH)_2 \rightarrow Ba^{2+} + 2OH^- \tag{7.12}$$

The reaction of a nonmetal oxide with a metal oxide (Equations 7.13 and 7.14) leads to salt formation.

$$BaO + SO_3 \rightarrow BaSO_4 \tag{7.13}$$

$$NH_3 + H_2SO_4 \rightarrow [NH_4]^+[HSO_4]^- \tag{7.14}$$

7.4.3 Amphoteric Oxides

Many oxides (e.g., BeO, Al_2O_3, Ga_2O_3) react with both acids and bases and are called **amphoteric** oxides (Equations 7.15 and 7.16; note the change in coordination for Al). Boron oxides are acidic but the other metal oxides in Group 13 are amphoteric. For transition metals, low oxidation state metal oxides (e.g., CrO) are normally basic. Intermediate oxidation states (e.g., Cr_2O_3) tend to be amphoteric, and high oxidation states are acidic (e.g., CrO_3).

$$Al_2O_3 + 6H_3O^+ + 3H_2O \rightarrow 2\ [Al(OH_2)_6]^{3+} \tag{7.15}$$

$$Al_2O_3 + 2OH^- + 3H_2O \rightarrow 2\ [Al(OH)_4]^- \tag{7.16}$$

7.5 Solvent System Definition

Water and other solvents autoionize by forming a cation (the acid) and an anion (the base) representative of that solvent. The following three reactions show H$^+$ transfer and the formation of cations and anions.

The reverse reaction is termed the **neutralization** reaction.

acid	base	conjugate base	conjugate acid	
$H_2O_{(l)}$	$+ H_2O_{(l)}$	$\leftrightarrow OH^-_{(aq)}$	$+ H_3O^+_{(aq)}$	$K_{diss} = 10^{-14}$
$NH_{3(l)}$	$+ NH_{3(l)}$	$\leftrightarrow NH_2^-_{(l)}$	$+ NH_4^+_{(l)}$	$K_{diss} = 10^{-30}$
$H_2SO_{4(l)}$	$+ H_2SO_{4(l)}$	$\leftrightarrow HSO_4^-_{(l)}$	$+ H_3SO_4^+_{(l)}$	$K_{diss} = 10^{-4}$

The **autoionization reaction** is an equilibrium reaction as shown in Equation 7.17 for water where K_w is 1.01×10^{-14} at 25 °C (neutral pH = 7) and 1.148×10^{-15} at the freezing point of water (neutral pH = 7.47 at 0 °C). In seawater, K_w is 6.166×10^{-14} at 25 °C (neutral pH = 6.605) due to the increased ionic strength [9, 10].

$$K_w = [H_3O^+]\,[OH^-] \tag{7.17}$$

Other solvents do not transfer an H^+, but their autoionization reaction still results in forming a cation (acid) and an anion (base) as in these solvents that transfer a chloride ion.

$$OPCl_{3(l)} + OPCl_{3(l)} \leftrightarrow OPCl_2^+_{(l)} + OPCl_4^-_{(l)}$$

$$BCl_{3(l)} + BCl_{3(l)} \leftrightarrow BCl_2^+_{(l)} + BCl_4^-_{(l)}$$

7.5.1 Leveling Effect

The strength of an acid has only been discussed in the context of water as a solvent, but it can vary depending on the solvent. The dissolution of acetic acid into water, liquid ammonia, and sulfuric acid gives much different products, indicating that it can even be a base in sulfuric acid. This phenomenon is known as the leveling concept, which states that all acids and bases stronger than the characteristic cation and anion of a given solvent be leveled to the latter. Note that the conjugate base HSO_4^- of H_2SO_4 and the conjugate acid NH_4^+ of NH_3 are formed in the acetic acid reactions.

$CH_3COOH + H_2O \leftrightarrow H_3O + CH_3COO^-$	weak acid
$CH_3COOH + NH_3 \rightarrow NH_4^+ + CH_3COO^-$	strong acid
$CH_3COOH + H_2SO_4 \rightarrow HSO_4^- + CH_3COOH_2^+$	base

7.6 Gas Phase Acid–Base Strength

Many reactions involving the hydrogen ion occur in the atmosphere or gas phase where solvent effects are negligible so another measure of acid–base strength is needed. The **proton affinity** (PA) of a base in the gas phase is the enthalpy or energy released (negative thermodynamic quantity) during reactions as defined in Equations 7.18a and 7.18b. The PA value is defined as the negative value of the $\Delta H_{reaction}$. It is a positive quantity as shown for neutral and negative ions in Table 7.3, and can be determined experimentally using ion cyclotron resonance spectroscopy.

$$B_{(g)} + H^+_{(g)} \rightarrow BH^+_{(g)} \tag{7.18a}$$

$$PA = -\Delta H_{reaction} \text{ or } \Delta H_{PA}$$

$$B^-_{(g)} + H^+_{(g)} \rightarrow BH_{(g)} \tag{7.18b}$$

Table 7.3 Proton affinity (PA) data for selected chemical species from Refs. [11, 12]. The values are arranged to increase for the neutral species (blue column) for comparison with H_2

Neutral species	IP (eV)	PA (eV)	PA (kJ mol^{-1})	Ionic species	PA (kJ mol^{-1})	Species	PA (kJ mol^{-1})
He	24.581	1.84	178				
H	13.595	2.65	256	H$^-$	1675		
N	14.545	3.39	327	N^{3-}	3084		
O$_2$	12.071	4.38	423	O$_2^-$	1476	HO$_2$	661
H$_2$	15.426	4.39	424				
O	13.615	5.04	486				
N$_2$	15.581	5.13	495	N$_3^-$	1439		
NO	9.264	5.51	532	NO$^-$	1519		
CO$_2$	13.769	5.68	548				
CH$_4$	12.99	5.72	552	CH$_3^-$	1745		
CO	14.014	6.15	593				
OH	12.9	6.2	598	OH$^-$	1635	HO$_2^-$	1573
C	11.265	6.42	619			H$_2$O$_2$	678
HC≡CH	11.41	6.65	642				
S	10.357	6.86	662				
C$_2$	12.15	6.9	666				
H$_2$O	12.62	7.22	697	OH$^-$	1635	O^{2-}	2318
H$_2$S	10.47	7.38	712	HS$^-$	1469	S^{2-}	2300
H$_2$Se		7.43	717	HSe$^-$	1466	Se^{2-}	2200
HCN	13.91	7.43	717	CN$^-$	1469		
CH	10.64	7.7	743				
PH$_3$		8.18	789	PH$_2^-$	1552		
NH$_3$	10.15	8.85	854	NH$_2^-$	1689	NH^{2-}	2565

The data listed in Table 7.3 can be understood when the proton affinity Equation 7.18 is broken into the sum of three separate processes via a **Born–Haber** cycle as in Figure 7.2 resulting in Equation 7.19 (the + sign for each process indicates that the enthalpy of the process costs energy as in the IP for the H atom, and the − sign indicates that the process releases energy). The first reaction [IP(H)] does not vary, but the bond dissociation energy and the electron affinity vary for each base considered. Note the minus sign in Equation 7.19 indicating the reverse of each process. The EA for B is the same as the IP for B$^-$ [IP(B$^-$)]. When the PA is known, the cycle can be used to estimate one of the other parameters ($\Delta H_{IP(B-)}$ or $\Delta H_{EA(B)}$), which is similar to Equation 6.13 for lattice energy.

$$H^+_{(g)} + e^- \rightarrow H_{(g)} \qquad \text{reverse of the ionization potential (IP) for the H atom [IP(H)] (+)}$$

$$B_{(g)} + H_{(g)} \rightarrow BH_{(g)} \qquad \text{reverse of the bond dissociation energy (BDE) (+)}$$

$$B^-_{(g)} \rightarrow B_{(g)} + e^- \qquad \text{reverse of the electron affinity (EA) for B [EA(B)] (−); IP for B$^-$ (+)}$$

$$B^-_{(g)} + H^+_{(g)} \rightarrow BH_{(g)} \qquad \text{proton affinity as } \Delta H_{PA} (−)$$

$$\Delta H_{PA} = -[\Delta H_{IP(H)} + \Delta H_{EA(B)} + \Delta H_{BDE(BH)}] = -[\Delta H_{IP(H)} - \Delta H_{IP(B-)} + \Delta H_{BDE(BH)}] \qquad (7.19)$$

The higher the value of PA in Table 7.3, the more basic the chemical species. Comparison of neutral and ionic species for the same central atom (e.g., H_2O, OH^-, O^{2-}) shows that ionic species have larger PA values as

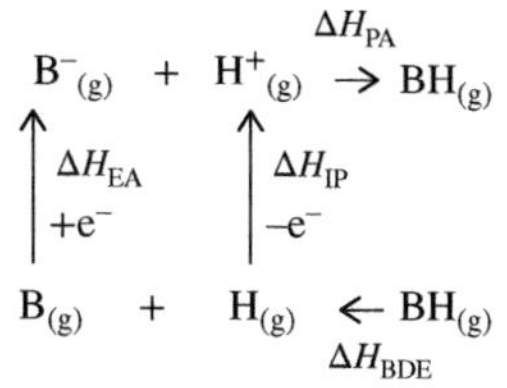

$$\begin{array}{ccc} & & \Delta H_{PA} \\ B^-_{(g)} & + \quad H^+_{(g)} & \rightarrow BH_{(g)} \\ \uparrow \Delta H_{EA} & \quad \uparrow \Delta H_{IP} & \\ +e^- & \quad -e^- & \\ B_{(g)} & + \quad H_{(g)} & \leftarrow BH_{(g)} \\ & & \Delta H_{BDE} \end{array}$$

Figure 7.2 *Born–Haber cycle for the determination of the ΔH_{PA}*

they are more basic. For H_2O, H_2S, and H_2Se, the PA values increase slightly because the bond dissociation energy decreases on descending the periodic table more than the electron affinity increases (e.g., 1.1 eV for H_2S versus 0.75 eV for H_2O).

The proton affinity of NH_3 is greater than that of H_2O because the bond dissociation energy of O–H bonds is higher than that of N–H bonds; i.e., the inverse of the bond dissociation energy is a higher positive enthalpy so decreases the PA for H_2O. Also the electron affinity for H_2O (0.75 eV) is greater than that for NH_3 (0.16 eV) and contributes to decreasing the PA value.

For NH_3, the PA value is larger than that of PH_3 as it is more basic. In this case, the N–H bond is stronger than the P–H bond; however, the electron affinity of NH_3 (0.16 eV) is substantially smaller than that for PH_3 of 1.263 eV. Thus, the larger EA for PH_3 results in its lower PA value. In many other instances, the EA is the dominant term in calculating PA.

7.6.1 H_3^+ as a Reactant

As noted in Section 5.5.1, the reaction of H_2 with H_2^+ forms the cation H_3^+, which is the most common cation in the universe. Table 7.3 shows that it should be an excellent hydrogen ion donor as all species with a larger PA value than H_2 will accept H^+ from H_3^+ as in Equation 7.20.

$$H_3^+ + X \rightarrow H_2 + HX^+ \tag{7.20}$$

Among the chemical species (X) in Table 7.3 that can accept H^+ are $\mathbf{N_2}$, the carbon compounds not containing hydrogen (CO_2, CO, C, and C_2) as well as those already containing hydrogen (CH_4 and HCN). These carbon reactions in interstellar space can lead to the formation of small organic molecules that are the building blocks for simple biochemicals such as amino acids [13].

7.7   Lewis Definition

In 1923, Lewis also defined an acid as an electron pair acceptor and a base as an electron pair donor. Inspection of all the acid–base reactions above shows that they meet the criterion of this definition. However, the Lewis definition can also account for neutral species and transition metal reactions in addition to ionization reactions as in the examples below. The reaction of a neutral acid and neutral base results in the formation of a Lewis acid–base adduct or **complex**, and the bond has been referred to as a **coordinate covalent** or **dative** bond. Planar BF_3 and $:N(CH_3)_3$ which is pyramidal, react to form such a bond. The arrow between N and B indicates donation of the electron pair from N to B. Both the B and N atoms assume tetrahedral geometry in the complex.

$$BF_3 + :N(CH_3)_3 \rightarrow (CH_3)_3N - BF_3 \text{ or } (CH_3)_3N \rightarrow BF_3$$

These reactions are important for all metal ions in aqueous solution as free metal ions have bound water molecules, which are displaced by another base, which is also called a **ligand**.

$$Ag^+_{(aq)} + 2 :NH_3 \rightarrow Ag(NH_3)_2^+; \text{ which is formally ammonia substitution for water}$$

$$Fe + 5 :C\equiv O: \rightarrow Fe(CO)_5$$

The Lewis acid–base definition has some similarity with redox chemistry (Chapter 2). An oxidizing agent accepts electrons; therefore, it is formally an acid whereas a reducing agent donates electrons and is formally a base.

7.7.1   MOT

The Lewis definition of an acid–base reaction corresponds well with the frontier molecular orbital theory (FMOT) approach, which indicates donation of an electron or electron pair from the HOMO of a base to the LUMO of an acid. If the symmetry characteristics of the HOMO and LUMO match and their energies are appropriate, the reaction should proceed quickly. This is the case for the hydroxide ion (−1.825 eV HOMO) donating a pair of electrons to H^+ (−13.6 eV LUMO) to form H_2O as discussed in Figure 5.22 and HS^- donating electrons to reduce I_2 to form I^- in Figures 5.27 and 5.28.

7.7.2   Molecular Iodine Adducts or Complexes as Examples

Many solvents act as electron pair donors to iodine to form a Lewis acid–base complex according to the following reaction, which can be monitored by UV–Vis spectroscopy (see Figure 1.2). I_2 absorbs at a wavelength (λ) of 500 nm (violet color is reflected), and the acid–base complexes absorb at wavelengths that are shifted toward the blue or UV region (giving a brownish color) as in Table 7.4. The shift in wavelength can be rationalized using MOT as these are examples of **charge transfer** transitions.

$$D: + I_2 \rightarrow D - I_2$$

On the left of Figure 7.3, the upper orbitals are shown for I_2 including the stable LUMO ($\sigma* = -2.55$ eV). The absorption at 500 nm (blue dashed arrow in Figure 7.3) is due to a $\pi * \rightarrow \sigma *$ transition between the I_2 orbitals. On reaction, the donor HOMO combines with the $I_2 \sigma*$ LUMO to form two new molecular orbitals (the bonding σDI_2 and antibonding $\sigma *DI_2$). Two absorption peaks are now possible; the first is the $\pi *I_2 \rightarrow \sigma *DI_2$ transition (blue arrow in Figure 7.3), which is of higher energy and lower wavelength than the original I_2 transition and which is responsible for the absorptions in Table 7.4. This transition is a direct measure of the bond strength; the greater the ΔE (also smaller the λ) of this transition, the greater the $D - I_2$ bond strength. The second possible absorption is due to the σDI_2 to $\sigma *DI_2$, but it is of high enough energy to be observed in the UV region. Both of these absorptions result from the mixing of electron donor and acceptor orbitals

Table 7.4   *Spectroscopic data for oxygen and nitrogen donor compounds to form a Lewis acid–base complex with I_2 [14]*

Donor compound (D)	λ (nm)	Donor compound (D)	λ (nm)
CH_3CN	464	$(CH_3)_2SO$	436
1,4-dioxane $[O(CH_2CH_2)_2O]$	447	pyridine	407

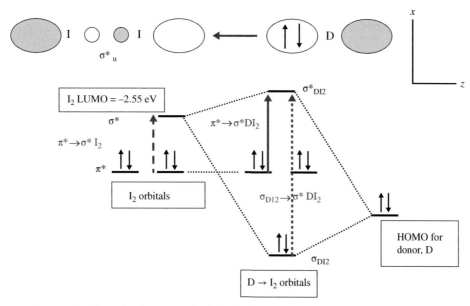

Figure 7.3 *Iodine molecular orbitals are on the left, the donor HOMO is on the right, and the LCAO of these orbitals to form the molecular orbitals of the complex is in the center (compare with Figure 5.46). The upper figure is a possible orbital combination between the donor and* I_2

so are considered **charge transfer** transitions from the donor to the acceptor. The frontier MO diagram for complex formation between I_2 and Lewis bases in Figure 7.3 will be similar for formation of other Lewis acid–base complexes.

7.7.3   Thermodynamics of Lewis Acid–Base Reactions

Drago and his group [14, 15] proposed Equation 7.21 to calculate the enthalpy of a Lewis acid–base reaction in nonpolar and nonbasic solvents; i.e., they do not solvate or coordinate with the metal. Here, A and B refer to the acid and base, respectively. **Nonsolvating** solvents do not exhibit cation (acid) or anion (base) character as in the solvent system definition of acids and bases; thus, their reactions resemble gas-phase-like behavior.

$$- \Delta H = E_A E_B + C_A C_B \tag{7.21}$$

The parameter E indicates the electrostatic contribution (combination of ionic or dipole–dipole) and C the covalent contribution to bonding. Equation 7.21 bears similarity to the Heitler–London and Mulliken wave function equations (Section 5.2.1) for H_2 as there are ionic and covalent terms for the total energy. E_A and C_A were defined with a value of 1 for I_2. Base parameters then were calculated from known enthalpy data collected via calorimetric and spectroscopic data between the reactions of I_2 with neutral bases. Table 7.5 shows E and C parameters for a few neutral acids and bases from [15]. Note that the C term is larger than the E term for all species other than I_2, indicating that the complexes formed exhibit more covalent bonding. A general rule from these data indicates that acids tending to bind covalently will bind to bases that bind covalently; the same is true for acids and bases that tend to bind as ionic species. Note that Table 7.4 showed that the wavelengths for these complexes shift to the blue region in accordance with these E and C data.

Table 7.5 E and C parameters for select acids and bases in kcal mol^{-1} [15]

Acid	E_A	C_A	Base	E_B	C_B
I_2	1.0	1.0	Pyridine	0.88	6.92
SO_2	1.12	0.726	$(CH_3)_2SO$	0.969	3.42
$(CH_3)_3B$	5.77	1.76	1,4 dioxane	0.68	2.82
ICl	4.15	1.61	$CH_3C{\equiv}N$	0.533	1.77

The calculations using Equation 7.21 work well for neutral acids and bases; e.g., the calculated and experimental values for the reaction of pyridine with I_2 are within 1 kJ mol^{-1}.

$$-\Delta H_{calc} = (1 \times 6.92) + (1.0 \times 0.88) = -7.80 \text{ kcal mol}^{-1} \text{ or } -32.6 \text{ kJ mol}^{-1}$$

$$\Delta H_{expt} = -33.3 \text{ kJ mol}^{-1}$$

However, to accommodate charged species, which transfer electron density from negative ions to positive ions or neutral molecules, additional t_A and t_B terms were later added. Overall, this method developed parameters for hundreds of molecules and allowed the prediction of reactivity between molecules for which no experimental data were available.

7.7.4 Lewis Acid–Base Reactions of CO_2 and I_2 with Water and Hydroxide Ion

7.7.4.1 CO_2

The carbon dioxide acid–base system is the most important to biological and geochemical processes on earth. Natural waters on most of the earth's surface have a pH in the range of 4–8. The value of the Henry's Law constant at 25 °C for the dissolution of gaseous CO_2 in pure water is low (Equation 7.22a) as the final pH = 5.5. In contrast, the Henry's Law constant for CO_2 in seawater at 25 °C (pH = 8.10), which has significant base content, is five orders of magnitude higher (Equation 7.22b). The kinetics of the reaction of CO_2 with H_2O to form H_2CO_3 (Equation 7.22c) is also over five orders of magnitude slower than the reaction of CO_2 with OH^- to form HCO_3^- (Equation 7.22d). Equations 7.22e and 7.22f give the acid dissociation constants of H_2CO_3 in freshwater at 25 °C.

$$CO_{2(g)} \leftrightarrow CO_{2(aq)} \leftrightarrow H_2CO_3 \qquad K_H = 3.30 \times 10^{-7} \text{ mol l}^{-1} \text{ Pa}^{-1} = [CO_{2(aq)}]/P_{CO_2(g)} \qquad (7.22a)$$

$$CO_{2(g)} \leftrightarrow CO_{2(sw)} \leftrightarrow HCO_3^- \qquad K_H = 3.24 \times 10^{-2} \text{ mol l}^{-1} \text{ Pa}^{-1} \qquad (7.22b)$$

$$CO_{2(aq)} + H_2O \leftrightarrow H_2CO_3 \qquad k = 3.0 \times 10^{-2} \text{s}^{-1} \qquad (7.22c)$$

$$CO_{2(aq)} \leftrightarrow OH^- \leftrightarrow HCO_3^- \qquad k = 8.5 \times 10^3 \text{ M}^{-1} \text{ s}^{-1} \qquad (7.22d)$$

$$H_2CO_3 + H_2O \leftrightarrow H_3O^+ + HCO_3^- \qquad K_{a1} = 4.50 \times 10^{-7} \qquad (7.22e)$$

$$HCO_3^- + H_2O \leftrightarrow H_3O^+ + CO_3^{2-} \qquad K_{a1} = 4.70 \times 10^{-11} \qquad (7.22f)$$

Although the oceans cover 70% of the earth's surface, over time increased CO_2 dissolution into seawater has made the ocean more acidic so it will absorb less CO_2. Thus, as gaseous CO_2 emissions to the atmosphere continue to increase, the ocean is less able to uptake the CO_2.

The difference in the Henry's Law constant for CO_2 in pure water (Equation 7.22a) versus seawater (Equation 7.22b) can be explained by the unfavorable energy for the LUMO of CO_2 (+0.6 eV); i.e., CO_2 is

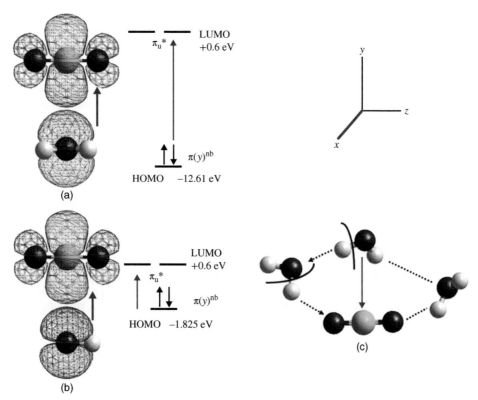

Figure 7.4 *Frontier orbitals for the reaction of (a) CO₂ and H₂O. (b) CO₂ and OH⁻. (c) MM+ rendition of hydrogen bonding, which permits hydrogen ion exchange during the reaction of water with CO₂. Blue arrows indicate direction of electron transfer, and black dashed arrows indicate hydrogen ion movement. Models produced with HyperChem™ version 8.0.10*

a poor electron acceptor. The HOMO energies for H_2O and OH^- are -12.61 and -1.825 eV, respectively. There is a substantial uphill energy barrier of 13.21 eV for the HOMO of H_2O to donate a pair of electrons to the LUMO of CO_2. However, the uphill barrier for OH^- to donate electrons is only 2.425 eV; thus, seawater can dissolve more CO_2 than waters at pH $\leq$ 7. Figure 7.4 a and b shows the orbitals and energetics involved in the dissolution of CO_2 in water and base, respectively. Data in Equation 7.22a indicate that CO_2 still dissolves in pure water despite the prediction of the orbital analysis energetics. In Figure 7.4c, hydrogen bonding between adjacent water molecules allows formation of OH^- that can then attack CO_2 [16, 17]. Thus, the dissolution of CO_2 in pure water is very slow. Other gases such as SO_2, which has a LUMO energy of -1.107 eV and a $K_H = 1.8 \times 10^{-5}$ mol L^{-1} Pa^{-1}, also do not quickly dissolve in water. The uphill HOMO–LUMO barrier is 11.5 eV for SO_2.

7.7.4.2   Lewis Acid–Base Reaction of I_2 with Water and Hydroxide Ion

In a similar way, I_2 also does not react with H_2O but does react with OH^- to form HOI. The energy of the I_2 LUMO is -2.55 eV whereas the HOMO energies for H_2O and OH^- are -12.61 eV and -1.825 eV, respectively.

7.7.5   Lewis Acid–Base Competitive Reactions

Another way to measure acid–base strength is to perform competitive reactions such as displacement reactions or metathesis reactions. **Displacement reactions** occur when one Lewis base or acid replaces or substitutes for another as in the reaction where $N(CH_3)_3$ donates electrons more readily than ammonia to BF_3. Transition metal ions are also well known to participate in substitution or displacement reactions (Chapter 9). Both of the gases CO and HCN (which dissociates in the blood to H^+ and CN^-) are toxic to animals as CO and CN^- readily displace O_2 from Fe(II) in hemoglobin leading to asphyxiation and death.

$$H_3N - BF_3 + :N(CH_3)_3 \rightarrow (CH_3)_3N \rightarrow BF_3 + :NH_3$$

$$Fe(H_2O)_6{}^{2+} + 6CN^- \rightarrow Fe(CN)_6{}^{4-} + 6H_2O$$

Formally, the hydrogen ion transfer reaction of a Bronsted acid is a displacement reaction as water acts as a base to accept a H^+ from the acid.

$$H_2S_{(aq)} + H_2O_{(l)} \rightarrow HS^-{}_{(aq)} + H_3O^+{}_{(aq)}$$

A **metathesis reaction** results in an interchange of the individual acid and base components from two compounds so is a double displacement reaction. It frequently leads to a precipitate when metal ions are involved. In this case, the methyl anion $[:CH_3]$ displaces Cl from the Si–Cl bond. The reaction is formally a competition reaction. The formation of solid LiCl drives the reaction energetics.

$$CH_3 - Li + (CH_3)_3Si - Cl \rightarrow LiCl_{(s)} + (CH_3)_3Si - (CH_3)$$

7.8   Classification of Acids and Bases

7.8.1   Irving–Williams Stability Relationship for the First Transition Metal Series

The competitive nature of chemical reactions has been well studied, and several classifications have been produced over the years. The Irving–Williams series [18] showed that the stability constant (Equation 7.23; where {} indicate activity; see Section 13.1.1) of a series of stable $M(H_2O)_6{}^{2+}$ high spin ions with a given ligand to form ML complexes varies as $Ba^{2+} < Sr^{2+} < Ca^{2+} < Mg^{2+} < Mn^{2+} < Fe^{2+} < Co^{2+} < Ni^{2+} < Cu^{2+} > Zn^{2+}$.

$$M(H_2O)_6 + L \leftrightarrow ML(H_2O)_4 + 2H_2O \qquad K_1 = \frac{\{ML\}}{\{M\}\{L\}} \qquad (7.23)$$

As shown in Table 7.6, the **ligands** contain two atoms (combination of N, O, and S), which donate one electron pair each to the metal ion (Lewis acid–base covalent bonding). The increase in stability of the ML complex follows the charge to size (Z/r or Z_{eff}/r) ratio for the metal ion. On going across the periodic table, Z_{eff}/r and stability increase whereas on descending the periodic table, Z_{eff}/r and stability decrease. For $^-OOC - COO^-$ (oxalate) and $H_2NCH_2COO^-$ (glycine) there is an increase in log K_1 from Ba^{2+} to Mg^{2+}, then an increase from Mn^{2+} to Cu^{2+} and a sharp decrease to Zn^{2+}. The discontinuity for Cu^{2+} is due to the Jahn–Teller effect (Section 8.5.1.5). The strength of the different ligands binding the metals also varies in a regular manner. Group 2 metals bind better with O atoms from carboxylic acids than with ligands containing N and S atoms. Ligands with N and S atoms bind better with the transition metal ions in an increasing manner across the periodic table. Cu^{2+} reduces to Cu^+ with ligands containing S atoms. The Zn^{2+} data show that the S and N atoms in $HSCH_2CH_2NH_2$ (mercaptoethylamine) bind better to Zn^{2+} than the S and O atoms in

Table 7.6   *The log of K_1 for the divalent cations of Group 2 and selected first row transition metals from [19–22] (thermodynamic constants Section 13.1.2). Atoms in bold font are those binding to the metal ions. Empty cells indicate data are not available for that complex*

Ligand	Ba^{2+}	Sr^{2+}	Ca^{2+}	Mg^{2+}	Mn^{2+}	Fe^{2+}	Co^{2+}	Ni^{2+}	Cu^{2+}	Zn^{2+}
$^-\mathbf{O}\mathbf{O}C - C\mathbf{O}\mathbf{O}^-$	2.31	2.54	3.00	3.43	3.95	4.31	4.72	5.16	6.23	4.87
$H_2\mathbf{N}CH_2C\mathbf{O}\mathbf{O}^-$		0.91	1.39	2.08	3.19	4.31	5.07	6.18	8.57	5.38
$H\mathbf{O}CH_2CH_2\mathbf{N}H_2$								2.98	5.7	3.7
$H_2\mathbf{N}CH_2CH_2\mathbf{N}H_2$				0.37	2.74	4.34	5.5	7.31	10.5	5.7
$H\mathbf{S}CH_2CH_2C\mathbf{O}\mathbf{O}^-$					4.38		5.84			7.86
$H\mathbf{S}CH_2CH_2\mathbf{N}H_2$	1.37	1.55	2.21	2.30			7.68	10.0		10.2
$H\mathbf{S}(CH)(C\mathbf{O}\mathbf{O}^-)\mathbf{N}H_2$					4.7		8.00	9.82		9.17

$HSCH_2CH_2COO^-$ (mercaptoactetate). The amino acid cysteine contains N, S, and O atoms, but the stability constant data indicate that the N and S atoms preferentially bind Zn^{2+}.

7.8.2   Class "a" and "b" Acids and Bases

Ahrland, Chatt, and Davies [23] characterized metals as Class "a" and "b" according to their preference in bonding to certain ligands or donor atoms, which follows from the Irving–Williams series. Class "a" metals are the alkali and alkaline earth metals and the first transition series metals with higher oxidation states (e.g., Ti^{4+}, Fe^{3+}, Co^{3+}) along with H^+. The Class "b" metals include the second and third transition series metals with lower oxidation states and usually have six or more "d" electrons (e.g., Cu^{2+}, Ag^+, Hg^{2+}, Hg_2^{2+}). In general, Class "a" cations and anions are smaller and at the top of a group in the periodic table whereas Class "b" metals and ligands are larger and near the bottom of a group in the periodic table. The ligands that prefer to bind to these classes of metals are in Table 7.7. For example, sulfide (and thioethers) prefer to bind with Hg^{2+} and Ag^+ so displace other metals from these ligands during reactions as shown in natural waters [24] whereas water, hydroxide, ammonia, and amines bind better with Fe^{3+}, Mn^{4+}, and Ti^{4+}.

7.8.3   Hard Soft Acid Base (HSAB) Theory

Pearson [25] coined the terms "hard" and "soft" to describe Class "a" and Class "b," respectively. Thus, a hard acid is a Class "a" metal and a hard base is a Class "a" ligand; these tend to be smaller chemical species. The Class "b" metals and ligands are larger and more **polarizable** as their electron clouds are more easily distorted.

Table 7.7   *Tendency for metals to bind with different donor atoms*

Tendency to bind to Class "a" metals	Tendency to bind to Class "b" metals
$N \gg P > As > Sb$	$N \ll P < As < Sb$
$O \gg S > Se > Te$	$O \ll S < Se \sim Te$
$F \gg Cl > Br > I$	$F < Cl < Br < I$

Table 7.8   *Classification of chemical species as acids and bases according to HSAB theory*

Hard	Borderline	Soft
	Acids	
H^+, Li^+, Na^+, K^+, Be^{2+}, Mg^{2+}, Ca^{2+}, Sc^{3+}, Ti^{4+}, Cr^{3+}, Cr^{6+}, Mn^{2+}, Mn^{7+}, Fe^{3+}, Al^{3+}, Ga^{3+}, Be^{2+}, Be^{2+}, La^{3+}, Sn^{4+}, SO_3, BF_3	Fe^{2+}, Co^{2+}, Ni^{2+}, Cu^{2+}, Zn^{2+}, Sn^{2+}, Pb^{2+}, Bi^{3+}, Ru^{3+}, Sb^{3+}, Rh^{3+}, Ir^{3+}, Os^{2+}, BBr_3, SO_2, $B(CH_3)_3$, R_3C^+, NO^+	Metal atoms, Cu^+, Au^+, Ag^+, Tl^+, Hg^{2+}, Hg_2^{2+}, CH_3Hg^+, Cd^{2+}, Pd^{2+}, Pt^{2+}, Pt^{4+}, I_2, I^+, I, Br_2, BH_3
	Bases	
F^-, Cl^-, O^{2-}, OH^-, H_2O, ROH, RO^-, NH_3, ClO_4^-, SO_4^{2-}, PO_4^{3-}, NO_3^-, CO_3^{2-}, acetate, R_3N	Br^-, NO_2^-, SO_3^{2-}, N_2, N_3^-, SCN^-, pyridine, S_2^{2-}, HS^-	H^-, I^-, SCN^-, CO, CN^-, R_3P, R_3As, R_2S, RS^-, $S_2O_3^{2-}$, benzene, ethylene

Many soft acids are π acids as they can accept electron density into their π or π^* orbitals ("backbonding," Section 8.6.3). He also added several other ligands or molecules to the list and noted acids and bases that have borderline behavior (Table 7.8).The base SCN^- can bind through S or N so it is a soft base when binding via the N atom and a borderline base when binding through the S atom.

Pearson suggested two rules for HASB theory. First, "Hard acids prefer to bind to hard bases and soft acids prefer to bind to soft bases." Second, the hard acid–hard base interaction is stronger than the soft acid–soft base interaction. Based on these rules, it is possible to predict the reactivity between a wide variety of compounds as in the following competitive and displacement reactions. For the competitive reactions below (bond energies from Appendix 5.1, Chapter 5), the lattice energy dominates the energetics of attractive forces.

Competitive reactions	Best hard acid – hard base interaction
$LiI + CsF \rightarrow LiF + CsI$	Li^+ and F^-
$-347 \quad -481 \rightarrow -573 \quad -315$	Δ Bond energies $= -60 \, kJ \, mol^{-1}$
$HgF_2 + BeI_2 \rightarrow BeF_2 + HgI_2$	Be^{2+} and F^-
$-368 \quad -289 \rightarrow -632 \quad -145$	Δ Bond energies $= -240 \, kJ \, mol^{-1}$
Displacement reactions	
$R_2S - BF_3 + R_2O \rightarrow R_2O - BF_3 + R_2S$	F is more electron withdrawing so B is harder
$R_2O - BH_3 + R_2S \rightarrow R_2S - BH_3 + R_2O$	H is less electron withdrawing so B is softer

Pearson was able to quantify **hardness** and **polarizability** (softness) using the ionization potential of the base (HOMO) and the electron affinity (LUMO) of the acids (see Section 3.8.7). Thus, hard molecules have a large HOMO–LUMO gap and soft molecules have a small HOMO–LUMO gap. For example, the hardness values for gas phase HI, HCl, and HF are 5.3, 8.0, and 8.8 eV, respectively. These values indicate the following order for dissociation HI > HCl > HF, which reflects the pK_a values for these strong acids in water (Table 7.1).

HSAB theory gives insights into the earth's mineral composition. In oxygenated environments, metal ions with high oxidation states form oxides and oxyhydroxides, e.g., SiO_2, Al_2O_3, Fe_2O_3, $FeOOH$, MnO_2, $FeTiO_3$, TiO_2. These minerals can also include other high oxidation state metal ions into their crystal structure; e.g.,

Table 7.9   The $pH_{(ZPC)}$ for several common minerals from Ref. [26]

Mineral	$pH_{(ZPC)}$	Mineral	$pH_{(ZPC)}$	Mineral	$pH_{(ZPC)}$
Corundum, Al_2O_3	9.4	Spinel, $MgAl_2O_4$	9.0	Pyrite, FeS_2	1.4
Quartz, SiO_2	2.9	Magnetite, Fe_3O_4	6.5	Pyrrhotite, FeS	3.0
Anatase, TiO_2	5.8	Hematite, Fe_2O_3	8.6	Sphalerite, ZnS	1.7
Ilmenite, $FeTiO_3$	6.3	Goethite, FeOOH	9.7	Galena, PbS	1.4
Zincite, ZnO	8.8	Birnessite, MnO_2	4.6	Chalcopyrite, $FeCuS_2$	1.8

small amounts of Cr^{3+} in Al_2O_3 lead to the gemstone ruby. In sulfidic environments, metal ions with lower oxidations states form sulfides and persulfides (e.g., FeS, FeS_2, ZnS, PbS, $FeCuS_2$), and they can include other low oxidation state metal ions into their crystal structure. Formally all these minerals are composed of covalent bonds between the acid and base so are insoluble or sparingly soluble in water.

7.9   Acid–Base Properties of Solids

The focus in Chapter 6 was on the energetics and packing of ions and atoms in solids. However, all solids have surfaces (e.g., Figure 6.21), which have an interface with the atmosphere or a solution. The surfaces of common solids and minerals also have acid–base character in aqueous solutions. The **zero point of charge** (ZPC) is used as an indicator of acid–base strength, and occurs when half of the surface sites are protonated and the other half are not; thus, the surface is neutral. The pH where the solid surface is neutral is defined as the $\mathbf{pH_{(ZPC)}}$. The following equation represents the surface acid–base character for δ-MnO_2 (birnessite) where > indicates the solid. The other symbol, $*K^S_{a2}$, indicates the apparent dissociation constant of H^+, and the subscript 2 indicates the second dissociation constant as $> MnOH_2^+$ exists at very low pH.

$$> Mn - OH \rightarrow MnO^- + H^+ \quad pH_{(ZPC)} = *K^S_{a2} = 10^{-4.6}$$

When the solution pH $<$ $pH_{(ZPC)}$, the surface is positive (and acts as an acid) and when the solution pH $>$ $pH_{(ZPC)}$, the surface is negative (and acts as a base). Table 7.9 lists common oxide and sulfide minerals along with their $pH_{(ZPC)}$ values. Sulfide minerals have smaller values of $pH_{(ZPC)}$ than oxide minerals because sulfide is a **softer base** than oxide, which binds better with the hydrogen ion, a hard acid. Thus, the surfaces of sulfide minerals are not protonated except at very low pH values; i.e., loss of hydrogen ions is easier as in the hydrogen halide acids noted in Section 7.8.2 and Table 7.1. Likewise, the pK_1 value for H_2S is 7.0 and compares with the pK_w for H_2O of 14 in fresh water. The result is that sulfide minerals have a negative surface at lower pH rendering them more basic and more reactive to dissolved metal ions, which can absorb to their surface.

References

1. Rickard, D. (2015) *Pyrite: A Natural History of Fool's Gold*, Oxford University Press, pp. 320.
2. Schoonen, M. A. A. and Barnes, H. L. (1988) An approximation of the second dissociation constant for H_2S. *Geochimica et Cosmochimica Acta*, **52**, 649–654.
3. Stumm W. and Morgan J. J. (1996) *Aquatic Chemistry* (3rd ed.), John Wiley and Sons, Inc., New York, pp. 1022.

4. Morel, F. M. M. and Hering, J. G. (1993) *Principles and Applications of Aquatic Chemistry*, John Wiley and Sons, Inc., New York, pp. 588.

5. Baes, C. F. and Mesmer, R. E. (1976) *The Hydrolysis of Cations*, Kreiger Publishing Co., Malabar, Fl. pp. 489.

6. Sisley, M. J. and Jordan, R. B. (2006) First hydrolysis constants of hexaaquacobalt(III) and –manganese(III): longstanding issues resolved. *Inorganic Chemistry*, **45**, 10758–10763.

7. Shannon, R. D. (1976) Revised effective ionic radii and systematic studies of interatomic distances in halides and chalcogenides. *Acta Crystallographica A*, **32**, 751–767.

8. Ricci, J. E. (1948) The aqueous ionization constants of inorganic oxygen acids. *Journal of the American Chemical Society*, **70**, 109–113.

9. Millero, F. J. (2007) The Marine Inorganic Carbon Cycle. *Chemical Reviews*, **107**, 308–341.

10. Millero, F. J. (2013) *Chemical Oceanography* (4th ed.), CRC Press, Boca Raton, FL, pp. 571.

11. Lias, S. G., Bartmess, J. E., Liebman, J. F., Holmes, J. L., Levin R. D. and Malllard, W. G. (1988) Gas-phase ion and neutral thermochemistry. *Journal of Physical and Chemical Reference Data,* **17**, Supplement No. 1. American Chemical Society, American Institute of Physics, U.S. National Bureau of Standards.

12. Ervin, K. M. and Lineberger, W. C. (2005) Photoelectron spectroscopy of phosphorus hydride anions. *The Journal of Chemical Physics*, **122**, 194303.

13. Oka T. (2013) Interstellar H_3^+. *Chemical Reviews*, **113**, 8738–8761.

14. Drago, R. S. and Vogel, G. C. (1992) Interpretation of spectroscopic changes upon adduct formation and their use to determine E and C parameters. *Journal of the American Chemical Society*, **114**, 9527–9532.

15. Drago, R. S. and Wayland, B. B. (1965) A double-scale equation for correlating enthalpies of Lewis acid-base interactions. *Journal of the American Chemical Society*, **87**, 3571–3577.

16. Lewis M. and Glaser R. (2003) Synergism of catalysis and reaction center rehybridization: a novel mode of catalysis in the hydrolysis of carbon dioxide. *Journal of Physical Chemistry A*, **107**, 6814–6818.

17. Luther, III, G. W. (2004) Kinetics of the reactions of water, hydroxide ion and sulfide species with CO_2, OCS and CS_2: frontier molecular orbital considerations. *Aquatic Geochemistry*, **10**, 81–97.

18. Irving, H. and Williams R. J. P. (1948) Order of Stability of metal complexes. *Nature*, **162**, 746–747.

19. Martell, A. E. and Smith, R. M. (1977) *Critical Stability Constants* Vol. 3, Other organic ligands, Plenum Press, New York.

20. Smith, R. M. and Martell, A. E. (1976) *Critical Stability Constants* Vol. 4, Inorganic ligands, Plenum Press, New York.

21. Martell, A. E. and Smith, R. M. (1982) *Critical Stability Constants* Vol. 5, First supplement, Plenum Press, New York.

22. Smith, R. M. and Martell, A. E. (1989) *Critical Stability Constants* Vol. 6, Second supplement, Plenum Press, New York.

23. Ahrland, S., Chatt, J. and Davies, N. R. (1958) The relative affinities of ligand atoms for acceptor molecules and ions. *Quarterly Reviews of the Chemical Society*, **12**, 265–276.

24. Luther, III, G.W. and Rickard, D. T. (2005) Metal sulfide cluster complexes and their biogeochemical importance in the environment. *Journal of Nanoparticle Research*, **7**, 389–407.

25. Pearson, R. G. (1988) Absolute electronegativity and hardness: application to Inorganic Chemistry. *Inorganic Chemistry*, **27**, 734–740.

26. Xu, Y. and Schoonen, M. A. A. (2000) The absolute energy positions of conduction and valence bands of selected semiconducting minerals. *American Mineralogist*, **85**, 543–556.

8

Introduction to Transition Metals

8.1 Introduction

In Section 7.8, transition metals were shown to accept a pair of electrons from another molecule, which is known as a ligand. The bond is a covalent bond between the metal and the ligand. There is a large assortment of ligands that can donate one or more pairs of electrons to a single metal ion. Those that only donate one pair of electrons are called monodentate ligands whereas those that donate more than one pair of electrons are called multidentate or polydentate ligands (also known as **chelates;** from the Greek word khele for "claw" of a crab). For organometallic compounds, there are other possibilities; e.g., the five carbon atoms in the cyclopentadiene anion donate six electrons to a metal center, and the six carbon atoms in benzene also donate six electrons to a metal center. Table 8.1 shows some typical monodentate and multidentate ligands with their names, common abbreviations, and the atoms that donate the pair(s) of electrons to the metal (such atoms are known as the **ligating atoms**). Figure 8.1 shows several examples of multidentate ligands. Organisms in nature produce a wide variety of multidentate chelates to bind metals; for example, terrestrial and aquatic plants synthesize chlorophyll, which binds magnesium and other metals, and bacteria synthesize **siderophores** such as desferrioxmaine-B to bind and acquire iron (III).

8.2 Coordination Geometries

The study of transition metal or coordination chemistry demonstrated the need to understand which ligands actually bind the metal in the **inner coordination sphere** in addition to understanding the primary oxidation state or valence of the metal. Transition metals display a wide range of coordination geometries starting with one ligating atom or electron pair donor bonding to a metal. Figure 8.2 shows common coordination geometries for metals when 4–7 ligands bind or complex the metal, and Figure 8.3 shows some 8 coordinate geometries (coordination numbers up to 12 are possible as shown for pure metals, Section 6.2.7.1). Some of the geometries are identical to what was discussed using the valence shell electron pair repulsion theory (see Figure 5.3); e.g., there are no additional geometries for coordination two (linear) and three (trigonal planar). Note that four coordination includes a square planar geometry in addition to the tetrahedral geometry.

Inorganic Chemistry for Geochemistry and Environmental Sciences: Fundamentals and Applications, First Edition. George W. Luther, III.
© 2016 John Wiley & Sons, Ltd. Published 2016 by John Wiley & Sons, Ltd.
Companion Website: www.wiley.com/go/luther/inorganic

Table 8.1 *Common monodentate and multidentate ligands and nomenclature*

Name (abbreviation or old name)	Formula	Donor atom(s)
Acetate	CH_3COO^-	O
Aqua (aquo)	H_2O	O
Amido	NH_2^-	N
Ammine	NH_3	N
Azido	N_3^-	N
Bromido (bromo)	Br^-	Br
Carbonato	CO_3^{2-}	O
Carbonyl	CO	C
Chlorido (Chloro)	Cl^-	Cl
Cyanido (cyano)	CN^-	C
(iso)cyano	CN^-	N
Fluoride (fluoro)	F^-	F
Hydrido	H^-	H
Hydroxido	OH^-	O
Iodido (Iodo)	I^-	I
Nitride	N^{3-}	N
Nitrato	NO_3^-	O
Nitrito – κN (nitro)	NO_2^-	N
Nitrito – κO (nitrito)	NO_2^-	O
Oxido (oxo)	O^{2-}	O
Pyridine (py)	C_5H_5N	N
Sulfido	S^{2-}	S
Thiocyanato – κN	SCN^-	N
Thiocyanato – κS	SCN^-	S
Thiolato	RS^-	S
Triphenylphosphine	$(C_6H_5)_3P$	P

Multidenate ligands

Ethylenediamine (en; 1,2-diaminoethane)	$H_2NCH_2CH_2NH_2$	N (2)
1,3-Diaminopropane (tn)	$H_2NCH_2CH_2CH_2NH_2$	N (2)
Glycinato (gly)	$H_2NCH_2CO_2^-$	N(1), O(1)
Oxalato (ox)	See figure $^-O_2C - CO_2^-$	O (2)
2,4-Pentanediono or acetylacetonato (acac)	See figure $CH_3COCHCOCH_3$	O (2)
o-Dihydroxybenzene (catechol; cat)	See figure	O (2)
8-Hydroxyquniolinato or oxinate (oxine)	See figure	N(1), O(1)
2,2'-Bipyridine or bipyridyl (bipy)	See figure	N (2)
1,10-Phenanthroline (phen)	See figure	N (2)
Diethylenetriamine (dien)	$NH(CH_2CH_2NH_2)_2$	N (3)
Cyclam	See figure	N (4)
Tet-b	See figure	N (4)
Triaminotriethylamine (tren)	See figure $N(CH_2CH_2NH_2)_3$	N (4)
Porphyrin macrocycle	See figure	N (4)
Corrin macrocycle	See figure	N (4)
Nitrilotriacetic acid (NTA)	See figure	N(1), O(3)
Ethylenediaminetetraacetic acid (EDTA)	See figure	N(2), O(4)
Desferrioxamine-B (DFOB)	See figure	O(6)

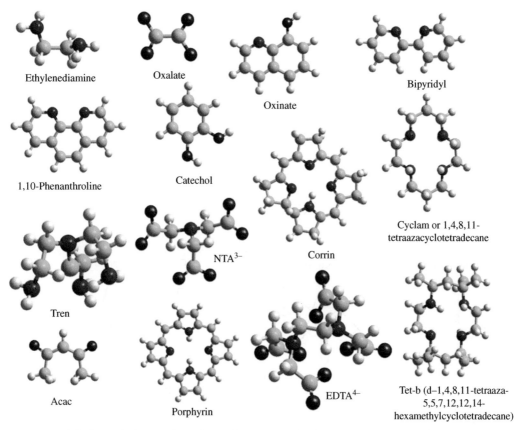

Figure 8.1 *Models for multidentate ligands. Carbon (aqua); hydrogen (white); oxygen (black), and nitrogen (dark blue). The corrin group has one C atom less than the porphyrin macrocycle ring (one C atom is missing between two of the cyclic pyrroles). Models created with the Hyperchem™ program (version 8.0.10)*

Five coordination includes a square pyramidal geometry in addition to the trigonal pyramidal geometry. Six coordination includes trigonal prismatic geometry (D_{3d}; note the overlap of atoms of the two trigonal planes along the C_3 axis) in addition to the octahedral geometry (distortions will be discussed later), which is formally a trigonal antiprism looking down the C_3 axis as the atoms for the trigonal planes do not overlap. Looking down the C_4 axis shows a square plane capped at the top and the bottom. Seven coordination includes the trigonal-face capped octahedron, the square-face capped trigonal prismatic geometry, and the pentagonal bipyramidal geometry (see Figure 5.3).

Eight coordination includes the cube (square prism), square antiprism (D_{4d}), and the trigonal dodecahedron (D_{2d}) geometries. The cube and trigonal dodecahedron are related to sliding of the atoms from a cube on the C_4 axis; note that atoms D, F, and H move upward as A and C move downward and B, E, and G move inward so that the original planes (ABCD and DEFG) of the cube are lost. Atoms A, B, C, and D form a wing-like geometry as do atoms E, F, G, and H.

All of these geometries can display a range of geometrical and/or optical isomers. In the following discussion, the isomers of transition metal complexes will center on four and six coordination.

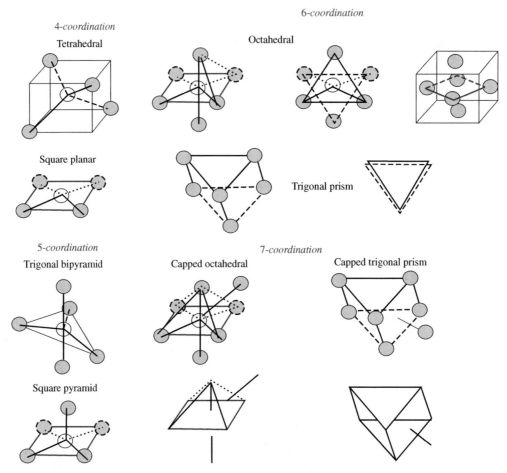

Figure 8.2 *Geometries for four, five, six, and seven ligating atoms attached to a central metal atom. The octahedron is shown in three separate arrangements to show its relation to the cube and as a trigonal antiprism. Two arrangements are given for the trigonal prism; the central metal atom is not drawn for this structure. The seven coordinate structures are related to the octahedron and the trigonal prism*

8.3   Nomenclature

In the examples below, the chemical formula will be specified and the proper name provided. The International Union of Pure and Applied Chemistry (IUPAC) has provided extensive rules for the nomenclature of inorganic complex compounds [1] that are briefly outlined here.

First, the cation name is given followed by the anion name. For cations, the metal name is given without a suffix whereas anionic ligands are given an "o" suffix and neutral ligands retain their normal name. To indicate the number of ligands bound to a given metal, the prefixes in Table 8.2 are used.

Second, the oxidation states and the inner sphere coordination number of the metal in the complex metal ion are explicitly considered; brackets [ ] are used to enclose the metal with its inner sphere ligands in the metal complex. For **chemical formulae**, the central atom symbol is given first and the ligand symbols listed alphabetically as in $[CoBrCl(NH_3)_4]^+$.

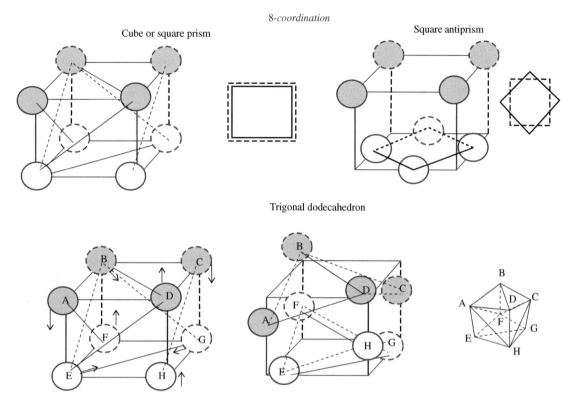

Figure 8.3 *Possible eight coordinate geometries; the central metal atom is not drawn in these structures. Top down views are also given for the cube and square antiprism*

Third, as noted in Table 8.1, the ligating atom(s) of the ligand need to be described. When naming compounds where more than one ligating atom can bind the central metal, κ (kappa) notation is used before the ligating atom to specify that it is the electron donor to the metal.

Fourth, the type of ligand bonding to the metal or metals needs to be described – e.g., terminal ligand versus bridging ligand (μ is used explicitly) versus capping ligand (η-**Hapto**) for organometallic compounds (Section 9.10).

Fifth, geometrical isomers are given a prefix – e.g., cis or trans (see below).

Sixth, optical isomers are given a symbol prefix for left (Λ) or right (Δ) handedness (see Section 8.4.6).

Nomenclature examples are now provided. In the following discussion of bonding and isomers (Section 8.4), the proper IUPAC name will be provided for the compounds discussed.

8.3.1   Complex Ion is Positive

$[Cu(NH_3)_4]SO_4$ Tetraamminecopper(II) sulfate
                Tetraamminecopper(2+) sulfate
                Tetraamminecupric sulfate

Table 8.2   *The names/prefixes used in the nomenclature for multiple numbers of the same ligand bound to a metal. The second column of prefixes are used with composite ligand names [e.g., (bis)ethylenediamine indicates two ethylenediamine molecules bound to a metal center] or to avoid ambiguity*

Number	Prefix I	Prefix II	Number	Prefix I	Prefix II
1	Mono	Mono	7	Hepta	Heptakis
2	Di	Bis	8	Octa	Octakis
3	Tri	Tris	9	Nona	Nonakis
4	Tetra	Tetrakis	10	Deca	Decakis
5	Penta	Pentakis	11	Undeca	Undeca
6	Hexa	Hexakis	12	Dodeca	Eicosa

The complex ion is $[Cu(NH_3)_4]^{2+}$ as four ammonia molecules bind to the copper metal center. Note that the Roman numeral II or the number 2+ can be used to signify copper's oxidation state or the complex ion's total charge, respectively. The use of the Roman numeral system is called the Stock system whereas the charge on the complex is known as the Ewing–Bassett system.

The use of a number before the plus or minus sign as in 2+ for the complex ion $[Cu(NH_3)_4]^{2+}$ indicates charge whereas the use of +2 is reserved for exponents in mathematical equations. Tetra from Table 8.2 is used to indicate the total number of ammonia molecules bound to Cu. The use of cupric for the 2+ oxidation state of copper was once common but is not commonly used now.

8.3.2   Complex Ion is Negative

$$K_3[Fe(CN)_6]^{3-}$$ Potassium hexacyanidoferrate(III)
Potassium hexacyanidoferrate(3−)
Tripotassium hexacyanidoferrate

Cyano has been replaced by cyanido but is still commonly used. Note that the Roman numeral III or the number 3− can be used to signify iron's oxidation state or the complex ion's total charge, respectively. The suffix "ate" is used for the metal to indicate that the complex ion is negative (note that iron is not negative as it is 3+). The Latin words with the "ate" suffix are still used for iron (ferrate), silver (argentate), lead (plumbate), tin (stannate), and gold (aurate) as the words are the source of their symbols: Fe, Ag, Pb, Sn, Au.

8.3.3   Complex Ion with Multiple Ligands

$$[CoBr(NH_3)_5]Cl_2$$ Pentaamminebromidocobalt(III)chloride

The ligands are named in alphabetical order but the prefix is not considered part of the ligand name. The complex ion is $[CoBr(NH_3)_5]^{2+}$.

8.3.4   Complex Ion with Ligand that can Bind with More Than One Atom (Ambidentate)

	with N bonding
[CoCl(NH$_3$)$_4$(NO$_2$)]Cl	Tetraamminechloridonitrito-κN-cobalt(III)chloride
Old name	Tetraamminechloridonitrocobalt(III)chloride
	with O bonding
[CoCl(NH$_3$)$_4$(NO$_2$)]Cl or [CoCl(NH$_3$)$_4$(ONO)]Cl	Tetraamminechloridonitrito-κO-cobalt(III)chloride
Old name	Tetraamminechloridonitrito-cobalt(III)chloride

The complex ion in each case is [CoCl(NH$_3$)$_4$(NO$_2$)]$^+$.

8.3.5   Complex Ion with Multidentate Ligands

[Co(en)$_3$]Cl$_3$ tris(ethylenediamine)cobalt(III)chloride

The complex ion is [Co(en)$_3$]$^{3+}$.

8.3.6   Two Complex Ions with a Bridging Ligand

[Co(NH$_3$)$_5$–O–Co(NH$_3$)$_5$]$^{4+}$
[{Co(NH$_3$)$_5$}$_2$ (μ − O)]   μ-Oxido-bis(pentaamminecobalt(III))

The complex ion is [(NH$_3$)$_5$Co − O − Co(NH$_3$)$_5$]$^{4+}$. Here one oxide ion, O^{2-} ligand, bridges two Co atoms by donating a pair of electrons to each Co.

[Co(NH$_3$)$_5$(μ-OH)$_2$Co(NH$_3$)$_5$]
[(Co(NH$_3$)$_5$)$_2$(μ-OH)$_2$]   di-μ-Hydroxido-bis(pentaamminecobalt(III))

The complex ion is [(NH$_3$)$_5$Co − (OH)$_2$ − Co(NH$_3$)$_5$]$^{4+}$. Here two hydroxide ions, OH$^-$ ligands, bridge two Co atoms by donating a pair of electrons to each Co.

8.4   Bonding and Isomers for Octahedral Geometry

Alfred Werner in a detailed series of experiments in the late 19th and early 20th century demonstrated how to distinguish between the inner and outer coordination sphere of a metal. He was able to do this for square planar and octahedral coordination geometries to the metal using the aqueous metal compounds of Co(III), Cr(III), and Pt(II) (d^6, d^3, and d^8 electron configurations, respectively), which are stable with long half-lives to dissociation and ligand exchange (Table 9.1). Transition metal complexes exhibit a wide range of isomers. When metals and ligands have the same overall chemical formula but have different ligands bound to the central metal ion or atom, the isomers are called structural or constitutional isomers, and there are four types (hydrate isomers, ionization isomers, coordination isomers, and linkage isomers). When metals and ligands have the same overall chemical formula and the ligands bound to the central metal ion or atom are the same, these isomers are called **geometrical isomers** and may be **optical isomers**.

8.4.1 Ionization Isomerism

Ionization isomerism is shown for compounds with the same chemical formula but which give different ions upon dissolution and dissociation; hydrate isomers below can be included in this definition. Again ligands attached to the metal in the inner coordination sphere remain bound to a metal ion. An example of ionization isomerism is given by the following pairs of compounds when ligand spectator ions exchange places.

$[Co(NH_3)_5SO_4]Cl$ or pentaamminesulfatocobalt(III) chloride dissolves with chloride being the spectator ion, which can be readily precipitated with the addition of Ag^+ from a solution of silver nitrate but not a solution of Ba^{2+}.

$$[Co(NH_3)_5SO_4]Cl + Ag^+ \rightarrow AgCl \downarrow + [Co(NH_3)_5SO_4]^+$$

$$[Co(NH_3)_5SO_4]Cl + Ba^{2+} \rightarrow no\ reaction$$

$[Co(NH_3)_5Cl]SO_4$ or pentaamminechloridocobalt(III) sulfate dissolves with sulfate being the spectator ion, which can be readily precipitated with Ba^{2+} from a solution of barium chloride but not a solution of Ag^+.

$$[Co(NH_3)_5Cl]SO_4 + Ag^+ \rightarrow no\ reaction$$

$$[Co(NH_3)_5Cl]SO_4 + Ba^{2+} \rightarrow BaSO_4 \downarrow + [Co(NH_3)_5Cl]^{2+}$$

8.4.2 Hydrate (Solvate) Isomers

The stoichiometry for the solid compound, $Cr(H_2O)_6Cl_3$, has three different solid chemical species that are called hydrate isomers as the number of water molecules around the coordination sphere of the metal vary; those not in the coordination sphere hydrate the molecule. The first hydrate isomer, which is violet in color, has six water molecules directly binding the Cr^{3+} in the inner coordination sphere; the three chloride ions are spectator ions in the outer coordination sphere. On dissolving this molecule in water, the three chloride ions would react with Ag^+ from silver nitrate to form AgCl precipitate, which can be collected and analyzed. This molecule is given the following chemical notation, $[Cr(H_2O)_6]Cl_3$, where as noted in the nomenclature discussion, the brackets indicate that this is a complex ion with a positive charge of three and that all waters bind the inner sphere of chromium. The complex ion is represented as $[Cr(H_2O)_6]^{3+}$, and the coordination number for the metal is six due to the six water molecules directly donating a pair of electrons to chromium. The three chloride ions are in the outer sphere part of the complex to balance or associate with the positive charge of the complex ion (Equation 8.1) – a 1 cation:3 anion or 1:3 electrolyte. The proper name for this compound is hexaaquachromium(III) chloride.

$$[Cr(H_2O)_6]Cl_3 \rightarrow [Cr(H_2O)_6]^{3+} + 3\ Cl^- \tag{8.1}$$

The second molecule, which is light green in color, has five water molecules and one chloride ion directly binding the Cr^{3+} in the inner coordination sphere; thus, only two Cl^- are spectator ions and would react with Ag^+ once the molecule is dissolved in water (Equation 8.2; 1:2 electrolyte). This molecule is represented as $[CrCl(H_2O)_5]Cl_2 \cdot H_2O$ where the one water in the outer coordination sphere is a water of hydration or a water of crystallization (other solvent molecules can also do this). The complex ion is represented as $[CrCl(H_2O)_5]^{2+}$. The proper name for this compound is pentaaquachloridochromium(III) chloride.

$$[CrCl(H_2O)_5]Cl_2 \cdot H_2O \rightarrow [CrCl(H_2O)_5]^{2+} + 2\ Cl^- \tag{8.2}$$

Similarly the third molecule, which is dark green, has four water molecules and two chloride ions binding the Cr^{3+} in the inner coordination sphere with two waters of hydration in the solid or $[CrCl_2(H_2O)_4]Cl_2 \cdot 2H_2O$; thus, only one Cl^- is a spectator ion and would react with Ag^+ once the molecule is dissolved in water (Equation 8.3; 1:1 electrolyte). The complex ion is represented as $[CrCl_2(H_2O)_4]^+$. The proper name for this compound is tetraaquadichloridochromium(III) chloride.

$$[CrCl_2(H_2O)_4]Cl \cdot 2H_2O \rightarrow [CrCl_2(H_2O)_4]^+ + Cl^- \qquad (8.3)$$

The latter also has two geometrical isomers, cis and trans, which are discussed below.

These three molecules are an example of **hydrate isomerism**. Another experimental test to determine the number of ions in solution is to measure the ionic conductivity in solution as each of these dissolutions of the solid material leads to a different number of ions in solution.

8.4.3 Coordination Isomerism

Coordination isomerism occurs when the cation and the anion are both complex metal ions with different ligands attached to the metals even though the chemical formula is identical. Examples with different metal ions for the cation and anion on switching the ligands are shown here.

$[Co(NH_3)_6] [Cr(CN)_6]$ hexaamminecobalt(III) hexacyanidochromate(III)
versus
$[Cr(NH_3)_6] [Co(CN)_6]$ hexaamminechromium(III) hexacyanidocobaltate(III)

An example with the same metal ion as both cation and anion but having different inner sphere coordination is Pt.

$[Pt(NH_3)_4] [PtCl_6]$ tetraammineplatinum(II) hexachloridoplatinate(IV) – the Pt(II) complex cation is square planar whereas Pt(IV) complex anion is octahedral.
Versus
$[Pt(NH_3)_4Cl_2] [PtCl_4]$ tetraamminedichloroplatinum(IV) tetrachloridoplatinate(II) – the Pt(IV) complex is octahedral whereas the Pt(II) complex anion is square planar.

8.4.4 Linkage Isomerism

Linkage isomerism occurs when a ligand can bind a metal from either of two different atoms (also known as **ambidentate**), and two different examples are given for the nitrite ion and the thiocyanate ion. Nitrite can bind a metal either from the terminal oxygen atom or the central nitrogen atom. Thiocyanate, SCN^-, can bind either from the terminal sulfur or the terminal nitrogen atom. Hard soft acid base theory can be used to predict the mode of binding when different metals are involved. For example in thiocyanate, the sulfur atom is softer than the nitrogen atom so sulfur would be predicted to combine with softer metals and nitrogen would be predicted to combine with harder metals.

The nitrite ion produces two different isomers with the following chemical formula, $[Co(NH_3)_5NO_2]^{2+}$, during the aqueous reaction of nitrite with $[Co(NH_3)_5Cl]^{2+}$ as the chloride ion undergoes ligand exchange with the nitrite ion.

$$[Co(NH_3)_5Cl]^{2+} + NO_2^- \rightarrow Cl^- + [Co(NH_3)_5(ONO)]^{2+} \ (red)$$

$$[Co(NH_3)_5Cl]^{2+} + NO_2^- \rightarrow Cl^- + [Co(NH_3)_5(NO_2)]^{2+} \ (yellow)$$

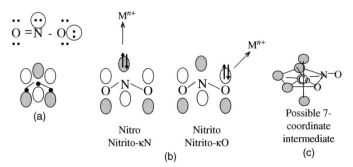

Figure 8.4 *(a) Circled electron pairs indicate that bonding of nitrite from the nitrogen or oxygen atom to a metal cation can occur. (b) Possible N and O bonding modes using the $2b_2{}^*$ orbital (see Figure 5.50,π^*), respectively. (c) Possible pentagonal bipyramid intermediate with both N and O binding Co ("**side on**" bonding)*

When the reaction is performed at colder temperatures, a red compound is formed first and on subsequent warming it is converted to the yellow compound. In the red compound, the nitrite binds via the oxygen atom (nitrito) whereas in the yellow compound, the nitrite ion binds via the nitrogen atom (nitro) to the cobalt. The use of O-18 radiolabeled nitrite [2] showed that nitrite did not dissociate from the metal (was not released to solution) during the transformation even under photochemical excitation, confirming that this conversion between isomers is an intramolecular rearrangement. This process can be explained recalling that the HOMO of nitrite is a π^* orbital (Figure 5.50) that is delocalized across all three atoms as shown in Figure 8.4. Although the HOMO has slightly more nitrogen that oxygen character, statistically the cobalt complex ion will encounter oxygen twice as many times as nitrogen in the transition state. Figure 8.4 indicates that the binding between cobalt and nitrite could occur via the oxygen atom and then slide to the nitrogen atom. Alternately, bonding to Co could occur via N and O simultaneously in a possible seven-coordinate intermediate.

The proper nomenclature for each of these complex ions is

$$[Co(NH_3)_5(ONO)]^{2+} \quad \text{Pentaamminenitrito-}\kappa\text{O-cobalt(III)}$$
$$[Co(NH_3)_5(NO_2)]^{2+} \quad \text{Pentaamminenitrito-}\kappa\text{N-cobalt(III)}.$$

Because nitrite is an intermediate in NH_3 oxidation to $NO_3{}^-$ in environments with low oxygen, and in $NO_3{}^-$ reduction to N_2 or NH_3, the linkage bonding scenario is very important in describing $NO_2{}^-$ reactivity in the environment.

These ambidentate ligands also have the capability of bridging between two metal complex ions. In this case, two different atoms from the ambidentate ligand will bind the two metal centers. For example, the S in SCN^- will bind metal center one whereas the N in SCN^- will bind to the second metal center $(M_1\text{-S-C}{\equiv}\text{N-M}_2)$.

8.4.5 Geometrical Isomerism – Four Coordination

Geometrical isomerism is very important in transition metal complexes. For four coordination, tetrahedral geometries do not allow for geometrical isomers, but optical isomers are possible as in carbon chemistry. However, square planar complexes readily accommodate geometrical isomerism. Figure 8.5 shows the square planar complexes for platinum(II) (d^8 electron configuration) in the neutral compound diamminedichloridoplatinum(II) [or diamminedichloridoplatinum(0)] with the chemical formula $[PtCl_2(NH_3)_2]$. In the cis compound, the two chloride anions (and the two ammonia molecules) are cis to each other with an angle of $90°$

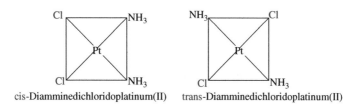

Figure 8.5   *Cis and trans geometries for diamminedichloridoplatinum(II), a neutral square planar molecule*

for the chloride–platinum–chloride bond angle; this molecule has a dipole moment. In the trans compound, the chloride ions are trans to each other with a chloride–platinum–chloride bond angle of 180°; this molecule does not have a dipole moment. In naming these compounds, the prefixes cis and trans are placed before the complex's name.

Square planar complexes have a plane of symmetry; therefore, optical isomers are not possible.

8.4.5.1   *Geometrical Isomerism – Octahedral Coordination*

There are more possibilities for geometrical isomerism in octahedral complexes. The top two structures in Figure 8.6 show cis and trans isomers for the cation, $[CoCl_2(NH_3)_4]^+$ or tetraamminedichloridocobalt(III). Here, cis and trans are defined similarly as above for the neutral diamminedichloridoplatinum(II). If one of the ammonia molecules were exchanged for a third ligand, the chloride ions would still be placed in either the cis or trans position with a similar naming of the resulting complex ion. For example, the complex ion, $[CoBrCl_2(NH_3)_3]^0$, would be named cis or trans triamminebromidodichloridocobalt(III).

In the case when there are three identical ligands attached to the metal center, there are two possible arrangements. The first geometrical structure is to have all three similar ligands occupying positions of the octahedron that share a triangular face. This is known as a facial or "fac" isomer, and for the example on

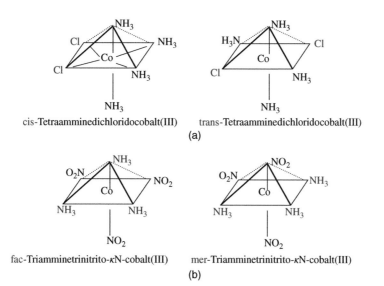

cis-Tetraamminedichloridocobalt(III)   trans-Tetraamminedichloridocobalt(III)

(a)

fac-Triamminetrinitrito-κN-cobalt(III)   mer-Triamminetrinitrito-κN-cobalt(III)

(b)

Figure 8.6   *Geometrical isomers in octahedral complexes using (a) cis, trans $[Co(NH_3)_4(Cl_2)]^+$ and (b) fac, mer $[Co(NH_3)_3(NO_2)_3]^0$; the N atom from nitrite binds to the cobalt*

the bottom left of Figure 8.6b, the name for this neutral molecule is fac-triamminetrinitrito-κN-cobalt(III), $[Co(NH_3)_3(NO_2)_3]^0$, as the nitrogen group binds to the cobalt. The second geometrical arrangement is for all three similar ligands to occupy three positions of the octahedron that are perpendicular to the C_4 axis; this is called a meridional isomer and is given the prefix "mer." The example on the bottom right of Figure 8.6b is named mer-triamminetrinitrito-κN-cobalt(III), $[Co(NH_3)_3(NO_2)_3]^0$. For multidentate ligands that have three ligating atoms binding to a metal, the ligand can bind in a facial or meridional manner, and the complex is named accordingly.

8.4.6   Optical Isomerism in Octahedral Geometry

In octahedral geometry, there are a significant number of possibilities for optical isomerism so discussion will be limited to just a couple of these. Figure 8.7 demonstrates optical isomerism for the two cis geometrical isomers of the complex ion $[CoCl_2(en)_2]^+$, which has two monodentate ligands (chloride) and two bidentate ligands (en = ethylenediamine). The trans isomer would have a plane symmetry and thus not exhibit optical isomerism. Figure 8.7a shows two cis isomers, which are mirror images of each other but which are not superimposable upon each other. These two structures are drawn to explicitly show the square plane or pseudo C_4 axis of the molecules. The two structures in Figure 8.7b are drawn to explicitly show the trigonal plane or pseudo C_3 axis of the molecules. The triangular face with the solid line for each lower complex is identical to the darker triangular face of the structure directly above it whereas the triangular face with the dashed lines for each lower complex is identical to the triangular face with the dashed lines directly above it. The structures in Figure 8.7b are also drawn with arrows indicating the direction of rotation about the C_3 axis. Here, the bottom left structure screws *counterclockwise* from the top triangular face to the bottom triangular face so is laevo or left handed, and the bottom right structure screws *clockwise* from the top triangular face to the bottom triangular face so is dextro or right handed. As shown in Figure 8.7, the naming for each of these complexes is the same except for the Greek symbol Λ (lambda) indicating left handedness and the Greek symbol Δ (delta) indicating right handedness.

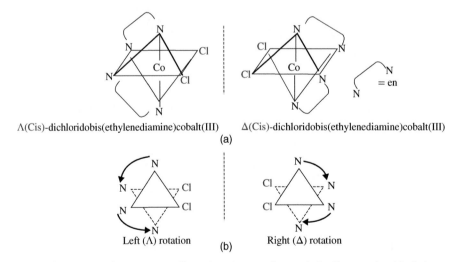

Λ(Cis)-dichloridobis(ethylenediamine)cobalt(III)   Δ(Cis)-dichloridobis(ethylenediamine)cobalt(III)

(a)

Left (Λ) rotation   Right (Δ) rotation

(b)

Figure 8.7   *Optical isomerism for two optically active cis complexes of* $[CoCl_2(en)_2]^+$ *with their proper nomenclature. The* CH_2CH_2 *in ethylenediamine is represented by the bracket. (a) Square plane explicitly drawn. (b) Rotation about trigonal plane (∼ C_3) axis*

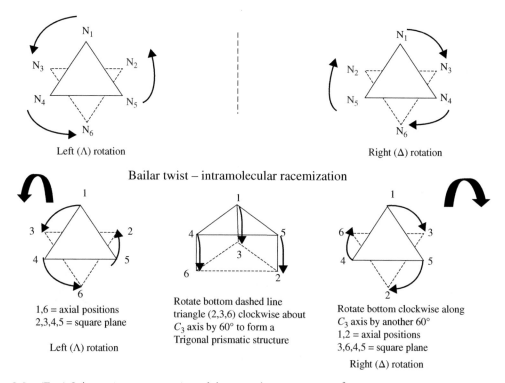

Left (Λ) rotation   Right (Δ) rotation

Bailar twist – intramolecular racemization

1,6 = axial positions
2,3,4,5 = square plane

Left (Λ) rotation

Rotate bottom dashed line
triangle (2,3,6) clockwise about
C_3 axis by 60° to form a
Trigonal prismatic structure

Rotate bottom clockwise along
C_3 axis by another 60°
1,2 = axial positions
3,6,4,5 = square plane

Right (Δ) rotation

Figure 8.8 *(Top) Schematic representation of the complex ion* $[Co(en)_3]^{3+}$ *looking down the* C_3 *axis. (Bottom) Bailar twist showing racemization*

Metal complexes, which have three bidentate ligands, also have optical isomers. The complex ion, $[Co(en)_3]^{3+}$, is shown in a schematic form in the upper part of Figure 8.8. The two nitrogen atoms with blue color have now been replaced by the two chloride ions that were shown in Figure 8.7b. The dark solid and dashed triangular faces have the same meaning as in Figure 8.7; counterclockwise and clockwise rotations about the trigonal plane or pseudo C_3 axis are identical to that shown in Figure 8.7. The bottom part of Figure 8.8 shows a possible pathway for intramolecular racemization known as the Bailar twist [3], which does not have any bond breaking or dissociation of the ligating atoms from the cobalt atom; note the direction of the arrows. Here, clockwise rotation by 60° of the bottom triangular face with positions numbered 2, 3, and 6 forms an intermediate with trigonal prismatic structure so that positions 1,3; 5,2; 4,6 now overlap. Another clockwise rotation by 60° of the bottom triangular face (120° full rotation) results in the conversion of the left-handed complex ion into the right-handed complex ion; note the change in the direction of the arrows representing the chelate, ethylenediamine. The Bailar twist is also known as the trigonal twist because of the rotation of the trigonal face about C_3 the axis.

Figure 8.9 shows another twist mechanism for the racemization of optical isomers that is known as the Ray–Dutt twist [3] (also known as the rhombic or tetragonal twist). The example shown is for the $[M(Cl)_2(en)_2]$ system, and the two chloride ions will remain in the same position throughout the twist. Molecules A1 and B1 are the Λ and Δ optical isomers, respectively, as displayed earlier in Figure 8.7 for perspective. Molecules A2 and B2 are the same optical isomers rotated by 90° and show the four nitrogen atoms in a roughly square or tetragonal arrangement so that the twist mechanism can be demonstrated.

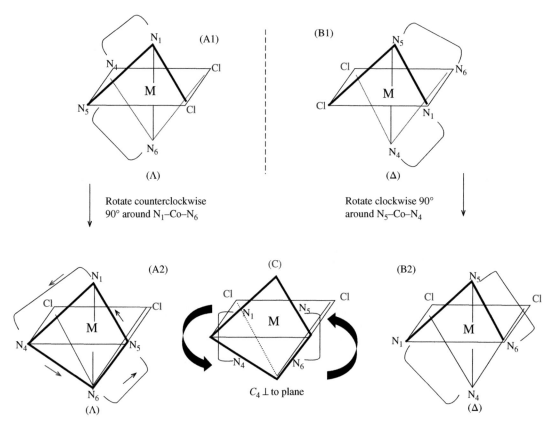

Figure 8.9 *(Top) Λ and Δ optical isomers for a cis-octahedral complex. (Bottom) Ray–Dutt twist showing racemization*

Molecule C is the intermediate formed during a concerted counterclockwise movement of molecule A2 along the edges of the triangular faces such that N_1 moves toward the original N_4 as N_4 moves toward the original N_6; likewise, N_6 moves toward N_5 as N_5 moves toward N_1. Continued counterclockwise motion results in molecule B2. The overall counterclockwise rotation from molecule A2 to molecule B2 is a pseudo-C_4 rotation (the four N atoms are not in a square arrangement in A2 or B2); thus, it is called a rhombic or tetragonal twist.

Both the Bailar twist and the Ray–Dutt twist are examples of intramolecular rearrangements without any bond breaking between the ligands and the central metal atom in the complex. Because of the fluid motion during these intramolecular processes, these molecules are called **fluxional**, and these twist mechanisms are examples of **Berry pseudorotation**; that is, the atoms exchange positions with each other. Pseudorotation can be enhanced by increasing the temperature, and it can be decreased or stopped by decreasing the temperature.

8.5   Bonding Theories for Transition Metal Complexes

There are three major theories that have been used to describe the bonding and coordination geometries for transition metal complexes. The first theory that found favor with inorganic chemists was valence bond theory, but it was unable to provide the information that inorganic chemists desired concerning the color and spectroscopy of transition metal complexes. Its use was supplanted by Bethe's [4] crystal field theory (CFT), which centered on the metal ion in the complex but not the ligands. Ligand field theory (LFT) developed when van Vleck [5] added a description of the bonding between the metals in the ligands to crystal field theory; frequently, the names LFT and CFT are used interchangeably. Eventually, molecular orbital theory (MOT), which describes all the bonding possible between the metal and the ligands, gave a thorough understanding of the bonding for transition metal complexes. The major points of crystal field theory and valence bond theory are part of molecular orbital theory. A brief history of these theories has been provided by Ballhausen in a series of three papers [6–8] and Figgis and Hitchman [9].

8.5.1 Valence Bond Theory

Many of the first transition metal complexes that were studied exhibited tetrahedral, square planar, or octahedral geometries. Because valence bond theory had reasonable success in describing the structure of many nonmetal compounds, its use also found great favor with inorganic chemists particularly before World War II. Major questions that inorganic chemists were interested in revolved around the paramagnetic or diamagnetic nature of transition metal complexes. Figure 8.10 shows how valence bond theory dealt with the octahedral complexes of Co(III) with a d^6 electron configuration with different ligands. With halide ions such as fluoride, the five 3d orbitals would be occupied as expected using the **Aufbau** principle where there would be four electrons unpaired (a **high spin** complex). To have an octahedral geometry with sp^3d^2 hybridization, the cobalt would accept six pairs of electrons from ligands to form bonds using s, p, and d orbitals from the fourth energy level. However, when NH_3 formed an octahedral complex with Co(III), the cobalt d^6 electrons are all paired so that the complex is diamagnetic (a **low spin** complex). In this case, the 3d orbitals could be used to provide sp^3d^2 hybridization as they would be vacant and available for accepting lone pairs of electrons from the ligands.

A similar procedure was used to describe many other transition metal complexes – e.g., the four coordinate structures of Ni(II) [d^8] with different ligands. For $NiCl_4^{2-}$, the complex is paramagnetic with two electrons unpaired as expected when using the Aufbau principle; thus, all of the 3d orbitals are occupied or partially occupied and a tetrahedral complex (sp^3 hybridization) results. In contrast, the $Ni(CN)_4^{2-}$ complex is diamagnetic so that four of the 3d orbitals are completely occupied, and one 3d orbital is empty and available for accepting a lone pair of electrons from a cyanide ligand. The resulting complex is square planar with sp^2d hybridization.

One important principle that is amply demonstrated by valence bond theory is that before understanding the bonding and the coordination geometry of a transition metal complex, it is important to know the oxidation

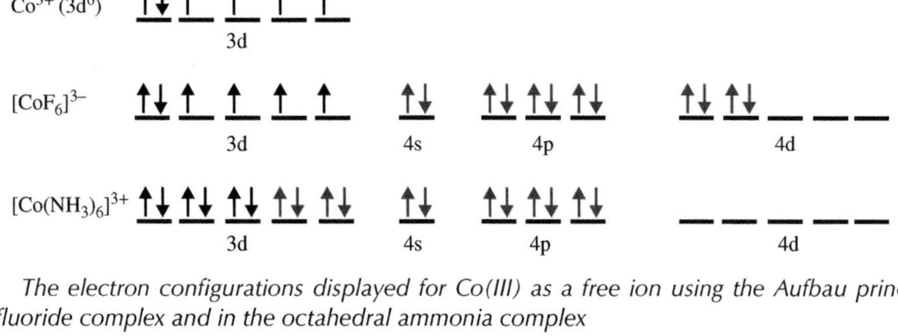

Figure 8.10 *The electron configurations displayed for Co(III) as a free ion using the Aufbau principle, in the octahedral fluoride complex and in the octahedral ammonia complex*

state of the metal and its electron configuration for the oxidation state. This information is essential regardless of the bonding theory (CFT, LFT, MOT) that is used to describe the metal complex.

8.5.2   Crystal Field Theory

Crystal theory was first developed by Hans Bethe for ions in crystals [4] and improved by van Vleck [5] to describe the electronic structure of transition metal complexes. These theories were a major improvement over valence bond theory because they could explain a wide variety of chemical and physical properties for metal complexes. Crystal field theory describes how the five d orbitals and the electrons in these orbitals interact with ligands, which donate a pair of electrons but which are considered a negative and spherical *point charge*. The five d orbitals will lose their spherical symmetry and degeneracy from the free ion based upon the geometry of the transition metal complex as the ligands will experience a repulsive force due to electrons in some of the d orbitals. The main geometries for discussion are octahedral and tetrahedral as well as square planar geometry and tetragonal distortions to the octahedral geometry.

Prior to discussing the ligand and metal d orbital interactions, it is necessary to take a further look at the d_{z^2} orbital (Figure 8.11). This orbital can have electron density on all three axes, and Figure 8.11 shows that it can be looked at as a combination of two orbitals, $d_{z^2-x^2}$ and $d_{z^2-y^2}$. As shown in the O_h character table (Table 4.9), the d_{z^2} orbital is an abbreviation for $2d_{z^2} - d_{x^2} - d_{y^2}$.

8.5.2.1   Octahedral Transition Metal Complexes

Figure 8.12 shows the five d orbitals for the central metal ion in an octahedral complex and the approach of the six ligands as point charges to the metal ion along the Cartesian coordinate axes, which are the bond axes. It is clear from the spatial arrangement that the d_{z^2} orbital and the $d_{x^2-y^2}$ orbital have a strong repulsion with the six ligand negative point charges. As shown in Table 4.9, these two d orbitals have the symmetry e_g (and are involved in the formation of bonding and antibonding orbitals as will be discussed in Section 8.6 on molecular orbital theory). The other three orbitals (d_{xy}, d_{xz}, d_{yz}) are in between the ligands that are on the bond axes and are not repelled by the negative point charges of the six ligands. Inspection of the octahedral character table shows that these three orbitals have the symmetry t_{2g}.

As a result of these ligand and d orbital interactions, there is a splitting of the two sets of d orbitals, which is represented in Figure 8.13 as an orbital energy diagram. On the left are the five degenerate d orbitals in a spherical environment (as if they were in a free atom) whereas on the right these d orbitals are now split into the e_g and t_{2g} sets. Because of the repulsion on the bond axes, the e_g orbitals (d_{z^2} and $d_{x^2-y^2}$) are at a higher energy in the energy level diagram, and the t_{2g} orbitals (d_{xy}, d_{xz}, d_{yz}) lie at a more stable energy. The difference in energy between these two sets of orbitals is the crystal field splitting parameter, Δ_{oct} or $10D_q$. The energy of the d orbitals in the spherical environment is the barycenter (or the midpoint in energy) for the

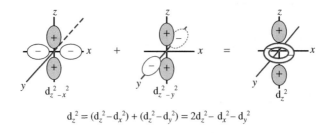

$$d_z^2 = (d_z^2 - d_x^2) + (d_z^2 - d_y^2) = 2d_z^2 - d_x^2 - d_y^2$$

Figure 8.11   *More information on the nature of the d_{z^2} orbital*

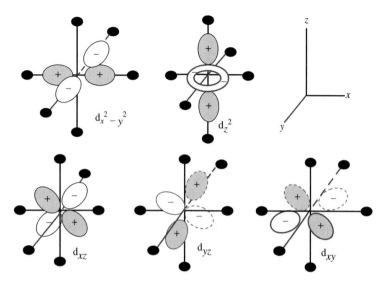

Figure 8.12   *The approach of six ligand point charges in an octahedral field to the five d orbitals*

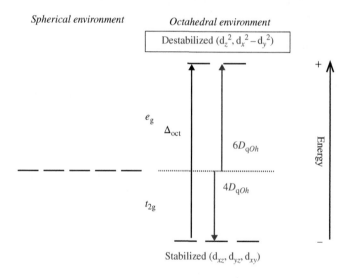

Figure 8.13   *Splitting of the d orbitals in an octahedral field*

orbitals split in an octahedral environment. Because the maximum occupancy of electrons in the t_{2g} set is six and in the e_g set it is four, each electron in the t_{2g} set has a splitting parameter of 0.4 Δ_{oct} or four D_{qO_h} units whereas each electron in the e_g set has a splitting parameter of 0.6 Δ_{oct} or six D_{qO_h} units. The actual value for the difference in energies between these two orbital sets is a function of the ligand and the oxidation state of the metal and will be described shortly. The energy gap between these two sets of orbitals can be measured by UV–Vis spectroscopy (the electromagnetic spectrum is given in Figure 1.2). It is important to note that the t_{2g} *orbitals are closer to the nucleus because of their enhanced stability*, and the e_g orbitals are farther away from the nucleus.

For a simple d^1 or t_{2g}^1 octahedral metal complex ion, the absorption of energy by the electron results in a transition from the t_{2g} orbital to one of the e_g orbitals for the purple $Ti(H_2O)_6^{3+}$ complex ion (Figure 8.14a). The transition of the electron between the two energy states is described with one of the following equations.

$$t_{2g}^1 e_g^0 \rightarrow t_{2g}^0 e_g^1 \quad \text{or} \quad e_g^1 \leftarrow t_{2g}^1$$

The maximum absorbance for this complex is at a wavelength of 493 nm ($20,300 \text{ cm}^{-1}$ or 243 kJ mol^{-1}) so that yellow and green light are absorbed. The wavelength(s) of light absorbed by the complex when white light is shone on the complex indicates the wavelength(s) of light reflected by the complex. The light reflected is the color actually observed and is the complement of the color absorbed; in this case, red and blue mix to give purple. The visible spectrum consists of the following primary colors in increasing energy and decreasing wavelength; red (620–750 nm), orange (590–620 nm), yellow (570–590 nm), green (495–570 nm), blue (450–495 nm), indigo (420–450 nm), and violet (380–420 nm) (ROYGBIV). Complementary colors are on the opposite edge of a color wheel; e.g., red and green; orange and blue; and yellow and violet. Thus, when yellow and green light are absorbed by the $Ti(H_2O)_6^{3+}$ complex, the complex reflects purple, and the splitting of these d orbitals can immediately be seen as a way to describe the color and spectroscopy of a metal complex. The simplicity of this model in explaining the spectroscopy of the complex is an improvement over the valence bond theory approach. The large absorption peak in the UV region is a ligand to metal charge transfer (LMCT) band that will be explained using the molecular orbital approach below.

$Fe(H_2O)_6^{2+}$ also gives a similar spectrum (Figure 8.14b). Octahedral d^6 ions have a similar electron configuration as Ti(III) because one of the t_{2g}^4 electrons (in the paired orbital) is excited to an e_g orbital. The transition can be represented as

$$t_{2g}^4 e_g^2 \rightarrow t_{2g}^3 e_g^3$$

Because the ionic charge is +2 for $Fe(H_2O)_6^{2+}$ the peak is at a longer wavelength (lower energy) in the near IR region than that for $Ti(H_2O)_6^{3+}$.

To describe high spin and low spin complexes, the filling of the electrons into the e_g and t_{2g} orbitals is provided in Figure 8.15 for the d^6 electron configuration [e.g., Fe(II)]. Because the t_{2g} orbitals are lower in energy (more stable than the e_g and the original d orbitals in a spherical environment), a concept known as the crystal field stabilization energy (CFSE) can be described. There are two energy level diagrams in Figure 8.15. The energy level diagram on the left is for the high spin complex, and the difference in the energy levels between the e_g and t_{2g} orbitals is smaller by a significant amount compared to the diagram on the right for the low spin complex. Because the d orbital splitting is smaller for the high spin complex, the complex is called **weak field**. In the weak field case, the extra stabilization energy for the t_{2g} orbitals is not high enough to overcome the **pairing energy** (P) of two electrons into each of the t_{2g} orbitals. Pairing energy is a positive energy term and results in a decrease in the stabilization of the complex because of the repulsive effect of two electrons in the same orbital and the loss of exchange energy related to the loss of maximum multiplicity as noted in Hund's rule (thus, $10D_q < P$).

When crystal field splitting results in significant stabilization of the t_{2g} orbitals as in the energy level diagram on the right of Figure 8.15, all of the electrons are paired in the t_{2g} orbitals as the stabilization of these orbitals is more than enough to compensate the pairing energy (thus, $10D_q > P$). This latter low spin case is known as the **strong field** case in crystal field theory terminology. Table 8.3 and Figure 8.16 summarize the relative CFSE for each d orbital electron configuration in both the weak and strong field cases.

A couple of important aspects of the pairing energy need to be described. First, the pairing energy decreases in magnitude on descending a given family in the periodic table because the 4d and 5d orbitals are much more diffuse. Thus, it becomes easier to put two electrons into the same orbital. Metals in the second and

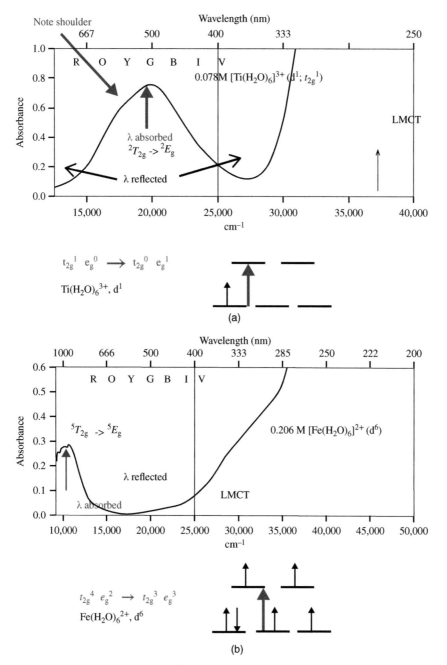

Figure 8.14 *Electronic spectra of (a) Ti(H$_2$O)$_6$$^{3+}$(d^1, t$_{2g}$^{1}) and (b) [Fe(H$_2$O)$_6$]$^{2+}$(d^6, t$_{2g}^{4}e_g$^{2}). The shoulders on the peaks of the Ti and the Fe complexes will be explained later. The vertical blue arrows represent the energies of the electronic transitions*

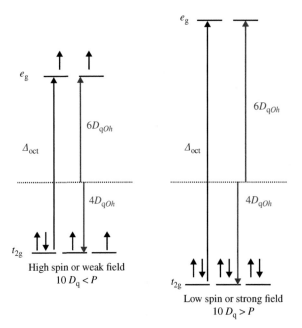

Figure 8.15 *Comparison of energy level diagrams for the weak field (high spin) and for the strong field (low spin) octahedral d^6 cases. $Fe(H_2O)_6^{2+}$ is an example of the weak field case and pyrite, FeS_2, is an example of the strong field case*

Table 8.3 *Simplified crystal field stabilization energies for weak and strong field octahedral complexes. Note that $10D_q = 1.0 \Delta_{oct}$*

Weak field (high spin)				Strong field (low spin)			
d^n	Configuration	Unpaired electrons	CFSE	d^n	Configuration	Unpaired electrons	CFSE
d^1	t_{2g}^1	1	$-4D_q$ ($-0.4\ \Delta_{oct}$)	d^1	t_{2g}^1	1	$-4D_q$
d^2	t_{2g}^2	2	$-8D_q$ ($-0.8\ \Delta_{oct}$)	d^2	t_{2g}^2	2	$-8D_q$
d^3	t_{2g}^3	3	$-12D_q$ ($-1.2\ \Delta_{oct}$)	d^3	t_{2g}^3	3	$-12D_q$
d^4	$t_{2g}^3 e_g^1$	4	$-6D_q$ ($-0.6\ \Delta_{oct}$)	d^4	t_{2g}^4	2	$-16D_q + P$
d^5	$t_{2g}^3 e_g^2$	5	$0D_q$	d^5	t_{2g}^5	1	$-20D_q + 2P$
d^6	$t_{2g}^4 e_g^2$	4	$-4D_q + P$	d^6	t_{2g}^6	0	$-24D_q + 3P$
d^7	$t_{2g}^5 e_g^2$	3	$-8D_q + 2P$	d^7	$t_{2g}^6 e_g^1$	1	$-18D_q + 3P$
d^8	$t_{2g}^6 e_g^2$	2	$-12D_q + 3P$	d^8	$t_{2g}^6 e_g^2$	2	$-12D_q + 3P$
d^9	$t_{2g}^6 e_g^3$	1	$-6D_q + 4P$	d^9	$t_{2g}^6 e_g^3$	1	$-6D_q + 4P$
d^{10}	$t_{2g}^6 e_g^4$	0	$0D_q + 5P$	d^{10}	$t_{2g}^6 e_g^4$	0	$0D_q + 5P$

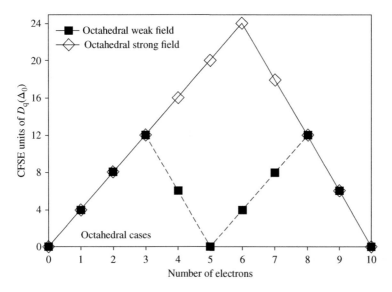

Figure 8.16 *Crystal field stabilization energies for octahedral weak and strong field complexes*

third transition series (rows five and six) of the periodic table exhibit almost exclusively strong field or low spin metal complexes whereas metals in the first row of the transition series (row four) of the periodic table can exhibit either strong field or weak field complexes. Second, some ligands have electron acceptor properties, which allow electrons in the same d orbital to interact or bond with a ligand orbital; this phenomenon is described in detail below using the molecular orbital theory approach and is called **backbonding** (Section 8.6.3). This type of ligand–metal bonding is important for the metals in the first transition metal row of the periodic table, and dictates whether the complex is a strong field or low spin complex. The difference between high spin and low spin complexes for the first row of transition metals results in unique chemical properties that occur in enzyme systems of a host of organisms (chapter 10).

There are numerous experimental data, which have been used to show that metal complexes have CFSE. One of the major outcomes of Figures 8.15 and 8.16 is that there is a significant energy difference between the t_{2g} and e_g orbitals for the weak field and strong field cases. The energy difference for an electron transition from a t_{2g} orbital to an e_g orbital is greater in the strong field case than in the weak field case. Thus, strong field complexes will absorb light toward the blue region and weak field complexes will absorb light toward the red region of the visible spectrum. Subsequent sections explore crystal field stabilization energies in octahedral and tetrahedral complexes using a variety of experimental data.

8.5.2.2 Evidence for Crystal Field Stabilization Energies (Visible Spectroscopy)

Ligand Effects on d Orbital Splitting. As noted previously, Cr(III) [d^3, t_{2g}^3] and Co(III) [d^6] octahedral complexes with a great variety of ligands have been well studied. Table 8.4 shows several complexes of these transition metal ions and the values of $10D_q$ in energy units of kJ mol^{-1} and cm^{-1} as well as the wavelength of the first major electronic transition (e.g., Figure 8.43a). For Co(III) complexes, the pairing energy (P) is about $21,000$ cm^{-1}; the splitting of the d orbitals for the ammonia complex exceeds the pairing energy. The d orbital splitting for the cyanide and ethylenediamine Co(III) complexes greatly exceeds the pairing energy. Additionally, Co(III) complexes typically have a larger splitting of the d orbitals than Cr(III) complexes with the same ligand because of the effective nuclear charge increase on going across the periodic table.

Table 8.4 Data on d orbital splitting and crystal field stabilization energy for various ligands in Cr(III) and Co(III) complexes

Cr^{3+} complex $(d^3; t_{2g}^{3})$	$\Delta_{oct} = 10D_q$		Co^{3+} complex (d^6)	$\Delta_{oct} = 10D_q$		
	kJ mol^{-1}	cm^{-1}(nm)		kJ mol^{-1}	cm^{-1}(nm)	
$[CrCl_6]^{3-}$	158	13,200 (757)				
$[CrF_6]^{3-}$	182	15,200 (658)	$[CoF_6]^{3-}$	182	13,100 (763)	$t_{2g}^{4}e_g^{2}$
$[Cr(H_2O)_6]^{3+}$	208	17,400 (574)	$[Co(H_2O)_6]^{3+}$	218	20,700 (483)	t_{2g}^{6}
$[Cr(NH_3)_6]^{3+}$	258	21,600 (463)	$[Co(NH_3)_6]^{3+}$	274	22,900 (437)	t_{2g}^{6}
$[Cr(en)_3]^{3+}$	262	21,900 (457)	$[Co(en)_3]^{3+}$	278	23,200 (431)	t_{2g}^{6}
$[Cr(CN)_6]^{3-}$	318	26,600 (376)	$[Co(CN)_6]^{3-}$	401	33,500 (299)	t_{2g}^{6}

Data for these and other ligands resulted in a tabulation of ligand effects on transition metal ions, which is known as the **spectrochemical series** in increasing field strength. The relative order of several important ligands in this series is as follows:

$$I^- < Br^- < S^{2-} < SCN^- < Cl^- < NO_3^- < F^- < OH^- < ox^{2-} < H_2O < NCS^-$$

$$< CH_3CN < \textbf{NH}_3 < \textbf{en} < \textbf{bipy} < \textbf{phen} < \textbf{NO}_2^- < \textbf{R}_3\textbf{P} < \textbf{R}_2\textbf{S} < \textbf{CN}^- < \textbf{CO} \sim \textbf{olefins} \sim \textbf{O}_2.$$

The ligands that are in bold font are the low spin/strong field ligands whereas the other ligands are the high spin/weak field ligands. An interesting point is that the halides are most representative of being negative **point charges**, but are weak field ligands with the least splitting of the d orbitals. The relative position of ligands in the spectrochemical series will be explained using molecular orbital theory below. Interestingly, $[Co(H_2O)_6]^{3+}$ is the only low spin water complex in the first transition series, but is unstable as it oxidizes water (Section 8.7.3).

Transition Metal Oxidation State Effects on d Orbital Splitting.   The splitting of the d orbitals is also affected by the oxidation state of the transition metal. Table 8.5 provides spectroscopic data for the same metal in different oxidation states bound to the same ligand. The difference in $10D_q$ on change in metal oxidation state can be substantial as in the cobalt and ruthenium complexes. The value of $10D_q$ increases with an increase in metal oxidation state as the metal–ligand bonds are shorter and thus stronger. Also the value of $10D_q$ increases

Table 8.5 Data on d orbital splitting for some +2 and +3 complexes of chromium, iron, cobalt, and ruthenium. The wavelength of absorption is in parentheses

Complex	$\Delta_{oct} = 10D_q$	M^{n+}	Complex	$\Delta_{oct} = 10D_q$	M^{n+}
$[Cr(H_2O)_6]^{3+}$	17,400 (574)	+3	$[Co(NH_3)_6]^{3+}$	22,900 (437)	+3
$[Cr(H_2O)_6]^{2+}$	14,000 (714)	+2	$[Co(NH_3)_6]^{2+}$	10,200 (980)	+2
$[Fe(H_2O)_6]^{3+}$	14,000 (714)	+3	$[Co(H_2O)_6]^{3+}$	20,700 (483)	+3
$[Fe(H_2O)_6]^{2+}$	10,400 (962)	+2	$[Co(H_2O)_6]^{2+}$	7,900 (1265)	+2
$[Fe(CN)_6]^{3-}$	35,000 (286)	+3	$[Ru(H_2O)_6]^{3+}$	32,200 (311)	+3
$[Fe(CN)_6]^{4-}$	32,200 (311)	+2	$[Ru(H_2O)_6]^{2+}$	19,800 (505)	+2

Table 8.6 *Data on d orbital splitting for +3 complexes of the chromium and cobalt families*

Complex	$\Delta_{oct} = 10D_q$		Complex	$\Delta_{oct} = 10D_q$	
	kJ mol^{-1}	cm^{-1}(nm)		kJ mol^{-1}	cm^{-1}(nm)
$[CrCl_6]^{3-}$	158	13,200 (757)	$[Co(NH_3)_6]^{3+}$	218	22,900 (437)
$[MoCl_6]^{3-}$	230	19,200 (521)	$[Rh(NH_3)_6]^{3+}$	408	34,100 (293)
			$[Ir(NH_3)_6]^{3+}$	490	41,000 (244)
$[RhCl_6]^{3-}$	243	20,300 (493)			
$[IrCl_6]^{3-}$	299	25,000 (400)	$[Co(en)_3]^{3+}$	278	23,600 (423)
			$[Rh(en)_3]^{3+}$	414	34,600 (289)
			$[Ir(en)_3]^{3+}$	495	41,400 (242)

on descending a given metal group (e.g., iron and ruthenium bound to water). The increase in d orbital splitting on descending a group is also shown in Table 8.6.

The Effects of Second and Third Transition Metals on d Orbital Splitting.   On the left of Table 8.6, there are metal complexes with the weak field ligand, chloride ion. On descending the periodic table from chromium to molybdenum and from rhodium to iridium, there is a substantial increase in $10D_q$. The $[CrCl_6]^{3-}$ and the $[MoCl_6]^{3-}$ complex ions have the electronic configuration of t_{2g}^3 whereas the rhodium and iridium chloride complexes are t_{2g}^6. The right-hand side of Table 8.6 shows the significant increase in $10D_q$ on descending the cobalt family for complexes with ammonia and with ethylenediamine, which are strong field ligands. For these complexes, the electron configuration is t_{2g}^6. From the data in Tables 8.4 through 8.6, it becomes obvious that the ligand dictates the spin state (high versus low) for the first transition metal series whereas in the second and third transition metal series, only the low spin case occurs because the 4d and 5d orbitals are more diffuse thus allowing the pairing of electrons to be easier in the same region of space.

Radii of Transition Metal Ions and CFSE.   The radii of M^{2+} and M^{3+} ions also provide further evidence for CFSE and d orbital splitting. Figure 8.17a shows that ionic radii decrease for high spin M^{2+} ions from d^0 to d^3 as the t_{2g} orbitals are filled sequentially to obtain a t_{2g}^3 electron configuration as these t_{2g} orbitals are of lower energy and thus closer to the nucleus. For d^4 and d^5 electron configurations, the e_g orbitals are filled. Because electrons in the e_g orbitals interact with the ligands' electrons, there is a repulsion and the ionic radius increases due to the d orbital splitting. Likewise from d^6 to d^8, the t_{2g} orbitals are filled until they all have a pair of electrons and the ionic radius again decreases. For d^9 and d^{10} electron configurations, the e_g orbitals are filled and the ionic radius increases again. For low spin M^{2+} ions from d^0 to d^6, the t_{2g} orbitals are filled sequentially to obtain a t_{2g}^6 electron configuration and the ionic radius decreases until electrons are added to the e_g orbitals from d^7 to d^{10}.

Figure 8.17b shows similar behavior for the low spin and high spin M^{3+} ions. Low spin cases are more prevalent for M^{3+} ions than for M^{2+} ions in the first transition metal series. Note that the radii for M^{3+} ions are shorter than the radii for M^{2+} ions, which correlates with the increased d orbital splitting for M^{3+} ions as in Table 8.5.

Lattice and Hydration Energies and CFSE.   Because the lattice energy (U_0; Section 6.3.5) is inversely proportional to ionic radius, an increase in lattice stability is observed on filling the t_{2g} orbitals as shown in Figure 8.18 for chloride and fluoride solids of the divalent cations. The characteristic double hump feature is

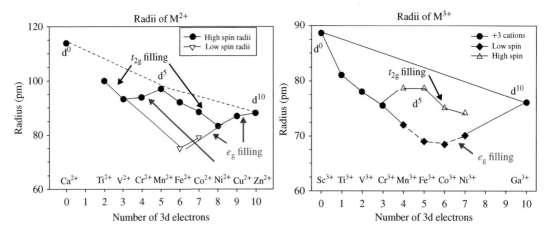

Figure 8.17 *(a) ionic radii for low spin and high spin M^{2+} ions. (b) ionic radii for low spin and high spin M^{3+} ions*

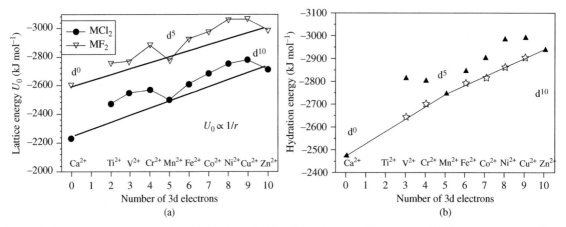

Figure 8.18 *(a) Lattice energy data for MCl_2 and MF_2 of the first transition series. (b) Hydration energy data (triangles) for the divalent ions of the first transition series; $\Delta H_{hydr} - \Delta_{oct}$ (stars)*

the inverse of what is observed in Figure 8.17. Likewise, the hydration energies of a gaseous +2 metal cation, which can be calculated from thermochemical data, yield the double hump feature.

$$M^{2+}_{(g)} + 6H_2O \rightarrow [M(H_2O)_6]^{2+}_{(aq)}$$

Data in Table 8.7 show that subtraction of the CFSE for the aqua complexes from the hydration energy ($\Delta H_{hydr} - \Delta_{oct}$) gives values that fall on the lines from d^0 to d^5 to d^{10} and indicate that d orbital splitting occurs.

Magnetic Properties and CFSE.   The magnetic properties of metal complexes also indicate the importance of d orbital splitting and CFSE. Table 8.8 provides the experimental and calculated magnetic moment for both high spin and low spin complexes of the first transition metal series. The agreement is quite good in

Table 8.7 Spectroscopic and hydration energy data for hexaaqua complexes for + 2 cations of the first transition series. Multiply $10D_q$ by relative CFSE in Table 8.3 to obtain CFSE in kJ mol^{-1}

+2 Ion	$10 D_q$ (cm^{-1})	$10 D_q$ kJ mol^{-1}	CFSE kJ mol^{-1}	Hydration E kJ mol^{-1}	$(\Delta H_{hydr} - \Delta_{oct})$ kJ mol^{-1}
Ca				2468	2468
V	12,100	144.7	173.6	2814	2640
Cr	14,000	167.4	100.4	2799	2698
Mn	8,500	101.7	0	2743	2746
Fe	10,400	124.4	49.8	2843	2793
Co	7,900	94.5	75.6	2904	2828
Ni	8,500	101.7	122.0	2986	2864
Cu	12,000	143.5	86.1	2989	2902
Zn				2936	2936

Table 8.8 Magnetic properties for some complexes of the first row transition metals. The number of electrons is indicated by #e$^-$. BM indicates Bohr magneton

Metal	# d e$^-$	# e$^-$ unpaired (high spin)	μ (BM) (expt.)	μ (BM) (spin only calc)	# e$^-$ unpaired (low spin)	μ (BM) (expt.)	μ (BM) (spin only calc)
Ti^{3+}	1	1, t_{2g}^1	1.73	1.73	–	–	–
V^{4+}	1	1, t_{2g}^1	1.68–1.78	1.73	–	–	–
V^{3+}	2	2, t_{2g}^2	2.75–2.85	2.83	–	–	–
V^{2+}	3	3, t_{2g}^3	3.80–3.90	3.88	–	–	–
Cr^{3+}	3	3, t_{2g}^3	3.70–3.9	3.88	–	–	–
Mn^{4+}	3	3, t_{2g}^3	3.80–4.0	3.88	–	–	–
Cr^{2+}	4	4, $t_{2g}^3 e_g^1$	4.75–4.90	4.90	2, t_{2g}^4	3.20–3.30	2.83
Mn^{3+}	4	4, $t_{2g}^3 e_g^1$	4.90–5.00	4.90	2, t_{2g}^4	3.18	2.83
Mn^{2+}	5	5, $t_{2g}^3 e_g^2$	5.65–6.10	5.92	1, t_{2g}^5	1.80–2.10	1.73
Fe^{3+}	5	5, $t_{2g}^3 e_g^2$	5.70–6.0	5.92	1, t_{2g}^5	2.0–2.5	1.73
Fe^{2+}	6	4, $t_{2g}^4 e_g^2$	5.10–5.70	4.90	0, t_{2g}^6	–	0
Co^{3+}	6	4, $t_{2g}^4 e_g^2$	–	4.90	0, t_{2g}^6	–	0
Co^{2+}	7	3, $t_{2g}^5 e_g^2$	4.30–5.20	3.88	1, $t_{2g}^6 e_g^1$	1.8	1.73
Ni^{3+}	7	3, $t_{2g}^5 e_g^2$	–	3.88	1, $t_{2g}^6 e_g^1$	1.8–2.0	1.73
Ni^{2+}	8	2, $t_{2g}^6 e_g^2$	2.80–3.50	2.83	–	–	–
Cu^{2+}	9	1, $t_{2g}^6 e_g^3$	1.70–2.20	1.73	–	–	–

almost every case. Equations 3.11 and 3.12 determine the magnetic moment, μ (units in Bohr magnetons [$eh/(4\pi m)$]), for the spin only case of the electrons in the atomic orbitals. It is also used for the transition metal cations in complexes.

$$\mu = [n(n + 2)]^{\frac{1}{2}} \text{ where } n = \text{\# of unpaired electrons} \tag{3.12}$$

As was observed in Section 3.8.2, orbital angular momentum of the electron can contribute to μ. The value of μ_{expt} normally increases over the value for the spin only case, and μ becomes larger for low spin d^5, and high spin d^6 and d^7 cases. In these cases, the electrons, for example, in the d_{xy} and/or $d_{x^2-y^2}$ orbitals, can be induced to circulate around the z-axis in the xy plane when a perpendicular field on the z-axis is applied to the xy plane.

Before discussing some other pertinent experimental data regarding CFSE, the crystal field splitting pattern for tetrahedral complexes is given.

8.5.2.3   Tetrahedral Transition Metal Complexes

Figure 8.19 shows the five d orbitals for the central metal ion at the center of the cube and the approach of the four ligands as point charges to the metal ion at alternate vertices of the cube.

It is clear from the spatial arrangement that all of the five d orbitals do not have a substantial interaction with the four point charges as in the octahedral field case. However, the d_{z^2} orbital and the $d_{x^2-y^2}$ orbital have the least interaction with the four ligands and are now stabilized relative to the other three d orbitals. These two d orbitals have the symmetry e from the character table for the tetrahedron, T_d (Table 4.8). The other three orbitals (d_{xy}, d_{xz}, d_{yz}) point toward the four ligands and are repelled to some extent by the negative point charges of the four ligands. They are involved in the formation of bonding and antibonding orbitals. Inspection of the tetrahedral character table shows that these three orbitals have the symmetry t_2.

Figures 8.20 and 8.21 summarize the relative CFSE for each d orbital electron configuration in tetrahedral geometry. Because the interaction of the ligand point charges is significantly less than what is found in octahedral geometry, the splitting of the d orbitals is much smaller in a tetrahedral geometry as shown in the energy level diagrams of Figure 8.20, and the relationship is given by the following equation (Equation 8.4; see AOM, 8.7.4.1).

$$\Delta_{\text{tet}} = 10D_{qT_d} = (4/9)\,\Delta_{\text{oct}} \tag{8.4}$$

Comparison of electronic spectra for the same metal cation with the same ligand in octahedral and tetrahedral geometries shows this relationship in many instances [9]; e.g., $MnCl_6^{4-}$ (7500 cm^{-1}) versus $MnCl_4^{2-}$ (3300 cm^{-1}); $FeCl_6^{4-}$ (7600 cm^{-1}) versus $FeCl_4^{2-}$ (4000 cm^{-1}).

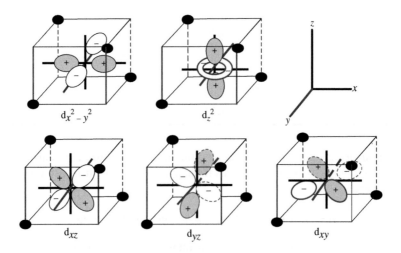

Figure 8.19   *The approach of four ligand point charges in a tetrahedral field to the five d orbitals*

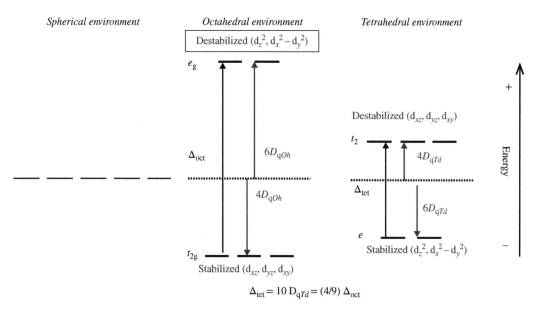

Figure 8.20 *Comparison of energy level diagrams for high spin or weak field octahedral complexes with high spin tetrahedral complexes*

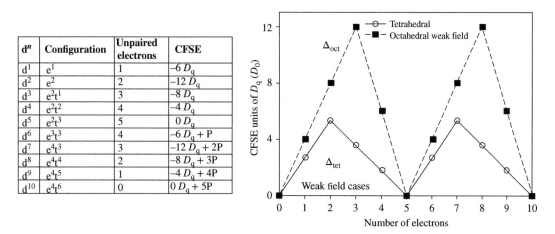

d^n	Configuration	Unpaired electrons	CFSE
d^1	e^1	1	$-6\,D_q$
d^2	e^2	2	$-12\,D_q$
d^3	e^2t^1	3	$-8\,D_q$
d^4	e^2t^2	4	$-4\,D_q$
d^5	e^2t^3	5	$0\,D_q$
d^6	e^3t^3	4	$-6\,D_q + P$
d^7	e^4t^3	3	$-12\,D_q + 2P$
d^8	e^4t^4	2	$-8\,D_q + 3P$
d^9	e^4t^5	1	$-4\,D_q + 4P$
d^{10}	e^4t^6	0	$0\,D_q + 5P$

Figure 8.21 *On the left is a table of simplified crystal field stabilization energies in tetrahedral geometry; on the right is a figure showing a comparison of crystal field stabilization energies in both tetrahedral and octahedral weak field cases*

Figure 8.21 provides the relative crystal field stabilization energies in a tetrahedral geometry on the left. These are then compared with crystal field stabilization energies in an octahedral weak field on the right; all values are normalized to values in octahedral field. The major outcome of this analysis is that weak field (high spin) complexes are found in tetrahedral geometry. In four coordinate geometry, strong field or low spin complexes exhibit square planar geometry.

8.5.2.4   Spinel Structures

The difference in crystal field stabilization energies between the tetrahedral and octahedral cases can be used to explain or to predict structures of the minerals known as spinels (Section 6.3.2.2). Spinels are solids with the stoichiometry $A^{II}B_2^{III}O_4$ where the Roman numerals indicate the oxidation states of A and B. ***Normal spinels*** have A^{2+} cations in T_d sites and both B^{3+} cations in O_h sites. ***Inverse spinels*** have A cations in O_h sites and half the B cations are in T_d sites and the other half in O_h sites. One of the most important minerals having an inverse spinel structure is magnetite, Fe_3O_4, or $Fe^{II}Fe_2^{III}O_4$ where the Fe^{II} cation occupies an O_h site so the formula is best represented as $Fe^{III}(Fe^{II}Fe^{III})O_4$. A simple calculation takes into account the CFSE for a given cation in both octahedral and tetrahedral geometries (normalized to octahedral CFSE units). The difference is known as the octahedral site stabilization energy (OSSE) and is given in the following equation.

$$OSSE = \text{Octahedral site stabilization energy} = CFSE(O_h) - CFSE(T_d \text{ in } O_h \text{ units})$$

Table 8.9 displays the results of the calculation for electron configurations d^1 to d^9 for both high spin and low spin complexes. Because tetrahedral symmetry leads to high spin complexes, only one calculation is required to convert CFSE T_d units to CFSE O_h units. Fe(III) is a d^5 electron configuration, and in high spin tetrahedral or octahedral geometries, the CFSE is zero (the same value is also obtained for the d^{10} electron configuration). Fe(II) is a d^6 electron configuration, and in a high spin octahedral geometry, the CFSE is 4. In a tetrahedral geometry, Fe(II) has a CFSE of 2.67; thus, the OSSE is 1.33. These data show that Fe(II) should occupy an O_h site as there is no stabilization for Fe(III) whether it would occupy an O_h site or a T_d site. Thus, magnetite is predicted to be more stable in the inverse spinel structure. In contrast, Mn_3O_4 has the normal spinel structure.

Using the data from Table 8.9, Table 8.10 predicts the structure of normal versus inverse spinels based on changes of the B or +3 cations with different A or +2 cations. Overall agreement is very reasonable.

8.5.2.5   Splitting of d Orbitals in a Tetragonal Field

In octahedral geometry, several metal electron configurations permit the placement of an electron into one or more of the t_{2g} or e_g orbitals. For example, in the case of $Ti(H_2O)_6^{3+}$, the d^1 electron could be placed in any one of the three degenerate t_{2g} orbitals. The **Jahn–Teller theorem** states that if the ground electronic configuration of a *nonlinear* molecule is degenerate, then the molecule *distorts* so as to remove the degeneracy

Table 8.9   *Octahedral site stabilization energies for the d^1 to d^9 electron configurations. These data indicate that the octahedral geometry is more stable than the tetrahedral geometry*

# d e⁻	CFSE, T_d in units of D_q for T_d	CFSE, T_d in units of D_q for O_h	CFSE for *octahedral high spin* in units of D_q for O_h	OSSE *high spin* in units of D_q for O_h	CFSE for *octahedral low spin* in units of D_q for O_h	OSSE *low spin* in units of D_q for O_h
d^1	$6 \times (4/9)=$	2.67	4	1.33	4	1.33
d^2	$12 \times (4/9)=$	5.33	8	2.67	8	2.67
d^3	$8 \times (4/9)=$	3.55	12	8.45	12	8.45
d^4	$4 \times (4/9)=$	1.78	6	4.22	16-P	14.22-P
d^5	$0 \times (4/9)=$	0	0	0	20-P	20-P
d^6	$6 \times (4/9)=$	2.67	4	1.33	24-P	21.33-P
d^7	$12 \times (4/9)=$	5.33	8	2.67	18-P	12.67-P
d^8	$8 \times (4/9)=$	3.55	12	8.45	12	8.45
d^9	$4 \times (4/9)=$	1.78	6	4.22	6	4.22

Table 8.10 *Use of OSSE data from Table 8.9 to predict normal versus inverse spinel structure. N= normal; I = inverse; NO = no prediction by OSSE*

	B cations					
	Al^{3+}		Fe^{3+}		Cr^{3+}	
Cation (A)	Exp.	Calc.	Exp.	Calc.	Exp.	Calc.
Mg^{2+}	0.881	NO	I	NO	N	N
Zn^{2+}	N	NO	N	NO	N	N
Cd^{2+}	N	NO	N	NO	N	N
Mn^{2+}	N	NO	I	NO	N	N
Fe^{2+}	N	I	I	I	N	N
Co^{2+}	N	I	I	I	N	N

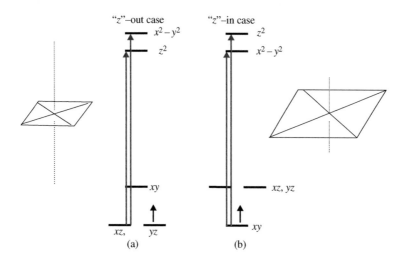

Figure 8.22 *Electronic energy level diagrams for z-out (a) and z-in (b) tetragonal distortions*

and become more stable. In an octahedral geometry, there are two possible distortions that can occur. The first and the one found most frequently is the tetragonal case where two ligands on the z-axis move away from the transition metal so that the bonds are longer; this is known as the z-out case. The second case is termed the z-in case as the two ligands on the z-axis move closer to the transition metal. Both cases are described in Figure 8.22, and both have the symmetry D_{4h} rather than O_h when all six ligands are identical in bond length.

As a result of the tetragonal distortion, the d orbitals in both the t_{2g} and e_g sets split further. For the z-out case, all the z type orbitals become more stable as there is less interaction with the ligands on the z-axis. Likewise, the d orbitals interact more with the ligands in the xy plane so that they become destabilized. For the z-in case, all the z-type orbitals become destabilized whereas the xy-type orbitals become stabilized. Figure 8.23 shows the relative stabilization of the tetragonal z-out case as compared to the octahedral case for both the d^1 and d^6 electron configurations. A major consequence of the d orbitals splitting is that the electron in the original t_{2g} orbital can now be excited to two energetically different e_g orbitals because of the splitting of the e_g set. In Figure 8.14a, the spectrum of the $Ti(H_2O)_6^{3+}$ complex showed a shoulder at longer wavelength

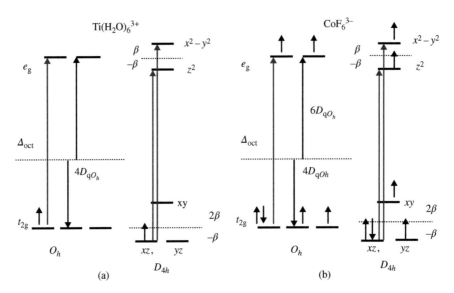

Figure 8.23 *Electronic energy level diagrams for* d^1 *(a) and* d^6 *(b) octahedral to z-out tetragonal distortion*

than the maximum at 493 nm; a shoulder at 8300 cm^{-1} can also be observed for Fe(H$_2$O)$_6$$^{2+}$. The shoulder is evidence that the complex gives two absorption peaks that overlap. The d orbital splitting is more substantial for salts of d^6 CoF$_6$$^{3-}$ with two peaks centered at 14,500 and 11,400 cm^{-1} [10] because of the ligand and the higher effective nuclear charge for Co(III) than Ti(III).

The only electron configurations in octahedral geometry that ***cannot*** exhibit tetragonal distortion are the d^3, d^5 high spin, d^6 low spin, and d^8 cases as the filling of the t_{2g} and e_g orbitals have only one possible combination. There are several other electron configurations in octahedral geometry that exhibit significant tetragonal distortions; e.g., d^4 high spin (see Mn(H$_2$O)$_6$$^{3+}$ in Figure 8.42a), d^7 low spin and d^9 (see Cu(H$_2$O)$_6$$^{2+}$ in Figure 8.42b). These electron configurations result in placing an electron in the e_g orbitals. For d^4 high spin and d^7 low spin, one electron is finally placed into an e_g orbital whereas for the d^9 case, there is a partially occupied orbital typically referred to as a hole as one electron is missing from the complete filling of the e_g orbitals. Because the e_g orbitals are formally antibonding σ orbitals (see Section 8.6) and partially occupied, the splitting of the orbitals is greater, and the resulting tetragonal distortion is more significant than in the d^1 and d^6 high spin cases where t_{2g} orbitals are filled last.

Some of the best experimental data for showing tetragonal distortion from an octahedral geometry are metal–ligand bond distances for d^4 and d^9 electron configurations, which are given for a few compounds of Mn(III), Cr(II), and Cu(II) in Table 8.11. Both z-in and z-out cases are given, and these exhibit two short and four long bonds, and two long bonds and four short bonds, respectively. The bond differences are substantial in both cases. Metal ions bound to porphyrin ligands (as shown in Figure 8.1), which have four nitrogen ligating atoms for binding in the xy plane, also exhibit tetragonal distortion.

8.5.2.6 *Splitting of d Orbitals in a Square Planar Field*

The ultimate tetragonal distortion is when the ligands on the z-axis are completely removed and there is no bonding on the z-axis resulting in a square planar geometry. Figure 8.24 shows the relative energy level diagrams for octahedral, tetragonal distortion, and square planar geometries. The diagram for tetragonal

Table 8.11 *Data for tetragonal distortion (the Jahn–Teller effect) based on metal–ligand bond distances from [11]*

Compound	Short distances (pm)	Long distances (pm)
"z" out case (long bonds on z axis)		
CuF$_2$	4 F at 193	2 F at 227
CuCl$_2$	4 Cl at 230	2 Cl at 295
CuBr$_2$	4 Br at 240	2 Br at 318
Cu(NH$_3$)$_6$$^{2+}$	4 NH$_3$ at 207	2 NH$_3$ at 262
CrF$_2$	4 F at 200	2 F at 243
MnF$_3$	2 F at 179	2 F at 191; 2 F at 209
MnOOH	4 O at 188	2 O at 230
"z" in case (long bonds on x and y axes)		
KCrF$_3$	2 F at 200	4 F at 214
KCuF$_3$	2 F at 196	4 F at 207

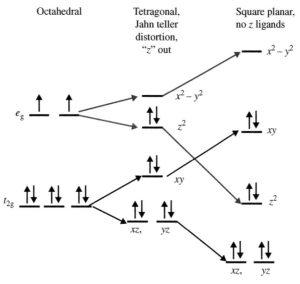

Figure 8.24 *Comparison of d-orbital splitting in six-coordinate versus four-coordinate structures; (d^8) octahedral, tetragonal distortion, and square planar complexes, respectively*

distorted and square planar geometries is similar except that in the square planar geometry all orbitals with a *z* component are lower in energy that those with *xy* components only. Krishnamurthy and Schaap [12] noted that d orbital splitting for square planar geometry could be either of the cases in Figure 8.24 based on potential energy considerations. As will be shown with the angular overlap model approach (Section 8.7), square planar complexes normally have the splitting shown for the tetragonally distorted case. The square planar geometry is a *low spin* four-coordinate structure as compared to tetrahedral geometry, which is a *high spin* four-coordinate structure.

8.5.2.7   Splitting of the Orbitals and Other Geometries

The crystal field splitting parameters have been calculated for a variety of geometries [3, 12], and the values of D_q are normalized to the octahedral case. Figure 8.25 gives details on the splitting of several important geometries, which will be used when discussing ligand substitution reactions in Section 9.2. The square pyramid (C_{4v}) and the trigonal bipyramid (D_{3h}) geometries can form when a square planar compound can gain a ligand to form a five coordinate intermediate, or when an octahedral compound can lose a ligand. The pentagonal bipyramidal geometry can form when an octahedral compound can gain a ligand to form a seven coordinate intermediate.

8.6   Molecular Orbital Theory

The discussion above only centered on the d orbitals of the metal, and did not explicitly consider the orbitals from the ligands. Thus, it was not possible to explain certain experimental data such as the **spectrochemical series** order of ligands binding to transition metals. The molecular orbital approach explicitly considers all the orbitals from the ligands and the transition metal. There are three cases, which can be used to explain the different bonding schemes between the ligands in the transition metal in an octahedral environment. They are (1) the sigma bonding only case between the ligands and the transition metal, (2) sigma bonding plus pi donor electron capability of the ligand to the metal and (3) sigma bonding plus pi acceptor capability of the ligand from the metal.

8.6.1   Case 1 – Octahedral Geometry (Sigma Bonding Only)

For case 1, the transition metal atom or ion has 1-s orbital, 3-p orbitals, and 5-d orbitals for a total of nine atomic orbitals. Table 4.9 gives the character table for the bonding of the central atom in octahedral geometry,

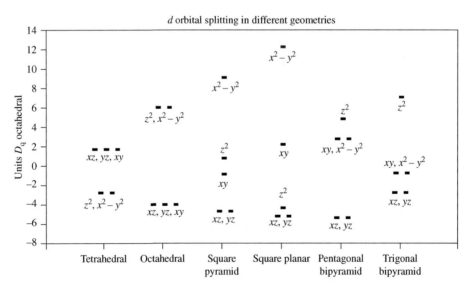

Figure 8.25   *Possible ways to split the metal d orbitals in several important four, five, six, and seven coordination number geometries (Source: Data from Ref. [12])*

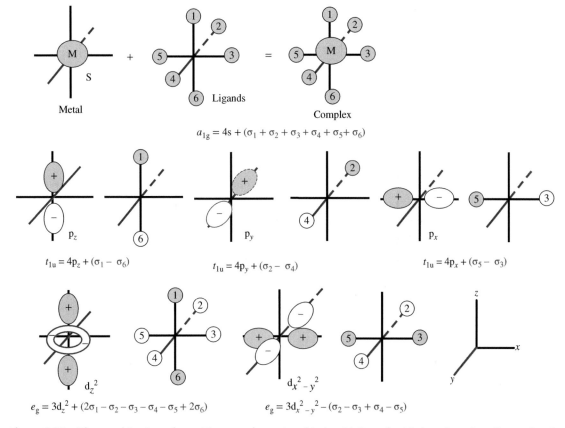

$$a_{1g} = 4s + (\sigma_1 + \sigma_2 + \sigma_3 + \sigma_4 + \sigma_5 + \sigma_6)$$

$$t_{1u} = 4p_z + (\sigma_1 - \sigma_6)$$

$$t_{1u} = 4p_y + (\sigma_2 - \sigma_4)$$

$$t_{1u} = 4p_x + (\sigma_5 - \sigma_3)$$

$$e_g = 3d_z^2 + (2\sigma_1 - \sigma_2 - \sigma_3 - \sigma_4 - \sigma_5 + 2\sigma_6)$$

$$e_g = 3d_{x^2-y^2} - (\sigma_2 - \sigma_3 + \sigma_4 - \sigma_5)$$

Figure 8.26 *The combination of transition metal atomic orbitals with ligand orbitals to form bonding molecular orbitals for case 1 sigma only bonding. The atomic orbitals or lobes of the atomic orbitals in blue color have a + sign of the wave function; white coloration indicates − sign of the wave function*

which shows that the 1-s orbital transforms as a_{1g}, the 3-p orbitals transform as t_{1u}, and the d_{z^2} and $d_{x^2-y^2}$ orbitals transform as e_g. These six atomic orbitals form six molecular orbitals when combined with six ligand orbitals, which donate electrons to the transition metal and are combined into group or SALC orbitals using the group theory approach. Figure 8.26 provides the six sigma bonding molecular orbitals (note that the overlap integral is everywhere positive for these combinations) in octahedral geometry along with the generalized wave functions for these bonding molecular orbitals. For simplicity, the ligand orbitals are drawn as "s"-like orbitals, and are given the symbol, σ, to represent sigma bonding. **Ammonia and simple amines** are examples of ligands that bind with transition metals in a sigma bonding only manner.

The d_{xz}, d_{yz}, and d_{xy} orbitals transform as t_{2g}, and are nonbonding orbitals as they have no electron density along the bond axes, which are also along the Cartesian coordinate axes. Clearly, the approaching ligands would have been an overlap integral (S) equal to zero (Figure 8.27).

There are six antibonding molecular orbitals, which are similar to the bonding molecular orbitals above, except for the sign of the wave function for the ligand orbitals. Figure 8.28 gives the combination of transition metal and ligand orbitals along with the generalized wave function for each resulting molecular orbital. On combination of the metal and ligand orbitals, the overlap integral gives a negative value ($S < 0$).

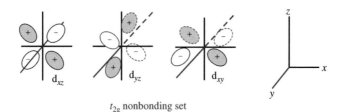

t_{2g} nonbonding set

Figure 8.27   *The t_{2g} nonbonding molecular orbitals for case 1; sigma only bonding*

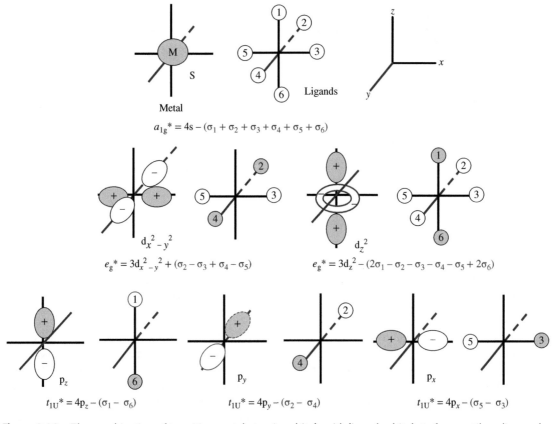

Figure 8.28   *The combination of transition metal atomic orbitals with ligand orbitals to form antibonding molecular orbitals for case 1; sigma only bonding*

The molecular orbital energy level diagram for these 15 molecular orbitals is shown in Figure 8.29. The atomic orbitals for the transition metal are on the left and the ligand-donating orbitals are on the right; the resulting molecular orbitals are in the center of the diagram. Because each ligand orbital donates two electrons to the transition metal complex, there are six bonding orbitals, which have both ligand and metal character, but have more ligand character. Likewise, the six antibonding orbitals also have ligand and metal character, but the contribution is mainly metal. Of particular interest is the fact that the metal e_g orbitals are no longer

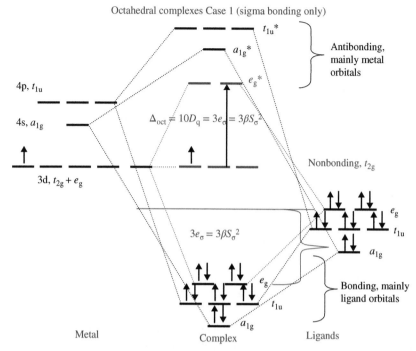

Figure 8.29   *Idealized molecular orbital energy diagram for a transition metal with a d^1 electron configuration when only sigma bonding occurs. For this figure and Figures 8.31 and 8.32, the metal d orbitals in the complex are drawn with blue lines to distinguish them from all other orbitals of the metal and the ligand. The energy terms $3e_\sigma$ and $3\beta S_\sigma^2$ will be discussed under the angular overlap model (Section 8.7)*

pure metal d orbitals and have antibonding character. The nonbonding t_{2g} orbitals only have metal character as there is no interaction with any ligand orbitals in a sigma bonding mode. The HOMO and LUMO orbitals depend on the orbital occupation; e.g., for octahedral high spin d^1, d^2, d^3, d^6, and d^7 the t_{2g} orbitals are both HOMO and LUMO (also SOMO); for octahedral high spin d^4, d^5, d^8, and d^9 the e_g orbitals are HOMO and LUMO (also SOMO); for low spin d^6, the HOMO are the t_{2g} orbitals and the LUMO are the e_g orbitals.

8.6.2   Case 2 – Octahedral Geometry (Sigma Bonding Plus Ligand π Donor)

When ligands have more than one lone pair of electrons to donate to the transition metal, π bonding can result between the metal t_{2g} orbitals and the additional lone pairs of electrons from the ligand. Figure 8.30 shows π bonding overlap for ligand pure p_x and p_z orbitals with the d_{xz} orbital of the metal atom; there are two other sets with similar overlap for the d_{xy} and d_{yz} metal orbitals. In this case, each ligand has two ligand π orbitals, which are perpendicular to the sigma bond axis, possible for bonding. The six ligand atoms have 12 π bonding vectors shown on the right side of Figure 8.30, and these give the reducible representation $\mathbf{\Gamma}_{\pi\,\text{oct}}$, which reduces to $\mathbf{\Gamma}_{\pi\,\text{oct}} = t_{1g} + t_{2g} + t_{1u} + t_{2u}$. Thus, the t_{2g} ligand orbital set can combine with the t_{2g} metal orbitals set.

	E	$8C_3$	$6C_2$	$6C_4$	$3C_2\,(=C_4^{\,2})$	i	$6S_4$	$8S_6$	$6\sigma_h$	$6\sigma_d$
$\mathbf{\Gamma}_{\pi\,\text{oct}}$	12	0	0	0	−4	0	0	0	0	0

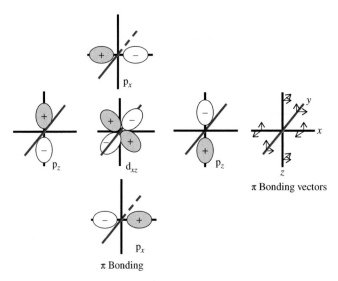

Figure 8.30   *The possible π bonding overlap of four p ligand orbitals with a d_{xz} metal orbital. The bonding vectors for six sets of ligand p orbitals bound to a central metal atom*

Figure 8.31 shows the molecular orbital energy level diagram for the π donor ligand case 2. The sigma bonding orbitals for the metal ligand complex are again identical to case 1, and the energy of the e_g* orbitals is unchanged. However, there are three more filled ligand t_{2g} orbitals that now interact with the empty or partially empty metal t_{2g} orbitals, which now are less stable than in the sigma bonding only case 1. The ligand t_{2g} orbitals are more stable as shown in the π donor complex part of the figure. The separation in energy between the metal t_{2g} and e_g* orbitals (Δ_{oct}) is less than in case 1; thus, the electron transition between these orbitals is of less energy and thus higher wavelength. Representative ligands for case 2 are the halides, hydroxide, oxide, sulfide, and water as shown in the spectrochemical series above. Another consequence of lowering the stability of the metal t_{2g} orbitals is that ionization of the metal ion becomes easier; this effect will be shown for the environmentally significant oxidation of the hexaaqua complexes of Fe(II) and Mn(II) as pH increases in aqueous solutions (Section 10.4.1).

The energy level diagram also shows the possibility of two more electron transitions due to the excitation of ligand t_{2g} electrons. These are called ligand to metal charge transfers (LMCT), which are very intense peaks toward the UV region as shown in Figure 8.14 and Section 8.8.1. The first transition is a low energy transition of a ligand t_{2g} electron to a metal t_{2g} orbital. The second is a higher energy transition of a ligand t_{2g} electron to a metal e_g* orbital.

8.6.3   Case 3 – Octahedral Geometry (Sigma Bonding Plus Ligand π Acceptor)

Case 3 results when a filled or partially filled metal t_{2g} orbital overlaps with empty t_{2g} orbitals of the ligand and results in π bonding (**backbonding**). There are several possible ligand t_{2g} orbitals that exhibit this type of π bonding. As shown in Figure 8.30, ligand molecular orbitals must have a positive and negative component of the wave function to interact with a metal t_{2g} orbital. Three empty molecular orbitals other than p orbitals that satisfy this requirement include the σ* orbital for H_2, CH_4 (CH_3X), and other alkanes; the π* orbitals found in carbon–carbon double bonds (C=C) and in triple bonds of C≡N, C≡N⁻, C≡O; and the d_{xz}, d_{yz}, and d_{xy} orbitals of organic sulfur (R_2S) and phosphorus compounds (R_3P). The resulting π bonding orbitals

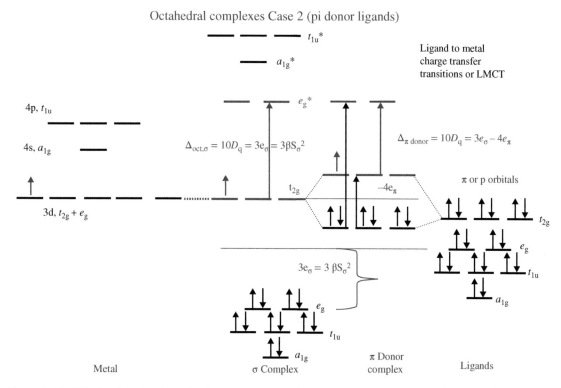

Figure 8.31 *Idealized molecular orbital energy diagram for a transition metal with a d^1 electron configuration when both sigma and pi donor bonding occur. CoF_6^{3-} (d^6) and $Ti(H_2O)_6^{3+}$ (d^1) are good case 2 examples. The metal d orbitals are shown with blue lines*

(signifying the metal to ligand electron flow) are given the following shorthand notation $d_\pi \to \sigma^*$, $d_\pi \to \pi^*$, and $d_\pi \to d_\pi$, respectively, where d_π after the arrow indicates d_{xz}, d_{yz}, and d_{xy} orbitals to R_2S and R_3P.

Figure 8.32 shows the molecular orbital energy level diagram for the π acceptor case 3. The sigma bonding orbitals for the metal–ligand complex are again similar to case 1, and the energy of the e_g^* orbitals is unchanged. However, there are three more empty ligand t_{2g} orbitals that now interact with the filled or partially filled metal t_{2g} orbitals, which become more stable than in the sigma bonding only case 1. The ligand t_{2g} orbitals are less stable as shown in the π acceptor complex part of Figure 8.32. The separation in energy between the metal t_{2g} and e_g^* orbitals (Δ_{oct}) is more than in the sigma bonding only case 1; thus, the electron transition between these orbitals is of higher energy and thus lower wavelength. Representative ligands for case 3 are olefins, dioxygen, and carbon monoxide as shown in the spectrochemical series above. Another consequence of increasing the stability of the metal t_{2g} orbitals is that ionization of the metal ion becomes more difficult. However, as will be shown in Section 9.10, this type of bonding interaction is extremely important in transition metal catalysis when **oxidative addition** occurs.

The energy level diagram also shows the possibility of another higher energy electron transition due to the excitation of the metal t_{2g} electrons to a ligand t_{2g} orbital (the black arrow for the π acceptor complex in Figure 8.32). These are called metal to ligand charge transfer transitions (MLCT).

Figure 8.33a shows a π donor bonding example for case 2 ($p \to d_\pi$) and Figure 8.33b–e shows π acceptor examples for case 3 ($d_\pi \to \sigma^*$, $d_\pi \to \pi^*$ and $d_\pi \to d_\pi$). Figure 8.33b shows that carbon monoxide binds

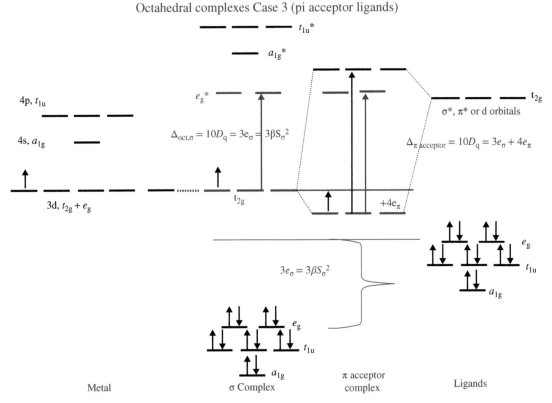

Figure 8.32   *Idealized molecular orbital energy diagram for a transition metal with a d¹ electron configuration when both sigma and pi acceptor bonding occur. The metal d orbitals are shown with blue horizontal lines*

a metal "end on." Figure 8.33d and e shows that ethylene in Zeise's salt binds Pt in a "side on" manner; here, there is a σ bond formed from the $C = C$ π bond of ethylene to a d^8 Pt^{2+} ($d_{x^2-y^2}$) orbital (not shown) and then a π acceptor bond from the d_{xz} orbital of Pt to the π* orbital of ethylene (the plane of the paper is the xz plane). O_2 can form a σ bond in either end on or side on attack of the metal (Figures 10.3, 10.7, 10.8, 10.9). The side on bonding formally has two carbons bonding equally to the Pt forming a trigonal planar arrangement, and is given the symbol η^2-C_2H_4 (hapto nomenclature for 2 C atoms binding; see Section 8.3).

8.7   Angular Overlap Model

The discussion above did not explicitly describe the energetics of the d orbital splittings as only the changes in $10D_q$ were noted. The angular overlap model (AOM, [13]) provides semi-quantitative understanding of the d orbital splittings in Figures 8.29, 8.31, and 8.32. For orbitals other than s, the extent of the overlap is dependent on the angle between the orbitals. Angular overlap was briefly discussed and described for p orbitals in Figure 5.20. The overlap integral S for an ML complex is given the symbol S_{ML} and for two p orbitals combining on the same axis $S_{ML} = S_\sigma \cos\theta$. When θ is 0°, maximum bonding occurs.

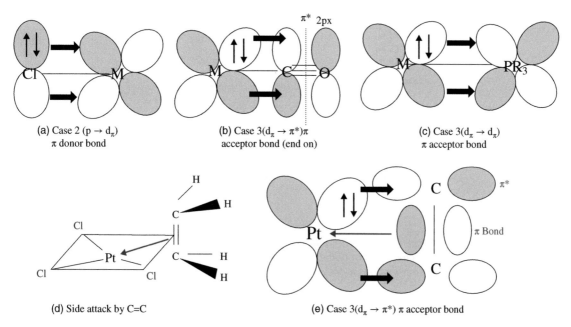

(a) Case 2 (p → d$_\pi$)
π donor bond

(b) Case 3(d$_\pi$ → π*)π
acceptor bond (end on)

(c) Case 3(d$_\pi$ → d$_\pi$)
π acceptor bond

(d) Side attack by C=C

(e) Case 3(d$_\pi$ → π*) π acceptor bond

Figure 8.33 *Examples of π bonding in metal complexes. (a) Case 2 π donor ligand; (b) case 3 π acceptor ligand end on attack by C in CO (d$_\pi$ → π*) ; (c) case 3 π acceptor ligand (d$_\pi$ → d$_\pi$); (d) case 3 π acceptor ligand side on attack by C = C (π → d$_{x^2-y^2}$ and d$_\pi$ → π *) in a square planar complex; (e) orbital overlap for case 3 π acceptor ligand side on attack (d$_\pi$ → π*)*

The energy of interaction, e, between the metal and ligand orbitals is given as

$$e_j = \sum_{L=1}^{N} \beta S_{ML}^2 = \beta S_\sigma^2 f(\delta, \phi)$$

where S_{ML} is the overlap integral between orbitals on the metal and each ligand and β is a constant inversely proportional to the difference in energy between the orbitals. Only the angular dependence need be considered because the radial portions of the wave functions are constant for a given set of ligands bound to a given metal ion. The total σ bond molecular orbital stabilization energy (**MOSE**) is the number of electrons (n_j) occupying each orbital summed over all five e_j orbitals.

$$\Sigma(\sigma) = \sum_{j=1}^{5} n_j e_j = \sum_{j=1}^{5} n_j \sum_{L=1}^{N} \beta S_{ML}^2$$

For bonding orbitals, the value is positive and for antibonding orbitals the value is negative. The S_{ML} functions for each of the five "d" orbitals are given in Figure 8.34.

The value of S_{ML} for each d orbital can be evaluated, and as an example, the value for the d$_{z^2}$ orbital is solved using S_{ML} for $S(d_{z^2}, \sigma) = [(1 + 3\cos 2\theta)/4] S_\sigma$. For the L_1 and L_6 positions on the z axis, the solutions are

$$L_1 \ \theta = 0° \text{ so } [(1 + 3(1))/4] S_\sigma = 1 S_\sigma; \text{ squaring gives } (1 S_\sigma)^2 = 1 \ S_\sigma^2$$

$$L_6 \ \theta = 180° \text{ so } [(1 + 3(1))/4] S_\sigma = 1 S_\sigma; \text{ squaring gives } (1 S_\sigma)^2 = 1 \ S_\sigma^2$$

Thus, L_1 and L_6 contribute $2 S_\sigma^2$ to the energy value of the d$_{z^2}$ orbital.

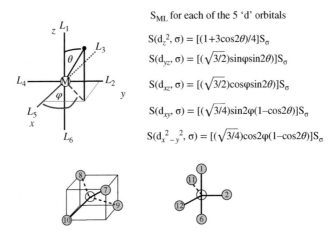

S_{ML} for each of the 5 'd' orbitals

$$S(d_{z^2}, \sigma) = [(1+3\cos2\theta)/4]S_\sigma$$

$$S(d_{yz}, \sigma) = [(\sqrt{3/2})\sin\varphi\sin2\theta)]S_\sigma$$

$$S(d_{xz}, \sigma) = [(\sqrt{3/2})\cos\varphi\sin2\theta)]S_\sigma$$

$$S(d_{xy}, \sigma) = [(\sqrt{3/4})\sin2\varphi(1-\cos2\theta)]S_\sigma$$

$$S(d_{x^2-y^2}, \sigma) = [(\sqrt{3/4})\cos2\varphi(1-\cos2\theta)]S_\sigma$$

Figure 8.34 *Octahedral arrangement of ligands and the Cartesian coordinate axes representing the ML$_6$ complex along with the angular functions for sigma bonding, S_{ML}, for each d orbital. Ligand positions 7 through 12 are also provided to calculate energy contributions for other geometries*

For the *xy* planar positions (L_2, L_3, L_4, L_5) on the *x* and *y* axes, the solution of S_{ML} for each position is the same as all *x,y* ligands have $\theta = 90°$. Thus, each position has the solution of

$$\cos 2\theta = \cos 180 = -1$$

and substituting "−1" into

$$S(d_{z^2}, \sigma) = [(1 + 3\cos 2\theta)/4]\, S_\sigma \text{ gives } ([1 + 3(-1)]/4)\, S_\sigma = (-2/4)\, S_\sigma$$

On squaring the value $(-2/4\, S_\sigma)$, it becomes $1/4\, S_\sigma^2$ for each position. Thus, $L_2, L_3, L_4,$ and L_5 each contribute this amount to the total S_{ML}^2 of the d$_{z^2}$ orbital.

Summing the values of all six positions gives a value of the square of the overlap integral as

$$S_{ML}^2 = [1 + 1 + 4(1/4)]\, S_\sigma^2 = 3\, S_\sigma^2$$

Multiplying this total value by β yields the total molecular orbital stabilization energy contribution of the d$_{z^2}$ orbital as

$$E = \beta\, S_{ML}^2 = 3\beta S_\sigma^2$$

The value of $3\beta S_\sigma^2$ is also expressed as $3e_\sigma$. Note that bonding is more important on the z axis than the x and y axes for d$_{z^2}$. The same energy value of $3\beta S_\sigma^2$ is also obtained for the d$_{x^2-y^2}$ orbital (as it should because it is an e_g orbital also); here bonding is exclusively on the x and y axes. The energy value of zero is obtained for each of the t_{2g} orbitals (d$_{xz}$, d$_{yz}$, d$_{xy}$) as they are not involved with sigma bonding.

Figure 8.35a shows the splitting of d orbitals using CFT and the angular overlap method of MOT. For the sigma bonding only case, the splitting of d orbitals in octahedral geometry gives

$$\Delta_{oct}\,(10D_q) = 3\beta S_\sigma^2 = 3e_\sigma$$

The MOT method explicitly considers the overlap of the ligand orbitals with the e_g metal orbitals. Thus, the e_g bonding orbitals in Figure 8.35 (and 8.29) are mainly ligand bonding orbitals whereas the metal e_g orbitals

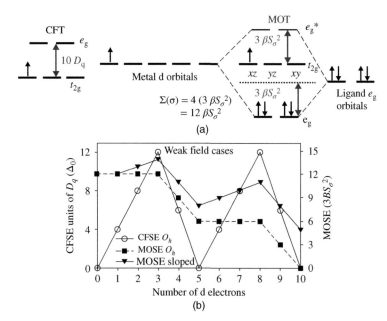

Figure 8.35 *(a) Comparison of d orbital splitting for CFT and MOT (angular overlap method) and their respective energy terms. (b) Comparison of the relative stabilization energies in CFT and MOT*

are mainly antibonding orbitals (e_g^*). The decrease in energy for each electron in a bonding orbital is $3\beta S_\sigma^2$ and the total contribution from two ligands donating two electrons each is $12\beta S_\sigma^2$. The energy contribution of each electron in an antibonding orbital is $-3\beta S_\sigma^2$; thus, these cancel the effect of bonding. The sum of the energies for the bonding and antibonding electrons of a metal–ligand complex is known as the molecular orbital stabilization energy (MOSE).

A comparison of the stabilization energies between these two approaches is given in Table 8.12 with a plot in Figure 8.35b. Table 8.12 shows that the MOSE for electron configurations d^0 through d^3 are constant; however, the Irving–Williams relationship [14; Section 7.8.1] indicates that the bonding strength of the complex increases going across the periodic table because of the increase in the charge to size ratio. It is common to plot the MOSE with a slope as in Figure 8.35b to account for the increase in bond strength.

8.7.1   AOM and π Ligand Donor Bonding

The metal t_{2g} orbitals also can accept electrons from lower energy t_{2g} ligand orbitals in π donor complexes. The interaction stabilizes the ligand orbitals but destabilizes the metal t_{2g} orbitals as they become weakly antibonding (see Figure 8.31) and more able to lose the electron. Without any development, the energy for the t_{2g} destabilization of each metal t_{2g} orbital is denoted as $-4e_\pi$ [9]. The overall d orbital splitting is then

$$\Delta_{oct} = 3\beta S_\sigma^2 - 4e_\pi = 3e_\sigma - 4e_\pi$$

The π donor bonding has a much smaller value than that for $3\beta S_\sigma^2$, and ranges from 10% to 50% of the $3\beta S_\sigma^2$ value. Figure 8.31 explicitly shows that there is a loss in MOSE of $-4e_\pi$ for each electron added to a metal t_{2g} orbital.

Table 8.12 Comparison of the stabilization energies using the angular orbital method (MOT, $3\beta S_\sigma^2$, or $3e_\sigma$) versus stabilization energies using the crystal field ($10D_q$) approach

Configuration	MOSE $3\beta S_\sigma^2$ (high spin)	MOSE $3e_\sigma$ (low spin)	CFSE $10 D_q$ (high spin)	CFSE $10 D_q$ (low spin)
d^0	12	12	0	0
d^1	12	12	4	4
d^2	12	12	8	8
d^3	12	12	12	12
d^4	9	12	6	16
d^5	6	12	0	20
d^6	6	12	4	24
d^7	6	9	8	18
d^8	6	6	12	12
d^9	3	3	6	6
d^{10}	0	0	0	0

8.7.2   AOM and π Ligand Acceptor Bonding

Electrons in metal t_{2g} orbitals also can be donated to higher energy empty t_{2g} ligand orbitals in π acceptor complexes as shown in Figure 8.32. The bonding interaction stabilizes the metal t_{2g} orbitals, which are π bonding orbitals, but destabilizes the ligand t_{2g} orbitals, which become antibonding orbitals. The energy for each t_{2g} is denoted as $+4e_\pi$ and the overall d orbital splitting is then

$$\Delta_{oct} = 3\beta S_\sigma^2 + 4e_\pi = 3e_\sigma + 4e_\pi$$

The π acceptor bond has a much smaller value than that for the sigma bonding value of $3\beta S_\sigma^2$, as overlap on the sigma bond axis is better. Figure 8.32 explicitly shows a gain in molecular orbital stabilization energy of $+4e_\pi$ for each electron added to an empty metal t_{2g} orbital.

8.7.3   MOT, Electrochemistry, and the Occupancy of Electrons in d Orbitals in O_h

The energies derived for the electrons in the t_{2g} and e_g^* orbitals from the angular overlap method are very useful in explaining oxidation and reduction reactions as adding electrons to an antibonding orbital loses MOSE or stability ($-3\beta S_\sigma^2$ for each electron added). Adding electrons to t_{2g} nonbonding orbital does not result in loss of stability. The following Co^{3+} reactions with their reduction potentials [15] indicate the usefulness of the MOT approach and MOSE data in explaining the redox potentials of metal complexes. The first three reactions have ligands bound to Co^{3+} normally considered as weak field and the other four reactions have strong field ligands.

$$[Co(H_2O)_6]^{3+} + e^- \rightarrow [Co(H_2O)_6]^{2+} \qquad E° = 1.92 \text{ V}$$
$$[Co(ox)_3]^{3-} + e^- \rightarrow [Co(ox)_3]^{2-} \qquad E° = 0.55 \text{ V}$$
$$[Co(edta)]^- + e^- \rightarrow [Co(edta)]^{2-} \qquad E° = 0.38 \text{ V}$$
$$[Co(phen)_3]^{3+} + e^- \rightarrow [Co(phen)_3]^{2+} \qquad E° = 0.42 \text{ V}$$
$$[Co(NH_3)_6]^{3+} + e^- \rightarrow [Co(NH_3)_6]^{2+} \qquad E° = 0.11 \text{ V}$$
$$[Co(en)_3]^{3+} + e^- \rightarrow [Co(en)_3]^{2+} \qquad E° = -0.18 \text{ V}$$
$$[Co(CN)_6]^{3-} + e^- \rightarrow [Co(CN)_5(H_2O)]^{3-} + CN^- \qquad E° = -0.83 \text{ V}$$

8.7.3.1   Weak Field Ligands

Even though water is considered a weak field ligand, $[Co(H_2O)_6]^{3+}$ is low spin (t_{2g}^6) due to the high Z_{eff}/r for Co^{3+}. Nevertheless, $[Co(H_2O)_6]^{3+}$ oxidizes water and reduces to high spin $Co^{2+}(t_{2g}^5 e_g^{2*})$ over time so $[Co(H_2O)_6]^{3+}$ is not stable. Formation of Co^{2+} may seem at odds with MOSE data (energy loss of 6 βS_σ^2) as in the equation below.

$$Co^{3+} [t_{2g}^6 e_g^{*0}] (12\beta S_\sigma^2) + e^- \rightarrow Co^{2+} [t_{2g}^5 e_g^{*2}] (6\beta S_\sigma^2) \Delta E_{loss} = 6\beta S_\sigma^2$$

However, when bound to weak field π donor ligands such as ox and edta, Co(III) reduction occurs even though the electron initially enters an e_g^* orbital. Rearrangement of a low spin Co(II) complex to a high spin Co(II) complex does not cost as much energy (see **spin crossover**, Section 8.8.4) as the metal t_{2g} and e_g^* orbitals become closer in energy (MOT case 2; Section 8.6.2). Co^{2+} has a lower Z_{eff}/r so cannot be stabilized as low spin with these ligands. The following two equations represent a two-step process showing reduction followed by spin state conversion.

$$Co^{3+} [t_{2g}^6 e_g^{*0}] (12\beta S_\sigma^2) + e^- \rightarrow Co^{2+} [t_{2g}^6 e_g^{*1}] (9\beta S_\sigma^2) \Delta E_{loss} = 3\beta S_\sigma^2$$

$$Co^{2+} [t_{2g}^6 e_g^{*1}] (9\beta S_\sigma^2) + e^- \rightarrow Co^{2+} [t_{2g}^5 e_g^{*2}] (6\beta S_\sigma^2) \Delta E_{loss} = 3\beta S_\sigma^2$$

On increasing the number of lone pairs of electrons on the ligating atom(s) and the negative charge, Co(III) reduction becomes more difficult than for $[Co(H_2O)_6]^{3+}$ because more π bonding between the metal and ligand t_{2g} orbitals pushes electron density into the metal t_{2g} orbitals. Oxalate and EDTA are negatively charged and have more π donor electron pairs than does water; thus, their reduction potentials are lower. Also, oxalate and edta are multidentate ligands that can stabilize higher oxidation states due to the favorable thermodynamics associated with the chelate effect (see Section 9.1.2).

8.7.3.2   Strong Field Ligands

Co(III) becomes increasingly more stable to reduction (reduction is less favorable) when bound to strong field (low spin) ligands. Reduction is difficult because the added electron again must enter an antibonding e_g^* orbital, and rearrangement to a high spin Co(II) complex costs more energy than with weak field ligands so the product is Co(II) in the low spin state.

$$Co^{3+} [t_{2g}^6 e_g^{*0}] (12\beta S_\sigma^2) + e^- \rightarrow Co^{2+} [t_{2g}^6 e_g^{*1}] (9\beta S_\sigma^2) \Delta E_{loss} = 3\beta S_\sigma^2$$

Phenanthroline's (phen) aromatic character in $[Co(phen)_3]^{3+}$ aids reduction as the Co(III) t_{2g} electrons are already delocalized into the aromatic ring due to π acceptor bonding so there is less electron density at Co(III). This electron delocalization is an example of the **nephelauxetic** or cloud expanding effect (see Section 8.8.2.4). In contrast, cyanide ion has a negative charge, which hinders delocalization of the Co(III) t_{2g} electrons into CN^-; thus, Co(III) in $[Co(CN)_6]^{3-}$ is electron rich and reduction is not favorable. Because NH_3 and en are neutral σ only electron donors, their Co(III) complexes have E° values between $[Co(phen)_3]^{3+}$ and $[Co(CN)_6]^{3-}$.

The above interpretation using MOSE data is more instructive than using crystal or ligand field stabilization energies because MOSE values are directly related to the electrons being placed in either stable bonding or unstable antibonding orbitals.

8.7.4   AOM and Other Geometries

The angular orbital method is useful in describing other metal–ligand geometries, and a brief discussion of tetrahedral and cubic geometries are provided. First, using the equations in Figure 8.34, it is possible to

Table 8.13 *Angular overlap energy contributions, $S_{ML}{}^2$, of d orbitals for a variety of ligand geometries and positions (L# = ligand position in Figure 8.34) for metal–ligand sigma bonds*

CN	Geometry	Positions	L #	z^2	$x^2 - y^2$	xz	yz	xy
2	Linear	1, 6	1	1	0	0	0	0
3	Trigonal planar	2, 11, 12	2	0.25	0.75	0	0	0
3	T-shape	1, 3, 5	3	0.25	0.75	0	0	0
4	Tetrahedral	7, 8, 9, 10	4	0.25	0.75	0	0	0
4	Square planar	2, 3, 4, 5	5	0.25	0.75	0	0	0
5	Trigonal bipyramid	1, 2, 6, 11, 12	6	1	0	0	0	0
5	Square pyramid	1, 2, 3, 4, 5	7	0	0	0.333	0.333	0.333
6	Octahedral	1 through 6	8	0	0	0.333	0.333	0.333
8	Cube	$2 \times (7, 8, 9, 10)$	9	0	0	0.333	0.333	0.333
			10	0	0	0.333	0.333	0.333
			11	0.25	0.1875	0	0	0.5625
			12	0.25	0.1875	0	0	0.5625

calculate the relative energy values for a variety of positions in different geometries. Table 8.13 gives values of energy for each of the 12 positions for σ bonds. For O_h, the sum of positions 1 through 6 under z^2 and $x^2 - y^2$ gives a value of 3. The other geometries exhibited by metal–ligand complexes can be calculated in a similar way.

8.7.4.1   Tetrahedral Geometry

Tetrahedral geometry is dominated by high spin complexes in the first transition metal series as noted above in the discussion on crystal field theory. The atomic orbitals used for the central atom (metal) and the terminal atoms (ligands) bound to it have been described in Sections 4.6.3 and 5.6.1 for methane. In the case of the metal–ligand complex, the metal also has the d orbitals to be considered. The metal atom's p orbitals as well as the d orbitals with t_2 symmetry can interact with the appropriate ligand orbitals to form a tetrahedral complex, and Figure 8.36 shows the molecular orbital energy level diagram for a metal in tetrahedral geometry. In this case, the metal d orbitals of t_2 symmetry are used to form σ bonds (sd^3) as well as σ antibonding orbitals with the ligand orbitals of t_2 symmetry; the metal p orbitals are considered not to be involved in bond formation. Because halides, hydroxide, water, and other ligands with extra lone pairs of electrons can induce tetrahedral geometry, there are additional filled orbitals of t_2 symmetry as discussed in the octahedral case above; in addition, there are filled orbitals of t_1 symmetry of slightly higher energy than the t_2 set. The primary electronic transition, which represents the d orbital splitting, is from the nonbonding e orbital to the antibonding t_2 orbital (the solid blue arrow in Figure 8.36). The dashed arrows in the figure are known as ligand to metal charge transfer (LMCT) transitions (see below).

The energy separation between the e and t_2 orbitals is obtained from the angular overlap model data in Table 8.13. The sum of the $S_{ML}{}^2$ values for the tetrahedral ligand positions 7, 8, 9, and 10 gives a value of 1.333, which is 0.444 (or 4/9) that of the value for the octahedral ligand positions 1 through 6.

8.7.4.2   Cubic Geometry

In Chapter 4, octahedral, ML_6, and cubic, ML_8, structures were shown to have the same symmetry point group, O_h, even though the metal coordination numbers differ. The cubic structure has two tetrahedra that

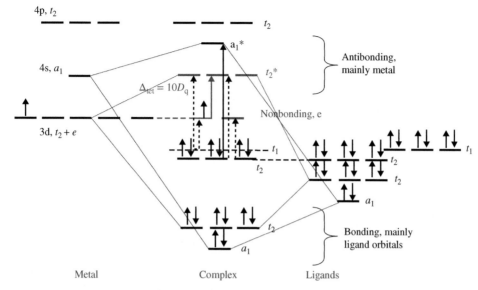

Figure 8.36 *Idealized molecular orbital energy diagram for a transition metal with a d^1 electron configuration in tetrahedral geometry (sd^3). The metal d orbitals are shown with blue lines. The t_1 molecular orbitals are filled but are not drawn that way in order to simplify the diagram. LMCT bands are the horizontal black dashed arrows*

interlock so that all eight vertices of the cube in Figure 8.34 are occupied. The data in Table 8.13 indicates that the molecular orbital stabilization energy is equal to two times that of the tetrahedral structure or 2.667, which is 8/9 of the MOSE for an octahedral structure [$D_{q(cube)} = (8/9) D_{q(oct)}$]. The splitting of the d orbitals is similar to the tetrahedral geometry so that the e_g orbitals are of lower energy. Because the cube has a center of inversion, the g subscript can again be used.

8.8 More on Spectroscopy of Metal–Ligand Complexes

The spectrochemical series indicates that Δ_{oct} increases in the following relative order (the bold font indicates ligands in complexes that are low spin, and the others are high spin).

$$I^- < Br^- < S^{2-} < SCN^- < Cl^- < NO_3^- < F^- < OH^- < ox^{2-} < H_2O < NCS^-$$

$$< CH_3CN < \mathbf{NH_3} < \mathbf{en} < \mathbf{bipy} < \mathbf{phen} < \mathbf{NO_2^-} < \mathbf{R_3P} < \mathbf{R_2S} < \mathbf{CN^-} < \mathbf{CO} \sim \mathbf{olefins} \sim \mathbf{O_2}.$$

This order can be roughly broken down into ligands with the following π bonding properties: π donor < weak π donor < no π effects (σ donor) < π acceptor. The softer halides (I^-, Br^-) with a total of four lone pairs of electrons exhibit strong π donor effects so do not act as point charges as initially predicted when using the crystal field approach. The weak π donors are those with only two lone pairs of electrons such as water and hydrogen sulfide with one available for sigma bonding and the other for π bonding. Cyanide has a negative charge so is less able to participate in π acceptor bonding than neutral isoelectronic species such as CO.

8.8.1   Charge Transfer Electronic Transitions

So far, the discussion has centered on relatively simple d orbital splitting of the metal. However, the number of peaks observed in a metal–ligand complex is more complicated and depends on the number of metal d electrons in the metal–ligand complex as well as the interactions of the metal d orbitals and electrons with the π donor electrons and π acceptor ligand orbitals. The latter are known as charge transfer complexes and two limiting cases are noted. Ligand to metal charge transfer (LMCT) complexes describe the excitation of ligand σ and π electrons to metal t_{2g} and e_g* orbitals for π donor complexes as in Figure 8.37. Metal to ligand charge transfer (MLCT) complexes describe the excitation of metal t_{2g} and e_g* electrons to π acceptor orbitals. In both cases, Figure 8.37 shows a solid blue arrow for the t_{2g} to e_g* spectrochemical series electron transition (or Δ_{oct}). It is possible to have both types of transitions occur in the same complex; e.g., especially for second and third row transition metals with chloride, bromide, iodide, and organosulfur ligands, which have two or more π donor orbitals and empty d or π acceptor orbitals.

For tetrahedral complexes, Figure 8.36 demonstrated four potential LMCT transitions from the ligand t_2 and $t_1(p\pi)$ orbitals with the dashed arrows. There are many examples of these types of transitions in tetrahedral geometry, and the compounds in Table 8.14 have been used by artists over the centuries when painting their masterpieces.

Thus, high oxidation state metal oxides and sulfides (weak field ligands) are common components of pigments. Permanganate (Figure 8.38a) and chromate (Figure 8.38b) exhibit color for ligand $p\pi \rightarrow$ metal 3d transitions. In Figure 8.38a, all four of the permanganate peaks are assigned according to the black dashed arrows in Figure 8.36. Because these metal anions have no d electrons in the e orbitals (see Figure 8.36), there

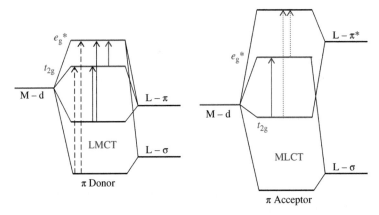

Figure 8.37   *LMCT (black lines on left) and MLCT (dashed blue lines) electron transitions in octahedral metal–ligand complexes*

Table 8.14   *Some chemical compounds use pigments in their charge transfer reactions and transitions [16]*

Pigment	Charge transfer reaction	Primary orbitals
PbCrO$_4$ (chrome yellow)	Cr^{6+}O^{2-} $\rightarrow$ Cr^{5+}O$^-$	ligand pπ $\rightarrow$ metal 3d
HgS (vermilion)	Hg^{2+}S^{2-} $\rightarrow$ Hg$^+$S$^-$	ligand pπ $\rightarrow$ metal 6s
Iron oxides (red/yellow ochres)	Fe^{3+}O^{2-} $\rightarrow$ Fe^{2+}O$^-$	ligand pπ $\rightarrow$ metal 3d

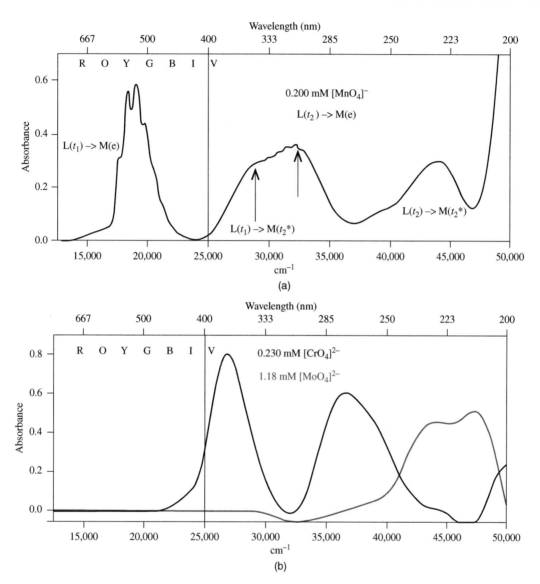

Figure 8.38   *Electronic spectra of (a) MnO$_4^-$; (b) CrO$_4^{2-}$ and MoO$_4^{2-}$*

is no d → d electronic transition. On descending a family in the periodic table, the energy gap between the pπ → metal d orbitals increases; thus, the complex or anion does not absorb light in the visible region. For example, CrO$_4^{2-}$ is colored but the MoO$_4^{2-}$ and WO$_4^{2-}$ anions are not as they absorb light in the UV region (blue shift of the peak as in Figure 8.38b).

8.8.2   Electronic Spectra, Spectroscopic Terms, and the Energies of the Terms for d → d Transitions

In the above discussion on the UV–Vis spectroscopy of d^1 (and d^6, d^9) high spin octahedral complexes, only one absorption peak or electronic transition was noted unless there was tetragonal distortion to remove the degeneracy of the t_{2g} and e_g orbitals. For other electron configurations, more than one absorption peak is possible, but every possible transition may not be observed in a spectrum.

8.8.2.1   Selection Rules for Electronic Transitions

Spectroscopic selection rules are a statement about which transitions are allowed and which are forbidden. The electric dipole model is used to understand absorption of electromagnetic radiation because the electric dipole moment of the molecule changes upon receiving incident radiation, and the intensity of the electronic transition is proportional to the strength of the dipole. The equation below gives the integral of the wave functions of the ground and excited states using the dipole moment vector operator, μ, where f and i are final (excited) and initial (ground) states, respectively, and μ is the electric dipole moment operator. The transition dipole moment, μ_{fi}, is a measure of the impulse that a transition imparts to the electromagnetic field. A large impulse gives an intense transition; a zero impulse is *forbidden*. The intensity of an impulse is proportional to the *square* of μ_{fi}. Two major selection rules result in the following equation.

$$\text{transition dipole moment} = \mu_{fi} = \int \psi_f{}^* \, \mu \, \psi_i \, d\tau$$

First, the **Laporte** selection rule states that in a centrosymmetric molecule or ion, the only allowed transitions are those accompanied by a change *in parity*; i.e., change from gerade to ungerade ($g \to u$ or $u \to g$) but not $g \to g$ or $u \to u$. Thus, for octahedral complexes, d $\to$ d transitions are forbidden as are s $\to$ s; p $\to$ p; f $\to$ f. Such transitions are weak when the selection rule is relaxed by the following.

(a) The complex may not be perfectly centrosymmetric in its ground state (e.g., ligands may be polyatomic or the metal has some distortion).
(b) The complex may undergo an asymmetrical vibration destroying its center of inversion.
(c) Most importantly, the $e_g{}^*$ orbitals in octahedral geometry are a mix of ligand and metal character so are not purely d orbitals (Laporte allowed).
(d) Tetrahedral complexes are not centrosymmetric (there is no assignment of g or u subscripts for the orbitals), and show more intense d $\to$ d absorptions.

The second spin selection rule states that transitions can only occur between energy states with the same spin multiplicity. Thus, promotion of an electron from one energy state to another cannot result in a change of spin for the electron; i.e., $\Delta S = 0$. Spin forbidden transitions are significantly less intense than spin allowed transitions. For example, $Mn(H_2O)_6{}^{2+}$ has a $t_{2g}{}^3 e_g{}^2$ (five unpaired electrons) ground state, and the transition of an electron from the ground state to the $t_{2g}{}^2 e_g{}^3$ excited state results in three unpaired electrons. As a result, $Mn(H_2O)_6{}^{2+}$ has a very weak pink color even in concentrated solutions (see Figure 8.48).

The relative intensities of spectroscopic absorption peaks in metal complexes with 3d orbitals comes from Beer's law, $A = \varepsilon_{max} bC$, where ε_{max} is the molar absorptivity, b is the path length of the spectroscopic cell in centimeters, and C is the molar concentration of the metal complex. The list in Table 8.15 shows the relative intensities for different types of spectroscopic peaks or bands. The charge transfer transitions are the most intense as they are spin allowed and symmetry (parity) allowed (e.g., ligand pπ $\to$ metal d or ligand pπ $\to$ metal s).

8.8.2.2   Spectroscopic Terms for Atoms

In the initial discussion of electronic absorption spectroscopy using the crystal field theory approach (Figure 8.14), the focus of attention was on a single absorption band between the t_{2g} and $e_g{}^*$ orbitals except

Table 8.15   *Types of spectroscopic peaks and their molar absorptivity, which indicates the intensity of the electronic transition*

Band type	ε_{max} (dm^3 mol^{-1} cm^{-1})
Spin forbidden	<1
LaPorte forbidden (d → d)	20–100
LaPorte allowed (d → d)	~250
Symmetry allowed (e.g., Charge transfer)	1,000–50,000

when splitting could occur due to Jahn–Teller distortions. However, many transition metal complexes exhibit two or more peaks as in Figure 8.39 depending upon the number of electrons occupying the d orbitals. The spectra of the free metal octahedral ions, $[M(H_2O)_6]^{n+}$, of the first transition metal series [15] provide excellent examples to describe the principles of spectroscopy.

In order to describe the possible d → d absorption peaks, it is necessary to describe the variety of ways that electrons can be arranged into atomic orbitals using spectroscopic term symbols. This is done first for the electron configuration of the free atom or ion in spherical geometry. When the atom or ion is in a molecular complex of lower geometry or symmetry such as octahedral, the atomic spectroscopic term symbol can split into other spectroscopic terms or states. This additional splitting of states can result in additional peaks in the UV–Vis spectrum. Fortunately, the maximum number of allowed peaks for octahedral geometry in the other electron configurations is three, but they may not all be observed due to overlap with charge transfer absorptions.

In Section 3.8.2, coupling of electron spin with the electron's orbital angular momentum (Russell–Saunders or **LS** coupling) resulted in two energy states (**microstates**) that were detected by X-ray or UV photoelectron spectroscopy for the noble gases other than helium. This coupling considers the ***electron–electron repulsions*** that occur when there is more than one electron that can occupy a given set of orbitals as well as when one electron can be placed into different degenerate orbitals.

In order to describe all the **microstates** and the energy states or term symbols (spectroscopic terms) that they represent, the following systematic approach is provided for the free metal ion.

1. Determine the possible values of both m_l and m_s for the given electron configuration. For the d^2 electron configuration, m_l values equal −2, −1, 0, +1, +2. The maximum value of $L = \Sigma m_1 = 4$ when two electrons are paired in the same orbital, which has a m_l value of 2.
2. Using the **Pauli** exclusion principle, determine the number of electron configurations or microstates that are allowed. Each microstate has one value for m_l and one value for m_s. For the d^2 electron configuration, the two electrons can occupy the same orbital when the spins are opposite (Table 8.16A); the two electrons can be in different orbitals but the spins can be aligned up or down ($m_s = 1$ or −1) (Table 8.16B); the two electrons can be in different orbitals when the spins are opposite ($m_s = 0$) (Table 8.16C).
3. Produce a chart of all the microstates possible. Below are the 45 microstates possible for the d^2 electron configuration.

The total number of microstates, N, for a given electron configuration, l^x, can be calculated using the following formula where $N_l = 2(2l + 1)$ [the number of m_l and m_s combinations for a single electron in the orbital set], x is the number of electrons, and l is the azimuthal quantum number.

$$N = \frac{N_1!}{x!(N_1\text{-}x)!}$$

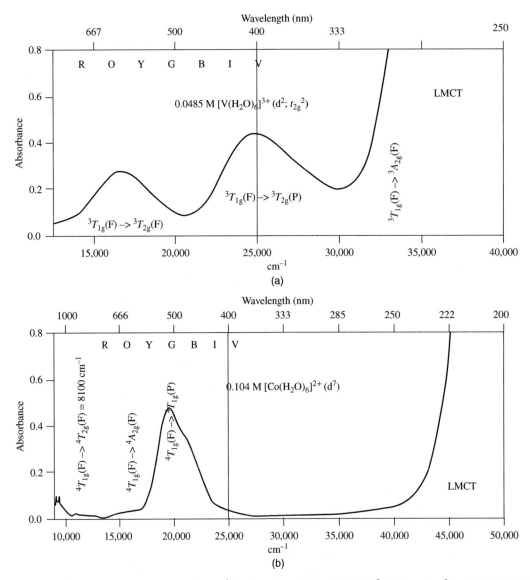

Figure 8.39 *Electronic spectra for (a) V(H$_2$O)$_6$$^{3+}$ (d^2, t$_{2g}$^{2}) and (b) Co(H$_2$O)$_6$$^{2+}$ (d^7, t$_{2g}$^{5}e$_g$^{*2}) in 3.0 M HCl. The $^3T_{1g}$(F) → $^3A_{2g}$(F) transition for V(H$_2$O)$_6$$^{3+}$ is hidden under the LMCT band*

These 45 microstates can be put into a table, which shows every microstate with its m_l and m_s values. The 45 microstates can be reduced to the five Russell–Saunders term symbols [of form ^xL in the absence of spin orbit coupling (J)]: ^{1}G, ^{3}F, ^{3}P, ^{1}D, and ^{1}S, which represent rectangular matrices shown in Table 8.17 (recall that S = 0, P = 1, D = 2, F = 3, and G = 4; and the multiplicity = $2\Sigma m_s + 1$). The left and right sections of Table 8.17 show the matrices and how the microstates produce the appropriate term symbols.

The ground state of a particular electron configuration may be chosen using **Hund's rules**. The first rule states that the ground state will be the one of maximum multiplicity ($\sum \mathbf{m_s}$) as the state will have

Table 8.16A   *Electrons paired in the same orbital so that $\sum m_s = 0$ spins paired; (five individual microstates distributed vertically with a different value of $\sum m_l$)*

			$\sum m_s = 0$		
m_l	2	1	0	−1	−2
	↑↓	↑↓	↑↓	↑↓	↑↓
$\sum m_l =$	4	2	0	−2	−4

Table 8.16B   *No electrons paired in the same orbital; all spins aligned (20 individual microstates distributed vertically)*

$\sum m_s = 1$ (all spins up)					$\sum m_s = -1$ (all spins down)				
m_l					m_l				
2	↑	↑	↑	↑	2	↓	↓	↓	↓
1	↑				1	↓			
0		↑			0		↓		
−1			↑		−1			↓	
−2				↑	−2				↓
$\sum m_l =$	3	2	1	0	$\sum m_l =$	3	2	1	0
m_l					m_l				
1	↑	↑	↑		1	↓	↓	↓	
0	↑				0	↓			
−1		↑			−1		↓		
−2			↑		−2			↓	
$\sum m_l =$	1	0	−1		$\sum m_l =$	1	0	−1	
m_l					m_l				
0	↑	↑			0	↓	↓		
−1	↑				−1	↓			
−2		↑			−2		↓		
$\sum m_l =$	−1	−2			$\sum m_l =$	−1	−2		
m_l					m_l				
−1	↑				−1	↓			
−2	↑				−2	↓			
$\sum m_l =$	−3				$\sum m_l =$	−3			

the maximum number of parallel spins with electrons in different regions of space, which results in **less electron–electron repulsion**. If two states have the same multiplicity, then the state with the higher value of L will lie at lower energy. Although this is true for the term symbols of atoms, there can be differences for molecules. For the d^2 electron configuration, the following order is predicted for the five states: $^3F < {}^3P < {}^1G < {}^1D < {}^1S$.

Table 8.18 shows the atomic terms that arise from each of the d^n electron configurations. Fortunately, the total number of terms is the same for the pairs with d^1, d^9; d^2, d^8; d^3, d^7; and d^4, d^6 electron configurations.

Table 8.16C   *No electrons paired in the same orbital but $\sum m_s = 0$ or spins opposed (20 individual microstates distributed vertically)*

Left group ($\sum m_s = 0$):

m_l				
2	↑	↑	↑	↑
1	↓			
0		↓		
−1			↓	
−2				↓
$\sum m_l =$	3	2	1	0
m_l				
1	↑	↑	↑	
0	↓			
−1		↓		
−2			↓	
$\sum m_l =$	1	0	−1	
m_l				
0	↑	↑		
−1	↓			
−2		↓		
$\sum m_l =$	−1	−2		
m_l				
−1	↑			
−2	↓			
$\sum m_l =$	−3			

Right group ($\sum m_s = 0$):

m_l				
2	↓	↓	↓	↓
1	↑			
0		↑		
−1			↑	
−2				↑
$\sum m_l =$	3	2	1	0
m_l				
1	↓	↓	↓	
0	↑			
−1		↑		
−2			↑	
$\sum m_l =$	1	0	−1	
m_l				
0	↓	↓		
−1	↑			
−2		↑		
$\sum m_l =$	−1	−2		
m_l				
−1	↓			
−2	↑			
$\sum m_l =$	−3			

Table 8.17   *The assignment of the term symbols from the individual microstates where the value "1" indicates each individual microstate of the 45 shown in Tables 8.16A–8.16C*

$\sum m_l = 4, \sum m_s = 0$ equals ^1G; 9 by 1 matrix = 9 microstates
$\sum m_l = 3, \sum m_s = 1$ equals ^3F; 7 by 3 matrix = 21 microstates
$\sum m_l = 1, \sum m_s = 1$ equals ^3P; 3 by 3 matrix = 9 microstates
$\sum m_l = 2, \sum m_s = 0$ equals ^1D; 5 by 1 matrix = 5 microstates
$\sum m_l = 0, \sum m_s = 0$ equals ^1S; 1 by 1 matrix = 1 microstate

	1	0	−1	m_s	1	0	−1
m_l							
4		1				G	
3	1	11	1		F	GF	F
2	1	111	1		F	GFD	F
1	11	1111	11		FP	GFPD	FP
0	11	11111	11		FP	GFPDS	FP
−1	11	1111	11		FP	GFPD	FP
−2	1	111	1		F	GFD	F
−3	1	11	1		F	GF	F
−4		1				G	

Table 8.18 *Terms for the d^n configurations*

Electron configurations	Atomic terms
d^1, d^9	^2D
d^2, d^8	$^3F, {}^3P, {}^1G, {}^1D, {}^1S$
d^3, d^7	$^4F, {}^4P, {}^2H, {}^2G, {}^2F, {}^2D$ (twice), ^2P
d^4, d^6	$^5D, {}^3H, {}^3G, {}^3F$ (twice), $^3D, {}^3P$ (twice), $^1I, {}^1G$ (twice), ^1F
d^5	$^6S, {}^4G, {}^4F, {}^4D, {}^4F, {}^2I, {}^2H, {}^2G$ (twice), ^2F (twice), ^2D (thrice), $^2P, {}^2S$

Table 8.19 *The splitting of atomic terms into terms in octahedral geometry*

Atomic term	Number of energy states	Group theory terms or states (O_h)
S	1	A_{1g}
P	3	T_{1g}
D	5	T_{2g} and E_g
F	7	T_{1g}, T_{2g} and A_{2g}
G	9	A_{1g}, E_g, T_{1g} and T_{2g}
H	11	E_g, T_{1g}, T_{1g} and T_{2g}
I	13	$A_{1g}, A_{2g}, E_g, T_{1g} T_{2g}$ and T_{2g}

In addition, the ground state terms are D for the d^1, d^9 and d^4, d^6 electron configurations and F for the d^2, d^8 and d^3, d^7 electron configurations.

8.8.2.3 *Spectroscopic Terms of Complexes*

In an octahedral metal–ligand complex, the five 3d orbitals are no longer equally degenerate with spherical geometry, and it is necessary to take into account the difference in energy between the t_{2g} and e_g orbitals. Thus, the atomic term symbols from Table 8.18 split into additional terms as in Table 8.19; i.e., the atomic terms correlate to terms in octahedral geometry using group theory approaches. The symbols for the terms have their usual meaning where A is one-dimensional representation, E is two-dimensional representation, and T is three-dimensional representation.

Thus, the ^3F atomic term for a d^2 electron configuration in an octahedral field transforms into three discrete terms ($^3T_{1g}, {}^3T_{2g}$, and $^3A_{2g}$) representing seven energy states consisting of 21 microstates. The ^3P atomic term transforms into one discrete term ($^3T_{1g}$) representing three energy states consisting of nine microstates. These d^2 terms will be used below to describe the absorption peaks expected in Figure 8.39a above.

8.8.2.4 *Splitting Diagrams for Octahedral Complexes*

Figure 8.40 demonstrates the types of d orbital splitting expected for all *high spin* electron configurations from the ground state atomic spectroscopic term symbol. The figure shows the ground state as well as the

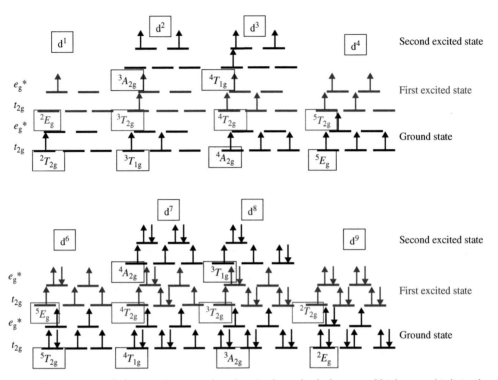

Figure 8.40 *Arrangement of electrons in ground and excited octahedral states of highest multiplicity for the d^1 through d^4 and d^6 through d^9 high spin electron configurations. The terms for each ground and excited electron configuration are also provided*

possible first and second excited states for each splitting based on Table 8.19. Note that the spin multiplicity in all electronic states for each electron configuration is identical so that the spin selection rule is not violated on electron excitation. The first excited state is a one-electron excitation and the second excited state has two electrons excited at the same time. Because the d^5 electron configuration can have only one electron configuration in the ground state, any excited electronic state must have a different spin multiplicity so the spin selection rule is violated. Thus, there are no spin allowed transitions or absorption peaks.

The d^1, d^4, d^6, d^9 Electron Configurations. The d^1 electron configuration has no electron repulsions to consider and is the easiest splitting to describe. The atomic term symbol is ^2D and splits into the terms T_{2g} and E_g in octahedral geometry. In Figure 8.40, the only possible spin allowed electron transition is from $t_{2g}^1 e_g^0 \rightarrow t_{2g}^0 e_g^1$ or $^2T_{2g} \rightarrow {}^2E_g$. Similarly, the d^6 electron configuration transforms the same way despite the electron–electron repulsions, and the only possible electron transition is from $t_{2g}^4 e_g^2 \rightarrow t_{2g}^3 e_g^3$ or $^5T_{2g} \rightarrow {}^5E_g$. Figure 8.14 showed the similarity in electronic spectra for $Ti(H_2O)_6^{3+}$ and $Fe(H_2O)_6^{2+}$. The similarity of the terms for different electron configurations in octahedral geometry are easily described with Orgel splitting or correlation diagrams (Figure 8.41). As the t_2 and e orbitals are inverted or reversed in octahedral and tetrahedral geometry, the diagram can be used to describe similar electron configurations in both geometries; e.g., d^1 octahedral is similar to d^9 tetrahedral as one electron or one hole can be placed into any one of the t_2 orbitals (a concept known as **hole formalism**). The diagram shows energy on the vertical axis and increasing field strength on the horizontal axis; the intersection of the T_2 and E lines indicates the ^2D atomic term, which has no splitting of the d orbitals.

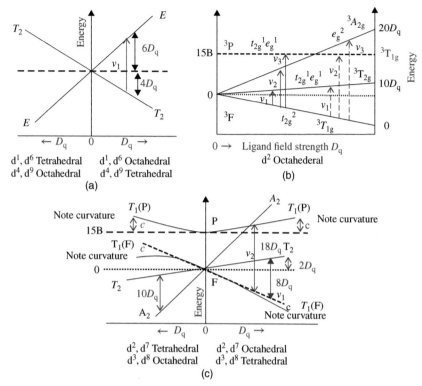

Figure 8.41 (a) Orgel correlation diagram demonstrating the splitting of a D term symbol as ligand field strength, D_q, increases in octahedral and tetrahedral geometries for d^1, d^4, d^6, and d^9. The subscript g is omitted as it does not apply to tetrahedral geometry. (b) Correlation diagram for the ^3F and ^3P term symbols of the d^2 electron configuration on going from the free atom to the octahedral state without bending due to mixing of states of the same term. (c) Correlation diagram for the ^3F and ^3P term symbols of the d^2, d^3, d^7, and d^8 electron configurations on going from the free atom to the octahedral state when bending due to mixing of states of the same term occurs. The subscript "g" is dropped in figures a and c, which also note electron configurations in tetrahedral geometry

The d^2, d^7 Electron Configurations. The d^2 case is the simplest electron configuration that has d orbital splitting and the effect of electron repulsion. As shown above, it has two terms of maximum spin multiplicity for the free ion (^3F is the ground state and ^3P is an excited state) as well as three singlet excited terms (^1S, ^1G, ^1D). The energy separations for spectroscopic terms are expressed with the empirically derived Racah parameters (A, B and C). A is a measure of the average of all interelectron repulsions possible whereas B and C are related to the repulsion energies between individual d electrons. All three Racah parameters are positive in energy as they represent electron–electron repulsions. For the d^2 case, the energies for each term have the following form.

$$E(^1S) = A + 14B + 7C \qquad E(^3P) = A + 7B$$

$$E(^1G) = A + 4B + 2C \qquad E(^3F) = A - 8B$$

$$E(^1D) = A - 3B + 2C$$

Because A is a measure of all interelectron repulsions, it is found in each of the equations. The parameter C is only found in equations for excited states that differ in spin multiplicity from the ground state. For terms of the same spin multiplicity separation is only a function of B. Thus, the difference in energy between the ^3F ground state and ^3P excited state equals 15 B as shown on the left of Figure 8.41b and c. This is also true for the d^3, d^7, and d^8 electron configurations. The parameters B and C differ for each atom or ion and are determined from atomic spectra. B indicates that the Racah parameter is for the free ion and is larger than B for the complex, which is given the notation B'. B' normally ranges from 0.7B to 0.9B because of mixing and delocalization of the d electrons with the ligand orbitals of the complex, which reduces electron–electron repulsions. All three molecular orbital theory cases (1, 2, and 3) provide ample information regarding the sharing of d electrons with ligand orbitals; the more covalent the bonding in the complex is, the lower the value of B' is. This phenomenon is known as the **nephelauxetic effect, β,** or "cloud-expanding" effect where $\beta = B'/B$ and has a value < 1. The effect is independent of the metal ion and a series of common ligands follows an ionic to covalent transition; $F^- > H_2O > NH_3 > en > Cl^- > Br^- > S^{2-} > I^-$. The effect is greatest for softer ligands that are highly polarizable.

As noted in Table 8.19, the ^3F ground state splits into $^3T_{2g}$, $^3T_{1g}$, and $^3A_{2g}$ terms; these and their d electron orbital populations are shown in Figure 8.40. Using the d orbital splitting calculations as in Table 8.3, the energies for these three terms are as follows at infinitely high and intermediate ligand field strength. The relative energies for these terms in infinitely high ligand field strength are thus $0D_q$, $10D_q$, and $20D_q$ as shown on the far right of Figure 8.41b.

	Strong field	Intermediate field (Figure 8.41c)
$E(t_{2g}^2 \text{ or } ^3T_{1g})$	$= (2 \times -4D_q) = -8D_q$	$= -6D_q$
$E(t_{2g}^1 e_g^1 \text{ or } ^3T_{2g})$	$= (-4D_q + 6D_q) = +2D_q$	$= +2D_q$
$E(e_g^2 \text{ or } ^3A_{2g})$	$= (2 \times 6D_q) = +12D_q$	$= +12D_q$

In the absence of any mixing of terms of similar symmetry, Figure 8.41b gives the relative energies of the ^3F and ^3P atomic terms for the d^2 case on the left, and the relative energies of the metal–ligand complex terms in infinitely strong field on the right. From table 8.19, the ^3P excited state transforms as a single term $[^3T_{1g}(P)]$ (these two terms are given in Figure 8.41b with blue color) and remains unchanged in energy on going from the atomic free ion state to the metal–ligand octahedral complex (horizontal line at 15B energy units on the left). Figure 8.41b shows that there are three *spin allowed* d → d electronic transitions (solid blue lines) for a weaker field as in the left part of Figure 8.41b and the right part of Figure 8.41c. The electronic transitions in increasing energy for a **very weak field** case are assigned using Equations 8.5 through 8.7 (the term symbols for the complex are indicated in parentheses with the atomic term that they came from).

$$\upsilon_1 = {}^3T_{1g}(F) \rightarrow {}^3T_{2g}(F) \quad \text{or} \quad t_{2g}^2 \rightarrow t_{2g}^1 e_g^1 = 8D_q \tag{8.5}$$

$$\upsilon_2 = {}^3T_{1g}(F) \rightarrow {}^3A_{2g}(F) \quad \text{or} \quad t_{2g}^2 \rightarrow e_g^2 = 18D_q \tag{8.6}$$

$$\upsilon_3 = {}^3T_{1g}(F) \rightarrow {}^3T_{1g}(P) \quad \text{or} \quad t_{2g}^2 \rightarrow t_{2g}^1 e_g^1 = 6D_q + 15B' \tag{8.7}$$

For this weaker field, the value of Δ_{oct} can be determined from $\upsilon_2 - \upsilon_1$ in both d^2 and d^7 cases (Figure 8.41c). The $[Co(H_2O)_6]^{2+}$ ion spectrum (Figure 8.39b) has been assigned these bands from Equations 8.5–8.7 [9]. The first transition $[^4T_{1g}(F) \rightarrow {}^4T_{2g}(F) \text{ or } t_{2g}^5 e_g^2 \rightarrow t_{2g}^4 e_g^3]$ occurs at 8100 cm^{-1}. The υ_2 peak $[^4T_{1g}(F) \rightarrow {}^4A_{2g}(F) \text{ or } t_{2g}^5 e_g^2 \rightarrow t_{2g}^3 e_g^4]$ near 16,000 cm^{-1} is weak as it is a two-electron jump whereas the υ_3 peak $[^4T_{1g}(F) \rightarrow {}^4T_{1g}(P) \text{ or } t_{2g}^5 e_g^2 \rightarrow t_{2g}^4 e_g^3]$ is at 19,400 cm^{-1}. Using these data and Equations 8.5–8.7, $10D_q$ is 7900 cm^{-1} and B' is 977 cm^{-1}.

As the ligand field strength increases (Figure 8.41b and c), the separation in energy between the terms also increases, and the energies for peaks v_2 and v_3 become inverted as the term $^3A_{2g}$ increases in energy whereas the $^3T_{1g}$ term remains constant in energy (note solid blue versus dashed blue arrows in Figure 8.41b). [The energy of the $^3T_{1g}(F) \rightarrow {}^3A_{2g}(F)$ or $t_{2g}{}^2 \rightarrow e_g{}^2$ peak now is twice that of the $^3T_{1g}(F) \rightarrow {}^3T_{2g}(F)$ or $t_{2g}{}^2 \rightarrow t_{2g}{}^1 e_g{}^1$ peak on increasing the ligand field strength.] The dashed blue arrows in Figure 8.41b show the possible peaks as the d orbital splitting approaches a **strong octahedral field** and the transitions are given in Equations 8.8–8.10.

$$v_1 = {}^3T_{1g}(F) \rightarrow {}^3T_{2g}(F) \quad \text{or} \quad t_{2g}{}^2 \rightarrow t_{2g}{}^1 e_g{}^1 = 8\,D_q \tag{8.8}$$

$$v_2 = {}^3T_{1g}(F) \rightarrow {}^3T_{1g}(P) \quad \text{or} \quad t_{2g}{}^2 \rightarrow t_{2g}{}^1 e_g{}^1 = 6D_q + 15B' \tag{8.9}$$

$$v_3 = {}^3T_{1g}(F) \rightarrow {}^3A_{2g}(F) \quad \text{or} \quad t_{2g}{}^2 \rightarrow e_g{}^2 = 18D_q \tag{8.10}$$

For this intermediate field strength, the value of Δ_{oct} can be determined from $v_3 - v_1$ in both d^2 and d^7 cases (Figure 8.41c). Fortunately, most weak field d^2 and d^7 cases have the type of electron transition sequence given in Equations 8.8–8.10. A convenient method to help assign the peaks is to calculate **the ratio** of v_3/v_1 or v_2/v_1; a value of near 2 indicates the lowest energy peak as the $^3T_{1g}(F) \rightarrow {}^3T_{2g}(F)$ transition and the other as the $^3T_{1g}(F) \rightarrow {}^3A_{2g}(F)$ transition.

For an **infinitely strong field** (right of Figure 8.41b), the electron transitions are described using Equations 8.11–8.13. The term "c" is related to mixing of the $^3T_{1g}(F)$ and $^3T_{1g}(P)$ states (see Section 8.8.2.5 and Figure 8.41c). Frequently "c" is omitted from these equations.

$$v_1 = {}^3T_{1g}(F) \rightarrow {}^3T_{2g}(F) \quad \text{or} \quad t_{2g}{}^2 \rightarrow t_{2g}{}^1 e_g{}^1 = 10D_q + c \tag{8.11}$$

$$v_2 = {}^3T_{1g}(F) \rightarrow {}^3T_{1g}(P) \quad \text{or} \quad t_{2g}{}^2 \rightarrow t_{2g}{}^1 e_g{}^1 = 6D_q + 15B' + 2c \tag{8.12}$$

$$v_3 = {}^3T_{1g}(F) \rightarrow {}^3A_{2g}(F) \quad \text{or} \quad t_{2g}{}^2 \rightarrow e_g{}^2 = 20D_q + c \tag{8.13}$$

The value of Δ_{oct} can be determined from $v_3 - v_1$ at infinite field strength for the d^2 case and the d^7 case (Figure 8.41c); also v_1 now equals $10D_q$. Note that the difference in energy for the two transitions to determine Δ_{oct} is from the same transitions as above for the very weak field case because the $^3T_{1g}(P)$ term remains constant in energy as the $^3A_{2g}(F)$ term increases in energy.

Aqueous solutions of V^{3+} (d^2 case) salts form $V(H_2O)_6{}^{3+}$ and reflect green as the spectrum consists (Figure 8.39a) of two broad bands with one in the red/yellow region at $17,200$ cm^{-1} with ε_{max} of 6 assigned as v_1 $[^3T_{1g}(F) \rightarrow {}^3T_{2g}(F)]$ and the other peak in the blue/violet region at $25,600$ cm^{-1} with ε_{max} of 8 assigned as v_2 $[^3T_{1g}(F) \rightarrow {}^3T_{1g}(P)]$. The third v_3 absorption peak $[^3T_{1g}(F) \rightarrow {}^3A_{2g}(F)]$ at $34,500$ cm^{-1} is not detected in water but can be detected when V^{3+} is doped into Al_2O_3 [15]. The +3 charge on V in $[V(H_2O)_6]^{3+}$ enhances d orbital splitting so $[V(H_2O)_6]^{3+}$ tends to show stronger field character as the ratio of v_2/v_1 is less than 2 (for the very weak field case, the ratio should be >2 from Equations 8.5, 8.6), but v_3/v_1 is 2. Although v_1 and v_2 are spin allowed, they are formally Laporte or parity forbidden; v_3 is also parity forbidden and requires the excitation of two electrons simultaneously.

Spectroscopy of Octahedral High Spin d^n Configurations: d^1, d^4, d^6, d^9. Figures 8.40 and 8.41 show similarity among the high spin configurations so that the number of spin allowed d $\rightarrow$ d absorption peaks is the same for different electron configurations. The d^1 and d^9 as well as the d^4 and d^6 electron configurations are expected to have *one* d $\rightarrow$ d *absorption peak*, which may be split because of Jahn–Teller distortions. The single absorption peak gives the energy for the splitting of the d orbitals as $10D_q$ or Δ_{oct}. These four electron configurations have ground state D atomic terms of maximum multiplicity that split into T_{2g} and E_g octahedral terms.

As shown in Figure 8.40, the d^1 and d^6 octahedral electron configurations have a similar arrangement of energy levels with a single electron transition from the $T_{2g} \rightarrow E_g$ levels. Figure 8.14a and b shows the spectra for $[Ti(H_2O)_6]^{3+}$ and $[Fe(H_2O)_6]^{2+}$. Likewise, the d^4 and d^9 electron configurations have a similar arrangement of energy levels but are inverse to the d^1 and d^6 cases. Thus, the octahedral d^4 and d^9 cases have a single electron transition from the $E_g \rightarrow T_{2g}$ level as shown for $[Mn(H_2O)_6]^{3+}$ and $[Cu(H_2O)_6]^{2+}$ in Figure 8.42; the broadness or shoulders on the peaks indicate that there is Jahn–Teller distortion as expected. Because of the similarity of the tetrahedral and octahedral electron configurations shown in Figure 8.41a and discussed previously, the electron transitions for tetrahedral d^1 and d^6 and for d^4 and d^9 are inverse to the transitions for their octahedral counterparts. The energy for the d orbital splitting in tetrahedral geometry is Δ_{tet}.

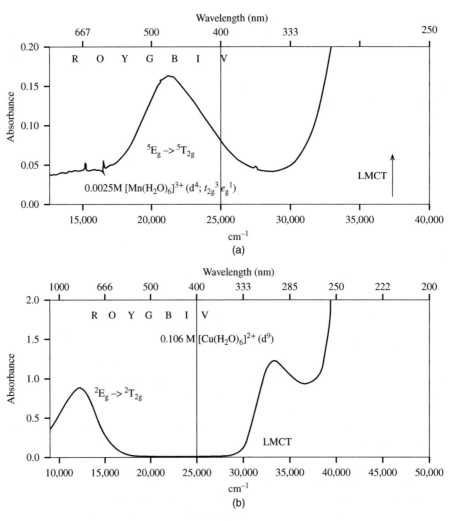

Figure 8.42 *Electronic spectra of (a)* $[Mn(H_2O)_6]^{3+}(d^4,\ t_{2g}{}^3e_g{}^{*1})$*; (b)* $[Cu(H_2O)_6]^{2+}(d^9,\ t_{2g}{}^6e_g{}^{*3})$

8.8.2.5   *Overview of d², d³, d⁷, d⁸ Electron Configurations*

Figure 8.40 shows that the d^8 electron configuration (or the two-hole case) has the same atomic terms as the d^2 electron configuration and thus the same splitting of the terms as the d^2 electron configuration. Note that the terms are reversed in energy; i.e, the ground state for the d^8 case is now the $^3A_{2g}$ term because there is only one possible combination to place the three unpaired electrons in the ground state (the value of D_q is also reversed in sign). The $^3T_{1g}$ term is now the second excited state of the ^3F term so it is possible to mix or interact with the $^3T_{1g}$ term of the ^3P term. When these two different $^3T_{1g}$ terms have near similar energy so they can interact or mix (much as two molecular orbitals of similar energy can interact), their energies will deviate from the predicted values in the absence of any interaction with one state increasing in energy and the other decreasing in energy; thus, the energy lines will no longer be linear and will bend so that they will not cross. This is known as the **non-crossing** rule and is shown in Figure 8.41c.

Similar to the d^2 case, the d^8 along with the d^3 and d^7 electron configurations are expected to have *three spin allowed d → d absorption peaks*. These four electron configurations have ground state F atomic terms of maximum multiplicity that split into T_{1g}, T_{2g}, and A_{2g} octahedral terms. The d^2 and d^7 electron configurations have the same relative energy level diagram as noted above, whereas the energy states for d^3 and d^8 electron configurations, which have the same relative energy level diagram as shown in Figures 8.40 and 8.41c, are inverse in order to the d^2 and d^7 cases.

The Orgel diagram in Figure 8.41c contains much information and differs to some extent from Figure 8.41b. Left of $0D_q$ in Figure 8.41c are the d^3 and d^8 octahedral cases and to the right of $0D_q$ are the d^2 and d^7 octahedral cases. The difference between the atomic F and P terms is 15B as above. However, the $T_1(P)$ term is no longer linear (the difference between the upward curve versus dashed horizontal line is assigned a value of "c"), and neither is the $T_1(F)$ term (downward curve versus the diagonal dashed line). These curvatures are most pronounced for the d^3 and d^8 octahedral cases on the left of the diagram because of the closeness in energy of the two T_1 terms; the higher value of "c" indicates more mixing between the states. For the d^2 and d^7 octahedral cases, there is less interaction.

8.8.2.6   *Spectroscopy of high spin d³, d⁸ Electron Configurations*

An excellent example of electron transitions starting with a $^4A_{2g}(F)$ ground state are aqueous solutions of Cr^{3+} (d^3 case) that produce the complex $[Cr(H_2O)_6]^{3+}$, which is light green (Figure 8.43). There are two broad absorption bands at $17,400$ cm^{-1} (yellow region) and $24,390$ cm^{-1} (blue region) with a very weak band at $37,000$ cm^{-1} in the UV region. Using Figure 8.40 and 8.41c for the octahedral d^3 electron configuration, these three absorption bands are assigned as Equations 8.14–8.16. Note that the equations for the absorption peaks have a correction of "c" for the interaction of the two T_1 terms. The value of Δ_{oct} ($10D_q$) is obtained from v_1 for both the d^3 and d^8 octahedral cases. Using Equations 8.14–8.16, $10\,D_q = v_1 = 17,400$ cm^{-1} and $D_q = 1740$ cm^{-1}; $c = 6930$ cm^{-1}; $B' = 613$ cm^{-1}.

$$v_1 = {}^4A_{2g}(F) \rightarrow {}^4T_{2g}(F) \quad \text{or} \quad t_{2g}{}^3 \rightarrow t_{2g}{}^2 e_g{}^1 = 17,400 \text{ cm}^{-1} = 10D_q \tag{8.14}$$

$$v_2 = {}^4A_{2g}(F) \rightarrow {}^4T_{1g}(F) \quad \text{or} \quad t_{2g}{}^3 \rightarrow t_{2g}{}^2 e_g{}^1 = 24,390 \text{ cm}^{-1} = 18D_q - c \tag{8.15}$$

$$v_3 = {}^4A_{2g}(F) \rightarrow {}^4T_{1g}(P) \quad \text{or} \quad t_{2g}{}^3 \rightarrow t_{2g}{}^1 e_g{}^2 = 37,000 \text{ cm}^{-1} = 15B' + 12D_q + c \tag{8.16}$$

The electronic spectrum of $[Ni(H_2O)_6]^{2+}$ (d^8) in Figure 8.43b is similar to that of $Cr(H_2O)_6{}^{3+}$. The v_1 transition $[{}^3A_{2g}(F) \rightarrow {}^3T_{2g}(F)$ or $t_{2g}{}^6 e_g{}^2 \rightarrow t_{2g}{}^5 e_g{}^3]$ is at 8700 cm^{-1}; the v_2 transition $[{}^3A_{2g}(F) \rightarrow {}^3T_{1g}(F)$ or

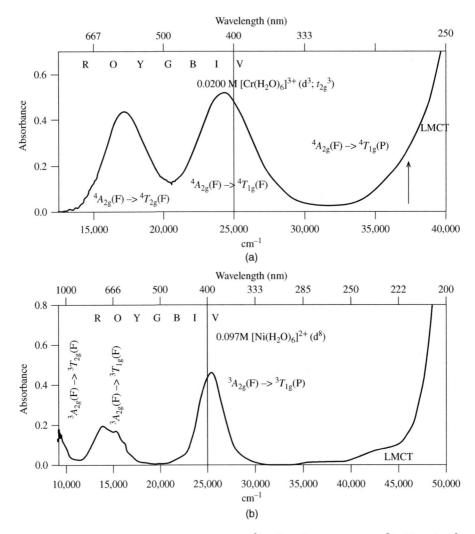

Figure 8.43   *Electronic spectra of (a) Cr(H₂O)₆³⁺ (d³; t₂g³); (b) Ni(H₂O)₆²⁺ (d⁸, t₂g⁶eg*²)*

$t_{2g}^6 e_g^2 \rightarrow t_{2g}^5 e_g^3$] is found at $14,500$ cm^{-1} and the v_3 peak [$^3A_{2g}$(F) $\rightarrow$ $^3T_{1g}$(P) or $t_{2g}^6 e_g^2 \rightarrow t_{2g}^4 e_g^4$] is at $25,300$ cm^{-1}. Using Equations 8.14–8.16, $10D_q = v_1 = 8700$ cm^{-1} and $D_q = 870$ cm^{-1}; $c = 1160$ cm^{-1}; $B' = 913$ cm^{-1}.

8.8.3   Energy and Spatial Description of the Electron Transitions Between t_{2g} and e_g^* Orbitals

Equations 8.14 and 8.15 both describe a $t_{2g}^3 \rightarrow t_{2g}^2 e_g^1$ electronic transition between the F atomic terms. The difference in energy for the absorption peaks v_1 and v_2 are related to the type of $t_{2g} \rightarrow e_g^*$ transitions possible. Figure 8.44 shows two possible transitions between t_{2g} orbitals and e_g^* orbitals. On the left is a transition from a d_{xy} to the d_{z^2} orbital; here the electron moves from the xy plane to the z-axis, which already has some electron density because of the occupancy of the d_{xz} and d_{yz} orbitals. Thus, there is significant increase in the

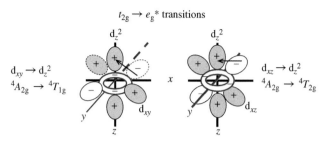

$t_{2g} \rightarrow e_g{}^*$ transitions

Figure 8.44 *Schematic diagram of the movement of t_{2g} electrons from different t_{2g} orbitals to the same $e_g{}^*$ orbitals; Left (v_2 transition); Right (v_1 transition)*

electron–electron repulsion from that in the ground state. In contrast, the transition from the d_{xz} orbital to the d_{z^2} orbital does not significantly increase electron–electron repulsions from the ground state. The absorption peaks v_1 and v_2 are thus different in energy because of these electron repulsions. The v_1 peak has the lower energy transition assigned as $^4A_{2g}(F) \rightarrow \, ^4T_{2g}(F)$ whereas the v_2 peak has the higher energy transition and is assigned as $^4A_{2g}(F) \rightarrow \, ^4T_{1g}(F)$. There are four other electron transitions; two are similar lower energy transitions to the $d_{xz} \rightarrow d_{z^2}$ orbital transition, and they are the $d_{yz} \rightarrow d_{z^2}$ and $d_{xy} \rightarrow d_{x^2-y^2}$ transitions. The two higher energy transitions similar to the d_{xy} to the d_{z^2} orbital transition are the $d_{xz} \rightarrow d_{x^2-y^2}$ and the $d_{yz} \rightarrow d_{x^2-y^2}$ transitions.

8.8.4 More Details on Correlation Diagrams

More complete correlation diagrams are given in Figures 8.45 through 8.47 for the d^3, d^6, and d^5 electron configurations to show the number of energy states available and how the d orbitals split on increasing field strength. In each figure, the left-hand free atom or ion term indicates that there is zero d orbital splitting and the splitting increases from high to low spin on going right. There is a significant crossover (indicated by the dashed blue lines) for the energy of states that have different terms, but which do not mix as they have different symmetry.

8.8.4.1 The d^3 Correlation Diagram

In the d^3 case for Cr^{3+} (Figure 8.45), the ground state energy does not change on increasing field strength as the three electrons can only be in one ground state of maximum multiplicity, the $^4A_{2g}$, for both weak and strong fields. Under very strong field conditions, the doublet states become more stable, and it is possible to have spin forbidden transitions from the $^4A_{2g}$ term to ^2E_g and other doublet states with very low ε_{max} values along with the three spin allowed transitions discussed above (Figure 8.43a) and noted in Figure 8.45 (three black arrows on left and right).

8.8.4.2 The d^6 Correlation Diagram

With increasing field strength, the correlation diagram for the d^6 electron configuration (Figure 8.46 and Fe^{2+} as example) shows that the ground state changes from a paramagnetic complex with maximum multiplicity ($^5T_{2g}$ state) to a diamagnetic complex with a ground state term of $^1A_{1g}$. The diagram indicates that there should be only one spin allowed d $\rightarrow$ d absorption peak from the $^1A_{1g}$ to $^1T_{1g}$ state for an infinitely strong field, and this gives Δ_{oct}. Other spin forbidden transitions are possible.

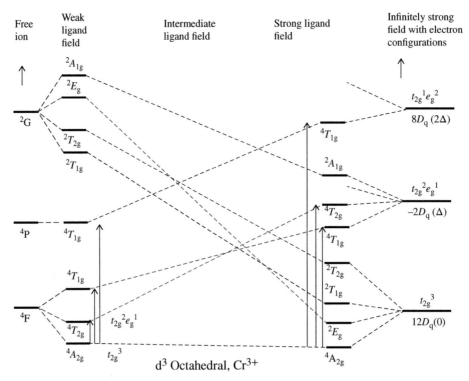

Figure 8.45 *Correlation diagram for the d^3 case in octahedral symmetry (d^7 in tetrahedral). The energies in units of D_q of the infinitely strong field case are given on the right with the relative energies in units of Δ in parentheses. The arrows at the top pointing up indicate other higher energy terms not drawn on the diagram (Source: From [9], Figure 6.5. Reproduced with permission from Wiley, Inc.)*

8.8.4.3 The d^5 Correlation Diagram

With increasing field strength, the correlation diagram for the d^5 electron configuration (Figure 8.47 with the hexaaqua ions of Mn^{2+} and Fe^{3+} as examples) shows that the ground state changes from a paramagnetic complex with maximum multiplicity ($^5T_{2g}$ state) to a slightly paramagnetic complex with a ground state term of $^2T_{2g}$. The left part of the diagram indicates that there should be NO spin allowed d → d absorption peak from the $^6A_{1g}$ to any other state for weak field cases (the six arrows indicate spin forbidden transitions).

$Mn(H_2O)_6^{2+}$ shows very weak spin forbidden transitions peaks and a slight pink coloration at about 1.0 molar concentration (Figure 8.48). Note that the $Mn(H_2O)_6^{2+}$ peaks are 10–100 times lower in absorbance than that of the other hexaaqua ions of the first transition series, which have spin allowed d → d transitions. For the strong field case, only one spin allowed transition is available in the visible region and it is the $^2T_{2g} \rightarrow {}^2E_g$, which gives Δ_{oct}.

Tanabe–Sugano [17] diagrams for each electron configuration provide a more quantitative approach than that shown above, but are only mentioned here. These diagrams plot the ratio of the energy of the states versus B, the Racah parameter, (E/B) versus D_q/B. The E/B ratio of a state can vary as a line and/or curve with D_q/B, and the spectral peaks are assigned as above. Using the energies from the peak assignments gives a measure of D_q/B for the complex.

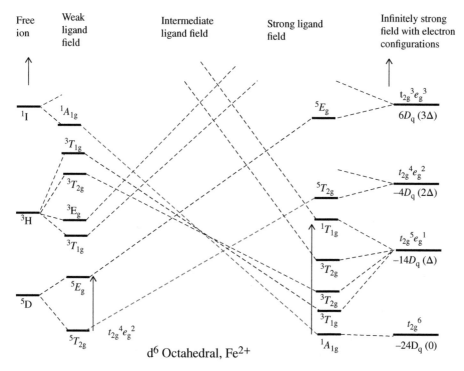

Figure 8.46 *Correlation diagram for the d^6 case in octahedral symmetry (d^4 in tetrahedral). The energies in units of D_q of the infinitely strong field case are given on the right with the relative energies in units of Δ in parentheses. The arrows at the top pointing up indicate other higher energy terms not drawn on the diagram (Source: From Ref. [9], Figure 6.8. Reproduced with permission from Wiley, Inc.)*

8.8.5 Luminescence

The discussion on spectroscopy has centered on the electronic transitions that result in absorption peaks, and the light that is reflected, which determines the color of the metal–ligand complex observed. Metal–ligand complexes can also exhibit luminescence, which is the emission of radiation after the complex has been excited by the absorption of radiation. There are two kinds of luminescence. The first is **fluorescence**, which is the decay of the radiation from an excited state to the ground state when both states have the same multiplicity. Fluorescence is a spin allowed transition and is quite fast with half-lives on the order of nanoseconds. **Phosphorescence** is the second form of luminescence and is the decay of the radiation from an excited state back to a ground state when both have different spin multiplicity; thus, phosphorescence is a spin forbidden transition, and is significantly slower than fluorescence.

Figure 8.49 shows the absorption transitions and the luminescence emissions for the mineral ruby, which is alumina (Al_2O_3) with up to 2% Cr^{3+} ions replacing and substituting for Al^{3+} ions. Both aluminum and chromium are surrounded by six oxide anions so have octahedral geometry, but only the Cr^{3+} d electrons are involved in electron transitions. More details on the lower energy states for Cr^{3+} ions are also provided in Figure 8.43. The $t_{2g} \rightarrow e_g^*$ visible absorption transitions in ruby are

$$^4A_{2g} \rightarrow {}^4T_{2g}(F), \text{ which is the green absorption peak and}$$

$$^4A_{2g} \rightarrow {}^4T_{1g}(F), \text{ which is the blue absorption peak.}$$

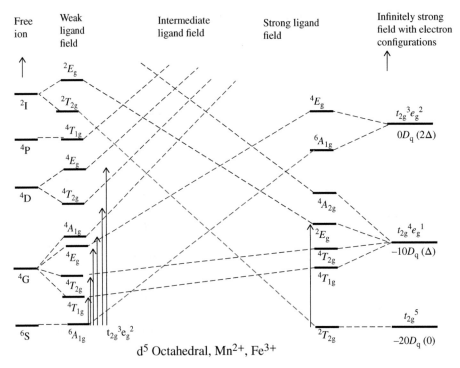

d^5 Octahedral, Mn^{2+}, Fe^{3+}

Figure 8.47 *Correlation diagram for the d^5 case in octahedral symmetry. The energies in units of D_q of the infinitely strong field case are given on the right with the relative energies in units of Δ in parentheses. The arrows at the top pointing up indicate other higher energy terms not drawn on the diagram (Source: From Ref. [9], Figure 6.9. Reproduced with permission from Wiley, Inc.)*

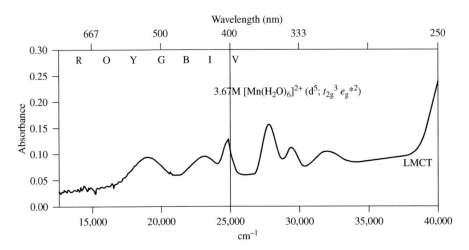

Figure 8.48 *Electronic spectrum of $Mn(H_2O)_6{}^{2+}$. The six peaks from left to right arise from transitions from the $^6A_{1g}$ ground state to the ^4P states ($^4T_{1g}$, $^4T_{2g}$, ^4E_g, $^4A_{1g}$) and ^4E states ($^4T_{2g}$, ^4E_g) as shown by the arrows in Figure 8.47*

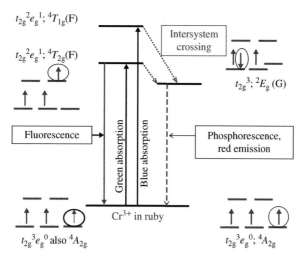

Figure 8.49 *The electronic transitions and energy states that occur during the absorption and the luminescence of light by Cr^{3+} ions in the mineral ruby. The difference in energy for the two excited states $t_{2g}^2 e_g^1$ (^4T) is shown on the left in Figure 8.45. Transition (loss of energy) from the $^4T_{1g}$(F) to $^4T_{2g}$(F) is known as internal conversion*

These lead to reflection of red light, which gives ruby its characteristic color. Fluorescence occurs by emission of the light absorbed, but is not a major feature for the luminescence of ruby. Phosphorescence occurs when an electron in one of the excited states from a spin allowed electron transition undergoes **intersystem crossing** to another excited state with a different multiplicity. This process is nonradiative decay via thermal loss of energy to the surroundings, and occurs on a much faster timescale than fluorescence; i.e., on the order of picoseconds compared to nanoseconds. For Cr^{3+}, the excited e_g* electron in a $^4T_{2g}$ (F) or $^4T_{1g}$(F) state ($t_{2g}^2 e_g^1$) crosses to a t_{2g}^3 excited ^2E_g state (see blue lines in Figure 8.45; there is a slight splitting of the t_{2g} orbitals in the ^2E_g state in Figure 8.49 due to a distortion from octahedral geometry) so that there is only one electron unpaired. Because this doublet state is of different multiplicity than the quartet ground state ($^4A_{2g}$), radiative decay or emission of light is now a spin forbidden process. Radiative decay back to the ground state is slow (microseconds or longer), and the phosphorescence emission of red light occurs at 627 nm. The red phosphorescence emission also adds to the reflected red color from the absorption of green and blue light. Emission of some phosphorescent materials can continue even when the light source is turned off.

8.8.6 Magnetism and Spin Crossover in Octahedral Complexes and Natural Minerals

An important feature of the correlation diagrams of Figures 8.45–8.47 is the crossover of the energy states (indicated by the dashed blue lines) as ligand field strength increases in octahedral complexes. Some first transition series metal complexes can exist as both high and low spin complexes simultaneously or can be changed from high to low spin with increasing pressure and/or decreasing temperature. This phenomenon is known as spin crossover and has been reviewed in *Topics in Current Chemistry* [18]. The largest number of examples are for metals with octahedral geometry and d^6 electron configurations with Fe(II) comprising the largest majority as low spin t_{2g}^6 has maximum CFSE and MOSE. The first report for a synthetic Fe(II) complex was [Fe(phen)$_2$(NCS)$_2$].

The changeover from high to low spin in octahedral geometry results in a decrease in the bond length for the metal–ligand bond as electrons move from the e_g* orbital to the t_{2g} orbitals, which are closer to the nucleus (see

Figure 8.17; Table 6.4). For Fe^{2+}, the decrease in bond length is 17 pm and for Fe^{3+} it is 9.5 pm. An increase in pressure would stimulate the high to low spin change so that the volume of the complex would decrease. In nature, extreme pressures are found at depth in the earth's mantle and core. Examples where both Fe(II) and Fe(III) undergo a change from high to low spin under high pressure and high temperature are found in the most abundant minerals of Earth's lower mantle (between about 670 and 2890 km depth). Two important deep earth minerals are **ferropericlase** [(MgFe)O with Mg and Fe as 6 coordinate] and the $MgSiO_3$ **perovskite** structure named bridgmanite [19], which has Fe substituting for Mg and Si as in $(MgFe)(Si, Fe)O_3$ with six coordination [20]. Fe_2O_3 [21] and FeS [22] also have six coordinate Fe. The spin crossover for ferropericlase and bridgmanite has been hypothesized to be responsible for seismic discontinuities [23, 24]. Note that the ligands (O,S) are normally considered high spin inducing ligands, but under high pressure can stabilize low spin metal ions.

An increase in temperature results in enhanced electron movement within the complex with higher vibrational energy levels being occupied (Figure 5.9) according to Equation 8.17 where i is the ground vibrational or electronic state and j is an excited state. Increased T leads to an increase in the bond length and the volume with a consequent shift from the low spin to the high spin complex occurring. A plot of the percentage of high spin species versus increasing absolute temperature will give an "s" shaped curve that increases in one of three ways: gradual, steep, or in steps depending on the compound or solid.

$$\frac{N_i}{N_j} = \left(e^{-(E_i - E_j)/(kT)} \right) \tag{8.17}$$

Figure 8.50 shows a plot of the molar magnetic susceptibility (χ_M) versus temperature for materials that do not show spin crossover. Diamagnetic species undergo no change in χ_M with increasing absolute T, but dilute solutions of paramagnetic species with a permanent paramagnetic moment (μ, in Bohr magnetons, Equation 3.12) follow the Curie Law (Equation 8.18) and show a decrease with increasing absolute T. The magnetic dipoles of neighboring atoms in paramagnetic species are randomly distributed.

$$\chi_M = \frac{C}{T} = \frac{N^2 \mu^2}{3RT} \tag{8.18}$$

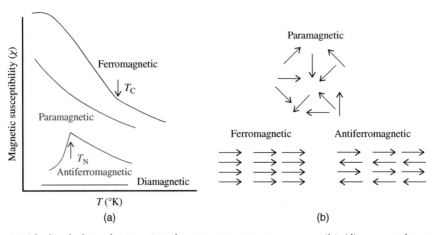

Figure 8.50   (a) Idealized plots of magnetic substances versus temperature. (b) Alignment of magnetic dipoles for paramagnetic, ferromagnetic, and antiferromagnetic species

Ferromagnetic substances have the magnetic dipoles on neighboring atoms aligned in parallel, but as temperature increases there is a tendency for random distribution of dipoles. Below the Curie temperature (T_C), ferromagnetism dominates. Antiferromagnetic materials have magnetic dipoles aligned in antiparallel, and below the Neel temperature (T_N), antiferromagnetism dominates.

Geochemists use the ferromagnetic properties of magnetite [Fe_3O_4 formed during the serpentization process, Equations 8.19 and 8.20 (not balanced)] and the diamagnetic properties of pyrite (FeS_2 formed at black smoker hydrothermal vents) to determine whether hydrothermal vents are present at mid-ocean spreading centers [25, 26]. Reduced crustal magnetization indicates the formation of pyrite (and other metal sulfides). Magnetization measurements are used by mining companies to search for hydrothermal deposits to mine precious metals.

$$Mg_{1.8}Fe_{0.2}SiO_4 + 1.37H_2O \rightarrow 0.5Mg_3Si_2O_5(OH)_4 + 0.3Mg(OH)_2 + 0.067Fe_3O_4 + 0.067H_2$$

Olivine   serpentine   brucite   magnetite   (8.19)

$$Olivine[(MgFe)_2SiO_4] + H_2O + CO_2 \rightarrow serpentine[Mg_3Si_2O_5(OH)_4] + Fe_3O_4 + CH_4 \qquad (8.20)$$

8.8.7 Note about f Orbitals in Cubic Symmetry (O_h)

The f_{x^3}, f_{y^3} and f_{z^3} orbitals are on the Cartesian coordinate axes and transform as t_{1u} in O_h symmetry similar to the metal "p" orbitals that are involved with σ bonding; these orbitals can also bind with the ligand orbitals used in π bonding (see Figure 8.30). The $f_{z(x^2-y^2)}$, $f_{x(z^2-y^2)}$, and $f_{y(z^2-x^2)}$ orbitals, which are similar in shape to the f_{xyz} orbital, transform as t_{2u} and can bind with ligand orbitals used in π bonding. The f_{xyz} orbital transforms as the nonbonding a_{2u} orbital.

References

1. Connelly, N. G., Damhus, T., Hartshorn, R. M. and Hutton, A. T. (2005) *Nomenclature of Inorganic Chemistry: IUPAC Recommendations 2005*. IUPAC, RSC Publishing, Norfolk, UK, pp. 366. (Also known as the red book).
2. Murmann, R. K. and Taube, H. (1956) The mechanism of the formation and rearrangement of nitrito-cobalt(III) amines. *Journal of the American Chemical Society*, **78**, 4886–4890.
3. Basolo, F. and Pearson R. G. (1967) *Mechanisms of Inorganic Reactions*, Wiley, New York, pp. 701.
4. Bethe, H. (1929) Termaufspaltung in Kristallen. *Annalen der Physik*, **5**, 133–208.
5. Van Vleck, J. H. (1935) The group relation between the Mulliken and Slater-Pauling theories of valence. *Journal of Chemical Physics*, **3**, 803–806.
6. Ballhausen, C. J. (1979) Quantum mechanics and chemical bonding in inorganic complexes. I. static concepts the bonding; dynamic concepts of Valency. *Journal of Chemical Education*, **56**, 215–218.
7. Ballhausen, C. J. (1979) Quantum mechanics and chemical bonding in inorganic complexes. II. Valency and inorganic metal complexes. *Journal of Chemical Education*, **56**, 294–297.
8. Ballhausen, C. J. (1979) Quantum mechanics and chemical bonding in inorganic complexes. III. The spread of the ideas. *Journal of Chemical Education*, **56**, 357–361.
9. Figgis, B. N. and Hitchman, M. A. (2000) *Ligand Field Theory and Its Applications*, Wiley-VCH, New York, pp. 354.
10. Cotton, F. A. and Meyers, M. D. (1960) Magnetic and spectral properties of the spin-free $3d^6$ systems Fe(II) and cobalt(III) in cobalt hexafluoride ion: Probable observation of dynamic Jahn-Teller effects. *Journal of the American Chemical Society*, **82**, 5023–5026.

11. Wells, A. F. (1986) *Structural Inorganic Chemistry* (5th ed.), Oxford University press, London.

12. Krishnamurthy, R. and Schaap, W. B. (1969) Computing ligand field potentials and the relative energies of d orbitals. *Journal of Chemical Education*, **46**, 799–810.

13. DeKock, R. L. and Gray, H. B. (1980) *Chemical Structure and Bonding*, Benjamin/Cummings Publishing Co, Menlo Park, CA, pp. 491.

14. Williams, R. J. P. and Frausto da Silva, J. J. R. (2006) *The Chemistry of Evolution: The Development of Our Ecosystem*, Clarendon Press, Oxford, pp. 481.

15. Richens, D. T. (1997) *The Chemistry of Aqua Ions*, John Wiley and Sons, Chichester, pp. 592.

16. Brill, T. B. (1980) *Light Its Interaction with Art and Antiquities*, Plenum Press, NY, pp. 287.

17. Tanabe, Y. and Sugano, S. (1954) On the absorption spectra of complex ions I. *Journal of the Physical Society of Japan*, **9**, 753–766. Two other papers are in the same journal. Vol. 9, 766–779 and Vol. 11, 864–877.

18. Gutlich, P. and Goodwin, H. A. (2004) Spin crossover – an overall perspective. *Topics in Current Chemistry*, **233**, 1–47, in Spin Crossover in transitions metal complexes (eds. Gutlich, P. and Goodwin, H. A), Springer-Verlag Berlin. See also volumes 234 and 235.

19. Tschaunner, T., Ma, C., Beckett, J. R., Prescher, C., Prakapenka, V. B. and Rossman, G. R. (2014) Discovery of bridgmanite, the most abundant mineral in the earth, in a shocked meteorite. *Science*, **346**, 1100–1102.

20. Shahnas, M. H., Peltier, W. R., Wu, Z. and R. Wentzcovitch (2011) The high-pressure electronic spin transition in iron: Potential impacts upon mantle mixing. *Journal of Geophysical Research*, **116**, B08205, doi: 10.1029/2010JB007965.

21. Hemley, R. J., Mao, H. K and Gramsch, S. A. (2000). Pressure-induced transformations in deep mantle and core minerals. *Mineralogical Magazine*, **64**(2), 157–184.

22. Rueff, J. P., Kao, C. C., Struzhkin, V. V., Badro, J., Shu, J., Hemley, R. J. and Mao, H. K. (1999) Pressure-induced high-spin to low-spin transition in FeS evidenced by X-ray emission spectroscopy, *Physical Review Letters*, **82**(16), 3284–3287.

23. Antonangeli, D., Siebert, J. , Aracne, C. M., Farber, D. L., Bosak, A., Hoesch, M., Krisch, M., Ryerson, F. J. , Fiquet, G. and Badro, J. (2011) Spin crossover in ferropericlase at high pressure: a seismologically transparent transition? *Science*, **331**, 64–67.

24. Lin, J.-F., Speziale, S., Mao, Z. and Marquardt, H. (2013) Effects of the electronic spin transitions of iron in lower mantle minerals: Implications for deep mantle geophysics and geochemistry. *Reviews of Geophysics*, **51**, 244–275.

25. Tivey, M. A., Rona, P. A. and Kleinrock, M. C. (1996) Reduced crustal magnetization beneath relict hydrothermal mounds: TAG hydrothermal field, Mid-Atlantic Ridge, 26 °N. *Geophysical Research Letters*, **23**, 3511–3514.

26. Tivey, M. A. and Dyment, J. (2010) The magnetic signature of hydrothermal systems in slow spreading environments, In *Diversity of Hydrothermal Systems on Slow Spreading Ocean Ridges*, Geophysical Monograph Series 188 (Eds. P. A. Rona, C. W. Dvey, J. Dyment and B. J. Murton), American Geophysical Union, Washington, DC, pp 43–65. doi: 10.1029/2008GM000773.

9

Reactivity of Transition Metal Complexes: Thermodynamics, Kinetics and Catalysis

9.1 Thermodynamics Introduction

Transition metal complexes undergo ligand substitution reactions in which the first coordination sphere of ligands bound to the metal are substituted by ligands of different composition. There are four main types of substitution reactions that describe metal complexes other than water (or solvent) exchange, that is, the water molecule replaced by a water solvent molecule. They are **metal complex formation** (Equation 9.1; where ligand replaces a water molecule), **aquation** or **solvolysis** (Equation 9.2; where a water or solvent molecule replaces another ligand that is not the solvent), **anation** (Equation 9.3; complex formation where a ligand(s) replaces a water molecule in the free metal ion complex, $M(H_2O)_6$, with other ligands bound to the metal), and **ligand exchange** (Equation 9.4; where one ligand replaces another ligand and neither is a solvent molecule).

$$M(H_2O)_6 + L \rightarrow M(H_2O)_5L + H_2O \qquad \text{complex formation} \qquad (9.1)$$

$$ML_5X + H_2O \rightarrow ML_5(H_2O) + X \qquad \text{aquation} \qquad (9.2)$$

$$ML_5(H_2O) + X \rightarrow ML_5X + H_2O \qquad \text{anation} \qquad (9.3)$$

$$ML_5X + Y \rightarrow ML_5Y + X \qquad \text{ligand exchange} \qquad (9.4)$$

9.1.1 Successive Stability Constants on Water Substitution

The simplest substitution or metal complex formation reaction is when a water molecule(s) is replaced by another ligand(s). For an octahedral complex, six water molecules can be replaced by six different ligands or six ligating atoms from multidentate ligands. The successive replacement of water molecules by another monodentate ligand (entering group) is given by the expressions 9.5a–9.5c, which describe the individual formation constants, K_f, for each substitution reaction. Water is not part of the mathematical expression since

Inorganic Chemistry for Geochemistry and Environmental Sciences: Fundamentals and Applications, First Edition. George W. Luther, III.
© 2016 John Wiley & Sons, Ltd. Published 2016 by John Wiley & Sons, Ltd.
Companion Website: www.wiley.com/go/luther/inorganic

it is the solvent and is incorporated in K. Species surrounded by [ ] indicate concentrations whereas species surrounded by { } indicate activities (see Sections 7.8.1, 13.1.1 and 13.1.2).

$$M(H_2O)_6 + L \leftrightarrow ML(H_2O)_5 + H_2O \qquad K_1 = \frac{[ML]}{[M][L]} \tag{9.5a}$$

$$ML(H_2O)_5 + L \leftrightarrow ML_2(H_2O)_4 + H_2O \qquad K_2 = \frac{[ML_2]}{[ML][L]} \tag{9.5b}$$

$$\vdots$$

$$ML_{n-1}(H_2O) + L \leftrightarrow ML_n + H_2O \qquad K_n = \frac{[ML_n]}{[ML_{n-1}][L]} \tag{9.5c}$$

The overall formation constant of the final product is given by Equation 9.6a and is the product of the stepwise formation constants (Equation 9.6b). The formation constant of the complex is also known as the stability constant, a term which is used interchangeably with formation constant.

$$\beta_n = \frac{[ML_n]}{[M][L]^n} \tag{9.6a}$$

$$\beta_n = K_1 K_2 K_3 \cdots K_n \tag{9.6b}$$

The inverse of K_f, the formation constant, is the dissociation constant, K_d (Equation 9.7).

$$ML \leftrightarrow M + L \quad K_d = \frac{[M][L]}{[ML]} \tag{9.7}$$

The formation constant, K_f, is related to the Gibbs free energy (ΔG) and the redox potential, E, of the substitution reaction via Equation 9.8.

$$\Delta G = -nFE = -(RT)\ln K_f = \Delta H - T\Delta S \tag{9.8}$$

Successive stepwise formation constants typically follow the order $K_1 > K_2 > K_3 > \cdots > K_n$.

Thus, each K value has a statistical contribution from entropy, ΔS, if the enthalpy, ΔH, is constant (a reasonable first approximation). The formation constants of Ni^{2+} amines, $[Ni(NH_3)_n(OH_2)_{6-n}]^{2+}$, display this behavior well [1].

n (ligand number)	1	2	3	4	5	6
log K_n	2.72	2.17	1.66	1.12	0.67	0.03

However, a reversal of the relation $K_n < K_{n-1}$ can occur and indicates a change in the electronic structure or coordination of the complex as more monodentate ligands are added. For Cd complexes with bromide, the successive equilibrium constants are log $K_1 = 1.56$, log $K_2 = 0.54$, log $K_3 = 0.06$, and log $K_4 = 0.37$ [1]. Aqua complexes are six coordinate (octahedral) whereas halide complexes are four coordinate (tetrahedral). The loss of three water molecules as the fourth bromide replaces the fourth water molecule results in an increase in entropy ($\Delta S > 0$; four product molecules vs two reactant molecules in Equation 9.9), a decrease in Gibbs free energy, and an increase in K_f (log $K_3 = 0.06$ and log $K_4 = 0.37$).

$$[CdBr_3(OH_2)_3]^- + Br^- \rightarrow [CdBr_4]^{2-} + 3H_2O$$

$$\text{C.N.} = 6 \qquad\qquad\qquad \text{C.N.} = 4 \tag{9.9}$$

S^{2-} and other anions in group 16 induce a similar increase in K_n; for example, as bisulfide displaces water from $Zn(H_2O)_6{}^{2+}$ and $Cu(H_2O)_6{}^{2+}$, there is a loss of the proton from HS^- and the resulting sulfide (S^{2-}) complexes of Cu and Zn become tetrahedral [2, 3]. For these d^{10} metals, there is no MOSE advantage to have an octahedral complex (Table 8.12).

9.1.2   The Chelate Effect

When bidentate or multidentate ligands ($edta^{4-}$, porphyrins, siderophores) substitute for monodentate ligands, an increase in entropy ($\Delta S > 0$) and hence K is also observed. This increase in stability for multidentate ligands over monodentate ligands is referred to as the **chelate effect**, and is most important for macrocyclic ligands such as the **porphyrins**, which are produced by phytoplankton, and **siderophores**, which are produced by bacteria. A few simple examples of the chelate effect follow. In the metal–ligand complex reactions of Equations 9.10a,b (at 25 °C; $\mu = 1.0$ M), replacement of two water molecules by ethylenediamine and two ammonia molecules in $[Cu(OH_2)_6]^{2+}$ gives different thermodynamic data [1].

$$[Cu(OH_2)_6]^{2+} + en \rightarrow [Cu(en)(OH_2)_4]^{2+} + 2H_2O$$

$$\log K_1 = 10.54 \qquad \Delta H = -54.8 \text{ kJ mol}^{-1} \qquad \Delta S = +25 \text{ J mol}^{-1} \text{ deg}^{-1} \tag{9.10a}$$

$$[Cu(OH_2)_6]^{2+} + 2NH_3 \rightarrow [Cu(NH_3)_2(OH_2)_4]^{2+} + 2H_2O$$

$$\log \beta_2 = 7.83 \qquad \Delta H = -46.4 \text{ kJ mol}^{-1} \qquad \qquad \Delta S = -4.2 \text{ J mol}^{-1} \text{ deg}^{-1} \tag{9.10b}$$

The bonds formed in each reaction are Cu–N so should be similar in bond strength or enthalpy. However, reaction 9.10a has three product molecules versus two reactant molecules whereas reaction 9.10b has the same number of reactant and product molecules. The increase in product molecules causes an increase in entropy, which increases the stability constant of the complex and lowers the ΔG of the complex as in Equation 9.8.

The chelate effect occurs regardless of the monodentate ligand initially bound to the transition metal as in Equation 9.11 for the substitution of two ammonia molecules by ethylenediamine (25 °C; $\mu = 1.0$ M). This reaction is formally a ligand exchange reaction (Equation 9.4) as a molecule other than water is replaced. The ΔH of the reaction is not large as the total number of Ni–N bonds is the same in the reactant and the product, but the ΔS value is quite positive contributing significantly to the $\log K_1$ value.

$$[Ni(NH_3)_2(OH_2)_4]^{2+} + en \rightarrow [Ni(en)(OH_2)_4]^{2+} + 2NH_3$$

$$\log K_1 = 2.27 \qquad \Delta H = -5.0 \text{ kJ mol}^{-1} \qquad \Delta S = +29 \text{ J mol}^{-1} \text{ deg}^{-1} \tag{9.11}$$

The formation of five-membered rings counting the atoms in the chelate and the metal are also more stable than six-membered rings as in the case of Equations 9.12a,b with en and tn(1,3-diaminopropane). Here, the ΔS value does not change but the ΔH value decreases with a consequent decrease in $\log K$ (25 °C; $\mu = 0.10$ M).

$$[Cu(OH_2)_6]^{2+} + en \rightarrow [Cu(en)(OH_2)_4]^{2+} + 2H_2O$$

$$\log K_1 = 10.48 \quad \Delta H = -52.7 \text{ kJ mol}^{-1} \qquad \Delta S = +25 \text{ J mol}^{-1} \text{ deg}^{-1} \tag{9.12a}$$

$$[Cu(OH_2)_6]^{2+} + tn \rightarrow [Cu(tn)(OH_2)_4]^{2+} + 2H_2O$$

$$\log K_1 = 9.68 \qquad \Delta H = -47.7 \text{ kJ mol}^{-1} \quad \Delta S = +25 \text{ J mol}^{-1} \text{ deg}^{-1} \tag{9.12b}$$

In the first transition series, the ligands attached to the metal dictate whether the transition metal exhibits a high or low spin d electron configuration. In the examples above, all metals exhibit a high spin electron configuration. Bipyridyl is an example of a bidentate ligand that can induce a low spin electron configuration. When replacing the first two water molecules in $Fe(H_2O)_6^{2+}$ to form $[Fe(bipy)(H_2O)_4]^{2+}$, $\log K_1 = 4.2$. When replacing the third and fourth water molecules to form $[Fe(bipy)_2(H_2O)_2]^{2+}$, $\log K_2 = 3.7$. However, when replacing the last two water molecules to form $[Fe(bipy)_3]^{2+}$, $\log K_3 = 9.3$. The total number of ligating atoms bound to Fe^{2+} is still six so an increase in stability is not due to a change in coordination number as in the cadmium bromide example above. The increase in stability for the tris complex, $[Fe(bipy)_3]^{2+}$, is due to Fe^{2+} now being a low spin (t_{2g}^6) ion with the d electrons closer to the nucleus whereas the bis complex, $[Fe(bipy)_2(H_2O)_2]^{2+}$, has Fe^{2+} in a high spin state $t_{2g}^4 e_g^2$. The two bound water molecules in the bis complex induce the weak field complex. The stability of the low spin complex is related to the molecular orbital stabilization energy (MOSE) as all d electrons are in nonbonding t_{2g} orbitals, whereas for the high spin complex, two electrons are in the antibonding e_g^* orbitals.

The stability of a metal ion with a multidentate chelate has been used in medicine to detoxify lead that has been ingested when children are inadvertently exposed to Pb. A solution of EDTA is administered intravenously (intramuscular administration can also be done), and the EDTA binds the lead ion to form a Pb–EDTA complex, which is then excreted in the urine. A solution of dimercaptosuccinic acid can be drunk with the same result. These are examples of **chelation therapy**. Many bacteria make hexadentate ligands called *siderophores*, which are excellent at binding Fe(III) [and other metals] to keep it in solution and prevent its precipitation as iron(III)oxyhydroxides (Table 10.1 and Section 13.1).

Substitution reactions occur not only in solution but also at the surface of solids and membranes found in single cell organisms such as phytoplankton (Section 13.2.2). Here, a soluble chelated metal must undergo *dechelation* at the surface so that the surface ligating atoms can bind the metal. The transport of a metal from the surface of the membrane into the inner portion of the cell is governed by the thermodynamics and kinetics of the substitution processes. In the next section, the kinetics of substitution processes is discussed.

9.2   Kinetics of Ligand Substitution Reactions

The thermodynamic stability constants indicate how stable a complex is but do not indicate anything about how fast or slowly the complex may form. The kinetics of complex formation can vary substantially; as an example, the rate of water exchange for a metal in different oxidation states (Cr^{2+} vs Cr^{3+}) can vary by 15 orders of magnitude (Table 9.1). The terms **inert** and **labile** are used to indicate how slowly and fast ligand substitution can occur, respectively. Taube [10] suggested that complexes that react to completion in about 1 min (half-life, $t_{1/2}$, less than 30 s) are substitution labile, and those with longer half-lives are substitution inert. Inert complexes are considered to have kinetic stability, which is unrelated to thermodynamic stability described above. Complexes that are extremely labile require the use of **stopped flow**, P-jump, or T-jump experimental techniques.

The simplest ligand substitution reaction is the replacement of one water molecule by another for fully hydrated metal ions (the **"free" ion**). The exchange rate of one water molecule by another is governed by the nature of the coordinated metal, its charge, size and CFSE or LFSE (MOSE) energy. These rates have been determined using a variety of physical instrumentation, in particular by nuclear magnetic resonance spectroscopy. Equation 9.13 is a formal way of describing the water exchange reaction where * indicates a bulk water molecule that exchanges with one water molecule bound to the metal. In NMR experiments, ^{17}O labeled water is the solvent. Table 9.1 gives the water exchange rate constants for several metal +1, +2, +3, and +4 ions as well as some monohydroxo complexes of +3 metal ions (Equation 9.14).

$$M(H_2O)_6 + H_2O^* \leftrightarrow M(H_2O)_5(H_2O^*) + H_2O \tag{9.13}$$

$$[M(H_2O)_5(OH)]^{2+} + H_2O^* \leftrightarrow [M(H_2O)_4(H_2O^*)(OH)]^{2+} + H_2O \tag{9.14}$$

Table 9.1   Water exchange rate constants (k) for several metal ions with their half-lives and d orbital splitting primarily in octahedral field. Bolded ions show the range of values

ION (+1)	k (s^{-1})	$t_{1/2}$ (s)	ION (+2)	k (s^{-1})	$t_{1/2}$ (s)	
$Li(H_2O)_4^+$	1×10^8	6.9×10^{-9}	$Be(H_2O)_4^{2+}$	7.3×10^2	3.30×10^{-4}	
$Na(OH_2)_6^{2+}$	5×10^8	1.4×10^{-9}	$Mg(OH_2)_6^{2+}$	6.7×10^5	1.03×10^{-6}	
$K(OH_2)_6^{2+}$	1×10^9	6.9×10^{-10}	$Ca(OH_2)_6^{2+}$	3.2×10^8	2.16×10^{-9}	
$Rb(OH_2)_6^{2+}$	2×10^9	3.4×10^{-10}	$Sr(OH_2)_8^{2+}$	3.5×10^8	1.98×10^{-9}	
$Cs(OH_2)_6^{2+}$	5×10^9	1.4×10^{-10}	$Ba(OH_2)_8^{2+}$	7.2×10^8	9.6×10^{-10}	

ION (+3)	k (s^{-1})	$t_{1/2}$ (s)		ION (+2)	k (s^{-1})	$t_{1/2}$ (s)	
$Ti(OH_2)_6^{3+}$	1.8×10^5	3.85×10^{-6}	t_{2g}^1	$V(OH_2)_6^{2+}$	8.7×10^1	7.97×10^{-3}	t_{2g}^3
$V(OH_2)_6^{3+}$	5.0×10^2	1.39×10^{-3}	t_{2g}^2	$\mathbf{Cr(OH_2)_6^{2+}}$	1×10^9	6.9×10^{-10}	$t_{2g}^3 e_g^1$
$Cr(OH_2)_6^{3+}$	2.4×10^{-6}	2.89×10^5	t_{2g}^3	$Mn(OH_2)_6^{2+}$	2.1×10^7	3.30×10^{-8}	$t_{2g}^3 e_g^2$
$Fe(OH_2)_6^{3+}$	1.6×10^2	4.33×10^{-3}	$t_{2g}^3 e_g^2$	$Fe(OH_2)_6^{2+}$	4.4×10^6	1.58×10^{-7}	$t_{2g}^4 e_g^2$
$Ru(OH_2)_6^{3+}$	3.5×10^{-6}	1.98×10^5	t_{2g}^5	$Co(OH_2)_6^{2+}$	3.2×10^6	2.17×10^{-7}	$t_{2g}^5 e_g^2$
$Rh(OH_2)_6^{3+}$	2.2×10^{-9}	3.15×10^8	t_{2g}^6	$Ni(OH_2)_6^{2+}$	3.2×10^4	2.17×10^{-5}	$t_{2g}^6 e_g^2$
$\mathbf{Ir(OH_2)_6^{3+}}$	1.1×10^{-10}	6.30×10^9	t_{2g}^6	$Cu(OH_2)_6^{2+}$	4.4×10^9	1.58×10^{-10}	$t_{2g}^6 e_g^3$
$Al(OH_2)_6^{3+}$	1.29	5.37×10^{-1}		$Zn(OH_2)_6^{2+}$	7×10^7	9.90×10^{-9}	$t_{2g}^6 e_g^4$
$Ga(OH_2)_6^{3+}$	4.0×10^2	1.73×10^{-3}		$Cd(OH_2)_6^{3+}$	3×10^8	2.31×10^{-9}	$t_{2g}^6 e_g^4$
$In(OH_2)_6^{3+}$	4×10^4	1.73×10^{-5}		$Hg(OH_2)_4^{2+}$	2×10^9	3.47×10^{-10}	$t_{2g}^6 e_g^4$
$La(OH_2)_8^{3+}$	1×10^8	6.93×10^{-9}		$Ru(OH_2)_6^{2+}$	1.8×10^{-2}	3.85×10^1	t_{2g}^6
$\mathbf{Eu(OH_2)_7^{3+}}$	5.00×10^9	1.39×10^{-10}		$Pd(OH_2)_4^{2+}$	5.6×10^2	1.24×10^{-3}	$t_{2g}^6 e_g^2$
$Gd(OH_2)_8^{3+}$	8.30×10^8	8.35×10^{-10}		$Pt(OH_2)_4^{2+}$	3.9×10^{-4}	1.78×10^3	$t_{2g}^6 e_g^2$
$Tb(OH_2)_8^{3+}$	5.58×10^8	1.24×10^{-9}		$\mathbf{Pb(OH_2)_6^{2+}}$	7×10^9	9.90×10^{-11}	
$Dy(OH_2)_8^{3+}$	4.34×10^8	1.6×10^{-9}					
$Ho(OH_2)_8^{3+}$	2.14×10^8	3.24×10^{-9}		ION (+4)	k (s^{-1})	$t_{1/2}$ (s)	
$Er(OH_2)_8^{3+}$	1.33×10^8	5.21×10^{-9}		$Th(OH_2)_{10}^{4+}$	5.00×10^7	1.39×10^{-8}	
$Tm(OH_2)_8^{3+}$	9.10×10^7	7.62×10^{-9}		$U(OH_2)_{10}^{4+}$	5.00×10^6	1.39×10^{-7}	
$Yb(OH_2)_8^{3+}$	4.70×10^7	1.47×10^{-8}					

$\mathbf{M(OH)^{2+}}$	k (s^{-1})	$t_{1/2}$ (s)	
$Cr(OH_2)_5(OH)^{2+}$	1.8×10^{-4}	3.85×10^3	t_{2g}^3
$Fe(OH_2)_5(OH)^{2+}$	1.2×10^5	5.78×10^{-6}	$t_{2g}^3 e_g^2$
$Ru(OH_2)_5(OH)^{2+}$	5.9×10^{-4}	1.17×10^3	t_{2g}^6
$Rh(OH_2)_5(OH)^{2+}$	4.2×10^{-5}	1.65×10^4	t_{2g}^6
$Ir(OH_2)_5(OH)^{2+}$	5.6×10^{-7}	1.24×10^6	t_{2g}^6
$Al(OH_2)_5(OH)^{2+}$	3.10×10^4	2.23×10^{-5}	t_{2g}^0
$Ga(OH_2)_5(OH)^{2+}$	4.00×10^2	1.73×10^{-3}	t_{2g}^0

Source: Data compiled from several sources [4–9].

The rate of water exchange is considered first order in the metal–aqua complex as water is the solvent and no other ligands are present. Table 9.1 shows that the rate of water exchange for these metal complexes varies over a range of 20 orders of magnitude (from $Ir(OH_2)_6^{3+}$ to $Pb(OH_2)_6^{2+}$) as do their first order half-lives. Values larger than 10^8 s^{-1} indicate a diffusion-controlled process. Values larger than $10^{1.5}$ s^{-1} indicate a labile complex whereas those less than that indicate an inert complex as suggested by Taube [10]. Using water exchange rates, Langford and Gray [11, 12] divided metal ions into four broad classes as follows.

9.2.1 Kinetics of Water Exchange for Aqua Complexes

CLASS I: $k \geq 10^8$ s^{-1} Diffusion-controlled ions are very fast and LABILE.
Ions in this class are the alkali and alkaline earths (except for Be^{2+} and Mg^{2+}), Cd^{2+}, Hg^{2+}, Cr^{2+}, Mn^{3+}, Cu^{2+}, and many of the lanthanide +3 ions. Note that these ions have noble gas or pseudo-noble gas configurations or are Jahn–Teller distorted ions [(Cr^{2+}, Mn^{3+} both d^4) and Cu^{2+}, d^9]. The lanthanides have higher coordination numbers so that water loss is statistically more possible. The ratio of the charge squared to the radius is small ($Z^2/r \leq 10$ C^2 m^{-1}).
CLASS II: $k \sim 10^4$–10^8 s^{-1} water exchange is fast and the ions are LABILE.
In this class are all divalent first row transition metal ions (except for V^{2+}, Cr^{2+}, Cu^{2+}), Mg^{2+}, and several trivalent lanthanide ions (M^{3+}). Z^2/r ranges from 10 to 30 C^2 m^{-1}.
CLASS III: $k \sim 1$–10^4 s^{-1}
This class contains Be^{2+}, V^{2+}, Al^{3+}, Ga^{3+}, and several trivalent first row transition metal ions. $Z^2/r >$ 30 C^2 m^{-1}. Ions (e.g., V^{2+}) may exhibit some LFSE.
CLASS IV: $k \sim 10^{-9}$–10^{-1} s^{-1}
Water exchange is INERT as originally defined by Taube.
This class is dominated by octahedral ions that exhibit very high LFSE such as Cr^{3+}, Rh^{3+}, Ir^{3+}, and Mn^{4+} along with the low spin cases of Co^{3+} and Fe^{2+}. Note that the octahedral d^3(t_{2g}^3) and d^6(t_{2g}^6) as well as the square planar d^8 configurations of Ni^{2+}, Co^+, and Pt^{2+} have very high LFSE. In tetrahedral geometry, first row transition elements are high spin and labile whereas in square planar geometry, all transition elements are low spin and inert.
When LFSE is not a factor, there are some straightforward comparisons based on size and charge of the free ions.

1. An increase in the rate constant corresponds with an increase in the radius of the metal; for example,

$$Ca(OH_2)_6^{2+} > Mg(OH_2)_6^{2+} > Be(OH_2)_6^{2+}$$

2. An increase in the rate constant corresponds with a decrease in ionic charge; for example,

$$Na(OH_2)_6^+ > Mg(OH_2)_6^{2+} > Al(OH_2)_6^{3+}$$

9.2.2 Intimate Mechanisms for Ligand Substitution Reactions

The term intimate describes the details of the ***activation and the energetics of formation of the activated complex*** in the rate-determining step. For octahedral complexes, Langford and Gray [11] described two limiting mechanistic cases, dissociative and associative mechanisms.

A **dissociative** (D) mechanism results in the loss of a ligand (**the leaving group**) from an octahedral complex to form a five-coordinate intermediate [ML_5] in Equation 9.15 and Figure 9.1a, and is referred to as an

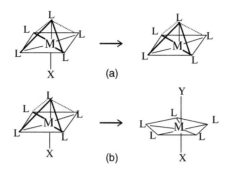

Figure 9.1 *Schematic diagrams representing (a) a dissociative mechanism with formation of a five-coordinate square pyramid intermediate and (b) an associative mechanism with formation of a seven-coordinate pentagonal bipyramid intermediate*

S_N1 limiting reaction. Bond breaking of the M–X bond is the rate-controlling factor for the activation energy; thus, *the reaction rate is more sensitive to variation of the leaving group*. The **entering group** is Y.

$$ML_5X + Y \rightarrow [ML_5] + X + Y \rightarrow ML_5Y + X \tag{9.15}$$

The rate equation for formation of the product, ML_5Y, is given in Equation 9.16, and shows that the rate is only proportional to the concentration of the starting reactant, ML_5X. Thus, this is a first-order reaction.

$$\text{Rate} = \frac{d}{dt}[L_5MY] = k_{obs}[L_5MX] \tag{9.16}$$

An **associative** (A) mechanism results in the gain of a ligand for an octahedral complex to form the seven-coordinate intermediate $[ML_5XY]$ in Equation 9.17 and Figure 9.1b, and is referred to as an S_N2 limiting reaction. Bond making (M–Y bond) is the rate-controlling factor for the activation energy; thus, *the reaction rate is sensitive to variation of both the leaving and the entering groups*. In this process, as the M–Y bond begins to form and increase in bond strength, the M–X bond must gradually weaken and then be broken to form the product.

$$ML_5X + Y \rightarrow [ML_5 \ XY] \rightarrow ML_5Y + X \tag{9.17}$$

The rate equation for formation of the product, ML_5Y, is given in Equation 9.18, and shows that the rate is proportional to the concentration of each reactant. Thus, this is a second-order reaction.

$$\text{Rate} = \frac{d}{dt}[L_5MY] = k_{obs}[L_5MX][Y] \tag{9.18}$$

9.2.3   Kinetic Model and Activation Parameters

The description of dissociative and associative reaction mechanisms and the formation of their intermediates indicate that there is a significant change in potential energy from the initial energy states of the reactants to the formation of the intermediate. In **transition state theory**, the energy maximum required to form the activated complex is the **activation energy**, E_a, as shown in Figure 9.2a. This reaction progress along the **reaction coordinate** (or pathway) requires collisions of the reactants (left of the maximum in Figure 9.2a) with enough energy to form the activated complex in the transition state (maximum in the curve for Figure 9.2a). During this

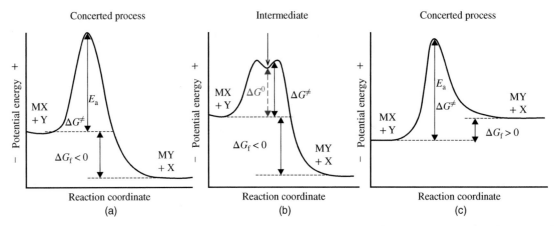

Figure 9.2   *Representative reaction coordinate diagrams showing concerted processes (a, c) that do not result in an intermediate and (b) showing the formation of an intermediate*

process, the initial reactants undergo bond breaking and making, which require bond stretching and angular distortions to form the activated complex. A large E_a decreases the reaction rate so that the reaction progress will be slow. Then, the activated complex undergoes another set of internal atomic movements to form (or relax to) the products (right of Figure 9.2a). Because the products are of lower energy than the reactants, the Gibbs free energy of formation or reaction, ΔG_f or ΔG_r, is negative and the equilibrium constant, K_f, is positive indicating that the reaction is favorable. The smooth curve in Figure 9.2a indicates a concerted process such that an intermediate does not form or has a very small but constant concentration. Although $\Delta G_f < 0$, the intermediate could decompose back to the reactants; if the reaction would proceed backward from the intermediate (or the products) to reactants, the same reaction coordinate is followed. This is known as **the principle of microscopic reversibility**, and requires that when the reaction is at equilibrium, the forward and reverse reactions proceed at equal rates.

Figure 9.2b shows a double maximum for the curve, and the energy valley or potential well indicates that an intermediate forms over the reaction coordinate. If the potential well is very deep, then the intermediate may be (meta)stable and thus be detected. Again, this reaction shows a negative value for ΔG_f and is a favorable reaction. ΔG^0 is the free energy of formation for the intermediate from the reactants.

Figure 9.2c shows another concerted process without formation of an intermediate. However, the products have a higher potential energy than the reactants so formation of products is not favored; that is, $\Delta G_f > 0$.

As temperature and pressure increase, reactions generally proceed faster, and it is possible to determine the activation energies associated with the reaction. These are important in determining whether a reaction undergoes dissociative or associative mechanisms.

9.2.3.1   *Arrhenius Equation*

There are two approaches to determining the energetics along the reaction coordinate. The first of these is the Arrhenius equation 9.19, which relates the reaction rate constant to the $E_a^{\ddagger}$, the activation energy. Here, k is an abbreviation for k_{obs}, which is used to describe the rate constant in Equations 9.15 and 9.18.

$$k = A\,e^{\frac{-E_a^{\ddagger}}{RT}} \tag{9.19}$$

Arrhenius behavior indicates that as the temperature increases, the number of collisions, A, increases, and the reaction rate increases. The approach to determining the E_a is to determine the rate constant at three or more temperatures, and then plot the $\ln k$ versus $1/T$.

$$\ln k = \frac{-E_a^{\ddagger}}{RT} + \ln A$$

The slope of this plot is $E_a^{\ddagger}/R$ where $R = 8.314$ J (mol K)$^{-1}$. If $E_a < 42$ kJ mol^{-1}, then the reaction is a diffusion-controlled process. If $E_a > 42$ kJ mol^{-1}, then the reaction is a chemically controlled reaction process.

9.2.3.2 Eyring Equation

In the Eyring equation 9.20, the reaction rate constant is related to the Gibbs free energy of activation, $\Delta G^{\neq}$, which is graphically portrayed in Figure 9.2a–c. The ΔG^0 is the Gibbs free energy for the activated complex.

$$k = \frac{k'T}{h}e^{\frac{-\Delta G^{\neq}}{RT}} = \frac{k'T}{h}e^{\frac{-\Delta H^{\ddagger}}{RT}}e^{\frac{\Delta S^{\ddagger}}{R}} \tag{9.20}$$

where k' is the Boltzmann constant and $\ddagger$ indicates activation. Rearranging and taking the logarithm gives the following equation.

$$\ln\frac{k}{T} = \frac{-\Delta H^{\ddagger}}{RT} + \ln\frac{k'}{h} + \frac{\Delta S^{\ddagger}}{R}$$

A plot of $[\ln(k/T)]$ versus $1/T$ yields a slope of $-\Delta H^{\ddagger}/R$ and an intercept of $[\ln(k'/h) + \Delta S^{\ddagger}/R]$.

If $\Delta S^{\ddagger} < -10$ entropy units (e.u., J mol^{-1} °K^{-1}), an associative mechanism occurs as a metal complex undergoes bond making in the transition state. If $\Delta S^{\ddagger} > +10$ entropy units (e.u. or J mol^{-1} °K^{-1}), a dissociative mechanism occurs as the metal complex undergoes bond breaking. If the value is in between these two limits, then it is not possible to indicate a particular mechanism.

Note that $\ln k$ is proportional to $\Delta G^{\neq}$, and ΔG_f may be proportional to $\Delta G^{\neq}$. Thus $\ln k$ can be related to ΔG_r, or the equilibrium constant, K, or the standard reduction potential ($E°$) for a series of metal complexes (same metal reacting with different ligands). The following linear free energy relationship (LFER) then results.

$$\ln k = n \ln K + C \tag{9.21}$$

Swaddle [13] showed that the coefficient n gives information on dissociative versus associative mechanisms in octahedral substitution for a series of reactions with the same Y. If $n = 1.0$, a dissociative mechanism is likely because there is little difference in the transition states for the different complexes studied; that is, the kinetic parameter (k or $\Delta G^{\neq}$) varies the same as the thermodynamic parameter (K or $\Delta G_{reaction}$). If $n = 0.5$, an associative mechanism occurs for a series of complexes.

9.2.3.3 Pressure-Dependent Reactions

In addition to the rate depending upon temperature, the reaction rate may depend upon variation in pressure (e.g., every 10 m of water depth is another atmosphere of overpressure so the reaction rate can increase with ocean depth). The **volume of activation**, $\Delta V^{\ddagger}$, is complex because it is a component of volume changes in the reactants and volume changes in the surrounding solvent, which is known as solvent electrostriction ($\Delta V^{\ddagger}_{solvent}$).

$$\Delta V^{\ddagger} = \text{volume of activation} = \Delta V^{\ddagger}_{intrinsic} + \Delta V^{\ddagger}_{solvent}$$

$\Delta V^{\ddagger}(P)$ and k are related to the rate constant via the ideal gas law (Equation 9.22).

$$k = e^{\frac{-P\Delta V^{\ddagger}}{RT}} \tag{9.22}$$

A plot of $\ln k$ versus P gives a slope of $-\Delta V^{\ddagger}/(RT)$. Similar to ΔS, a negative value indicates an associative mechanism, and a positive value indicates a dissociative mechanism unless solvent effects are important.

Electrostriction ($\Delta V^{\ddagger}_{solvent}$) deals with charges in the activated complex and how H_2O (or the solvent) affects that complex. Any process that creates charge (two charges of the same sign come together) leads to solvent electrostriction and gives a "negative" contribution to $\Delta V^{\ddagger}$ as H_2O around each ion counteracts repulsion. $\Delta V^{\ddagger}$ is often large and negative when anionic ligands leave cationic complexes in D mechanisms (solvent H_2O coordinates each ion). Neutralization of charge in the intermediate (two charges of opposite sign come together) gives a "positive" contribution to $\Delta V^{\ddagger}$ as H_2O leaves each ion. $\Delta V^{\ddagger}$ is large and positive when anionic ligands leave anionic complexes as H_2O is less tightly bound to ions of smaller charge.

9.2.4   Dissociative Versus Associative Preference for Octahedral Ligand Substitution Reactions

A possible way of predicting whether associative or dissociative mechanisms are preferred for octahedral metal complexes is to calculate the crystal field activation energy (CFAE) on going from an octahedral starting geometry to a square pyramid (C_{4v}) intermediate using CFSE for each geometry [14, 15]. The same calculation is done on going from an octahedral starting geometry to an intermediate with the pentagonal bipyramid (D_{5h}) geometry. Tables 9.2A and 9.2B give the CFAE values for both high spin and low spin cases on going from an octahedral geometry to square pyramid and pentagonal bipyramid geometries, respectively.

The CFAE low spin data assume no change in spin multiplicity on going from an electron configuration in octahedral geometry to one of the intermediate geometries. The CFAE data for low spin d^3, d^4, and d^8 electron configurations are more stable in square pyramid as the d^3 and d^4 electron configurations in pentagonal bipyramidal have electrons that go into orbitals of higher energy to maintain electron spin (see Figure 8.25).

Negative CFAE values in Tables 9.2A and 9.2B refer to loss of crystal field stabilization energy and the destabilization of the complex intermediate. When comparing CFAE for high spin complexes in square pyramid and pentagonal bipyramidal geometries, dissociation is favored or preferred over association for d^3 to d^5 and d^8–d^9 electron configurations, but not for d^1, d^2, and d^7 electron configurations. When comparing

Table 9.2A   *CFAE data for a square pyramid (S_N1) and pentagonal bipyramidal (S_N2)* **high spin** *intermediates. Note that positive energies are considered to be stable*

d^n	O_h CFSE high spin	C_{4v} CFSE high spin C.N. = 5	CFAE high spin C.N. = 5	O_h CFSE high spin	D_{5h} CFSE high spin C.N. = 7	CFAE high spin C.N. = 7
d^0	0	0	0	0	0	0
d^1	4	4.57	0.57	4	5.28	1.28
d^2	8	9.14	1.1	8	10.56	2.56
d^3	12	10.0	−2.0	12	7.74	−4.26
d^4	6	9.14	3.14	6	4.92	−1.08
d^5	0	0	0	0	0	0
d^6	4	4.57	0.57	4	5.28	1.28
d^7	8	9.14	1.14	8	10.56	2.56
d^8	12	10	−2.0	12	7.74	−4.26
d^9	6	9.14	3.14	6	4.92	−1.08
d^{10}	0	0	0	0	0	0

Table 9.2B   *CFAE data for a square pyramid (S_N1) and pentagonal bipyramidal (S_N2)* **low spin** *intermediates. Note that positive energies are considered to be stable*

d^n	O_h CFSE low spin	C_{4v} CFSE low spin C.N. = 5	CFAE low spin C.N. = 5	O_h CFSE low spin	D_{5h} CFSE low spin C.N. = 7	CFAE low spin C.N. = 7
d^0	0	0	0	0	0	
d^1	4	4.57	0.57	4	5.28	1.28
d^2	8	9.14	1.14	8	10.56	2.56
d^3	12	10.00	**−2.00**	12	7.74	**−4.26**
d^4	16	14.57	**−1.43**	16	13.02	**−2.98**
d^5	20	19.14	−0.86	20	18.3	−1.70
d^6	24	20	**−4.00**	24	15.48	−8.52
d^7	18	19.14	1.14	18	12.66	−5.34
d^8	12	10.00	**−2.00**	12	9.84	−2.16
d^9	6	9.14	3.14	6	4.92	−1.08
d^{10}	0	0	0	0	0	0

square pyramid and pentagonal bipyramidal geometries in low spin complexes, dissociation is favored for d^3–d^9 electron configurations, but not for d^1 and d^2. Thus, a dissociative mechanism is preferred more often than an associative mechanism. However, octahedral complexes, which have the inert electron configurations [d^3, d^6 (low spin), d^8], have negative CFAE values so ligand substitution reactions in octahedral geometry are considered "inert" to both types of mechanisms.

Figure 9.3 shows plots of the water exchange rate constant, CFAE, and MOSE versus high spin d electron configuration for the [M(H$_2$O)$_6$]$^{2+}$ and [M(H$_2$O)$_6$]$^{3+}$ free ions. Positive values for CFAE and MOSE indicate a stable energy configuration. The water exchange rate shows a significant decrease for the t_{2g}^3 and the $t_{2g}^6 e_g^2$ electron configurations. For these electron configurations, these decreases correspond to the increase in MOSE (CFSE) and the decrease in CFAE when octahedral metal complexes undergo dissociation to five-coordinate square pyramid geometry. These data indicate that a dissociative mechanism is viable for the water exchange reaction of many metal ions.

9.2.5   Stoichiometric Mechanisms

Langford and Gray [11] described three classes for stoichiometric mechanisms: dissociative (**D**), associative (**A**) and interchange (**I**), which can be either I_d (for dissociative activation) or I_a (for associative activation). Representative changes in potential energy versus reaction coordinate profiles are shown in Figure 9.4. In determining stoichiometric mechanisms, it is necessary to determine the sequence of elementary steps leading from reactants to products. Mathematically, the observed rate expressions in Equations 9.15 and 9.18 are related to the elementary steps so that k_{obs} is found to be a function of other kinetic and/or thermodynamic constants related to these elementary steps as well as the concentrations of L$_5$MX, X, or Y. In each of these three cases, the rate law is written as a function of the rate limiting or determining step. Because the lifetime of the intermediate is considered negligible, its concentration is assumed to be zero; thus, the **steady state approximation** can be used to solve for the rate law using the mathematical expressions for the elementary steps. Figure 9.4 shows representative reaction coordinate diagrams for these three main classes.

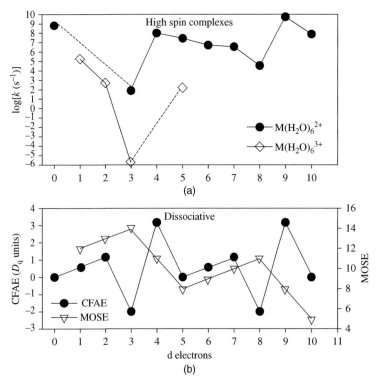

Figure 9.3   *Figure (a) shows a variation of the water exchange rate constant with d electron configuration. Figure (b) shows the variation of MOSE and CFAE with high spin electron configuration. The MOSE data are plotted with a slight slope (see discussion of Figure 8.35)*

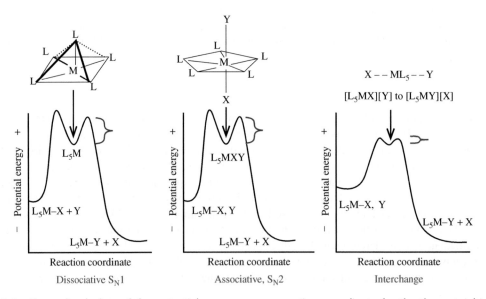

Figure 9.4   *Generalized plots of the potential energy versus reaction coordinate for the three stoichiometric mechanisms of octahedral complexes. The dashed lines for interchange indicate the formation of an ion pair* [L_5MX][Y] *that interchanges to* [L_5MY][X] *prior to forming products*

9.2.5.1 Dissociative Mechanism (S_N1 Limiting)

A dissociative mechanism must follow dissociative activation and energetics. The mathematical treatment for the sequence of elementary steps leading from reactants to products for S_N1 dissociative reactions starts with the equilibrium and rate-limiting step that involves breaking the M–L bond to form the intermediate, L_5M, as in Equation 9.23. This represents the dissociation of the reactant. The intermediate also has a second coordination sphere or an outer sphere of loosely held water (solvent) molecules, entering group (Y), and other spectator ions around it.

$$L_5MX \underset{k_{-1}}{\overset{k_1}{\rightleftharpoons}} L_5M + X \tag{9.23}$$

where

$$K = \frac{k_1}{k_{-1}}$$

The entering group, Y, then bonds with the five-coordinate intermediate, L_5M, to form the product L_5MY (Equation 9.24).

$$L_5M + Y \xrightarrow{k_2} L_5MY \tag{9.24}$$

The rate law for the formation of L_5MY (from Equation 9.24) is Equation 9.25.

$$\frac{d}{dt}[L_5MY] = k_2[L_5M][Y] \tag{9.25}$$

Applying the **steady-state approximation** to these equations for the intermediate L_5M gives Equation 9.26, which indicates that there is only one formation term for L_5M with the two loss terms.

$$\frac{d}{dt}[L_5M] = 0 = k_1[L_5MX] - k_{-1}[L_5M][X] - k_2[L_5M][Y] \tag{9.26}$$

Rearranging Equation 9.26 to solve for L_5M gives Equation 9.27.

$$[L_5M] = \frac{k_1[L_5MX]}{k_{-1}[X] + k_2[Y]} \tag{9.27}$$

Substituting Equation 9.27 into Equation 9.25 gives the complete rate law (Equation 9.28) for the formation of the product L_5MY, which indicates second-order kinetics under certain conditions.

$$\frac{d}{dt}[L_5MY] = \frac{k_1k_2[L_5MX][Y]}{k_{-1}[X] + k_2[Y]} = k_{obs}[L_5MX] \tag{9.28}$$

Thus, the rate is a function of k_{obs} and $[L_5MX]$, which is measurable with time. The intermediate, $[L_5M]$, needs to have a long lifetime for detection and "proof" of the dissociative mechanism (**D or S_N1 limiting**). The full form of k_{obs} is Equation 9.29.

$$k_{obs} = \frac{k_1k_2[Y]}{k_{-1}[X] + k_2[Y]} \tag{9.29}$$

If $k_2[Y]$ is very large (or $k_{-1}[X]$ is very small, the intermediate does not react readily with X to reform the reactant), then Equation 9.29 reduces to $k_{obs} = k_1$, and the rate depends on $[L_5MX]$ and M–X bond breaking or dissociation.

9.2.5.2   *Interchange Mechanism (I_a, I_d)*

The interchange mechanism is an example of a concerted process, and it is not possible to trap the intermediate. In an I_d or dissociative interchange process, any bonding of the metal to the entering and leaving groups is weak so it resembles a dissociation process. In an I_a or associative interchange process, bonding of the metal to the entering and leaving groups is more substantial so it resembles an association process. The mathematical treatment for the sequence of elementary steps for an interchange reaction is also known as the ***Eigen–Wilkins*** mechanism [16]. In this mechanism the two reactants form an ion pair, $[L_5MX, Y]$, which is an outer sphere (OS) complex and can be represented by an equilibrium constant (Equation 9.30; K or K_{OS} which is sometimes referred to as a pre-equilibrium or pre-association constant). In the ion pair, the leaving group X is still bound to M, but the entering group Y is not.

$$L_5MX + Y \rightleftharpoons (L_5MX, Y) \text{ and } K \text{ or } K_{OS} = \frac{[L_5MX, Y]}{[Y][L_5MX]} \tag{9.30}$$

The rate determining step is the interchange between X in the inner coordination sphere of M and Y in the second or outer coordination sphere of M so that Y becomes bound to the inner coordination sphere of M (Equation 9.31), which has a rate constant, k, and a rate law as in Equation 9.32.

$$(L_5MX, Y) \xrightarrow{k} (L_5MY, X) \text{ rate limiting step} \tag{9.31}$$

$$\frac{d}{dt}[L_5MY] = k[(L_5MX, Y)] \tag{9.32}$$

Using the expression for K_{OS} and Equation 9.32 gives a rate law, Equation 9.33, for the interchange ligand substitution reaction.

$$\frac{d}{dt}[L_5MY] = k[(L_5MX, Y)] = kK_{OS}[L_5MX][Y] \tag{9.33}$$

Once the exchange is complete, the product, L_5MY, is formed and X can go into the bulk solution (Equation 9.34). The activated complex in this process has a very small potential energy well so a ***metastable intermediate cannot be detected*** as shown in Figure 9.4.

$$(L_5MY, X) \xrightarrow{\text{fast}} L_5MY + X \tag{9.34}$$

The initial concentration of the reacting complex, $[L_5MX]_0$, is the sum of the remaining $[L_5MX]$ and the ion pair, $[L_5MX, Y]$, as in Equation 9.35.

$$[L_5MX]_0 = [L_5MX] + [(L_5MX, Y)] \tag{9.35}$$

Using the equilibrium expression for the ion pair (Equation 9.30), Equation 9.35 becomes Equation 9.36.

$$[L_5MX]_0 = [L_5MX] + K_{OS}[L_5MX][Y] \tag{9.36}$$

Rearranging this expression to solve for $[L_5MX]$ gives Equation 9.37.

$$[L_5MX] = \frac{[L_5MX]_0}{1 + K_{OS}[Y]} \tag{9.37}$$

Substituting $[L_5MX]$ from Equation 9.37 into Equation 9.33 gives the full rate law (Equation 9.38) for the interchange ligand substitution reaction, which again shows second order kinetics.

$$\frac{d}{dt}[L_5MY] = k[(L_5MX, Y)] = kK_{OS}[L_5MX][Y] = \frac{kK_{OS}[L_5MX]_0[Y]}{1 + K_{OS}[Y]} \tag{9.38}$$

Thus the rate of formation of the product is a function of the initial concentration of $[L_5MX]$. There are two *limiting cases* for the interchange reaction. First, if $K_{OS}[Y] \gg 1$, then $k_{obs} = k$ where k is the rate-limiting step for the interchange of X and Y (Equation 9.31); the mechanism is considered I_d **or** S_N1. Second, if $K_{OS}[Y] \ll 1$, then $k_{obs} = kK_{OS}$ (see Equation 9.33), and the mechanism is considered I_a **or** S_N2. If the equilibrium constant cannot be determined experimentally, it can be calculated using the following equilibrium expression known as the **Fuoss–Eigen** equation 9.39.

$$K_{OS} = \frac{4}{3}\pi a^3 N_A \left(e^{-\frac{V}{kT}}\right) \tag{9.39}$$

where a is the distance of closest approach for the ion pair; N_A is the Avogadro constant; V is the coulombic potential energy (Equation 9.40) where z is the charge on the entering and leaving groups, e is the electrostatic charge, and ε is the dielectric constant of the solvent (78 for water).

$$V = \frac{z_X z_Y e^2}{4\pi\varepsilon a} \tag{9.40}$$

If one of the reactants is neutral or uncharged as in the case of ammonia and the amines, $V = 0$ and the exponential term for K_{OS} vanishes.

9.2.5.3 Associative Mechanism (S_N2 Limiting)

An associative mechanism must follow associative activation and energetics. The mathematical treatment for the sequence of elementary steps leading from reactants to products for S_N2 associative reactions starts with the equation for the formation of the seven coordinate intermediate, ML_5XY, which leads to the equilibrium expression in Equation 9.41.

$$ML_5X + Y \underset{k_{-1}}{\overset{k_1}{\rightleftharpoons}} ML_5XY \tag{9.41}$$

Equation 9.42 leading to the formation of products from the seven-coordinate intermediate, ML_5XY, is considered *irreversible* and leads to the rate expression in Equation 9.43.

$$ML_5XY \xrightarrow{k_2} ML_5Y + X \quad \text{rate limiting step} \tag{9.42}$$

$$\frac{d[L_5MY]}{dt} = k_2[ML_5XY] \tag{9.43}$$

Applying the **steady-state approximation** to the formation of the seven-coordinate intermediate gives the rate expression 9.44.

$$\text{rate} = \frac{d[L_5MXY]}{dt} = 0 = -k_2[L_5MXY] + k_1[L_5MX][Y] - k_{-1}[L_5MXY] \tag{9.44}$$

Solving for $[ML_5XY]$ gives Equation 9.45.

$$[L_5MXY] = \frac{k_1}{k_2 + k_{-1}}[L_5MX][Y] \tag{9.45}$$

The rate of formation of the product, ML_5Y, can be written from combining Equations 9.43 and 9.45 by substituting for $[L_5MXY]$ into Equation 9.43. The result is the complete associative rate expression in Equation 9.46.

$$\frac{d[L_5MY]}{dt} = k_2[ML_5XY] = \frac{k_2k_1}{k_2 + k_{-1}}[L_5MX][Y] = k_{obs}[L_5MX][Y] \tag{9.46}$$

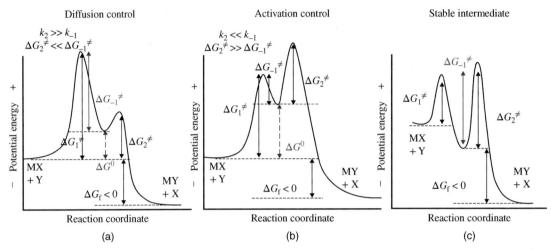

Figure 9.5 *Schematic reaction coordinate diagrams for (a) diffusion and (b) activation controlled limits for associative reactions (S_N2). The diagrams are also useful for indicating S_N1 reactions when the three kinetic constants can be evaluated (see also Equation 9.62 and Table 9.11). (c) Schematic shows a stable energy well for an intermediate*

The associative mechanism has two limiting cases (Figure 9.5a and b) for the determination of k_{obs}, which are calculated from the kinetic constants in Equation 9.46. The ***diffusion-controlled limit*** ($k_2 \gg k_{-1}$) indicates that the formation of the intermediate ML_5XY (and the products) is more favored than its breakup to reform the reactants. Thus, k_{-1} vanishes in the denominator and k_{obs} depends only on the rate of bond formation leading to the seven-coordinate intermediate (Equation 9.47).

$$\frac{k_2 k_1}{k_2} = k_{obs} = k_1 \tag{9.47}$$

The ***activation-controlled reaction*** ($k_2 \ll k_{-1}$) leads to a linear free energy relationship relating k_{obs} to the equilibrium formation of the seven-coordinate intermediate (Equation 9.48). Here, bond rupture of the intermediate is the rate-determining step.

$$\frac{k_2 k_1}{k_{-1}} = k_{obs} = k_2 K \tag{9.48}$$

If a ***stable intermediate*** can be formed and detected, the energy of the intermediate will be between the energies of the reactants and the products as in Figure 9.5c. In this case, k_2 and k_{-1} are slow and the intermediate has significant activation energies to overcome and form products or reform reactants.

9.2.6   Tests for Reaction Mechanisms

To determine the possible ***intimate reaction mechanism*** (dissociative or associative), it is useful to compare the rates of reactions for a series of complexes of the same metal cation where (1) the entering group is the same but the leaving group varies and (2) the entering group varies and the leaving group is the same. The limiting rate constant is calculated from the stoichiometric mechanisms above. Our understanding of mechanisms for reactions with water as solvent comes mainly from the published data on Co(III), Cr(III), Fe(III), Ni(II), and Pt(II) complexes in the period 1950–1970 because their substitution reactions are sufficiently slow or "inert" to follow reaction progress.

9.3   Substitution in Octahedral Complexes

9.3.1   Examples of Dissociative Activated Mechanisms

9.3.1.1   Aquation or Acid Hydrolysis Reaction (Leaving Group Varies)

The kinetics of the aquation reaction 9.49 for $Co(NH_3)_5X$ complexes shows that the rate constant varies five orders of magnitude as the bound ligand, X, varies (Table 9.3). This large change in k indicates that the leaving group bond strength (Co–X) is significant so the reaction undergoes dissociative or d activation. Using Equation 9.21, the $\log k$ versus $-\log K$ of formation (K_{diss}) of the $Co(NH_3)_5X$ leads to a linear free energy relationship with a value of $n = 1.00 \pm 0.03$ indicating d activation [11] for $[Co(NH_3)_5X]^{2+}$ complexes with a charge of -1 for X.

$$Co(NH_3)_5X + H_2O \rightarrow Co(NH_3)_5(H_2O) + X \tag{9.49}$$

9.3.1.2   Anation Reaction (Entering Group Varies)

The anation reaction of $[Co(NH_3)_5(H_2O)]^{3+}$ with different entering groups, Y, (Equation 9.50) shows that the rate constant varies less than one order of magnitude as the entering ligand, Y, varies (Table 9.4). Here the small change in k indicates that the entering group is not significant so undergoes *dissociative activation*. Although the Co–Y bond strength varies, it has no influence on the rate. The key to substitution here is that

Table 9.3   Rate constants (s^{-1}) for aquation of $[Co(NH_3)_5X]^{n+}$ complexes at $25\,°C$

Complex	k	K_{eq} (M^{-1})
$[Co(NH_3)_5(NO_3)]^{2+}$	2.7×10^{-5}	0.077
$[Co(NH_3)_5(I)]^{2+}$	8.3×10^{-6}	0.21
$[Co(NH_3)_5(H_2O)]^{3+}$	5.8×10^{-6}	0.018
$[Co(NH_3)_5(Cl)]^{2+}$	1.8×10^{-6}	1.25
$[Co(NH_3)_5(Br)]^{2+}$	6.3×10^{-6}	0.37
$[Co(NH_3)_5(H_2PO_4)]^{2+}$	2.6×10^{-7}	7.4
$[Co(NH_3)_5(SO_4)]^{+}$	8.9×10^{-7}	12.4
$[Co(NH_3)_5(F)]^{2+}$	8.6×10^{-8}	25
$[Co(NH_3)_5(N_3)]^{2+}$	2.1×10^{-9}	
$[Co(NH_3)_5(NCS)]^{2+}$	3.7×10^{-10}	470

Source: Data from [14, 17].

Table 9.4   Rate constants (note units) at $45\,°C$ for anation of $[Co(NH_3)_5(H_2O)]^{3+}$ with different Y ligands from [14, 18, 19]

Y	k (s^{-1}) $45\,°C$	k (M^{-1} s^{-1}) $25\,°C$
H_2O	100×10^{-6}	6.6×10^{-6}
N_3^{-}	100×10^{-6}	
SO_4^{2-}	24×10^{-6}	15×10^{-6}
Cl^{-}	21×10^{-6}	0.1×10^{-6}
NCS	16×10^{-6}	1.3×10^{-6}

there be a significant excess of Y. Because water is the solvent and is in great excess, the rate constant for other entering ligands *cannot exceed* the rate of water exchange when *d* activation occurs.

$$[Co(NH_3)_5(H_2O)]^{3+} + Y \rightarrow [Co(NH_3)_5(Y)] + H_2O \tag{9.50}$$

9.3.1.3   *Change in Oxidation State of the Complex (Associative vs Dissociative Activation)*

The anation reactions of $[Fe(H_2O)_6]^{3+}$ appear to proceed by an interchange mechanism because both five- and seven-coordinate intermediates are not favored due to low CFAE (Tables 9.2A and 9.2B). In the anation reaction for $[Fe(H_2O)_6]^{3+}$ and its first hydrolysis $[Fe(OH)(H_2O)_5]^{2+}$ complex the charge of the complex changes from +3 to +2. Comparing the rate constants for these reactants with the same entering ligand, Y, (Table 9.5) shows that there is significant increase in the rate constant for the +2 complex over the +3 complex. A higher positive charge on the complex indicates associative activation as the L–O bonds are stronger and more difficult to break. In contrast, the +2 complexes appear to undergo dissociative activation as the OH^- ion can help stabilize the intermediate by donating electron density to the metal via π donor bonding (see Figure 8.31 and see Cis in Figure 9.6).

9.3.2   Associative Activated Mechanisms

9.3.2.1   *Change in the Metal Atom of the Complex*

The anation reactions of $[Rh(H_2O)(NH_3)_5]^{3+}$ and $[Ir(H_2O)(NH_3)_5]^{3+}$ show a decrease going down the periodic table (Table 9.6), which correlates with an increase in the M–OH_2 bond energy. Also, the rate constant for Cl^- as well as other entering ligands, which are lower in concentration than water, exceeds the rate of water exchange, indicating that *a* activation occurs.

9.3.2.2   *Aquation or Acid Hydrolysis Reaction*

Swaddle [13] used Equation 9.21 and data in Table 9.7 for a series of $[Cr(H_2O)_5X]^{2+}$ complexes to show that a plot of the rate constants, *k*, for the aquation reaction of $[Cr(H_2O)_5X]^{2+}$ (reverse of Equation 9.51)

Table 9.5   *Limiting rate constants (l_a; $mol^{-1}\,s^{-1}$) for anation of $[Fe(H_2O)_6]^{3+}$ and $[Fe(OH)(H_2O)_5]^{2+}$ with different Y ligands from [20]*

Y (entering ligand)	$[Fe(H_2O)_6]^{3+}$	$[Fe(OH)(H_2O)_5]^{2+}$
SO_4^{2-}	2.3×10^3	1.14×10^5
Cl^-	4.8	5.5×10^3
Br^-	1.58	2.8×10^3
NCS^-	90	5.0×10^3

Table 9.6   *Limiting rate constants (s^{-1}) for anation of $[Rh(H_2O)(NH_3)_5]^{3+}$ at 65°C and $[Ir(H_2O)(NH_3)_5]^{3+}$ at 85°C with different Y ligands*

Y^{x-}	$[Rh(H_2O)(NH_3)_5]^{3+}$ [21]	$[Ir(H_2O)(NH_3)_5]^{3+}$ [22]
Cl^-	4.2×10^{-3}	9.2×10^{-4}
H_2O	1.6×10^{-3}	2.2×10^{-4}

versus the $-\log K$ of the forward reaction (K_{diss}) for Equation 9.49 gave a value of $n = 0.56$, indicating an *associative* activation. For the hydrolyzed $[CrOH(H_2O)_5]^{2+}$ complex, the rate constants are much faster for the anation reaction (Equation 9.52) suggesting a dissociative mechanism with OH^- stabilizing a five-coordinate intermediate via π donor bonding as for the Fe(III) species in Table 9.5 (see Figure 9.6).

$$[Cr(H_2O)_6]^{3+} + X \leftrightarrow [Cr(H_2O)_5X]^{2+} + H_2O \tag{9.51}$$

$$[CrOH(H_2O)_5]^{2+} + Y^{n-} \rightarrow [CrOH(H_2O)_4Y]^{-n+1} + H_2O \tag{9.52}$$

Table 9.7 *Second order limiting rate constant data for the anation of* $[Cr(H_2O)_6]^{3+}$

	$[Cr(H_2O)_6]^{3+}$		$[CrOH(H_2O)_6]^{2+}$
Y^{n-}	k_{obs} (M^{-1} s^{-1})	K_{eq} (M^{-1})	k_{obs} (M^{-1} s^{-1})
NCS$^-$	170×10^{-8}	136	9.7×10^{-5}
NO$_3^-$	73×10^{-8}	9.7×10^{-8}	15×10^{-5}
Cl$^-$	2.9×10^{-8}	1.1×10^{-1}	4.2×10^{-5}
Br$^-$	0.9×10^{-8}	2.2×10^{-3}	2.7×10^{-5}
I$^-$	0.08×10^{-8}	1.0×10^{-5}	0.28×10^{-5}
F$^-$	2000×10^{-8}	2.14×10^4	
SCN$^-$	0.4×10^{-8}	4.5×10^{-4}	

Source: Data from [13, 16, 23].

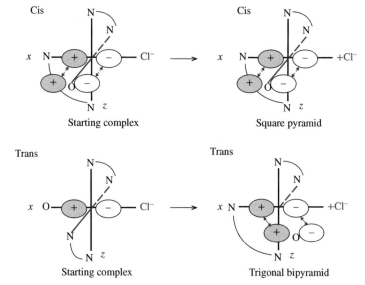

Figure 9.6   *(**Cis**) Stabilization of a square pyramid intermediate by donation of π electrons from OH^- to a t_{2g} (p) orbital of the metal ion. (**Trans**) Rotation of three atoms to form a trigonal bipyramid intermediate in order to donate π electrons from OH^- to a t_{2g} (p) orbital of the metal ion. The xz plane is the plane of the paper*

9.4 Intimate Mechanisms Affected by Steric Factors (Dissociative Preference)

Substitution of CH_3 groups for H atoms on carbon atoms in (bis)ethylenediamine complexes of Co(III) produces larger rate constants (Table 9.8) for the aquation of Co(III) complexes (Equation 9.53). Increased steric factors (as well as the electron donation capacity of CH_3 groups) indicate *dissociative activation* as the entering ligand is hindered from attacking and binding Co(III).

$$trans\text{-}Co(N\text{-}N)_2Cl_2^+ + H_2O \rightarrow trans\text{-}Co(N\text{-}N)_2Cl(H_2O)^{2+} + Cl^- \qquad (9.53)$$

Similar kinetic effects are exhibited for the **trans** complexes of the quadridentate macrocyclic ligand, cyclam (Equation 9.54), and its hexamethylated derivative, tet-b in Equation 9.55 [25], which reacts over 10 times faster than the cyclam complex.

$$Co(Cyclam)Cl_2^+ + H_2O \rightarrow Co(cyclam)Cl(H_2O)^{2+} + Cl^- \quad k\,(s^{-1}) = 1.1 \times 10^{-6} \qquad (9.54)$$

$$Co(tet\text{-}b)Cl_2^+ + H_2O \rightarrow Co(tet\text{-}b)Cl(H_2O)^{2+} + Cl^- \quad k\,(s^{-1}) = 9.3 \times 10^{-4} \qquad (9.55)$$

9.4.1 Intimate Mechanisms Affected by Ligands in Cis versus Trans Positions (Dissociative Preference)

For the aquation reactions of cis and trans (bis)ethylenediamine complexes of Co(III) with two different monodentate ligands ($Co(en)_2LX$, Equations 9.56 and 9.57), Table 9.9 shows that the cis complexes have higher rate constants than trans complexes for π-donor ligands (OH^-) and σ-donor ligands (NH_3). However, π-acceptor ligands (NO_2^-) have higher rate constants for trans versus cis complexes.

$$cis\text{-}Co(en)_2(L)Cl^+ + H_2O \rightarrow cis\text{-}Co(en)_2(L)(H_2O)^{2+} + Cl^- \qquad (9.56)$$

$$trans\text{-}Co(en)_2(L)Cl^+ + H_2O \rightarrow trans\text{-}Co(en)_2(L)(H_2O)^{2+} + Cl^- \qquad (9.57)$$

Also, the rate constants follow the order $k_{\pi\,donor} > k_{\pi\,acceptor} > k_{\sigma\,donor}$ whether the complex has cis or trans isomerism. All cis complexes react and form products that remain cis whereas the trans π donor complexes form complexes that undergo rearrangement to form cis complexes [14]. These results indicate that π donor ligands can stabilize a five-coordinate square pyramid intermediate as shown in Figure 9.6 and enhance M–L bond breaking, indicating that *dissociative activation* occurs for cis complexes. Figure 9.6 also demonstrates that the trans complex would have to undergo an intramolecular rotation of the O atom in OH^- along with the two N atoms in one ethylenediamine molecule (movement around a trigonal face similar to the Bailar twist, Section 8.4.6) to get to a trigonal bipyramid intermediate. This increased motion would slow down the reaction and lead to *isomeric rearrangement* (see Figure 9.7).

Table 9.8 Rate constants (s^{-1}) for aquation of Co(III) complexes with substituted ethylenediamine ligands from [24]

Ethylenediamine derivative (N-N)	$k\,(s^{-1})$
$H_2N\text{-}CH_2 \text{ - } CH_2 \text{ -}NH_2$	3.2×10^{-5}
$H_2N\text{-}CH_2 \text{ - } CH(CH_3) \text{ -}NH_2$	6.2×10^{-5}
$H_2N\text{-}CH_2 \text{ - } C(CH_3)_2 \text{ -}NH_2$	2.2×10^{-4}
$H_2N\text{-}C(CH_3)_2 \text{ - } C(CH_3)_2 \text{ -}NH_2$	Instantaneous

Table 9.9   *Acid hydrolysis of Co(III) complexes [14]*

Complex	$k\ (s^{-1})$	Ligand bonding
trans-Co(en)$_2$(OH)Cl$^+$	1.6×10^{-3}	π donor
cis-Co(en)$_2$(OH)Cl$^+$	1.2×10^{-2}	π donor
trans-Co(en)$_2$(NO$_2$)Cl$^+$	9.8×10^{-4}	π acceptor
cis-Co(en)$_2$(NO$_2$)Cl$^+$	1.1×10^{-4}	π acceptor
trans-Co(en)$_2$(NH$_3$)Cl$^+$	3.4×10^{-7}	σ donor
cis-Co(en)$_2$(NH$_3$)Cl$^+$	5.0×10^{-7}	σ donor

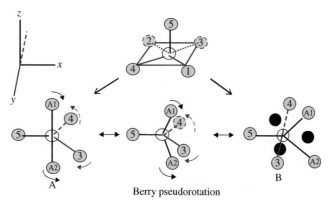

Berry pseudorotation

Figure 9.7   *Square pyramid rearranges to trigonal bipyramid A or B by in plane bend of positions 1, 2 or 3, 4. For structure A converting to B, each of the equatorial positions 3 and 4 move 30° away from each other in the xy plane as axial positions A1 and A2 move 30° toward each other in the xz plane. The three black spheres for structure B indicate three possible positions for attack by the entering ligand*

9.4.2   Base Hydrolysis

The overall reaction for replacement of a ligand by hydroxide ion is known as base hydrolysis (Equation 9.58), and the rate equation is second order in reactant and hydroxide ion (Equation 9.59).

$$[Co(NH_3)_5X]^{2+} + OH^- \rightarrow [Co(NH_3)_5OH]^+ + X^- \tag{9.58}$$

$$\text{rate} = k\{[Co(NH_3)_5X]^{2+}\}\{OH^-\} \tag{9.59}$$

However, the same rate equation is valid in basic solution when a ligand Y (rather than OH$^-$) replaces X$^-$ as in Equation 9.60 for metal complexes, which have the monodentate ligands NH$_3$, RNH$_2$, and H$_2$O, which can be deprotonated. These data indicate that a unique mechanism operates.

$$[Co(NH_3)_5X]^2 + Y \rightarrow [Co(NH_3)_5Y]^+ + X^- \tag{9.60}$$

The mechanism that was proposed by Garrick in 1937 [14] has three steps (Equation 9.61a–c). The first step is the deprotonation of one NH$_3$ to form NH$_2{}^-$, which is a π donor ligand, to form a conjugate base of the starting material. The second slow and rate-determining step is the dissociation of the leaving group from the metal center to form a five-coordinate intermediate, which is stabilized by the π donor ligand. Finally, OH$^-$

binds with the metal center, and the NH_2^- group accepts a H^+ from water, regenerating the base.

$$[Co(NH_3)_5Cl]^{2+} + OH^- \leftrightarrow [Co(NH_3)_4(NH_2)Cl]^+ + H_2O \quad \text{(fast, equilibrium, } H^+ \text{ removal)} \quad (9.61a)$$

$$[Co(NH_3)_4(NH_2)Cl]^+ \rightarrow [Co(NH_3)_4(NH_2)]^{2+} + Cl^- \quad \text{(slow, dissociation of } X = Cl^-) \quad (9.61b)$$

$$[Co(NH_3)_4(NH_2)]^{2+} + OH^- + H_2O \rightarrow [Co(NH_3)_5OH]^+ + OH^- \quad \text{(fast)} \quad (9.61c)$$

When the entering group is another ligand Y, Y replaces OH^- in Equation 9.61c; the first two steps are identical as the rate law is the same. This mechanism is termed $S_{N1}CB$ (substitution, nucleophilic, unimolecular, conjugate base). For the $[Co(CN)_5Cl]^{3-}$ complex where CN^- replaces all NH_3 ligands, the reaction does not occur as the rate is $> 10^6$ slower.

For metals in the first transition series, base hydrolysis reactions are about 10^6 times faster than acid hydrolysis or aquation reactions (Table 9.10; compare with data in Tables 9.3 and 9.4). The rate constants are slower when the leaving group X is more basic ($F^- > Cl^- > Br^- > I^-$) and the reactions show a large positive entropy of activation consistent with a *dissociative* mechanism. The $[Co(NH_3)_5(NO_2)]^{2+}$ complex reacts very slowly as the electron-donating capability of the NH_2^- group to Co is offset by the electron-accepting capability of the NO_2^- group; thus, the leaving group is less labile to dissociation. The cis and trans isomers of $[Co(en)_2(X)Cl]^{2+}$ complexes frequently have similar reaction rates (Table 9.10), indicating similar characteristics that often undergo stereochemical change. All these factors combined indicate the formation of a five-coordinate trigonal bipyramid intermediate for Equation 9.61b.

The top structure in Figure 9.7 shows formation of a square pyramid intermediate on losing a ligand trans to position 5. Rapid attack of ligand Y leads to retention of geometry. However, the square pyramid can rearrange to one of the trigonal bipyramid structures A or B if it has a significant lifetime, and then attack of the nucleophile Y (Equation 9.61c) can occur at any of the three trigonal planar sites (black spheres, B) as the L–M–L bond angles are 120° allowing Y to approach the metal with less steric constraints. This would result in a trans complex rearranging statistically to a product with a 2:1 ratio of cis to trans provided there are no electronic or steric factors in play that could stabilize one geometry over the other. The trigonal bipyramid A can also interconvert through a square pyramid to form trigonal bipyramid B, which is known as **Berry pseudorotation**. Here, one equatorial position remains constant as the other two equatorial positions exchange with the two axial positions via vibrational motions.

Table 9.10 *The rate constants ($M^{-1}s^{-1}$) and entropy of activation ($J\,mol^{-1}\,°K^{-1}$) for base hydrolysis of Co(III) complexes*

Complex	k_{OH} (25 °C)	$\Delta S^{\ddagger}$ (25 °C)	Complex	k_{OH} (0 °C)
$[Co(NH_3)_5I]^{2+}$	23	+42		
$[Co(NH_3)_5Br]^{2+}$	7.5	+40	cis-$[Co(en)_2(NO_2)Cl]^{2+}$	0.03
$[Co(NH_3)_5Cl]^{2+}$	0.85	+36	trans-$[Co(en)_2(NO_2)Cl]^{2+}$	0.08
$[Co(NH_3)_5F]^{2+}$	0.012	+20	cis-$[Co(en)_2OHCl]^{2+}$	0.37
$[Co(NH_3)_5(CH_3COO)]^{2+}$	0.0007		trans-$[Co(en)_2OHCl]^{2+}$	0.017
$[Co(NH_3)_5(NO_2)]^{2+}$	4.2×10^{-6}	+30		
cis-$[Co(en)_2Cl_2]^{2+}$	1,000		cis-$[Co(en)_2Cl_2]^{2+}$	15.1
trans-$[Co(en)_2Cl_2]^{2+}$	3,000		trans-$[Co(en)_2Cl_2]^{2+}$	85
trans-$[Co(en)_2F_2]^{2+}$	64		cis-$[Co(en)_2BrCl]^{2+}$	23
trans-$[Co(en)_2Br_2]^{2+}$	12,000		trans-$[Co(en)_2BrCl]^{2+}$	110
trans-$[Co(cyclam)Cl_2]^{2+}$	67,000	+49		

Source: Data from [14, 26].

9.5   Intimate Versus Stoichiometric Mechanisms

The discussion above tried to distinguish dissociative from associative mechanisms. However, purely associative mechanisms are not known as no seven-coordinate intermediate has been detected. The interchange reaction (Equation 9.38) provides tests to discriminate dissociative from associative activation. The anation reaction of $[Ni(H_2O)_6]^{2+}$ with various entering ligands (Equation 9.62; Table 9.11) has been studied by Wilkins [27]. The k_{obs} data show that there is a three order of magnitude difference on changing the entering group. However, when calculating K_{OS} from Equation 9.39, the ratio of k_{obs}/K_{OS}, which is k for the interchange of the entering group with water as the leaving group, is nearly constant. Thus, k_{obs} differences stem from K_{OS}, indicating that this is an I_d mechanism.

$$[Ni(H_2O)_6]^{2+} + Y^{n-} \rightarrow [Ni(H_2O)_5Y]^{-n+2} + H_2O \tag{9.62}$$

For water exchange reactions of the free $[M(H_2O)_6]^{n+}$ ions, the mechanism starts as A or I_a for Ti^{3+}. As the t_{2g} orbitals are filled on going across the periodic table, I_a becomes less effective. For the first hydrolysis products $[MOH(H_2O)_5]^{n-1}$, OH^- can stabilize a five-coordinate intermediate via π donor bonding (Figure 9.6) so I_d is important. This is observed in Table 9.1 for the rate constants of water exchange.

Measurement of k_1, k_{-1}, and k_2 from Equations 9.23 and 9.24 versus Equation 9.46 permit an assessment of *dissociative versus associative activation* as shown for the anation reaction shown in Equation 9.63. Both reactants are anions so formation of an outer sphere complex should be slow. Ratios of the product formation rate constant, k_2, to the rate constant for reformation of the reactant, k_{-1}, should be < 1 for an associative mechanism and > 1 for a dissociative mechanism. Table 9.12 shows that for the $[RhCl_5(H_2O)]^{2-}$ reaction that the ratios are < 1, and the reaction coordinate can be represented by the activation-controlled diagram in Figure 9.5b.

$$[RhCl_5(H_2O)]^{2-} + Y^- \rightarrow [RhCl_5Y]^{3-} + H_2O \tag{9.63}$$

Table 9.11   *Second order limiting rate constant data showing the Eigen–Wilkins mechanism*

Y^{n-}	k_{obs} (M^{-1} s^{-1})	K_{OS} (M^{-1})	$k = k_{obs}/K_{OS}$
CH_3COO^-	1×10^5	3	3×10^4
SCN^-	6×10^3	1	6×10^3
F^-	8×10^3	1	8×10^3
HF	3×10^3	0.15	2×10^4
H_2O			3×10^3
NH_3	5×10^3	0.15	3×10^4
$NH_2(CH_2)_2N(CH_3)_3^+$	4×10^2	0.02	2×10^4

Table 9.12   *Ratios of k_2/k_{-1} for the reaction in Equation 9.63 (35 °C) from [28]*

Y^-	k_2/k_{-1}	Y^-	k_2/k_{-1}
Cl^-	0.021	SCN^-	0.079
Br^-	0.016	NO_2^-	0.106
I^-	0.018	N_3^-	0.143

9.6   Substitution in Square Planar Complexes (Associative Activation Predominates)

Square planar geometry is important for metal ions with the d^8 [Pt^{2+}, Pd^{2+}, Ni^{2+}, Ir^+, Rh^+, Co^+] electron configuration, and their complexes are classified "inert" to substitution even though two axial positions are available for attack to form a six-coordinate complex. Group theory and the angular overlap method (Table 8.13) demonstrate that the bonding orbitals are a mix of the s, p_x, p_y, and $d_{x^2-y^2}$ orbitals with the ligand orbitals. These approaches also indicate that the p_z, d_{xz}, d_{yz}, d_{xy} orbitals are not stabilized or destabilized; however, the $d_{x^2-y^2}$ orbital and d_{z^2} orbital have overlap energy contributions (de)stabilized by 0.75 and 0.25, respectively. Figure 9.8 displays the molecular orbital energy level diagram for Ni(II) in square planar geometry (D_{4h}) for the sigma bond only case. The HOMO orbital is the d_{z^2} orbital and the LUMO is the $d_{x^2-y^2}$ orbital, although the 4s and $4p_z$ orbitals are also empty and have electron-accepting capability. In the case of significant mixing of the 4s and d_{z^2} orbitals (both with a_{1g} symmetry), the d_{z^2} orbital is lowered in energy and can be more stable than the d_{xy} orbital.

Pt^{2+} reactions have been studied extensively and Equation 9.64 describes the general substitution reaction. The ***trans effect*** is an important feature in substitution of these complexes.

$$MA_3X + Y^- \rightarrow MA_3Y + X^- \tag{9.64}$$

Here, ligands that are trans to a leaving group in square planar complexes have a stronger trans ligand–metal bond so influence the rate of substitution by enhancing bond breaking of the spectator ligand trans to it. Figure 9.9 shows the stereospecific formation of *cis*- and *trans*-diamminedichloridoplatinum(II). Addition of NH_3 to [$PtCl_4$]$^{2-}$ forms [$Pt(NH_3)Cl_3$]$^-$, and on addition of a second NH_3, the cis compound forms. Addition

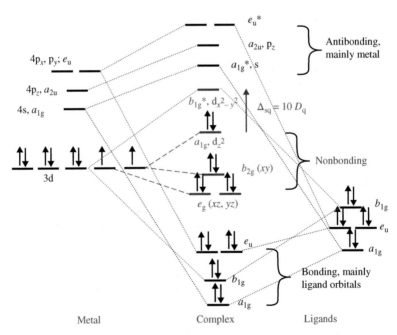

Figure 9.8   *Relative molecular orbital diagram for Ni(II) complexes with square planar geometry. The a_{2u} MO (p_z) is nonbonding.*

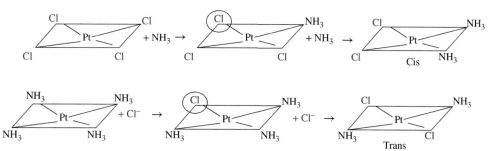

Figure 9.9   *The first displacement of a ligand in $[PtCl_4]^{2-}$ and $[Pt(NH_3)_4]^{2+}$ followed by the trans effect of Cl^- to give stereospecific displacement on addition of the second ligand*

of Cl^- to $[Pt(NH_3)_4]^{2+}$ forms $[Pt(NH_3)_3Cl]^-$, and on addition of a second Cl^-, the trans compound forms. In both cases, the Cl^- trans to another Cl^- or NH_3 directs substitution of either NH_3 or Cl^- to a trans position, respectively.

The ligands that are the best trans directors (T) are π acceptors and σ donors that form strong bonds with the metal ion; the decreasing order for trans direction is $CN^- \sim CO \sim NO \sim H^- \sim CH_3^- \sim SC(NH_2)_2 \sim R_2S \sim PR_3 > SO_3H^- > NO_2^- \sim I^- \sim SCN^- > Br^- > Cl^- > py > RNH_2 \sim NH_3 > OH^- > H_2O$. The stereospecific displacement of the leaving ligand (X) indicates that the mechanism involves a potential stabilization of an intermediate as the trans ligand (T) influences the weakening of the M–X bond.

Square planar complexes can undergo dissociative or associative mechanisms as well as interchange mechanisms, I_a and I_d. The rate law for substitution of square planar complexes is typically a combination of first order and second order pathways as shown in Equation 9.65. Running the experiment under pseudo-first order conditions gives a linear increase of k_{obs} with [Y], and k_{obs} is related to k_S (first-order pathway) and k_Y (second-order pathway) (note that the subscripts S and Y indicate the reaction order (first and second order, respectively) and not the stoichiometric steps in the reaction, which are numerals) as in Equation 9.66.

$$\text{Rate} = k_S[MA_3X] + k_Y[MA_3X]\,[Y] = (k_S + k_Y[Y])\,[MA_3X] \tag{9.65}$$

$$k_{obs} = k_S + k_Y[Y] \tag{9.66}$$

A linear plot of k_{obs} versus [Y] gives k_S as the intercept and k_Y as the slope where k_S represents the slow loss of X^- and its displacement by the solvent (solvent path) whereas k_Y represents direct nucleophilic displacement of X^- by Y^- (reagent path). Both paths are bimolecular, and the $\Delta S^{\ddagger}$ for these reactions is very negative, indicating associative activation and rate-determining steps. The solvent path has the following mechanistic steps (Equations 9.67a,b) where both steps have associative activation through formation of the five-coordinate intermediate $[MA_3X(H_2O)]$, which then reacts with Y^- to form $[MA_3Y(H_2O)]$.

$$MA_3X + H_2O \rightarrow MA_3X(H_2O) \rightarrow MA_3(H_2O)^+ + X^- \tag{9.67a}$$

$$MA_3X(H_2O)^+ + Y^- \rightarrow MA_3Y(H_2O) + X^- \rightarrow MA_3Y + (H_2O) \tag{9.67b}$$

The reaction path has the following stoichiometric mechanistic steps (Equation 9.68) indicating associative activation through the five-coordinate intermediate $[MA_3XY]^-$. Figure 9.5 gives possible reaction coordinate diagrams for these reactions (where k_{-2} is negligible for irreversible reactions).

$$MA_3X + Y^- \underset{k_{-1}}{\overset{k_1}{\rightleftarrows}} MA_3XY^- \underset{k_{-2}}{\overset{k_2}{\rightleftarrows}} MA_3Y + X^- \tag{9.68}$$

9.6.1 Effect of Leaving Group

Table 9.13 provides data for the displacement of X^- in $Pt(dien)X^+$ complexes by pyridine (py) (Equation 9.69) where dien = diethylenetriamine, $NH(CH_2CH_2NH_2)_2$. The rates vary by 10^6, indicating that Pt–X bond breaking is involved in the transition state and is comparable to Pt–Y bond making.

$$Pt(dien)X^+ + py \rightarrow Pt(dien)py^{2+} + X^- \tag{9.69}$$

9.6.2 Effect of Charge

The ligand exchange reaction of Cl^- by H_2O for the hydrolysis of various aminechlorido Pt^{2+} complexes, which have different charges, does not show a large difference in rate constant data (Table 9.14) even though the charge of the complex varies from -2 to $+1$. These data indicate that both bond making and bond breaking in the transition state are similar and that both the solvent path (Equation 9.67a) and the reagent path (Equation 9.67b) are operative.

9.6.3 Nature of the Intermediate – Electronic Factors

The reaction of chloridobis(triethylphosphine)(ligand)Pt(II) complexes with pyridine (Equation 9.70) have been investigated to understand the transition state where T indicates different trans director ligands. As above, the first-order and second-order pathways have the rate laws in Equation 9.71.

$$trans\text{-}[PtCl(PEt_3)_2T] + py \rightarrow trans\text{-}[Ptpy(PEt_3)_2T]^+ + Cl^- \tag{9.70}$$

$$\text{1st order rate} = k_S[PtCl(PEt_3)_2T]; \quad \text{2nd order rate} = k_Y[PtCl(PEt_3)_2T][py] \tag{9.71}$$

Ligands trans to each other interact with the same metal orbitals for bonding so a strong σ donor (or π acceptor) trans to a weak σ donor will weaken the bond of that weak σ donor ligand to the metal; for example, the metal

Table 9.13 *Data for reaction Equation 9.69 in water at 25 °C from [14]*

Ligand X	k_{obs} (s^{-1})	Ligand X	k_{obs} (s^{-1})
H_2O	1900×10^{-6}	N_3^-	0.83×10^{-6}
Cl^-	35×10^{-6}	SCN^-	0.30×10^{-6}
Br^-	23×10^{-6}	NO_2^-	0.050×10^{-6}
I^-	10×10^{-6}	CN^-	0.017×10^{-6}

Table 9.14 *Rate of hydrolysis for platinum(II)chloridoamine complexes at 20 °C from [14]*

Complex	k_{obs} (s^{-1})	Product
$PtCl_4^{2-}$	3.9×10^{-5}	$Pt(H_2O)Cl_3^-$
$Pt(NH_3)Cl_3^-$	0.62×10^{-5}	$cis\text{-}Pt(NH_3)(H_2O)Cl_2$ (replace Cl^- trans to NH_3)
$Pt(NH_3)Cl_3^-$	5.6×10^{-5}	$trans\text{-}Pt(NH_3)(H_2O)Cl_2$ (replace Cl^- cis to NH_3)
$cis\text{-}Pt(NH_3)_2Cl_2$	2.5×10^{-5}	$cis\text{-}Pt(NH_3)_2(H_2O)Cl$
$trans\text{-}Pt(NH_3)_2Cl_2$	9.8×10^{-5}	$trans\text{-}Pt(NH_3)_2(H_2O)Cl$
$Pt(NH_3)_3Cl^+$	2.6×10^{-5}	$Pt(NH_3)_3(H_2O)$

Table 9.15   *Rate data in ethanol for complexes in Equation 9.70 at 25 °C from [14]*

T	k_S (s^{-1})	k_Y (M^{-1} s^{-1})	Ligand bonding
H$^-$	1.8×10^{-2}	4.2	σ donor
P(Et)$_3$	1.7×10^{-2}	3.8	π acceptor and σ donor
CH$_3^-$	1.7×10^{-4}	6.7×10^{-2}	σ donor
C$_6$H$_5^-$	3.3×10^{-5}	1.6×10^{-2}	σ donor
Cl$^-$	1.0×10^{-6}	4.0×10^{-4}	π donor; weaker σ donor

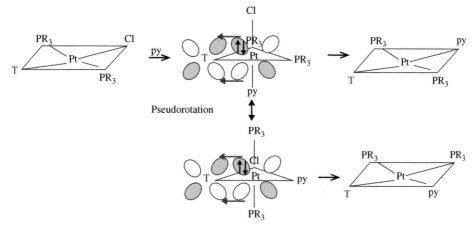

Figure 9.10   *Possible trigonal bipyramid intermediate formed for Equation 9.70 with a π acceptor orbital on T (e.g., P(Et)$_3$) trans to Cl$^-$. Note that the solvent path would have a similar intermediate. The xz plane is the plane of the paper*

p_z and d_{z^2} orbitals are localized and would interact with σ donor ligands. Table 9.15 shows that the rates are fastest for the strong σ donor and π acceptor ligands trans to Cl$^-$ and vary by 10^4.

Triethylphosphine, PEt$_3$, is an excellent example of a π acceptor as empty d orbitals on phosphorus accept electrons from the Pt d orbitals (e.g., d_{xz}; Figure 9.10) and strengthen the Pt–P bond while weakening the Pt–Cl bond. The top pathway in Figure 9.10 shows retention of configuration, which is exclusively found for reactions with square planar geometry. Note that the entering pyridine group and the Cl$^-$ leaving ligand occupy axial positions in the intermediate as the metal s and p_z orbitals can accept electrons; thus, axial attack by py (the entering Y group) is likely. The bottom pathway leads to the cis isomer, if the intermediate is long lived and can undergo Berry pseudorotation.

9.6.4   Nature of the Intermediate – Steric Factors

Convincing evidence, that trigonal bipyramid intermediates and associative activation control occur with square planar Pt(II) complexes, is presented in Table 9.16, which shows that k_1 decreases when methyl groups are added to the 2 positions of pyridine (the T ligand). The methyl groups on the ligand in either cis or trans positions to Y prevent axial attack of Y at the metal, thus inhibiting the associative mechanism

Table 9.16 *Rate data in ethanol for complexes in Equation 9.70 from [11] for py and methyl-substituted pyridine ligands*

Complex	T (cis) k_S (s^{-1}) 0 °C	T (trans) k_S (s^{-1}) 25 °C
py	8.0×10^{-2}	1.24×10^{-4}
2-Methylpyridine	2.0×10^{-4}	1.70×10^{-5}
2,6-Dimethylpyridine	1.0×10^{-6}	3.42×10^{-6}

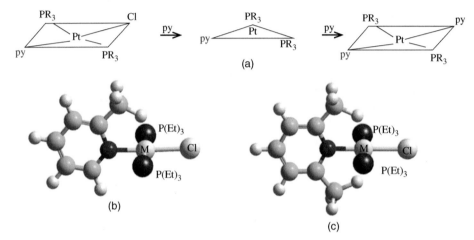

Figure 9.11 *(a) Possible trigonal planar intermediate formed for Equation 9.70 with T (as pyridine and its methylated derivatives) trans to Cl$^-$, (b) 2-methylpyridine complex, (c) 2,6-dimethylpyridine complex. Models created with the Hyperchem™ program (version 8.0.10)*

and favoring the dissociative mechanism with formation of a trigonal planar intermediate (Figure 9.11a). One CH_3 group (Figure 9.11b) on the T ligand would prevent axial attack at one position whereas two CH_3 groups (Figure 9.11c) would prevent axial attack at both axial positions. The reactions are 80,000 times slower for cis complexes on adding methyl groups to pyridine as the entering and leaving groups would be closer to ligand T and cause more steric hindrance for Y attack. The decrease in rate is only 36 for trans complexes as methyl groups are added to pyridine.

9.7 Metal Electron Transfer Reactions

Taube [29] defined two mechanisms of redox reactions for metal complexes: inner sphere and outer sphere. An ***inner sphere*** mechanism occurs when ligand exchange is faster than electron transfer; thus, ligand transfer between metal centers occurs. The ligand is an ion or molecule with two or more pairs of electrons to donate to both metal centers. In the transition state, the metal centers bind a common ligand as a bridge, which facilitates electron transfer in one direction from one metal d orbital to the other metal's d orbital as ligand exchange between the metals occurs in the opposite direction. ***Outer sphere*** reactions occur when electron transfer is faster than ligand exchange so no substitution or ligand exchange occurs. In the transition state, the intact metal complexes collide and come together without any bridging ligand, and the electron transfers between the two metal centers when their d orbitals overlap in a symmetry allowed combination. Both mechanisms are second-order reactions, first order in oxidant and reductant leading to the rate expression in Equation 9.72.

$$\text{rate} = k_2[\text{oxidant}][\text{reductant}] \tag{9.72}$$

9.7.1 Outer Sphere Electron Transfer

Marcus [30] derived quantitative expressions to model outer sphere electron processes, and two redox reactions are defined first. The **cross reaction** (Equation 9.73) occurs when two different metals (1 and 2) undergo electron transfer (rate constant $= k_{12}$) and there is a net chemical change ($\Delta G_r < 0$). The **self-exchange reaction** (Equation 9.74) occurs when the same metal in two different oxidation states undergoes electron transfer (rate constant $= k_{11}$ or k_{22}) with no net chemical exchange ($\Delta G_{\text{reaction}} = 0$; * indicates a radiolabeled Fe).

$$Fe(H_2O)_6{}^{2+} + Ce(H_2O)_6{}^{4+} \rightarrow Fe(H_2O)_6{}^{3+} + Ce(H_2O)_6{}^{3+}$$

$$Ox_1 \qquad + Red_2 \qquad \rightarrow Ox_2 \qquad + Red_1 \tag{9.73}$$

or

$$Fe*(H_2O)_6{}^{2+} + Fe(H_2O)_6{}^{3+} \leftrightarrow Fe*(H_2O)_6{}^{3+} + Fe(H_2O)_6{}^{2+} \tag{9.74}$$

The outer sphere mechanism has five steps. The rate-determining step could be step 2, 3 or 4.

1. Reactants diffuse together to form an outer sphere complex in which the first coordination spheres of both metals remain intact. No sigma bond breaking or making occurs in the reaction (this is very important for nonlabile/inert metal ions).
2. Bond distances around each metal change to become more like the products. There is shortening and elongation of metal–ligand bond distances due to vibrational motion.
3. The solvent shell around the outer sphere complex reorganizes. The metal orbitals can now overlap for eventual electron transfer; their nuclei are considered stationary during any electronic transitions as stated by the **Franck–Condon** principle.
4. The electron is transferred between the metal centers once a **symmetrical transition state** is achieved with little difference in the metal–ligand bond lengths due to vibrational motion.
5. After electron transfer occurs, a successor complex is formed that decomposes to the products, which diffuse away in a normally fast step.

The energetics of these features can be explained using potential energy curves. Figure 9.12a is similar to the Morse curve (Figure 5.9) and shows the ground electronic state (in black) with the first two vibrational levels for $Fe(H_2O)_6{}^{2+}$ and $Fe*(H_2O)_6{}^{3+}$. Each vibrational level has a range for the metal–ligand distance due to thermal motion, indicating that both reactants at some point in time can have the same metal–ligand distance to achieve a symmetrical transition state (the blue curves in Figure 9.12a indicate step 2 above). In this case, $Fe(H_2O)_6{}^{2+}$ must have some bond contraction as $Fe*(H_2O)_6{}^{3+}$ must have bond lengthening (arrows in Figure 9.12a).

Figure 9.12b shows energy wells (black color) for the initial state of both reactants in the outer sphere complex and the final state of the products after electron transfer for a self-exchange reaction. These potential energy curves overlap at the transition state and create intermediate states of higher energy. The dashed lines indicate the overlap without electronic interaction, but the splitting at the intersection of the curves lowers the barrier of electron transfer by $2H_{AB}$, which is a measure of the degree of mixing of the reactant and

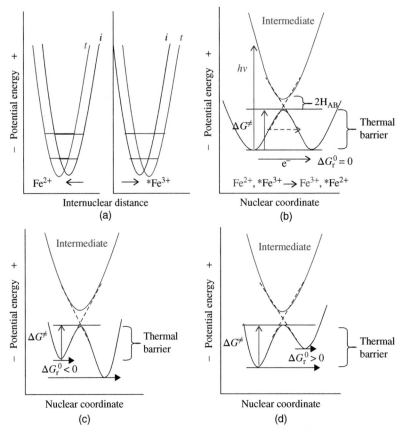

Figure 9.12   *(a) Simplified potential energy curves with vibrational levels (horizontal lines) versus distance for each metal complex; each vibrational level has a range for the metal–ligand bond distance [black curves for the initial or ground state (i); blue curves for the transition state (t)]. (b) Potential energy curves versus nuclear configuration of all atoms in the system for both reactants (left) and products (right) indicating $\Delta G^{\neq} = 0$ for the self-exchange redox reaction; the dashed blue line indicates tunneling and the $h\upsilon$ line indicates light-induced electron transfer. (c) Potential energy curves for both reactants (left) and products (right) indicating $\Delta G^{\neq} < 0$ for a heteronuclear redox reaction. (d) Potential energy curves for both reactants (left) and products (right) indicating $\Delta G^{\neq} > 0$ for a heteronuclear redox reaction*

product states. The Gibbs free energy of activation ($\Delta G^{\neq}$) is the **thermal barrier** to electron transfer and is the free energy needed to change the atomic coordinates from equilibrium values to the atomic coordinates in the activated complex (or intermediate) and then onto the products (the smooth black curve along the surface for the path of the electron is an adiabatic process). For $Fe(H_2O)_6{}^{2+}$ and $Fe*(H_2O)_6{}^{3+}$, both would achieve nearly the same nuclear configuration by thermal motion. The rate of electron transfer is greatest when the nuclear configurations are similar (the initial and final states in the reaction are similar), and $\Delta G^{\neq}$ is small.

In addition to thermal activation, there are two other pathways for electron transfer in Figure 9.12b [31]. First, light stimulation can promote an electron to higher electronic energy levels that permit ease of electron transfer down the well to the products. Second, electron tunneling, which allows electron transfer at greater distances beyond which collisions allow, is possible through t_{2g} orbitals. (Figure 3.7 shows the extent that hydrogen-like "d" orbitals can extend from the nucleus.) Figure 9.12c and d shows the reaction coordinate for the crossover reaction involving heteronuclear species for favorable and unfavorable reactions.

The Marcus equation states that the electron transfer (ET) rate constant, k_{ET}, is proportional to $\Delta G^{\neq}$ (Equation 9.75), which is related to the Gibbs free energy of the reaction, ΔG_r^0 (Equation 9.76), which can be determined from the standard potentials of the reactants. The parameter κ_e (Equation 9.75) is an electronic probability factor ranging from 0 to 1; it is 1 in adiabatic reactions (in non-adiabatic reactions, it is < 1) when the system remains on the lower energy surface. The parameter υ_n is the nuclear collision frequency factor (10^{13} $M^{-1}s^{-1}$ for uncharged ions in solution). $\Delta G^{\neq}$ is a measure of the reorganization energy required around the metal ions and is the sum of w_{11} (the non-electrostatic work required to bring the reactants together), $\Delta G_{ES}^{\neq}$ (the electrostatic interaction energy between the two reactants), $\Delta G_{IS}^{\neq}$ (the inner sphere rearrangement energy), and $\Delta G_{OS}^{\neq}$ (the outer sphere solvent reorganization energy). Many reactions are performed at high ionic strength so the w_{11} and $\Delta G_{ES}^{\neq}$ terms are negligible. Thus, the parameter λ (J mol^{-1}) is the reorganization energy required around the metal ions and is normally broken down into inner-shell vibration and outer-shell solvent components. The inner-shell component is a function of the change in M–L distance (Δd) on going from the ground to the transition state and, for simple solution reactions (e.g., Fe^{2+}, Fe^{3+}) ranges from 0 to 20 pm.

$$k_{ET} = k_{11} = \upsilon_N \kappa_e e^{\frac{-\Delta G^{\neq}}{RT}} \tag{9.75}$$

$$\Delta G^{\neq} = w_{11} + \Delta G_{ES}^{\neq} + \Delta G_{IS}^{\neq} + \Delta G_{OS}^{\neq} = \frac{1}{4}\lambda \left[1 + \frac{\Delta G_r^0}{\lambda}\right]^2 \tag{9.76}$$

For the $Fe(H_2O)_6^{2+,3+}$ self-exchange reaction, ΔG_r^0 is zero and $\Delta G^{\neq} = \frac{1}{4}\lambda$ (discussed further in Figure 9.16a). When λ is small and κ_e is one, the reaction is fast. For $t_{2g}(\pi)$ nonbonding orbitals in octahedral geometry, λ is small as there would be good overlap in the outer sphere precursor complex and changes in metal–ligand bond distances cost a minimum of energy. However, for $e_g^*(\sigma)$ antibonding orbitals, λ is high as these orbitals are directly involved with bonding and antibonding, and changes in metal–ligand bond distances cost more energy. For aqua metal complexes, λ is on the order of 1 eV or higher as the metal ions are not far from second coordination or outer sphere water molecules, but λ can be as low as 0.25 eV for metal centers that are shielded from water as in ***proteins and enzymes***. When π acceptor ligands (*o*-phen; bipy) are bound to metal ions such as Fe, $t_{2g}(\pi)$ electrons are delocalized to the ligand, thus lowering the reorganization energy and increasing κ_e when t_{2g} electrons are transferred. The rate constants for several self-exchange reactions are in Table 9.17 and show the effects of electron transfer between t_{2g} and e_g^* orbitals as well as metals bound to π acceptor ligands. The reactant and product spin states of each reactant pair show symmetrical electron arrangements except for the cobalt complexes. Clearly, the t_{2g} transitions are more favorable as they are not shielded by the ligand. Figure 9.13 demonstrates the symmetry allowed and forbidden overlap of d orbitals in outer sphere electron transfer reactions.

Co reactions are a special case because most Co(II) complexes are high spin and Co(III) complexes including water are low spin; thus, $Co^{2+,3+}$ self-exchange reactions are slow due to electronic factors. The $Co(NH_3)_6^{2+,3+}$ and $Co(H_2O)_6^{2+,3+}$ systems show a 10^9 difference in k_{11} indicating significant ligand effects as well. Reaction A in Figure 9.14 shows the unsymmetrical arrangement of the electron configurations for Co(II) and Co(III) hexaaqua complexes where a t_{2g} electron from $Co(H_2O)_6^{2+}$ could not transfer to an empty e_g^* orbital. Also, in the unsymmetrical reaction A, an e_g^* electron transfer would not occur as the reactants would not have a transition state similar to the products. However, the $Co(H_2O)_6^{3+}$ complex can rearrange via thermal motion or light induced activation from low spin $^1A_{1g}$ to high spin $^5T_{2g}$ (where the blue

Table 9.17   *Self-exchange rate constants ($M^{-1} s^{-1}$) for metals with octahedral coordination at 25°C from [14]*

Redox couple	k_{11}	Electron configuration	Transition
$Cr(H_2O)_6{}^{2+} + Cr(H_2O)_6{}^{3+}$	$\leq 2 \times 10^{-5}$	$t_{2g}{}^3 e_g{}^{*1}/t_{2g}{}^3 e_g{}^{*0}$	$e_g{}^* \rightarrow e_g{}^*$
$Co(NH_3)_6{}^{2+} + Co(NH_3)_6{}^{3+}$	$<10^{-9}$	$t_{2g}{}^5 e_g{}^{*2}/t_{2g}{}^6 e_g{}^{*0}$	(Unsymmetrical)
$Co(en)_3{}^{2+} + Co(en)_3{}^{3+}$	1.4×10^{-4}	$t_{2g}{}^5 e_g{}^{*2}/t_{2g}{}^6 e_g{}^{*0}$	(Unsymmetrical)
$Co(edta)^{2-} + Co(edta)^-$	1.4×10^{-4}	$t_{2g}{}^5 e_g{}^{*2}/t_{2g}{}^6 e_g{}^{*0}$	(Unsymmetrical)
$Co(o\text{-phen})_3{}^{2+} + Co(o\text{-phen})_3{}^{3+}$	1.1	$t_{2g}{}^5 e_g{}^{*2}/t_{2g}{}^6 e_g{}^{*0}$	(Unsymmetrical)
$Co(H_2O)_6{}^{2+} + Co(H_2O)_6{}^{3+}$	5	$t_{2g}{}^5 e_g{}^{*2}/t_{2g}{}^6 e_g{}^{*0}$	(Unsymmetrical)
$V(H_2O)_6{}^{2+} + V(H_2O)_6{}^{3+}$	0.01	$t_{2g}{}^3/t_{2g}{}^2$	$t_{2g} \rightarrow t_{2g}$
$Fe(H_2O)_6{}^{2+} + Fe(H_2O)_6{}^{3+}$	4	$t_{2g}{}^4 e_g{}^{*2}/t_{2g}{}^3 e_g{}^{*2}$	$t_{2g} \rightarrow t_{2g}$
$Fe(CN)_6{}^{4-} + Fe(CN)_6{}^{3-}$	740	$t_{2g}{}^5/t_{2g}{}^6$	$t_{2g} \rightarrow t_{2g}$
$Fe(o\text{-phen})_3{}^{2+} + Fe(o\text{-phen})_3{}^{3+}$	$>10^5$	$t_{2g}{}^5/t_{2g}{}^6$	$t_{2g} \rightarrow t_{2g}$
$Ru(NH_3)_6{}^{2+} + Ru(NH_3)_6{}^{3+}$	800	$t_{2g}{}^5/t_{2g}{}^6$	$t_{2g} \rightarrow t_{2g}$
$Os(bipy)_3{}^{2+} + Os(bipy)_3{}^{3+}$	5×10^4	$t_{2g}{}^5/t_{2g}{}^6$	$t_{2g} \rightarrow t_{2g}$
$IrCl_6{}^{3-} + IrCl_6{}^{2-}$	1.0×10^3	$t_{2g}{}^5/t_{2g}{}^4$	$t_{2g} \rightarrow t_{2g}$
$Cr(bipy)_3{}^{2+} + Cr(bipy)_3{}^{3+}$	2×10^9	$t_{2g}{}^4/t_{2g}{}^3$	$t_{2g} \rightarrow t_{2g}$

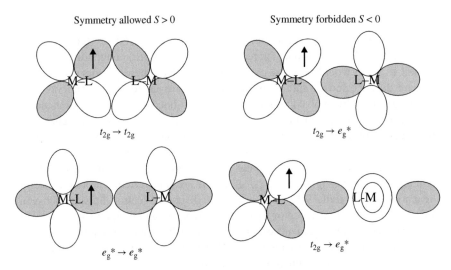

Figure 9.13   *(Top) Symmetry allowed and forbidden orbital overlap for outer sphere electron transfer. Blue indicates $+\psi$; white indicates $-\psi$*

lines cross in Figure 8.46). Figure 9.14 shows the low to high spin conversion with the blue arrow between reactions A and B as the ligand field strength for $Co(H_2O)_6{}^{3+}$ is low [32].

However, this is not the case for the other Co(III) complexes in Table 9.17 as the rearrangement from low to high spin for these Co(III) complexes to form a symmetrical system with the Co(II) complexes requires a reasonable energy or $\Delta G^{\neq}$ (toward the right of Figure 8.46). Although reaction B in Figure 9.14 is a t_{2g}

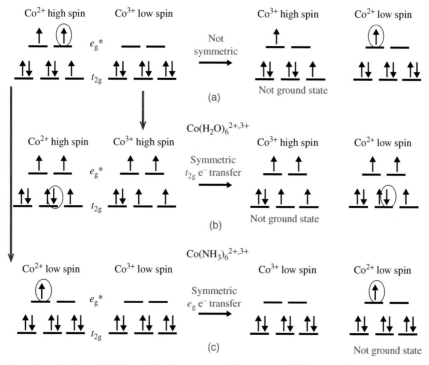

Figure 9.14 *Reactions A, B, and C show the electron configurations of the reactant and products after electron transfer. Reaction B is a symmetrical reactant state after $Co(H_2O)_6^{3+}$ converts from low to high spin. Reaction C is a symmetrical reactant state after $Co(NH_3)_6^{2+}$ converts from high to low spin. The circled electron is the electron being transferred. In every case, a product complex must undergo spin conversion to a ground state*

electron transfer and should be efficient, k_{ET} is still slower than other t_{2g} electron transfers in Table 9.17 due to the reorganization of cobalt's electronic structure.

The $Co(NH_3)_6^{2+,3+}$ system cannot undergo the $^1A_{1g}$ to $^5T_{2g}$ spin crossover because $Co(NH_3)_6^{3+}$ has a much higher $\Delta G^{\neq}$ than $Co(H_2O)_6^{3+}$ [32, 33]. In this case, $Co(NH_3)_6^{2+}$ would need to undergo a high spin to low spin conversion or crossover to form a symmetrical state (blue arrow between reactions A and C in Figure 9.14c). After electron transfer, the newly formed low spin $Co(NH_3)_6^{2+}$ would crossover back to high spin leading to an unfavorable $\Delta G^{\neq}$ for the product state. All these electronic factors can lead to a very small k_{11} for the $Co(NH_3)_6^{2+,3+}$ system, for which an inner sphere process is not available (NH_3 has only one lone pair of electrons).

9.7.2 Cross Reactions

For electron transfer between two different metals, Equation 9.76 depends on ΔG_r^0. Figure 9.15 shows a plot of the variation in k_{12} and $\Delta G^{\neq}$ as the ΔG_r^0 decreases when one metal ion reacts with several other metal ions with a constant λ of $50,000 \, \text{J mol}^{-1}$. As ΔG_r^0 decreases (becomes more negative), there is a short linear to slight exponential increase (blue dashed arrow Figure 9.15a) in k_{12} followed by a sharp exponential increase and then a decrease in k_{12}. This **inverted behavior** with k_{12} is important in photosynthesis [30] where there is a transfer of electronic excitation from antenna chlorophylls to a specialized chlorophyll molecule, which

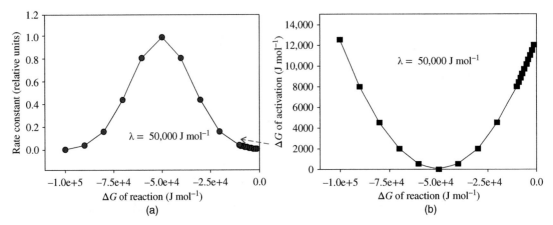

Figure 9.15   (a) k_{12} variation (from Equation 9.75) and (b) $\Delta G^{\neq}$ variation (from Equation 9.76) as ΔG_r^0 increases. The rate is a relative rate as v_n and κ_e are set to unity

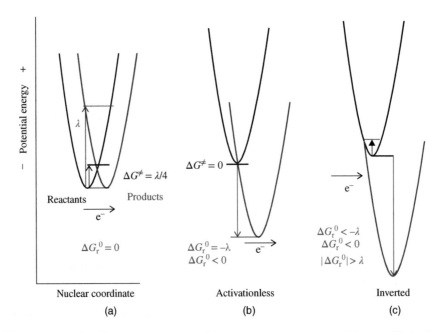

Figure 9.16   (a) Self-exchange process. (b) Activationless process. (c) Inverted behavior

then transfers the electron onward for the eventual reduction of CO_2 to organic carbon in photosystem I (PSI) center. To prevent the back reaction of the electron, $\Delta G^{\neq}$ and λ must be small, which occurs when the size of the reactants is large and the protein environment is largely nonpolar.

Figure 9.16 shows plots of nuclear coordinate as the ΔG_r^0 decreases for three conditions. Figure 9.16a is the self-exchange reaction where $\Delta G^{\neq} = \frac{1}{4}\lambda$ and when the energy state of the products crosses (mixes with) the energy state of the reactants to the right of the parabola representing the reactants; note the $\Delta G^{\neq} > 0$. When the energy of the products parabola intersects at the minimum energy of the reactants' parabola (Figure 9.16b), then $\Delta G^{\neq}$ is zero and electron transfer occurs readily and is **activationless** as $\Delta G_r^0 = -\lambda$. When the energy of the products parabola intersects to the left of the energy in the reactants' parabola (Figure 9.16c), then $\Delta G^{\neq} > 0$ again and inverted behavior is found; at this point, $\Delta G_r^0 < -\lambda$ or $|\Delta G_r^0| > \lambda$.

For the limiting case where $|\Delta G_r^0| \ll \lambda$, expanding the quadratic expression of Equation 9.76 leads to Equation 9.77 as $(\Delta G_r^0/\lambda)^2$ becomes small. Substitution into Equation 9.75 for k_{ET} yields Equation 9.78, and the dashed blue arrow in Figure 9.15a indicates the linear to exponential shape.

$$\Delta G^{\neq} = \frac{1}{4}\lambda\left[1 + \frac{\Delta G_r^0}{\lambda}\right]^2 \approx \frac{1}{4}\lambda\left[1 + \frac{2\Delta G_r^0}{\lambda}\right] = \frac{1}{4}(\lambda + 2\Delta G_r^0) \tag{9.77}$$

$$k_{ET} \approx \upsilon_N \kappa_e e^{-\frac{(\lambda + 2\Delta G_r^0)}{4RT}} \tag{9.78}$$

Marcus combined Equations 9.75 and 9.76 for two self-exchange reactions and formulated the cross relation Equation 9.79 to calculate k_{12} for the cross reaction (Equation 9.73). K_{12} is the equilibrium constant of the reaction, which can be experimentally determined or calculated from standard redox potentials or free energy data. The parameter f_{12} is normally 1 for simple ions in solution but is often calculated with Equation 9.80 when there is nonlinearity between $\Delta G^{\neq}$ and ΔG_r^0. The cross relation works best when ΔG_r^0 is not too large (e.g., blue dashed arrow in Figure 9.15a). Equation 9.79 has been used (1) to estimate k_{11} values and (2) to derive redox potentials when they are difficult to measure experimentally.

$$k_{12} = (k_{11}k_{22}K_{12}f_{12})^{1/2} \tag{9.79}$$

$$\log f_{12} = \frac{(\log K_{12})^2}{4\log(k_{11}k_{22}/Z^2)} \tag{9.80}$$

Agreement between experimentally determined and calculated rate constants is good when the electron transitions are symmetry allowed, but poor agreement occurs when transitions are symmetry forbidden as shown in Table 9.18. Rapid redox cross reactions are usually faster than each of the electron self-exchange reactions that are used to calculate the cross relation rate constant. Note that the ligand on each metal can be different.

9.7.3 Inner Sphere Electron Transfer

Inner sphere electron transfer reactions proceed with an additional step compared to outer sphere reactions as the ligand must also be transferred between the metal centers. There is substantial reorganization of the inner coordination sphere of each metal center to form a bridged M_1–L–M_2 complex as intermediate, which is similar to the **Eigen–Wilkins** mechanism.

The inner sphere mechanism has six main steps, which are outlined in Equations 9.81a–f.

1. Reactants diffuse together to form an outer sphere complex so that the bridging ligand on the reduced metal center 1 (M_1) is aligned to interact with the oxidized metal center 2 (M_2).
2. Once a ligand dissociates from M_1, the bridging ligand from M_2 attacks M_1.
3. Then the bridging ligand from M_2 binds with M_1 to form a ligand bridged intermediate (M_1–L–M_2) between both metal centers.
4. The electron is transferred from reduced M_1 to oxidized M_2 with the bridged intermediate (M_1–L–M_2) remaining intact.
5. After electron transfer occurs, ligand exchange occurs from M_2 to M_1, which is now the oxidized metal center.
6. A successor complex is formed that decomposes to the products, which diffuse away in a fast step.

Table 9.18   *Cross reaction and* k_{11} *(k_{22}) rate constants ($M^{-1} s^{-1}$) for octahedrally coordinated metals at 25°C from [34]. Agreement between experimental and calculated values within an order of magnitude is considered good. In each system the cross reactions are given in bold font*

Redox couple	k_{12} (k_{11})	k_{12} calc.	Transition
Symmetry forbidden			
(A) $Cr(H_2O)_6^{2+} + Fe(H_2O)_6^{3+}$	**2×10^3**	**1×10^6**	$e_g^* \to t_{2g}$ *(high spin metals)*
$Cr(H_2O)_6^{2+} + Cr(H_2O)_6^{3+}$	($\leq 2 \times 10^{-5}$)		$e_g^* \to e_g^*$
$Fe(H_2O)_6^{2+} + Fe(H_2O)_6^{3+}$	(4)		$t_{2g} \to t_{2g}$
(B) $Fe(H_2O)_6^{2+} + Co(H_2O)_6^{3+}$	**250**	**5.8×10^7**	$t_{2g} \to e_g^*$
$Fe(H_2O)_6^{2+} + Fe(H_2O)_6^{3+}$	(4)		$t_{2g} \to t_{2g}$
$Co(H_2O)_6^{2+} + Co(H_2O)_6^{3+}$	(5)		
(C) $Fe(o\text{-}phen)_3^{2+} + Co(H_2O)_6^{3+}$	**1.4×10^4**	**4×10^{-2}**	$t_{2g} \to e_g^*$
$Fe(o\text{-}phen)_3^{2+} + Fe(o\text{-}phen)_3^{3+}$	(3×10^7)		$t_{2g} \to t_{2g}$
$Co(H_2O)_6^{2+} + Co(H_2O)_6^{3+}$	(5)		
Symmetry allowed			
(D) $Fe(CN)_6^{4-} + Fe(o\text{-}phen)_3^{3+}$	**$>1 \times 10^8$**	**$>1 \times 10^8$**	$t_{2g} \to t_{2g}$ *(low spin metals)*
$Fe(o\text{-}phen)_3^{2+} + Fe(o\text{-}phen)_3^{3+}$	(3×10^7)		$t_{2g} \to t_{2g}$
$Fe(CN)_6^{4-} + Fe(CN)_6^{3-}$	(740)		$t_{2g} \to t_{2g}$
(E) $Fe(CN)_6^{4-} + IrCl_6^{2-}$	**3.8×10^5**	**1×10^6**	$t_{2g} \to t_{2g}$ *(low spin metals)*
$Fe(CN)_6^{4-} + Fe(CN)_6^{3-}$	(740)		$t_{2g} \to t_{2g}$
$IrCl_6^{3-} + IrCl_6^{2-}$	(10^3)		$t_{2g} \to t_{2g}$
(F) $V(H_2O)_6^{2+} + Ru(NH_3)_6^{3+}$	**1.3×10^3**	**2×10^3**	$t_{2g} \to t_{2g}$
$V(H_2O)_6^{3+} + V(H_2O)_6^{2+}$	(1.2×10^{-2})		$t_{2g} \to t_{2g}$
$Ru(NH_3)_6^{3+} + Ru(NH_3)_6^{2+}$	(4×10^3)		$t_{2g} \to t_{2g}$

Equation 9.81a was the first example demonstrating the mechanism [29]. $[Co(NH_3)_5Cl]^{2+}$ is an inert complex with a ligand substitution time scale on the order of hours in acid whereas $[Cr(H_2O)_6]^{2+}$ has a rate constant of water exchange on the order of 10^9s, yet the redox reaction occurs < 1 s. The mechanism has been explained as follows and is consistent with the fact that radiolabeled Cl^- in solution does not bind to Cr in the product. $[Cr(H_2O)_6]^{2+}$ loses a water and Cr(II) then forms a precursor complex (Equation 9.81b) and binds with the Cl attached to the Co(III) to form the binuclear metal complex as intermediate (Equation 9.81c). Electron transfer proceeds across the bridging Cl^- ligand to Co(III) (Equation 9.81d), and then the successor complex forms (Equation 9.81e) with transfer of Cl^- to form the Cr–Cl bond (Equation 9.81e). The successor complex decomposes (Equation 9.81f) to products with the complex, $[(NH_3)_5Co(H_2O)]^{2+}$, losing ammonia to form $[Co(H_2O)_6]^{2+}$.

$$Co(NH_3)_5Cl^{2+} + Cr(H_2O)_6^{2+} \to Co(H_2O)_6^{2+} + CrCl(H_2O)_5^{2+} + 5NH_4^+ \qquad (9.81a)$$
$$\underset{M_2 - t_{2g}^6}{} \quad \underset{M_1 - t_{2g}^3 e_g^{1*}}{} \quad \underset{M_1 - t_{2g}^5 e_g^{2*}}{} \quad \underset{t_{2g}^3}{}$$

$$[(NH_3)_5Co(d^6)\text{-}Cl\text{----}(d^4)Cr(H_2O)_5]^{4+} + H_3O^+ \text{ form precursor complex} \qquad (9.81b)$$

$$[(NH_3)_5Co(d^6)\text{----}Cl\text{----}(d^4)Cr(H_2O)_5]^{4+} + H_3O^+ \text{ form bridged activated complex} \qquad (9.81c)$$

$[(NH_3)_5Co(d^7)\text{----}Cl\text{----}(d^3)Cr(H_2O)_5]^{4+}$ electron transfer to Co(III)   (9.81d)

$[(NH_3)_5Co(d^7)\text{----}Cl\text{-}(d^3)Cr(H_2O)_5]^{4+}$ form successor complex   (9.81e)

$[(NH_3)_5Co(H_2O)]^{2+} + [Cr(H_2O)_5Cl]^{2+}$ bridged complex breaks up   (9.81f)

An inner sphere process is likely when there is a symmetry-forbidden mismatch between the metal orbitals as in the $Cr(H_2O)_6^{2+} + Fe(H_2O)_6^{3+}$ reaction (Table 9.18). Also, rate constant calculations with the Marcus cross relation (Equation 9.79) do not agree with the experiment.

The bridging ligand can be a π donor ligand, which has two or more electron pairs available or a π acceptor ligand as long as there are two lone pairs of electrons to bind with the two metal centers. In the case of cyanide ion ($:C\equiv N:$), both the C and N atoms form bonds with the two different metals. From an environmental standpoint, the halides, hydroxide ion, sulfate ion, acetate ion, and phosphate ion are excellent bridging ligands for inner sphere electron transfer reactions. Also, solid surfaces with hydroxide, oxide, and sulfide ions are excellent bridging ligands (Section 7.9).

9.8 Photochemistry

9.8.1 Redox

Figure 9.12b shows the electron transition for the reactants' ground state outer sphere complex to the intermediate state during a possible outer sphere transfer reaction. Once excited, the electron can proceed to the product state more rapidly than the thermal reaction in many cases. For a self-exchange symmetrical process, the reorganization energy λ (or $4\,\Delta G^{\neq}$) equals the energy of light $(h\upsilon)$.

The electron transfer rate can increase or decrease depending on the excited state versus the ground state after light activation. For example, Figure 9.14 describes the change in spin state for $Co^{2+,3+}$ complexes that results in faster electron transfer rates for the aqua complexes but not the ammonia complexes. For the self-exchange reaction of $Fe(bipy)_3^{2+,3+}$, k_{11} is 10^9 M^{-1} s^{-1}, but when light activated, k_{11} slows to 10^3 M^{-1} s^{-1} because the $Fe(bipy)_3^{2+}$ is converted from t_{2g}^6 to $t_{2g}^4 e_g^2$, leading to an unsymmetrical state for reaction with the t_{2g}^5 configuration for $Fe(bipy)_3^{3+}$ [35].

9.8.2 Photosubstitution Reactions d → d

Visible light is important in activating inert t_{2g}^3 and t_{2g}^6 octahedral metal complexes to ligand substitution, which is rather slow as noted in Tables 9.3–9.9. The light energy must match or exceed that of the d → d absorption peak(s). In t_{2g}^3 and t_{2g}^6 complexes, electrons transfer from t_{2g} to e_g^* orbitals (mainly metal orbital to orbitals with metal and ligand character as in Figure 8.49 for Cr^{3+}) resulting in bond lengthening that can lead to ligand dissociation and ligand substitution reactions. All $M(NH_3)_6^{3+}$ complexes for Cr, Co, Rh, and Ir react with water in the light to form $M(NH_3)_5(H_2O)^{3+}$ complexes at 25 °C (Equation 9.82a); for example, the $Co(NH_3)_6^{3+}$ is inert to substitution without visible light activation, and the reaction is instantaneous with proper light activation. Similar behavior is found for cyanide complexes (Equation 9.82b).

$$M(NH_3)_6^{3-} + H_2O + h\upsilon \rightarrow M(NH_3)_5(H_2O)^{3+} + NH_3 \qquad (9.82a)$$

$$Fe(CN)_6^{4-} + H_2O + h\upsilon \rightarrow Fe(CN)_5(H_2O)^{3-} + CN^- \qquad (9.82b)$$

However, the quantum efficiency for these $M(NH_3)_6{}^{3+}$ reactions is higher for Rh and Ir complexes by 10-fold or more and typically increases with ligand field strength. However, the aquation of $Co(CN)_6{}^{3-}$ to form $Co(CN)_5(H_2O)^{2-}$ has a quantum efficiency five times that of the Rh and Ir ammine complexes [36].

9.8.3   LMCT and Photoreduction

If enough light energy (blue into the UV region) is provided, ligand to metal charge transfer occurs, and the metal ion can be fully reduced and the ligand is oxidized (e.g., nitrite and oxalate in Equations 9.83a,b).

$$Co(NH_3)_5NO_2{}^{2+} + h\upsilon \rightarrow Co(H_2O)_6{}^{2+} + 5NH_3 + NO_2{}^+ \tag{9.83a}$$

$$2Fe(C_2O_4)_3{}^{3-} + h\upsilon \rightarrow 2Fe(C_2O_4)_2{}^{2-} + C_2O_4{}^{2-} + 2CO_2 \tag{9.83b}$$

9.8.4   MLCT Simultaneous Substitution and Photo-Oxidation Redox

In the UV region, metal to ligand charge transfer to π acceptor ligands results in oxidation of the metal. The following reaction shows Fe(II) oxidation and reduction of the ligand NO^+ to NO with water substitution.

$$Fe(CN)_5(NO)^{2-} + H_2O + h\upsilon \rightarrow Fe(CN)_5(H_2O)^{2-} + NO$$

Flash photolysis can lead to photo-oxidation of the metal as an electron is ejected to the solvent water forming the solvated electron as in these three examples. In acid, H_2 production occurs. Metal +3 ions can readily hydrolyze.

$$Ru(NH_3)_6{}^{2+} + H^+ + h\upsilon \rightarrow Ru(NH_3)_6{}^{3+} + \frac{1}{2}H_2$$

$$M(H_2O)_6{}^{2+} + h\upsilon \rightarrow M(H_2O)_5(OH)^{2+} + \frac{1}{2}H_2$$

$$Fe(CN)_6{}^{4-} + h\upsilon \rightarrow Fe(CN)_6{}^{3+} + e_{aq}^{-} \quad \text{Solvated electron with UV light}$$

9.9   Effective Atomic Number (EAN) Rule or the Rule of 18

The Lewis octet rule was important in describing simple bonding for the main group elements. Likewise, the transition series metals can achieve a completely filled $ns^2np^6(n-1)d^{10}$ electron configuration of 18 electrons, which is known as the rule of 18 or the **effective atomic number** (EAN) rule [37]. Although there are many compounds that can have less than or even more than 18 electrons and be stable, the EAN rule is very important for describing many stable compounds and catalytic cycles. These include neutral metal atoms reacting with carbon monoxide to form metal carbonyl compounds and with tri-organo-phosphine (R_3P) compounds to form metal phosphine compounds, both of which are important in metal catalysis (see Section 9.10). Figure 9.17 shows the simple metal carbonyl compounds for the first transition series, and includes the electron count to reach the rule of 18. Metal carbonyl compounds can be formed at ambient temperature and 1 atm of CO pressure with the powdered metal, but frequently are formed at higher temperatures and under higher CO pressures. They can also be formed by the reaction of a metal salt with a reducing agent in the presence of CO.

To reach 18 electrons, Cr has a total of 6 electrons in its outer shell and accepts 12 electrons when it reacts with six CO molecules (each of which donates 2 electrons to the Cr). Similarly, Fe and Ni react with five

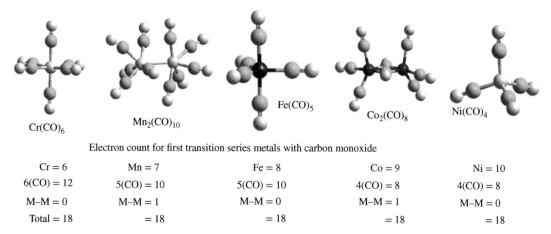

Electron count for first transition series metals with carbon monoxide

Cr = 6	Mn = 7	Fe = 8	Co = 9	Ni = 10
6(CO) = 12	5(CO) = 10	5(CO) = 10	4(CO) = 8	4(CO) = 8
M–M = 0	M–M = 1	M–M = 0	M–M = 1	M–M = 0
Total = 18	= 18	= 18	= 18	= 18

Figure 9.17 *Simple metal carbonyl compounds of the first transition series with the method for electron counting to reach the rule of 18. Models created with the Hyperchem™ program (version 8.0.10)*

and four CO molecules to reach an 18 electron count. However, Mn and Co have an odd number of electrons in their outer shells; thus, they form dimers where each metal supplies an electron to the other for an M–M bond. For $Mn_2(CO)_{10}$, there are five terminal CO ligands bound to each Mn for a total of 17 electrons, and the sharing of an electron between each Mn results in achieving the 18 electron count. For $Co_2(CO)_8$, two structures are known to exist in solution. The first is similar to $Mn_2(CO)_{10}$. The second has three terminal CO ligands bound to each Co plus two bridging (μ = bridging, Section 8.3) CO ligands, which still only donate two electrons per CO molecule to the total structure.

It is possible to distinguish the type of metal–carbon monoxide bonding using infrared spectroscopy. Free CO has a C≡O stretching frequency at 2143 cm^{-1} whereas terminal M–CO bonds have a C≡O stretching frequency ranging from 1852 to 2125 cm^{-1} indicating that the C–O bond is in between C≡O and C = O. The C = O stretching frequency for saturated ketones is 1715 cm^{-1}. Figure 8.33b demonstrated the π acceptor capability of CO as the π^* LUMO CO can accept electrons from the metal t_{2g} orbitals; placing electron density into the π^* orbital decreases the bond order and the C–O stretching frequency for CO. (Figure 8.33c shows similar **back bonding** for the M–P bond.)

Although bridging CO compounds have an M–C–M bond angle near 90° rather than the 120° bond angle (C–C–C) in organic compounds, the C–O stretching frequency in bridging metal carbonyl compounds ranges from 1850 to 1700 cm^{-1}. Thus, more electron density is placed into the CO π^* orbitals with a further decrease in the bond order and the C–O stretching frequency for CO.

It is possible to have more complex polymeric or cluster forms with CO and metal atoms. For example, in one structural form of $Fe_3(CO)_{12}$, the three Fe atoms form an equilateral triangle similar to the three C atoms in cyclopropane (C_3H_6). Each iron atom has three CO molecules, and there are three bridging CO molecules along with the three M–M bonds to reach a total electron count of 54 electrons ($54/3 = 18$ electrons for each Fe atom). Also, each Ir in $Ir_4(CO)_{12}$ has three terminal CO groups and the four Ir atoms bind to each other in a tetrahedron, which is similar to phosphorus in P_4 (Figure 4.6).

The electron counting of these and other cluster structures are similar to that used for the octet rule. $Mn(CO)_5$ has 17 electrons and is similar to a halogen atom (X•) and the CH$_3$•, which are 1 electron short of the

Table 9.19 *Carbon fragments, main group elements, and metal cluster fragments are each deficient by the same number of electrons*

C fragment	Main group analog	e^- count	Metal cluster fragment	e^- count
CH_4	Ar	8	$Cr(CO)_6$	18
CH_3	Cl	7	$Mn(CO)_5$	17
H_2C	S	6	$Fe(CO)_4$	16
HC	P	5	$Ir(CO)_3$	15

octet. Similarly, Cl^- has an octet whereas $[Co(CO)_4]^-$ has 18 electrons. Hoffmann [38] developed the concept of **isolobal fragments** to describe the parallel between the rule of 18 and the octet rule, "We will call two fragments isolobal if the number, symmetry properties, approximate energy and shape of the frontier orbitals and the number of electrons in them are similar – not identical, but similar." Table 9.19 shows some examples.

It is important to discuss the "weak" sigma bonding between the pair of electrons on the carbon atom in the CO to the neutral metal atom. The $3\sigma^{nb}$ orbital of CO has a very stable energy of $-14.1\,eV$ (see Figure 5.19) – one reason that higher pressures are needed to form some metal carbonyl compounds. Metal carbonyl compounds are typically diamagnetic and low spin, because of the low metal oxidation state, but the weak M–C σ bond indicates that **CO is labile** to dissociation. Thus, metal carbonyl compounds are very useful as transition metal catalysts (Section 9.13.1) where CO can be substituted readily.

9.10   Thermodynamics and Kinetics of Organometallic Compounds

Organic compounds also react with low oxidation state metal atoms to form M–C bonds. These can be direct donation of two electrons in a normal σ bond as in the Co–CH_3 bond of vitamin B_{12} where :CH_3 is formally an anion. There are also many **π donor ligands** that donate electrons to metals from the double bond of an unsaturated compound such as ethylene (two-electron donor), butadiene (two- or four-electron donor), and aromatic compounds (six-electron donor). An excellent example is the anion of Zeise's salt, $[PtCl_3(CH_2=CH_2)]^-$, or the trichlorido(η^2-ethylene)platinate(II) ion. Figure 8.33d and e shows ethylene ($CH_2=CH_2$) donating two electrons from its π bond to the Pt $d_{x^2-y^2}$ (formally a σ bond as the electron density is on the internuclear axis). Pt(II) also donates t_{2g} electrons to the π^* orbital of ethylene in a π bond. Because both C atoms in ethylene bind equally to the Pt, they are considered to be held or fastened to the Pt, which for nomenclature purposes is represented by the symbol η (eta) indicating **hapto**. The olefin compounds and the M–C (M–alkyl) bonds are exceptionally important in catalysis reactions as they lead to new products as described in Section 9.11.

Benzene and the cyclopentadiene anion (-1 charge) donate six electrons to a metal, and are aromatic ligands. The EAN rule is obeyed for the neutral molecule $Cr(CO)_3(C_6H_6)$ with the chemical name (η^6-benzene)(tricarbonyl)chromium (Figure 9.18) where η^6 indicates that all six C atoms bind to Cr equally. The best known cyclopentadiene anion compound is $Fe(\eta^5-C_5H_5)_2$, which is a Fe(II) compound that also obeys the EAN rule. Its chemical name is bis(η^5-cyclopentadienyl) iron, but its common name is ferrocene (Figure 9.18). It is an example of a **sandwich** compound as the Fe atom is between the two cyclopentadiene anion pentagons that are staggered to each other. It is the first example of the class of compounds called **metallocenes** (organometallic compounds where the metal binds between two planar ring systems).

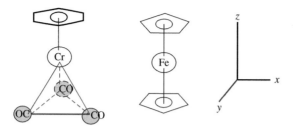

Figure 9.18   *Structure of $Cr(CO)_3(C_6H_6)$ with a vertical C_3 axis and of ferrocene, $Fe(\eta^5 - C_5H_5)_2$ with a vertical C_5 axis, and pentagonal antiprisms*

9.11   Electron Transfer to Molecules during Transition Metal Catalysis

A catalyst is a chemical species that increases the reaction rate for a thermodynamically favorable reaction ($\Delta G_r^0 < 0$). Although it is not consumed, it can form with the reactants a new metastable intermediate species, which then decomposes to regenerate the catalyst and the products. Figure 9.19a shows a reaction coordinate diagram indicating that the $\Delta G^{\neq}$ decreases from the uncatalyzed thermal reaction. In Figure 9.19b, the reaction coordinate diagram for the same reaction (note that ΔG_r^0 is the same) shows that a stable intermediate forms of lower energy than the products; thus, the reaction would stop at this point before products form as further reaction is unfavorable ($\Delta G_r^0 > 0$). In both figures, there are more elementary steps and intermediates formed along the reaction coordinate as shown by the additional lower energy potential wells. A thermodynamically unfavorable reaction cannot be enhanced via catalysts. However, there are chemical species, which are called catalytic poisons that retard a reaction; that is, they prevent one or more elementary steps along the reaction coordinate from occurring so that products cannot form.

For transition metals to be effective catalysts, a metal–ligand bond must be labile so that ligand substitution occurs readily at the various steps along the reaction path so that an inert/stable compound does not form and terminate the reaction. Transition metal catalysis is important for the production of many common chemicals used in industry and the laboratory such as acetic acid, aldehydes, and benzene. In the production of these

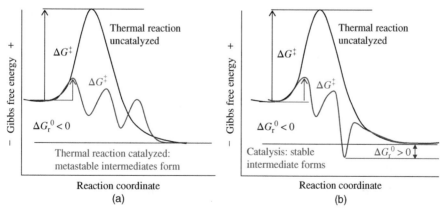

Figure 9.19   *Reaction coordinate diagrams for the thermal uncatalyzed reaction (black) (a) with catalyzed reactions leading to metastable intermediates (blue); (b) with catalyzed reactions leading to a stable intermediate (blue)*

chemicals, several steps occur in a **catalytic cycle** so that the reactants form the products and regenerate the catalyst. There are two key reactions for this process: *oxidation addition* (*OXAD*) *and reductive elimination* (*Redel*).

9.12   Oxidation Addition (OXAD) and Reductive Elimination (Redel) Reactions

In OXAD reactions, metals act as two electron donors (the reducing agent) and acceptors of two ligands from a reactant (the oxidizing agent). Thus, the metal loses two e^- and gains two ligands. Metals in lower oxidation states are more susceptible to losing electrons or acting as bases (**metal basicity**). In the $PtCl_4^{2-}$ example below (Equation 9.84b), Cl_2 is reduced to two chloride ions as square planar Pt(II) is oxidized to form octahedral Pt(IV); then the chloride ions bind to Pt(IV). Formally, the Pt loses two electrons and then obtains two lone pairs of electrons so Pt appears to act as an electron donor and acceptor in the same reaction. These reactions are important for square planar metal complexes (metal with a d^8 electron configuration and four ligands has a total of 16 electrons). Thus, the metal has two less electrons than the total 18 electrons required to achieve the rule of 18. OXAD reactions occur with electron acceptor molecules that have stable LUMOs as well as some of the strongest bonds known; for example, H_2, H–halogen (HX), halogen–halogen (X–X), CH_3–I, and O_2. These reactions lead to **activation of small molecules** for further chemistry.

$$ML_4 \quad + X–Y \rightarrow ML_4XY \tag{9.84a}$$

$$PtCl_4^{2-} \quad + Cl_2 \ \rightarrow PtCl_6^{2-} \tag{9.84b}$$

d^8, 16 e^- (unsaturated) d^6, 18 e^- (saturated)

square planar   octahedral (or tetragonal distorted)

Redel reactions are the reverse of OXAD reactions; that is, the metal gains two electrons and loses two ligands (reverse of Equation 9.84b). For some reactions, a combination of OXAD and Redel leads to a reversible reaction as in Vaska's complex *trans*$-Ir(CO)(P(C_6H_5)_3)_2Cl$, which reversibly binds both H_2 and O_2 (Equations 9.85a and 9.85b). In both cases, Ir(I) is oxidized to Ir(III), and H_2 and O_2 react **side on,** not end on, with Ir. On heating the product or passing inert gas into the solution, the original Ir complex is regenerated. Figure 9.20a shows that the weakly antibonding d_{z^2} HOMO orbital of square planar Ir(I) (Figure 9.8) donates two electrons (one each) into the π_x^* and π_y^* (not drawn) SOMO orbitals of O_2 so O_2 forms the peroxide ion. Figure 9.20b also shows that H_2 reacts to accept two electrons in its σ_u^* LUMO from the d_{z^2} HOMO orbital of Ir(I) so that hydride ions formally bind Ir(III) in the product. The two O (and H) atoms formally occupy cis positions around Ir(III). For H_2, the H–H bond is broken (**homolytic** cleavage) and the hydride ions are a source of reducing power. The following Equations 9.85a and 9.85b show these reaction types.

$$ML_4 + X_2 \rightarrow ML_4X_2 \text{ or } ML_4(X)_2$$

$$Ir(CO)(P(C_6H_5)_3)_2Cl + O_2 \quad \leftrightarrow IrO_2(CO)(P-phenyl_3)_2Cl \tag{9.85a}$$

$$Ir(CO)(P(C_6H_5)_3)_2Cl + H_2 \quad \rightarrow Ir(H)_2(CO)(P-phenyl_3)_2Cl \tag{9.85b}$$

Ir(I)d^8, 16 e^-   **Ir(III)d^6, 18 e^-**

Many biologically relevant reactions occur as shown for H_2. The types of sigma bonds that can be broken include dihalogen (Figure 9.20c) and CH_3–halide bonds (Figure 9.20d, the latter leads to methylation of metals; Section 5.6.3.3). The C–H bond in methane is possible, but its LUMO is not stable. The X–X (Chapter 4,

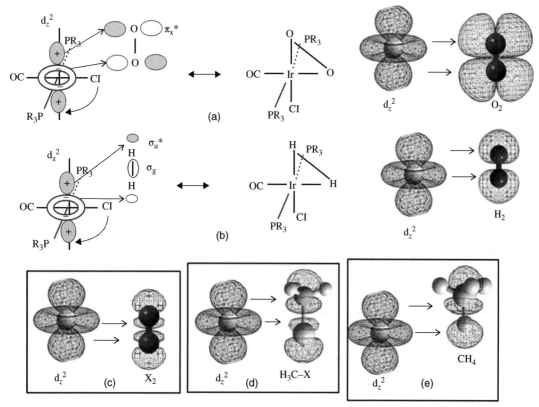

Figure 9.20 *Reactions showing electron donation from the d_{z^2} HOMO of Ir(I) to the LUMOs of (a) $O_2(\pi^*)$ and (b) $H_2(\sigma^*)$. Electron probability plots of the orbitals directly involved in electron transfer are provided on the right. Electron density contours (0.05) of (c) d_{z^2} HOMO to σ^* LUMO of dihalogens. (d) d_{z^2} HOMO to σ^* LUMO of C–X in methyl halides. (e) d_{z^2} HOMO to σ^* LUMO of C–H in methane. Orbitals created with the Hyperchem™ program (version 8.0.10)*

Figure 9.11) and C–X bonds (Figure 5.37) are broken as they accept electrons into the σ^* orbitals as shown in Figure 9.20. Figure 9.21 shows the orbital energy level diagrams for the metal, H_2 and Cl_2. The LUMO energies for H_2 and Cl_2 are stable so can accept electrons leading to bond cleavage and formation of ML_4HH and ML_4ClCl species. The HOMO energy for metals has been estimated by Pearson [39].

9.13 Metal Catalysis

Many catalytic reactions occur in homogeneous solutions and a key step is that Redel reactions are not truly reversible as new products rather than the original reactants are formed. Thus, the metal center is now different from the original material. On further reaction, the metal center may react with another chemical species, and the original metal center can be regenerated. In this case, the metal center is a catalyst and the catalytic cycle is restarted. For a catalyst to be effective and activate small molecules to do the chemistry and biochemistry, it must be regenerated at a high frequency (**turnover** rate). There are numerous examples of metal catalysis that are routes to chemicals and pharmaceuticals that are used daily throughout the world. Several of the scientists,

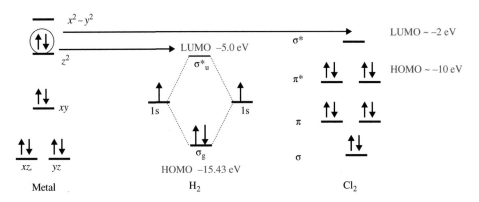

Figure 9.21 *Donation of electrons from a square planar metal to the LUMOs of* H_2 *and* Cl_2

who have designed these catalysts, have won Nobel prizes for their discovery. The following reactions show the synthesis of larger molecules from simple starting materials. Depending on the catalyst, these reactions frequently occur at room temperature for the second and third transition series metals but at higher pressures and temperatures for the first transition series.

$3\ HC \equiv CH \rightarrow C_6H_6$	Ziegler–Natta alkene polymerization ($TiCl_3$ or $AlCl_3$)
$CH_3OH + CO \rightarrow CH_3COOH$	Monsanto process (Co, Rh, Ir catalyst)
$3H_2 + N_2 \rightarrow NH_3$	Haber process (Fe at > 100 atm and $> 400\,°C$)

9.13.1   OXO or Hydroformylation Process

This is one of the most studied catalytic processes and it uses CO, H_2, and olefins to produce aldehydes in a chain extension reaction [40]. The starting compound is $Co_2(CO)_8$, a Co^0 compound. It reacts with H_2 to form the catalyst $HCo(CO)_4$, which has a C.N. of 5 with 18 electrons around the Co. The Co–H bond formally contains hydride (H^-) so both Co and H_2 act as electron acceptors and donors. Co in $HCo(CO)_4$ is labile as it forms $HCo(CO)_3$, which is Co^+ (d^8 and EAN = 16); thus, the Co becomes unsaturated. Figure 9.22a shows four major steps that are considered important in the catalytic cycle, and Figure 9.22b shows a schematic catalytic cycle.

The first step shows that $HCo(CO)_4$ loses a CO to form the unsaturated four-coordinate $HCo(CO)_3$ intermediate. In step 2, $HCo(CO)_3$ then reacts with the olefin to form a saturated complex (I) formed when the olefin double bond donates two electrons to the Co^+ center. In step 3, the $H:^-$ shifts from the Co^+ center to the olefin in an anti-Markownikoff addition to form a linear RCH_2CH_2Co species (II) with a Co–C σ bond where the Co^+ is four coordinate; another CO inserts into the C–Co bond to form a linear C–(CO)–Co arrangement (III) where the Co^+ is still four coordinate. In step 4, H_2 adds to the Co^+ in an OXAD reaction to form the Co^{3+} complex, which after another H^- or $(H:)^-$ shift to the CO bound to cobalt and a Redel step, produces the aldehyde and regenerates the $HCo(CO)_3$ catalyst. Under high CO pressures, the reaction slows down as the $HCo(CO)_3$ reforms the saturated $HCo(CO)_4$, which is excellent evidence for the formation of

(1) $Co_2(CO)_8 + H_2 \rightarrow \quad HCo(CO)_4 \quad \rightarrow \quad HCo(CO)_3 + CO$

(C.N. = 5; EAN = 18; saturated)   (Co^+, d^8, C.N. = 4; unsaturated)

(2) $HCo(CO)_3 +$ RCH=CH$_2$ Ligand addition $\begin{array}{c}H_2C \quad H \\ || \rightarrow Co(CO)_3 \\ HCR \end{array}$ (I)

(Co^+, d^8, C.N. = 4; unsaturated)   (Co^+, d^8; C.N. = 5; saturated)

(H⁻ shift)   (CO insertion)   O

(3) I $\Leftrightarrow$ RCH$_2$CH$_2$-Co(CO)$_3$ (II) $\rightarrow$ RCH$_2$CH$_2$-CCo(CO)$_3$ (III)

(Anti-Markownikoff)   (Chain extension)

(Co^+, d^8; C.N. = 4; unsaturated)   (Co^+, d^8; C.N. = 4; unsaturated)

(4) II $\begin{array}{c}H_2 \\ \Leftrightarrow \\ (OXAD)\end{array}$ $\begin{array}{c}O\ H \\ RCH_2CH_2\text{-}C\text{-}Co(CO)_3 \\ H\end{array}$ $\begin{array}{c}(H^- \text{ shift}) \\ \rightarrow \\ (Redel)\end{array}$ RCH$_2$CH$_2$CHO + HCo(CO)$_3$

(Co^{3+}, d^6; saturated)   (Catalyst regenerated)

(a)

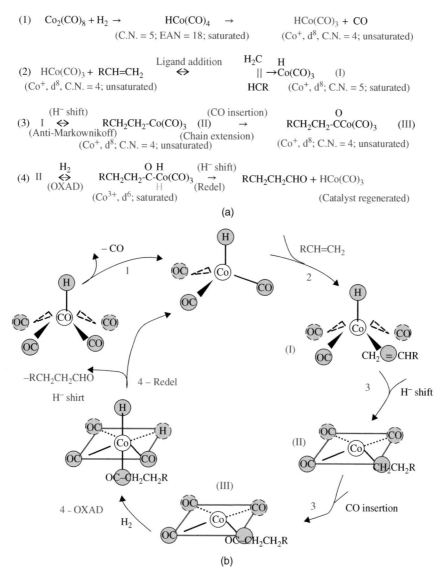

(b)

Figure 9.22   *(a) Proposed reaction sequence for the OXO process. (b)* **Tolman loop** *or catalytic cycle. The numbers indicate the steps in (a). The arrows show the regeneration of* HCo(CO)$_3$

the catalyst, $HCo(CO)_3$. Figure 9.22b shows a **Tolman loop** or catalytic cycle that describes the reactions in Figure 9.22a. Although $HCo(CO)_4$ and $HCo(CO)_3$ are drawn as square pyramid and tetrahedral, theoretical calculations indicate that these may be closer to trigonal bipyramid for $HCo(CO)_4$ and for $HCo(CO)_3$, which has a vacant site in the trigonal bipyramid structure. **Pseudorotation** between square pyramid and trigonal bipyramid structures also occurs.

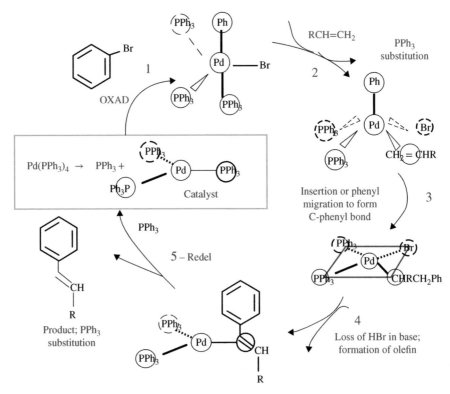

Figure 9.23   *The Heck reaction starting with Pd[P(Ph)$_3$]$_4$ in the box. Loss of a triphenylphosphine group leads to the catalyst Pd[P(Ph)$_3$]$_3$ Ph = phenyl or C$_6$H$_5$*

9.13.2   Heck Reaction

This reaction performs C–C coupling between two organic molecules [41] and uses the formation of a Pd–halogen bond to activate aromatic compounds (e.g., monobromobenzene). The general reaction is given below and Figure 9.23 shows the possible mechanism with a Tolman loop.

$$C_6H_5Br + RHC = CH_2 \rightarrow C_6H_5–CH = CHR + HBr$$

Figure 9.23 shows the formation of a five-coordinate Pd center when the catalyst, Pd[P(Ph$_3$)]$_3$, reacts with C$_6$H$_5$Br (1). The olefin displaces a P(Ph)$_3$ group in step 2. At step 3, a Pd-C σ bond [Pd − CH(R)CH$_2$Ph] forms when the phenyl group adds across the double bond. At step 4, HBr is lost with the H atom coming from the alkyl group, which results in a new π donor bond between Pd and the aromatic olefin compound. The olefin leaves when a P(Ph)$_3$ group displaces it (Redel step).

9.13.3   Methyl Transferases

An excellent biochemical example of an oxidative addition type reaction is the Co(I)–corrin complex (cobalamin) found in all organisms except plants [42]. Figure 9.24 shows the methionine synthase process, which cycles CH$_3$$^+$ between three chemical species. Co(I) is a d^8 ion and is a potent electron donor, which displaces

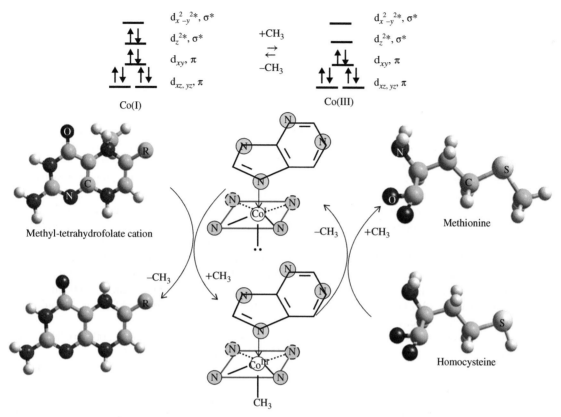

Figure 9.24 *The methionine synthase reaction mediated by* Co(I) ↔ Co(III) *transformations. (Upper) Electron configurations for Co(I) and Co(III). (Lower) Structures show the cycling of* CH_3^+ *via cobalamin from methylte-trahydrofolate cation to homocysteine. Models created with the Hyperchem™ program (version 8.0.10)*

a CH_3^+ group from methyl tetrahydrofolate cation (nucleophilic substitution reaction) to form methylcobal-amin [Co(III)]. A Co–CH_3 bond forms (Co^{3+}, d^6), and in a Redel type reaction, the CH_3^+ group is transferred to one of the two lone pairs of electrons on the S atom in homocysteine to form methionine (H^+ is released to solution). The Co(I) ↔ Co(III) transformation is an essential and efficient CH_3^+ shuttling reaction; however, Co needs to be consumed or eaten as vitamin B_{12} to replace any Co loss or excretion from the body. A signif-icant buildup of homocysteine in the blood indicates lack of vitamin B_{12} in humans, which increases the risk of cardiovascular diseases including blood clots, stroke, and heart attacks. Other methyl transfer reactions occur in the environment in a similar way with naturally produced methylating agents (CH_3X and DMSP) as noted in Section 5.6.3.3.

9.13.4   Examples of Abiotic Organic Synthesis (Laboratory and Nature)

Many catalytic reactions are **heterogeneous** as the catalyst is a solid with an activated surface, and the reac-tants are gaseous or dissolved that can adsorb to the surface. Table 7.9 showed that the surfaces of common materials or minerals can have acidic or basic sites depending on the solution pH. Thus, the catalyst's prop-erties can change as a function of pH. In the early 20th century, Fischer and Tropsch reacted CO from coal with H_2 (see Section 5.4.6) in the presence of metal catalysts to produce liquid hydrocarbons that were sold as fuels and lubricants. This commercial catalytic process allows CO to adsorb on the surface of solids containing Fe, Co, and Ni on alumina or high silica zeolites. Reaction of adsorbed CO and H_2 leads to generation of M–H bonds, with H^- migrating to bind CO similar to the oxo process. The reduction of CO produces simple alkanes as in the following reaction.

$$CO + 2H_2 \rightarrow CH_4, C_2H_6, C_3H_8, \text{etc.}$$

Figure 9.25a shows a top down view of how CO and H_2 can approach a metal surface. There are three ways that CO and H_2 can bind with the density of states of the Fe atoms: (1) across Fe atoms, (2) side on to an Fe atom in the plane or (3) side on perpendicular to the Fe plane (not shown). Figure 9.25b shows a side view with only CO across the Fe plane (xz). Figure 9.25c shows the Fe in the yz plane, and the possible side on orbital overlap of the Fe d_{yz} HOMO with the π^*_y LUMO of CO. The π^*_y LUMO of N_2 and the σ^*_u of H_2 are on the right and show that they have similar symmetry for the reaction with Fe. The s and partially filled d orbitals on Fe^0 clusters can interact (density of states for $E_{HOMO} = -5.2\,eV; E_{LUMO} = -2.5$ eV from [43]) with CO ($E_{LUMO} = -1.37$ eV) and H_2 ($E_{LUMO} = -5.0$ eV). The most favorably energetic interaction is that of the Fe HOMO with the H_2 LUMO to form H^-, but it likely reacts end on or perpendicular to one Fe atom (Fe $d_{z^2} \rightarrow \sigma^*_u$; not drawn) rather than across Fe atoms as the H–H bond length is only 74 pm. The CO (and N_2) bond lengths are 113 (110 pm N_2), respectively, and the nearest Fe–Fe distance is 260 pm (similar to mackinawite, Section 12.6.1). The radius of H^- is 139.9 pm in simple hydrides [44]; thus, it could bind across Fe atoms and then react with adsorbed CO and N_2.

The Fischer–Tropsch process may also occur at hydrothermal vents (water depths > 2000 m). These systems provide a rich source of raw materials (e.g., H_2, CO, H_2S, NH_3, reduced metals, metal sulfides, solid silicates, and aluminosilicates) as well as high temperatures (> 360 °C; Figure 1.1) and pressures (> 250 atm) to activate small molecules to adsorb to surfaces. These extreme conditions lead to abiotic synthesis of small organic molecules [45]. At vents, H_2 is produced via the serpentization reaction (Equation 8.19) or the formation of pyrite given below (Section 12.6.2). In the presence of excess H_2S and FeS (alone or with NiS), thiol (R-SH) compounds also form [46].

$$FeS + H_2S \rightarrow FeS_2 + H_2$$

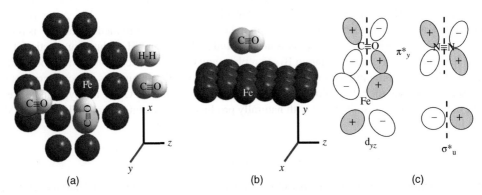

(a)　　　　　　　　　(b)　　　　　　　　　(c)

Figure 9.25 *(a) View of Fe atoms in the xz plane. (b) 90° rotation of the xz plane in (a) to show a side on view of the Fe atoms. (c) Possible side on orbital overlap with the Fe d_{yz} HOMO and π^*_y CO (N_2) LUMO; H_2 LUMO given for comparison. Models created with the Hyperchem™ program (version 8.0.10)*

9.13.5 The Haber Process Revisited

The **Haber** process to form NH_3 from N_2 and H_2, which uses a powdered Fe catalyst bound to SiO_2 and Al_2O_3 (Os and Ru can also be used), likely proceeds as a Fischer–Tropsch reaction. As noted in Section 5.4.6, there is a kinetic barrier to the reaction of N_2 and H_2, and the catalyst overcomes it. Although N_2 is isoelectronic with CO, it is a nonpolar molecule and does not readily donate a pair of electrons. Also, the E_{HOMO} of N_2 is -15.58 eV(σ_g), which is more stable than that of CO ($E_{HOMO} = -14.01$ eV), and its $E_{LUMO} = 2.2$ eV so is unstable. N_2 has a reasonable proton affinity (Table 7.3) so reaction with H^+ (and Fe) can lower the energy of its LUMO (e.g., Figure 5.17b for O species). For laboratory (and geochemical) reactions, high temperatures and pressures are necessary to activate the N_2 to bind the Fe catalyst with a σ bond. N_2 is also a π acceptor ligand as noted for the CO molecule above (see Figure 8.33b), and backbonding between the metal and N_2 likely aids its reduction to NH_3. Overall, the reaction of adsorbed N_2 and H_2 leads to generation of N–H bonds and NH_3 with hydride (H^-) or H• atoms migrating to the N_2 bound to Fe. The overall process is similar to the oxo process and to the ferredoxins in nitrogenase [47; Section 12.6.4.1]. Because H^- has an E_{HOMO} of $+0.754$ eV, it is an excellent reductant. H• will react as a free radical. Thus, both species can react with N_2.

References

1. Martell, A. E. and Hancock, R. D. (1996) *Metal Complexes in Aqueous Solutions*, Plenum Press, New York, pp. 253.
2. Luther, G. W., III, Theberge, S. M. and Rickard, D. T. (1999) Evidence for aqueous clusters as intermediates during zinc sulfide formation. *Geochimica et Cosmochimica Acta* **63**, 3159–3169.
3. Luther, III, G. W., Theberge, S. M., Rozan, T. F., Rickard, D., Rowlands, C. C. and Oldroyd, A. (2002) Aqueous copper sulfide clusters as intermediates during copper sulfide formation. *Environmental Science and Technology*, **36**, 394–402.
4. Eigen, M. and Mass, G. (1966) Über die kinetic der metallkomplexbildung der alkali- und erdalkaliionen in wässrigenlösungen. *Zeitschrift für Physikalische Chemie NeueFolge*, **Bg. 49**, S. 1633–177.
5. Margerum, D. W., Cayley, R., Weatherburn, D. C. and Pagenkopf, G. K. (1978) Kinetics and mechanisms of complex formation and ligand exchange. In *Coordination Chemistry* Vol. 2 (ed. Martell, A. E.) ACS Monograph 174 ACS, Washington, D.C. p. 1–221.
6. Richens, D. T. (1997) *The Chemistry of Aqua Ions*, pp. 592. John Wiley & Sons, Chichester.
7. Dunand, F. A., Helm, L., and Merbach, A. E. (2003) Solvent exchange on metal ions, in *Advances in Inorganic Chemistry*, Vol. 54, 1–69.
8. Helm, L., Nicole, G. M. and Merbach, A. E. (2005) Water and proton exchange processes on metal ions, in *Advances in Inorganic Chemistry*, Vol. 57, 327–379.
9. Helm, L. and Merbach, A. E. (2005) Inorganic and bioinorganic solvent exchange mechanisms. *Chemical Reviews*, **105**, 1923–1959.
10. Taube, H. (1952) Rates and mechanisms of substitution in inorganic complexes in solution. *Chemical Reviews*, **50**, 69–126.
11. Langford, C. H. and Gray, H. B. (1965) *Ligand Substitution Processes*, W.A. Benjamin, New York.
12. Langford, C. H. and Gray, H. B. (1968) Chemical & Engineering News, pp. 68–75.
13. Swaddle, T. W. (1974) Activation parameters and reaction mechanism in octahedral substitution. *Coordination Chemistry Reviews*, **14**, 217–268.
14. Basolo, F. and Pearson R. G. (1967) *Mechanisms of Inorganic Reactions*, Wiley, New York, p. 701.
15. Krishnamurthy, R. and Schaap, W. B. (1969) Computing ligand field potentials and the relative energies of d orbitals. *Journal of Chemical Education*, **46**, 799–810.

16. Wilkins, R. G. (1991) *Kinetics and Mechanism of Reactions of Transition Metal Complexes* (2nd ed.), Weinheim, VCH, pp. 465.

17. Langford, C. H. (1965) On the acid hydrolysis of $[Co(NH_3)_5X]^{+2}$ ions and the mechanism of interchange. *Inorganic Chemistry,* **4**, 265–266.

18. Wilkins, R. G. (1974) *The Study of Kinetics and Mechanism of Reactions of Transition Metal Complexes*, Allyn and Bacon Inc., Boston, MA, pp. 403.

19. Pearson, R.G. and Ellgen. (1975) Mechanism of inorganic reactions in solution, in *Physical Chemistry, An Advanced Treatise Vol. VII* (ed. H. Eyring), Academic Press Chapter 5.

20. Grant, N. and Jordan, R. B. (1981) Kinetics of solvent water exchange on Iron(III). *Inorganic Chemistry,* **20**, 55–60.

21. Monacelli, F. (1968) Anation reaction of $[Rh(H_2O)(NH_3)_5]^{3+}$ ion. Evidence for the nucleophilic attack of the entering group. *Inorganica Chimica Acta,* **2**, 263–268.

22. Borghi, E., Monacelli, F. and Prosperi, T. (1970) Anation and water exchange reactions of $Ir(NH_3)_5OH_2{}^{3+}$ cation. *Inorganic Nuclear Chemistry Letters,* **6**, 667–670.

23. Thusius, D. (1971) Rate constants and activation parameters for the formation of monosubstituted chromium(III) complexes. *Inorganic Chemistry,* **10**, 1106–1108.

24. Pearson, R. G., Boston, C. R. and Basolo, F. (1953) Mechanism of substitution reactions in complex ions. III. Kinetics of aquation of some cobalt(III) complex ions. *Journal of the American Chemical Society,* **75**, 3089–3092.

25. Chau, W. K. and Poon, C. K. (1971) Structural and mechanistic studies of coordination compounds. Part III. Steric effects in octahedral aquation of cobalt(III) Co complexes containing different quadridentate macrocyclic secondary amines. *Journal of the Chemical Society A*, 3087–3091.

26. Tobe, M. L. (1970) Base hydrolysis of octahedral complexes. *Accounts of Chemical Research,* **3**, 377–385.

27. Wilkins, R. G. (1970) Mechanisms of ligand replacement in octahedral nickel(II) complexes. *Accounts of Chemical Research,* **3**, 408–416

28. Robb, D., Steyn, M. M. and Krüger, H. (1969) Kinetics and mechanism of the anation by chloride, bromide, iodide, nitrite, azide and thiocyanate ions on the aquapentachlororhodate(III) anion. *Inorganica Chimica Acta,* **3**, 383–387.

29. Taube, H. (1983) Electron transfer between metal complexes – retrospective. *Nobel Lecture*, 8 December 1983.

30. Marcus, R. A. (1992) Electron transfer reactions in chemistry: theory and experiment. *Nobel Lecture*, 8 December 1992.

31. Brunschweig, B. S. and Sutin, N. (1999) Energy surfaces, reorganization energies, and coupling elements in electron transfer. *Coordination Chemistry Reviews***187**, 233–254.

32. Macartney, D. H. and Sutin, N. (1985) Kinetics of the oxidation of metal complexes by manganese(III) aquo ions in acidic perchlorate media: the $Mn(H_2O)_6{}^{2+} - Mn(H_2O)_6{}^{3+}$ electron-exchange rate constant. *Inorganic Chemistry,* **24**, 3403–3409.

33. Stynes, H. C. and Ibers, J. A. (1971) Effect of metal–ligand bond distances on rates of electron-transfer reactions. The crustal structures of hexaammineruthenium(II) iodide, $[Ru(NH_3)_6]I_2$, and hexaammineruthenium(III) tetrafluoroborate, $[Ru(NH_3)_6][BF_4]_3$. *Inorganic Chemistry,* **10**, 2304–2308.

34. Pennington, D. E. (1978) Oxidation–reduction reactions of coordination complexes. In *Coordination Chemistry* Vol. 2 (ed. Martell, A. E.), ACS Monograph 174. ACS Washington, D.C. p. 476–590.

35. Sutin, N. and Creutz, C. (1980) Light induced electron transfer reactions of metal complexes. *Pure and Applied Chemistry,* **52**, 2717–2738.

36. Ford, P. C., Wink, D. and Dibenedetto, J. (1983) Mechanistic aspects of the photosubstitution and photoisomerization reactions of d^6 metal complexes, in *Progress in Inorganic Chemistry*, Vol. 30, 213–271.

37. Tolman, C. A. (1972) The 16 and 18 electron rule in organometallic chemistry and homogeneous catalysis. *Chemical Society Reviews*, **1**, 337–353.

38. Hoffman, R. (1981) Building bridges between inorganic and organic chemistry. *Nobel Lecture*, 8 December 1981.

39. Pearson, R. G. (1988) Absolute electronegativity and hardness: application to inorganic chemistry. *Inorganic Chemistry*, **27**, 734–740.

40. Heck, R. F. and Breslow, D. S. (1961). The reaction of cobalt hydrotetracarbonyl with olefins. *Journal of the American Chemical Society*, **83**, 4023–4027.

41. Heck, R. F. (2006) Cobalt and palladium reagents in organic synthesis: the beginning. *Synlett*, **2006** (18), 2855–2860.

42. Brown, K. L. (2005) Chemistry and enzymology of vitamin B_{12}. *Chemical Reviews*, **105**, 2075–2149.

43. Wang, L.-S., Xi, L. and Zhang, H.-F. (2000) Probing the electronic structure of iron clusters using photoelectron spectroscopy.*Chemical Physics*, **262**, 53–63.

44. Lang, P. F. and Smith, B. C. (2010) Ionic radii for Group 1 and Group 2 halide, hydride, fluoride, oxide, sulfide, selenide and telluride crystals. *Dalton Transactions*, **39**, 7786–7791.

45. McCollom, T. M. and Seewald, J. S. (2007) Abiotic synthesis of organic compounds in deep-sea hydrothermal environments. *Chemical Reviews*, **107**, 382–401.

46. Rickard, D. and Luther, III, G. W. (2007) Chemistry of iron sulfides. *Chemical Reviews*, **107**, 514–562.

47. Hoffman, B. M., Lukoyanov, D., Yang, Z.-H., Dean, D. R. and Seefeldt, L. C. (2014). Mechanism of nitrogen fixation by nitrogenase: the next stage. *Chemical Reviews*, **114**, 4041–4062.

10

Transition Metals in Natural Systems

10.1 Introduction

Transition metals are very important in many geochemical and biological processes. There are some important differences between the family members of the first transition series and those of the second and third transition series. Recall that the "d" orbitals for the second and third transition series metals are more diffuse so pairing of electron spins occurs more readily. Also, shielding of nuclear charge by inner electrons leads to higher IP and EA values (Table 3.6); thus, they tend to have higher effective nuclear charges for congeners with comparable electron configurations.

In nature, the first transition series metals are typically characterized by +2 and +3 complexes, which exhibit tetrahedral, square planar, or octahedral geometry. In octahedral geometry, both high and low spin complexes are possible whereas square planar complexes are low spin and tetrahedral complexes are typically high spin. Although metal–metal bonds are known, they are much rarer than what is found in the second and third transition series metals. The metals are attacked or dissolved by dilute acids whereas the second and third transition series metals normally require concentrated oxidizing acids such as nitric acid alone or in combination with HCl (aqua regia). The second and third transition series metals form low spin complexes and exhibit coordination geometries of 4, 6, 7, and 8. There is a tendency for second and third transition series metals to form and remain in higher oxidation state compounds and ions than the first transition series metals. For example, molybdate as Mo(VI) is the dominant form of Mo in aquatic systems whereas chromate [with Cr(VI)] and Cr(III) exist in nature; Cr(III) converts to chromate when exposed to oxidants such as MnO_2 (an inner sphere electron transfer process).

10.2 Factors Governing Metal Speciation in the Environment and in Organisms

Many factors affect the reactivity of a metal ion in the environment. Section 1.8.2 indicated that the oxidation state of a metal [e.g., Fe(II) versus Fe(III)] depends on the redox condition of the environment. For example, once dissolved O_2 and NO_3^- are consumed, microbes decompose organic matter with other oxidants such as MnO_2 and $FeOOH$, resulting in reduction to Mn^{2+} and Fe^{2+}. Metal ions can exist as inorganic complexes

Inorganic Chemistry for Geochemistry and Environmental Sciences: Fundamentals and Applications, First Edition. George W. Luther, III.
© 2016 John Wiley & Sons, Ltd. Published 2016 by John Wiley & Sons, Ltd.
Companion Website: www.wiley.com/go/luther/inorganic

Table 10.1 *Fe(III)/Fe(II) redox couples for selected ligands bound to iron [1–4].*
E values for porphyrin compounds vary with substituents on the ring. E values for
$Fe(H_2O)_6^{3+}$ *and* $Fe(H_2O)_5(OH)^{2+}$ *are at pH = 0 and 4, respectively. All other E data*
are for pH = 7. The relationship between E and the ratio of the thermodynamic stability
constants (β, see also Equation 13.1.2) for the Fe(III)L and Fe(II)L complexes is at the
top of the table

$$E_{Fe(III)L/Fe(II)L} = E^0{}_{aq} - 0.05915 \log \left(\frac{[\beta]^{Fe(III)}}{[\beta]^{Fe(II)}} \right)$$

Redox reaction	E (Volts) at pH = 7
$Fe(phen)_3^{3+} + e^- \rightarrow Fe(phen)_3^{2+}$	1.07
$Fe(H_2O)_6^{3+} + e^- \rightarrow Fe(H_2O)_6^{2+}$	0.77
$Fe(H_2O)_5(OH)^{2+} + e^- \rightarrow Fe(H_2O)_5(OH)^+$	0.34
$Fe(H_2O)_4(asp)^{2+} + e^- \rightarrow Fe(H_2O)_4(asp)^+$	0.33
$Fe(NTA)^- + e^- \rightarrow Fe(NTA)$	0.32
$Fe(EDTA)^- + e^- \rightarrow Fe(EDTA)$	0.12
$Fe(porphyrin)^{3+} + e^- \rightarrow Fe(porphyrin)^{2+}$	0.100
$Fe^{3+}(Mb) + e^- \rightarrow Fe^{2+}(Mb)$	0.046
$Fe(ox)_3^{3-} + e^- \rightarrow Fe(ox)_3^{4-}$	0.02
$Fe(H_2O)_4(OH)_2^+ + e^- \rightarrow Fe(H_2O)_4(OH)_2$	−0.02
$Fe(OH)_3 + e^- \rightarrow Fe^{2+}(10^{-5}M)$	−0.07
$Fe(oxinate)_3^- + e^- \rightarrow Fe(oxinate)_3$	−0.20
$\alpha FeOOH + e^- \rightarrow Fe^{2+}(10^{-5}M)$	−0.25
$[4Fe - 4S]^{2+} + e^- \rightarrow [4Fe - 4S]^+$	−0.25 to −0.65 (see Table 12.2)
$Fe^{3+}(ferrioxamine\,B) + e^- \rightarrow Fe^{2+}(ferrioxamine\,B)$	−0.468
$Fe^{3+}(enterobactin) + e^- \rightarrow Fe^{2+}(enterobactin)$	−0.750

with the following common inorganic ligands: chloride, carbonate, sulfate, phosphate, and sulfide. However, there are a variety of organic ligands that are naturally produced and that outcompete the inorganic ligands for bonding with metal ions. Ligands also affect the redox and spin state of a metal couple such as Fe(III)/Fe(II) (Section 8.7.3 for Co; Table 10.1 for Fe), and nature uses ligand–metal bonding to affect reactivity and catalysis. The left of Figure 10.1 shows potentials for several reduction couples that are important in life processes (sometimes termed a **redox spectrum**). An oxidized partner of a redox couple at the top such as O_2/H_2O can be reduced by the reduced partner of any couple below it as $E_{reaction} = E_{red} + E_{ox}$. Thermodynamically, the O_2/H_2O couple is one of the most efficient, but it also shows the energy needed for water splitting (the reverse reaction that requires **photochemistry;** Section 10.7). The FeS proteins (see Section 12.6.4) have different oxidation states and ligand attachments that tune their redox potential over 1 volt; reduction of N_2 to NH_3 is possible for some Fe_4S_4 redox centers (ferredoxins). Selected aspects of the redox chemistry for the metal redox centers in Figure 10.1 are described below and in Chapter 12.

For the first transition metal series, the spin state of the metal ion is affected by ligand attachment, which affects metal reactivity; e.g., octahedral Fe^{2+} can exist in high spin $t_{2g}{}^4 e_g{}^{*2}$ in FeS (pyrrhotite) or low spin $t_{2g}{}^6$ in FeS_2 (pyrite). Metal ions and low spin states have stronger bonds because of the shorter metal bond distances and extra stability due to CFSE/MOSE. As a result, they are less labile and reactive; e.g., pyrite is inert to ligand dissociation in HCl, but can be oxidized with nitric acid whereas FeS does dissolve in HCl (because sulfide is released, FeS is termed **acid volatile sulfide** or AVS).

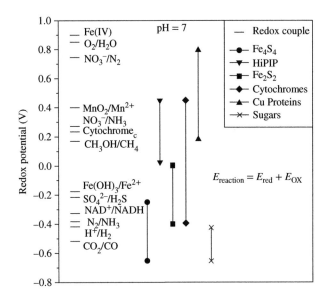

Figure 10.1 *Redox couples at pH 7 for important reactions in nature along with the potential ranges of sugars and some metal ligand redox centers. HiPIP indicates the Fe_4S_4 high potential ferredoxins (Fd). The N_2O/N_2 couple is highest at 1.35 V*

Environmental chemists and geochemists frequently ask the question "Is the metal dissolved or colloidal?" However, the physical state of the metal ion is frequently operationally defined by what passes through 0.2 or 0.4 μm filters. Interestingly, quantum-sized particles, nanoparticles, and **molecular clusters** (Chapter 12) can readily pass through these filters.

10.3 Transition Metals Essential for Life

In the first transition series, V, Cr, Mn, Fe, Co, Ni, Cu, and Zn have been found to be necessary for life in a variety of organisms [5]. In the second transition series metals, only Mo and Cd have been shown to be essential or capable of supporting life processes whereas only W in the third transition series has been found to be capable of supporting life processes. As different forms of life and chemical species occur in oxygenated environments versus sulfide-rich environments, the reactivity of a given metal ion with sulfide becomes critical in establishing that metal ion's availability to an organism. Before the Great Oxidation Event, which occurred between 2.0 and 2.4 billion years ago and resulted in the oxidation of H_2O to O_2 during photosynthesis, elements were in their reduced state and many metals formed insoluble metal sulfides. Based on solubility criteria of simple metal sulfides, Mn, Fe, Co, Ni, W, and Mo would be available for organisms whereas Cu and Zn would not be available for organisms. Also, any sulfide present would have reacted with Cu and Zn first because of their faster water exchange rates (Table 9.1). Although Mn^{2+} has a faster water exchange rate than Fe^{2+}, it does not form a sulfide mineral readily. Fe^{2+} forms FeS and FeS_2, but the total amount of Fe atoms is greater than that of S on earth by orders of magnitude (Figure 1.6). Thus, the ocean would have had significant concentrations of Mn^{2+} and Fe^{2+} for organisms to uptake to support life processes. A couple of brief examples follow.

Table 1.3 noted the importance of several metals in processes necessary for life. Many processes center around the binding and/or activation of small molecules or ions (e.g., N_2, NO_3^-, O_2, H_2, CH_4) so they can react with each other or another chemical. Organisms have taken advantage of transition metal chemistry because their "d" orbitals permit them to act as electron donors and acceptors at the same time (in low oxidation states), to be efficient redox species, and to be labile or inert to substitution reactions based upon the occupation of the "d" orbitals, which is a function of the ligand bound to the metal. These characteristics are exploited in nature for catalysis using metals in enzymes (**metalloenzymes**); i.e., biochemically mediated reactions using a metal center.

Nitrogen is an essential nutrient, and organisms (archaea, bacteria, plants) have developed chemical strategies using **metalloenzymes** to perform the eight-electron conversion between ammonia and nitrate. In most instances, metal enzymatic reactions involve a catalytic cycle that bears similarity to the hydroformylation reaction in the previous chapter. Some specific enzymatic processes that couple metal catalytic cycles with H^+, electron, and/or O atom transfer include the nitrogenases, reductases and oxidases, and cytochromes. Nitrogenases bind N_2 using Fe in an enzyme containing FeS, FeMoS, FeWS, or V complexes or **clusters** (Section 12.6.4); the overall reaction is given below and shows the addition of protons and electrons to N_2 to form NH_3 at ambient temperature. Although, assimilatory N_2 reduction to NH_3 only requires six electrons, the biochemical process uses eight electrons as H_2 is also a product. Nitrite reductase also uses FeS proteins and cytochromes to affect reduction of nitrite to ammonia (Section 12.7).

$$N_2 + 8e^- + 8H^+ + 16MgATP \rightarrow 2NH_3 + H_2 + 16MgADP + 16P_i(\text{or } PO_4^{3-})$$

Examples of oxidases are the Mo (and W) enzymes, which have Mo = O bonds and oxidize sulfite, nitrite, and dimethylsulfide (DMS) as below. They also perform dissimilatory sulfate and nitrate reduction (the reverse processes). These are considered O atom transfer reactions, and for the oxidations, the lone pair of electrons on the central atom attacks the labile Mo = O bond [6].

$$LMo^{6+}(O)(O)(\text{s-cysteine}) + SO_3^{2-} + OH^- \leftrightarrow LMo^{4+}(O)(OH)(\text{s-cysteine}) + SO_4^{2-} \quad \text{(reversible)}$$

$$L_2Mo^{6+}(O)(\text{s-serine}) + DMS \leftrightarrow L_2Mo^{4+}(\text{s-serine}) + DMSO \quad \text{(reversible)}$$

Before describing selected biological metal centers, the oxidation reactions of Mn^{2+} and Fe^{2+} by O_2 are first presented. These redox reactions are model examples that describe possible strategies that organisms need to employ for O_2 binding, storage, and transport within the organism.

10.4 Important Environmental Iron and Manganese Reactions

Iron and manganese are the two most important redox metals in the environment as the cycling between their reduced and oxidized phases occurs seasonally when environments change redox state (e.g., from oxic to sulfidic; Figure 2.8) due to increased organic matter decomposition (Figure 1.13a). At circumneutral pH [7], an interesting feature of these reactions is that $[Mn(H_2O)_6]^{2+}$ oxidation by O_2 is slow (5 yrs.), but $MnO_2[Mn(IV)]$ reduction by H_2S is fast (<10 min). However, $[Fe(H_2O)_6]^{2+}$ oxidation and Fe(III) mineral reduction occur at roughly "similar" rates (~30 min.). This reactivity feature of the oxidation and reduction reactions for each system suggests a common pathway in the electron transfer steps for the forward and reverse reactions in each system. However, the pathways for the high spin iron and manganese reactions are discreetly different from each other. For Fe reactions, the reactivity is based on electrons being added to or removed from $t_{2g}(\pi)$ orbitals whereas for Mn reactions, the reactivity is based on electrons being added to or removed from e_g^* (σ^*)

orbitals as in the following.

$$\text{Fe(II)} \; t_{2g}{}^4 e_g{}^{*2} \leftrightarrow \text{Fe(III)} \; t_{2g}{}^3 e_g{}^{*2}$$

$$\text{Mn(II)} \; t_{2g}{}^3 e_g{}^{*2} \leftrightarrow \text{Mn(IV)} \; t_{2g}{}^3 \; \text{or}$$

$$\text{Mn(II)} \; t_{2g}{}^3 e_g{}^{*2} \leftrightarrow \text{Mn(III)} \; t_{2g}{}^3 e_g{}^{*1}$$

The octahedral aqueous ions of Mn(II), Mn(III), Fe(II), and Fe(III) are all labile and undergo ligand exchange. However, octahedral Mn(IV) has an inert electron configuration so does not undergo ligand exchange readily even under acidic conditions. This inertness can retard chemical reactivity. With this brief overview, the reaction of the aqueous reduced metal ions with O_2 and their solid oxidized phases with H_2S (Chapter 11) over a range of pH values are described.

10.4.1   Oxidation of Fe^{2+} and Mn^{2+} by O_2 – Environmentally Important Metal Electron Transfer Reactions

In Chapter 9, electron transfer reactions were shown to be first order in the oxidant and reductant (Equation 9.72) for an overall second-order reaction. The experimentally determined rate laws are normally more complicated because of pH dependence for the chemical species in the reactions and the formation of solid products that autocatalyze the redox reaction. The oxidation reactions of Fe^{2+} and Mn^{2+} with O_2 are examples of these complications. They are also examples of **metal-centered** redox reactions as the metal loses an electron directly to the oxidant.

Figure 10.2a shows idealized kinetic plots obtained for the reactions of Fe(II) and Mn(II) with O_2. For Fe(II), pseudo first order plots are obtained, and the experimentally determined rate law over pH is given in Equations 10.1a and 10.1b where $k_H = 3 \times 10^{-12} \; \text{mol}^{-1} \; \text{L}^{-1} \; \text{min}^{-1}$ at 20 °C.

$$-\frac{d}{dt}[\text{Fe(II)}] = k_H[\text{Fe(II)}][\text{OH}^-]^2 O_2(\text{aq}) \tag{10.1a}$$

$$= (k_0[\text{Fe}^{2+}] + k_1[\text{Fe(OH)}^+] + k_2[\text{Fe(OH)}_2])[O_2(\text{aq})] \tag{10.1b}$$

[Fe(II)] indicates all forms of dissolved Fe(II) (Equation 10.1a) or total Fe(II) in solution. Equation 10.1b from [10] explicitly indicates the Fe(II) speciation; on increasing pH, OH^- replaces H_2O as a ligand and binds Fe^{2+}. Note the overall second-order dependence in Equation 10.1b where each Fe(II) species is considered in the parentheses. As pH increases the number of OH^- bound to Fe^{2+} increases.

However, for Mn(II), there is curvature in the kinetic plots (Figure 10.2a) indicating that more Mn(II) is lost than expected; this behavior suggests an autocatalytic reaction [11]. Although the initial oxidation is first order in Mn(II) and O_2, the experimentally determined rate law is Equation 10.2, and the second autocatalytic term indicates that Mn(II) adsorbs or binds to the surface of solid MnO_x ($MnOOH$, Mn_3O_4, or other solid-phase Mn compound), which forms in the reaction. In the first term of Equation 10.2, k_1 is a pseudo first order constant representing the loss of dissolved Mn(II), and it depends on [O_2]aq, pH, temperature, and Mn(II) species.

$$-\frac{d}{dt}[\text{Mn(II)}] = k_1[\text{Mn(II)}] + k_2[\text{Mn(II)}][\text{MnO}_x] \tag{10.2}$$

Recall that Figure 2.9a described the thermodynamics of $[\text{Mn(H}_2\text{O)}_6]^{2+}$ and $[\text{Fe(H}_2\text{O)}_6]^{2+}$ oxidation by O_2 to form H_2O and oxidized Fe and Mn solid phases in a multielectron process for O_2 to H_2O. However, that analysis is in contrast to the known reactivity and kinetics of these reactions that indicate that Mn^{2+} is mainly unreactive below a pH of 8–9, and Fe^{2+} is unreactive below a pH of 5. Figure 10.2b (also see Figure 2.9b)

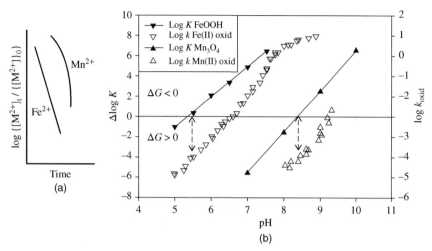

Figure 10.2 *(a) Idealized kinetics plots for Fe(II) and Mn(II) oxidation with O_2. (b) Calculated thermodynamic data with experimental kinetics data for the oxidation of $[Fe(H_2O)_6]^{2+}$ and $[Mn(H_2O)_6]^{2+}$ by O_2 to form O_2^-. Favorable reactions are those above the horizontal line. Hausmannite (Mn_3O_4) is the first formed oxide [8, 9]*

shows the available kinetic data for the reactions of Fe(II) species and O_2 compiled by [10] and of Mn(II) species and O_2 compiled by [11] over the pH range from 4–10 as well as the calculated thermodynamic data for the first electron transfer (O_2 to O_2^-) to form FeOOH and Mn_3O_4. The arrows indicate the predicted lowest pH at which the oxidation reactions are favorable. As noted in Chapter 2, these calculations are reasonable to one $\Delta \log K$ unit. The rate constants are very slow at the lower pH values; for $[Mn(H_2O)_6]^{2+}$, the abiotic reaction has not been observed to occur below pH = 8. These data indicate that these reactions are controlled by the first electron accepted by O_2 to form O_2^-, whose formation becomes thermodynamically more favorable as the pH increases [OH^- as a ligand binding to these M(II) species becomes important].The first electron transfer steps can be represented by these reactions.

$$[Fe(H_2O)_6]^{2+} + O_2 \rightarrow FeOOH + O_2^- + 3H^+ + 4H_2O$$

$$[Mn(H_2O)_6]^{2+} + O_2 \rightarrow MnOOH + O_2^- + 3H^+ + 4H_2O$$

An important geochemical point is that the banded iron formations (BIFs), which are alternating layers of silica and iron(III) oxyhydroxides, have very low Mn content. Thus, the oxidation of Fe^{2+} in the ancient ocean must have occurred at pH values less than 8 or 9.

The primary reason for the difference in reactivity of $[Mn(H_2O)_6]^{2+}$ and $[Fe(H_2O)_6]^{2+}$ with O_2 is related to the metal HOMOs and the O_2 LUMO or SOMO. $[Fe(H_2O)_6]^{2+}$ and O_2 have HOMO and LUMO orbitals of π character as shown in Figure 10.3. O_2 can interact with $[Fe(H_2O)_6]^{2+}$ in either an "end on" or "side on" position without a σ bond forming between the reactants. The Fe^{2+} d_{xz} and d_{yz} orbitals can interact with the O_2 π^*_x and π^*_y orbitals, respectively, for electron transfer. Thus, these reactants are ideal for an outer sphere electron transfer process with a low kinetic barrier [12]. Use of the Marcus cross relation for these reactants provides excellent agreement with experimental data [13]. For $[Fe(H_2O)_6]^{2+}$, deviation of the kinetics from linearity at pH > 8 suggests that an inner sphere electron transfer process may also occur, and that the chemical species is $[Fe(H_2O)_4(OH)_2]^0$. As pH increases, the OH^- ligand forms a σ bond with Fe(II) through its p_y orbital and a π bond to Fe(II) through its 1π p_x orbital (lower part of Figure 10.3). This enhances the metal's

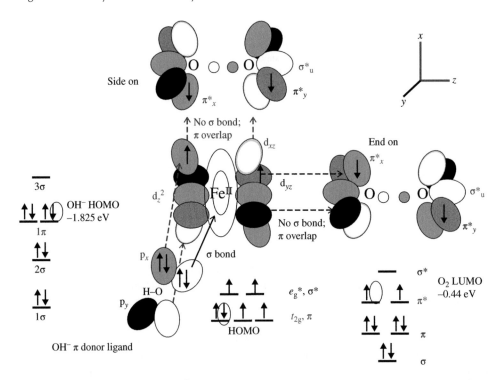

Figure 10.3 *Frontier MOs for* $[Fe(H_2O)_6]^{2+}$ *and* O_2*; both side on and end on attack by* O_2 *are shown. There are no* σ *bonds between Fe(II) and* O_2*, but there is* π *overlap for an outer sphere electron transfer. At higher pH, the bonding of* OH^- *to Fe(II) results in both* σ *and* π *overlap (dashed arrows), which enhances Fe(II) reducing power (the H–O bond is on the z axis). The circles in the energy level diagrams indicate that these* π *orbitals have continuous overlap between* OH^-*, Fe(II), and* O_2*. The black and gray lobes for* O_2 *and Fe(II) are in the yz plane*

ability to lose an electron (see Figure 8.31 for case 2 octahedral MO diagrams). In essence, the addition of π donor ligands to the metal center enhances its basicity (**metal basicity**) or reducing power. Also, the hydroxide stabilizes the Fe(III) product, which can hydrolyze further to oxides such as Fe_2O_3 (hematite). Because this outer sphere reaction is thermodynamically favorable and kinetically unhindered, it is possible to create O_2^- and other ROS in groundwater and deep waters, where light does not penetrate and there is a supply of Fe(II) and O_2 [14] (see also Figure 1.13b and c).

In contrast, Figure 10.4 shows that $[Mn(H_2O)_6]^{2+}$ has a HOMO orbital (d_{z^2}) with σ character that cannot overlap with the π^* LUMO of O_2 in an outer sphere electron transfer process as in Figure 10.3. Thus, this is a symmetry-forbidden process. Use of the **Marcus cross** relation for these reactants indicates that experimental kinetic data are 4–5 orders of magnitude faster than the calculated rate [13], further indicating that an outer sphere process is not accessible. The oxidation occurs at higher pH, indicating that the reactive species is not $[Mn(H_2O)_6]^{2+}$, but $[Mn(H_2O)_4(OH)_2]^0$. To overcome this barrier to an outer sphere electron transfer process, an inner sphere electron transfer process with a σ bond between (Mn(II) d_{z^2} with O_2 (π^*_x) must occur. Here, O_2 binds with Mn(II) in an "end on" or angular Mn–O–O arrangement. At higher pH, a similar inner sphere process is also possible for Fe(II).

Equation 10.2 indicated that autocatalytic behavior occurs in Mn(II) oxidation with O_2. Oxide ions on the MnO_x surface would serve to replace OH^- as a π donor ligand as they would bind dissolved Mn^{2+} (**surface**

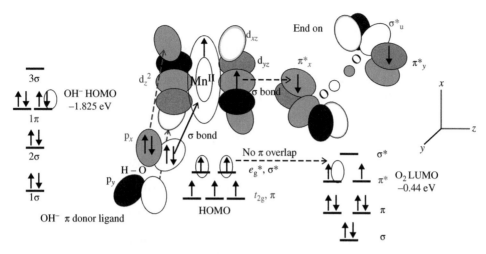

Figure 10.4 *Frontier MOs for* $[Mn(H_2O)_6]^{2+}$ *and* O_2. *There is no* π *overlap for an outer sphere electron transfer. An angular Mn–O–Oσ bond occurs between Mn(II) and* O_2. *At higher pH, both* σ *and* π *overlaps from* OH^- *to Mn(II) occur to enhance the reducing power of Mn(II). The black and gray lobes for* O_2 *and Mn(II) are in the yz plane*

complexation); thus, loss of an electron from Mn(II) should occur readily. Mn(II) oxidation can occur at lower pH values than shown in Figure 10.2b when MnO_x is present as in soils and sediments. The $pH_{(ZPC)}$ of MnO_2 is 4.6 (see Table 7.9; different Mn(III,IV) solid phases have values ranging from 2.8 to 4.6) indicating that Mn^{2+} binds readily to its surface at pH $\geq$ 4.6. The following reaction steps describe Mn(II) oxidation with O_2 and are consistent with the rate law. The Mn(II) adsorption step occurs at the water exchange rate of the individual Mn(II) species, which depends on the pH.

$$[Mn(H_2O)_4(OH)_2]^0 + O_2 \rightarrow MnO_2 \quad \text{(slow)}$$

$$MnO_2 + Mn(II) \rightarrow MnO_2 \cdot Mn(II) \quad \text{(adsorption is very fast)}$$

$$MnO_2 \cdot Mn(II) + O_2 \rightarrow 2\ MnO_2 \quad \text{(fast)}$$

In addition to surface complexation, ligands in soluble metal–ligand complexes can donate electrons to Fe^{2+} and Mn^{2+} and enhance the loss of electrons to O_2 and other electron acceptors so that Fe^{2+} and Mn^{2+} oxidation can occur at lower pH values [15–18]. Bacteria can also facilitate metal oxidation as both direct (enzymatic) and indirect (e.g., non-enzymatic bacterial spore coat mediation) processes are well documented for Mn^{2+} oxidation [19].

10.4.2 Redox Properties of Iron–Ligand Complexes

The redox potentials (Table 10.1 in Section 10.2) for the reduction of Fe(III) complexes to Fe(II) complexes or Fe^{2+} are instructive with regard to the stabilization of Fe(III) versus Fe(II). Fe(II) is more stable at higher E values, and Fe(III) is more stable when the E value is negative. Note that π acceptor ligands stabilize Fe(II), which is important for O_2 storage and transport discussed in the next section, and π donor ligands stabilize Fe(III). The increased number of ligating atoms (**denticity**) in a ligand binding to Fe also influences the Fe(III)/Fe(II) reduction potential [1, 2]. The reduction potentials for the last two entries in Table 1 are siderophore compounds, which have six ligating atoms per ligand bound to Fe. These complexes are so stable that free ligands secreted by organisms can solubilize Fe(III) solid phases to acquire Fe. Table 10.1 also shows that Fe–ligand chemistry spans the entire redox range found in Figure 10.1; thus, a variety of Fe–ligand compounds are found in organisms to affect biochemical processes (see Chapter 12).

Oxygen reduction to O_2^- and ultimately H_2O occurs via the following four one-electron steps with consumption of hydrogen ions (Equation 10.3a–10.3d; Half reaction potentials at standard state and pH = 7 from Table 2.2) so that four Fe(II) and Mn(II) are needed to fully reduce each O_2 molecule with formation of Fe and Mn solid oxidation products.

$$O_2 + e^- \rightarrow O_2^- \qquad\qquad p\varepsilon = -2.72; E_{w\,(pH\,7)} = 0.059\,(-2.72) = -0.16\,V \qquad (10.3a)$$

$$O_2^- + e^- + 2H^+ \rightarrow H_2O_2 \qquad p\varepsilon = 29.08\text{–}2pH; E_{w\,(pH\,7)} = 0.059(15.08) = +0.89\,V \qquad (10.3b)$$

$$H_2O_2 + e^- \rightarrow \text{•OH} + OH^- \qquad p\varepsilon = 16.71\text{–}pH; E_{w\,(pH\,7)} = 0.059(9.71) = +0.57\,V \qquad (10.3c)$$

$$\text{•OH} + OH^- + e^- + 2H^+ \rightarrow 2H_2O \qquad p\varepsilon = 42.92\text{–}pH; E_{w\,(pH\,7)} = 0.059(35.92) = 2.12\,V \qquad (10.3d)$$

Figure 10.5 shows that the ROS (O_2^-, H_2O_2, •OH) have almost no thermodynamic barrier from pH 0–14 [20] or significant kinetic barrier for reaction with Fe^{2+} and Mn^{2+}. Thus, it is possible to oxidize Fe^{2+} and Mn^{2+} at lower pH in environmental waters if H_2O_2 or other ROS are present. For example, the kinetics of the reactions of O_2^- with Mn(II) bound to sulfate, phosphate, and pyrophosphate has been shown to have high rate constants that exceed 10^7 M^{-1} s^{-1} [21]. Thus, these reactions are not rate limiting. The reaction of Fe^{2+} with H_2O_2 is known as the **Fenton** reaction, which has high oxidizing capability when the Fe(III) formed can be reduced back to Fe^{2+} *in situ*. Complexation of Fe^{3+} with siderophore-type ligands (Table 10.1) prevents the thermal reduction of Fe(III) and the Fenton reaction from proceeding. With light activation, ligand to metal charge transfer can occur, leading to Fe(II) complexes.

10.4.3   Metal Ions Exhibiting Outer Sphere Electron Transfer

In addition to Fe^{2+} [t_{2g}^{4} e_g^{*2}; octahedral] exhibiting outer sphere electron transfer reactions, Cu^+ [e^4 t_2^{6}; tetrahedral] and V^{4+} [$t_{2g}^{1}e_g^{0}*$; octahedral] lose electrons from t_2 orbitals during oxidation by O_2. A plot of the log $K_{reaction}$ versus log k_{et} for the oxidation of these ions, Fe^{2+}, $FeOH^+$, and $Fe(OH)_2$ by O_2 yields a linear free energy relationship as expected from **Marcus** theory [13].

10.5   Oxygen (O_2) Storage and Transport

From the above discussion on Mn(II) and Fe(II) oxidation in natural waters, O_2 is a potent oxidant, and OH^- and O^{2-} stabilize these metal ions in their higher oxidation states resulting in the formation of solid oxides or oxyhydroxides in the absence of ligands, which can bind and stabilize metals in solution. Vertebrate and invertebrate animals breathe or use O_2 for life processes so must prevent the oxidation of reduced metal ions and the formation of their higher oxidation state metal oxide or (oxy)hydroxide solids. The porphyrins are four-coordinate nitrogen-containing ligands that can complex a metal ion efficiently, and when housed within a protein, the metal center binds with histidine residues to form five-coordinate metal centers that then react with small molecules such as O_2, CO, CN^-, and NO while simultaneously preventing solid metal oxide formation. Formally, the metal center goes from 16 electrons to 18 electrons (**EAN rule**), and the reaction with O_2 is reversible.

Figure 10.6a shows the basic porphyrin building block without any substitution of hydrogen atoms on the carbon atoms. When a metal binds with a porphyrin, the two N–H hydrogen atoms are removed and all four

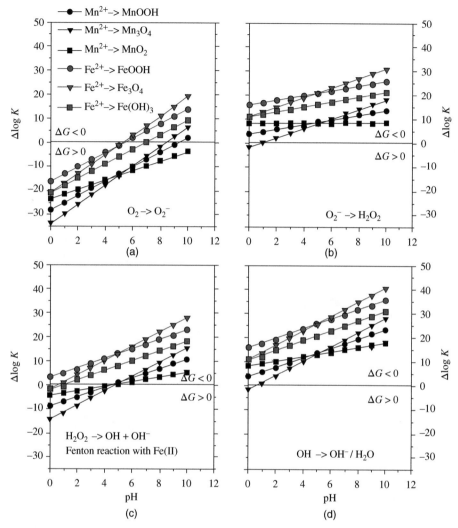

Figure 10.5 *Thermodynamics of the four successive one electron transfer reactions for* O_2 *to* H_2O *(Source: From [20]. Reproduced with permission from Springer.)*

nitrogen atoms bind with the metal; the resulting complex is known as a **metalloporphyrin** (Figure 10.6b). Figure 10.6c shows the primary structure of the heme group, which contains four methyl groups and two carboxylic acid groups, as well as two side chains (R_1 and R_2) that are hydrophobic. Figure 10.6d shows the bonding of O_2 to Fe(II), which is bound to the nitrogen atoms of the heme group and a histidine group. The bonding of O_2 to Fe(II) complexes found in blood is now described.

10.5.1   Hemoglobin

Hemoglobin (also spelled as haemoglobin) is the metalloprotein that contains iron for oxygen transport in the red blood cells of all vertebrates. In the lungs (or gills), hemoglobin binds O_2 and transports it to the rest of

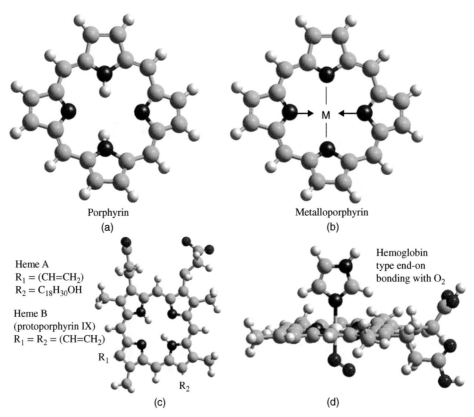

Figure 10.6 *(a) The porphyrin molecule. (b) Basic structure of a metalloprotein. (c) Structure of the heme group with representative porphyrin substitutions. (d) O_2 binding in a Fe(II) porphyrin. Models created with the Hyperchem™ program (version 8.0.10)*

the body (tissues, muscles) where it is then released in the process of respiration, which provides energy to run the metabolic machinery of the body, producing H_2O and CO_2 as products. Although hemoglobin must bind and transport O_2, it must be able to release it. In the tissues, O_2 is released to myoglobin, which has a stronger binding affinity for O_2 than hemoglobin **at lower pH and lower partial pressures of O_2**; thus, the CO_2 formed aids in the release of O_2 from the hemoglobin to myoglobin. The bonding and chemistry of hemoglobin has been well studied, and involves the stability and size of Fe(II) once it is in a low spin state.

In hemoglobin, there are four heme groups, which bind one Fe(II) each. Fe(II) is bound to the four nitrogen atoms in the heme group and an additional histidine nitrogen atom from a side chain of the protein (termed the **proximal** region) for a coordination number of five (Figure 10.7a).

In human lungs, the sixth coordination site has a water molecule loosely bound to the Fe(II) prior to binding with O_2, and this complex is known as deoxyhemoglobin. Assuming a localized octahedral geometry, the Fe(II) has the high spin electron configuration, $t_{2g}^4 e_g^{*2}$. Because high spin Fe(II) is 92 pm in size and has e_g^* electrons pointing to the ligands, it cannot fit easily into the porphyrin ring so it sits above the plane of the porphyrin ligand by about 40 pm. As O_2 enters the lungs, it displaces water and binds the Fe(II) as shown in the schematic diagram at the right of Figure 10.7a (also the structure in Figure 10.6d). The binding results in an angular Fe–O_2 bond (angular Fe–O–O; Figure 10.7b), and this complex is known as oxyhemoglobin. The $O_2 \pi_x^*$ orbital forms a σ bond with the Fe(II) d_{z^2} orbital similar to that noted in Figure 10.4; in addition,

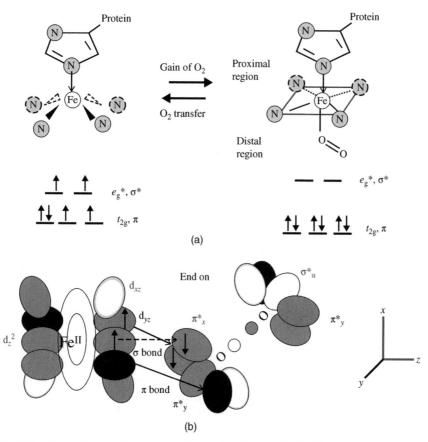

Figure 10.7 *(a) The change in coordination number and spin state for Fe(II) on going from deoxyhemoglobin (H$_2$O weakly bonds at the six site) to oxyhemoglobin. (b) The Fe(II) and O$_2$ frontier molecular orbitals involved in σ (dashed arrow) and π (solid arrows) bonding. The black and gray lobes for O$_2$ and Fe(II) are in the yz plane*

the O$_2$ π_y^* orbital forms a π bond with the Fe(II) d$_{yz}$ orbital as O$_2$ is a **π acceptor ligand**. However, the spin state of Fe(II) changes from high spin to low spin, t_{2g}^6, which reduces the size of Fe(II) to 75 pm so it can now fit into the square plane of the porphyrin ligand. The more stable bonding in the plane and the increase in MOSE arising from the low spin conversion further stabilizes the Fe–O$_2$ bond to dissociation. The overall result of Fe–O$_2$ bonding is a diamagnetic complex (formally one π* orbital of O$_2$ has two electrons so becomes ^{1}O$_2$ whereas the other π* orbital of O$_2$ accepts electrons from Fe^{2+}). At the tissues, O$_2$ is released from hemoglobin to deoxymyoglobin (initially high spin then converted to low spin on bonding with O$_2$), which contains only one heme group, and the hemoglobin returns to the lungs to repeat the process.

There are a couple of essential features to this O$_2$ exchange process at the tissues. First, at the tissues in the muscles where respiration occurs, the local pH value is approximately 6.8 due to the production of CO$_2$ compared to the 7.4 found in the blood. CO$_2$ is a waste product, which binds to histidine residues in the side chains of the protein. The lower pH weakens the Fe–O$_2$ bond so that hemoglobin can release O$_2$ to myoglobin, which has a higher affinity for O$_2$ at this lower pH [22]. Second, the proteins in hemoglobin and myoglobin help prevent the oxidation of Fe^{2+} to Fe^{3+} via steric factors; formation of Fe^{3+} renders the heme group incapable of oxygen transport. Formation of Fe^{3+} results in the loss of Fe from the body as in the equation below. Nevertheless, in the human body, the iron needs to be replaced on the timescale of 100 to 120 days in an adult.

$$4Fe^{2+} + O_2 \rightarrow 2 \ [Fe^{3+} - O - Fe^{3+}] \rightarrow \text{excretion or loss of Fe}$$

The size of red blood cells varies from 6 to 8 μm in adult humans, and the average molecular mass for hemoglobin is approximately four times that of myoglobin, which is ~17 kDa. Thus, the bonding and chemistry involved with oxygen transport by iron is not formally a solution process with dissolved chemical species. As a result of the size of the red blood cells and the molecular mass of the metalloproteins, only small diatomic molecules can easily get to the distal or ligand substitution site in the heme group. There are several small molecules and ions that are toxic to the system at the pH of the blood (7.4). HCN and H_2S are predominantly CN^- and HS^- at blood pH so can readily bind with Fe(II). High concentrations of these ions along with carbon monoxide result in asphyxiation and death.

Interestingly, the deep-sea tubeworm, *Riftia pachyptila*, has red blood (pH = 7.4) that contains hemoglobin similar to humans. However, it grows and survives by hosting endosymbiont chemoautotrophic bacteria that can perform chemosynthesis using the reaction between H_2S and O_2. X-ray data indicate that the hemoglobin has separate sites containing Zn^{2+} ions that bind HS^- so that both reactants are transported on the protein to the location where the endosymbionts perform chemosynthesis; thus, the Fe^{2+} centers do not react with sulfide [23].

10.5.2   Hemocyanin and Hemerythrin

There are many other known biological O_2 carriers, and two more are noted that have different bonding between the metal and O_2 than hemoglobin as there is no heme group (formally the word "heme" comes from the Greek work meaning blood). **Hemocyanin** (a blue copper protein) gives the blood a blue color and is found in snails (mollusks) and crabs (arthropods) [24]. The active site contains two Cu^+ ions with each bound to three histidine groups from the protein; thus, each Cu has 16 electrons. On reaction with O_2, both Cu^+ ions are oxidized to Cu^{2+} as O_2 is reduced to O_2^{2-} (Figure 10.8).The O_2 binds **side on** between the two copper atoms similar to the binding of O_2 in Vaska's complex (Figure 9.20a; see analogy to ethylene in Figure 8.33d) so that each Cu now has 18 electrons. The complex is diamagnetic, indicating that the spins are antiparallel for the two Cu^{2+} ions.

Hemerythrin is found in sea worms, and gives the blood a red color. The active site again contains two Fe^{2+} ions bound to donor atoms from histidine groups in the protein rather than a heme group similar to hemocyanin. However, there is one bridging OH group between the two Fe atoms with one Fe^{2+} six coordinate (bound to three histidine groups and two carboxyl bridging groups) whereas the other Fe^{2+} is five coordinate (bound to two histidine groups and the two carboxyl bridging groups); thus, the latter Fe^{2+} can accept 2 electrons from a donor ligand to obtain 18 electrons. Figure 10.9 shows that the O_2 binds **end on** to only the five-coordinate Fe^{2+} ion (similar to hemoglobin); also, the O_2 does hydrogen bond to the H atom from the OH group bridging the two Fe^{2+} ions [25].

10.6   Oxidation of CH_4, Hydrocarbons, NH_4^+

As noted in Equation 1.20, methane is produced in the environment during organic matter decomposition (methanogenesis; formally a disproportionation reaction) when all other electron acceptors have been consumed. In Section 5.6.1, the reaction of O_2 with CH_4 was shown to be kinetically hindered although thermodynamically favorable. Fortunately, CH_4 is oxidized by microbial processes [26] as it diffuses through the

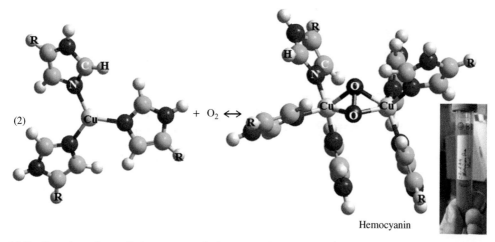

Figure 10.8 *Reaction of two Cu⁺ groups with O₂ to produce hemocyanin (structures prepared using Hyperchem™ version 8.0.10). (Lower right) Blood containing hemocyanin from the deep-sea vent snail* Alviniconcha *spp. collected at the Eastern Lau Spreading Center (2100 m water depth)*

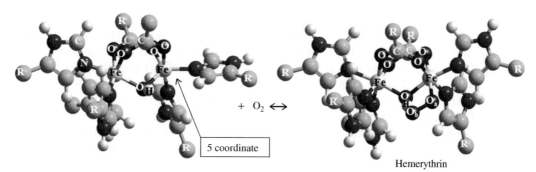

Figure 10.9 *Reaction of deoxyhemerythrin with O₂ to form oxyhemerythrin. O$_a$ and O$_b$ indicate the O atoms in O₂ that bind to Fe^{2+} and H⁺, respectively. Models created with the Hyperchem™ program (version 8.0.10)*

sediment and/or water column and before it reaches the atmosphere where it acts as a greenhouse gas that is 20 times more potent than CO_2. In marine systems, sulfate is an oxidant and in fresh water, oxidized Fe and/or Mn solid phases are possible oxidants used by bacteria. The thermodynamics for the reactions of CH_4 with these metal oxide phases is not always favorable at near neutral pH (Figure 10.10). Thus, organisms have produced efficient enzyme systems to activate both reactants using mainly iron and copper for metal catalysis and then to use the energy from the O_2 oxidation of CH_4 and other hydrocarbons, which are produced naturally (including oil and gas formation).

10.6.1 Cytochrome P450: An Example of Cytochrome (Heme – O₂) Redox Chemistry

Cytochrome P450 is a monooxidase enzyme or **metalloenzyme** system [27, 28] that has a Fe–heme complex that absorbs light at 450 nm (a sharp **Soret band**) as well as a redox potential at the lower end of cytochromes in Table 10.1. In this biochemical system, electron transfer reactions between redox partners (**NADH** to P450)

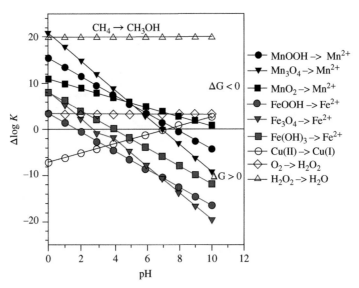

Figure 10.10   *Reaction of* CH_4 *with* O_2, H_2O_2, *Cu(II), and natural metal oxide and oxyhydroxide phases using the equations from Table 2.2*

within the protein occur as in photosynthesis (see below), but the reactions are not light activated as they are thermodynamically favorable (Figure 10.1). Although the P450 redox potential is not sufficient to oxidize hydrocarbons directly [e.g.; in Table 10.1, $E_{(pH\,7)}(CH_3OH \rightarrow CH_4) = 0.173$ V], cytochromes are capable of oxidizing hydrocarbons to alcohols as the Fe–porphyrin complex cycles Fe between the +2, +3 and +4 oxidation states. Also, the Fe center and its redox potential are tuned as in Figure 10.1. O_2 is reduced to form a $Fe^{IV} = O$ oxo intermediate (known as ferryl Fe), which also forms a free radical on the porphyrin ring that in turn abstracts a H• from the hydrocarbon allowing incoming water to form ROH. In this process (Equation 10.4), two H^+ and two one-electron transfers occur so formally O_2 is first reduced to H_2O_2, which then oxidizes the alkane. Figure 10.10 shows that the H_2O_2 reaction with CH_4 is more favorable than the O_2/CH_4 reaction.

$$CH_4(RH) + O_2 + 2e^- + 2H^+ \rightarrow CH_3OH \ (ROH) + H_2O \tag{10.4}$$

Figure 10.11 shows some of the known mechanistic details. The Fe in the heme (an iron protoporphyrin IX) is coordinated on the proximal side by a thiolate ion and by water on the distal side (**I**). The Fe^{3+} is d^5 (EAN = 17) and low spin. On addition of the hydrocarbon (RH) with H_2O displacement, Fe^{3+} becomes high spin (**II**, EAN = 15), and the Fe redox potential shifts 300 mV, enabling it to accept an electron from a redox partner (e.g., NADH). After reduction, Fe^{3+} becomes Fe^{2+} and binds O_2 in a manner similar to hemoglobin (**III**, oxy compound low spin Fe, EAN = 18). On addition of a second electron from a redox partner and a first H^+ transfer to the O_2, a peroxy, HO_2 Fe species (species **IV**), forms; Fe^{2+} also provides an electron to the HO_2 species reforming Fe^{3+} (EAN = 17). On additional H^+ transfer to the HO_2, water is lost and an O atom is left bound to the Fe^{3+}. Loss of another electron from Fe^{3+} to form Fe^{4+} as well as a second from the porphyrin to form a **π radical cation** (P•$^+$) on the delocalized porphyrin ring leads to the $Fe^{IV} = O$ ferryl or oxo intermediate (**V**, EAN = 18; both high spin and low spin complexes are possible). The porphyrin free radical ion abstracts a H• from the alkane to form R• and Fe^{IV}-OH species (**VI**). Addition of H_2O results in ROH and H^+ formation as Fe(IV) is reduced to Fe(III), reforming the starting species I. Because the redox process forms a π radical cation on the porphyrin (P•$^+$), which reacts with the alkane substrate, this can be

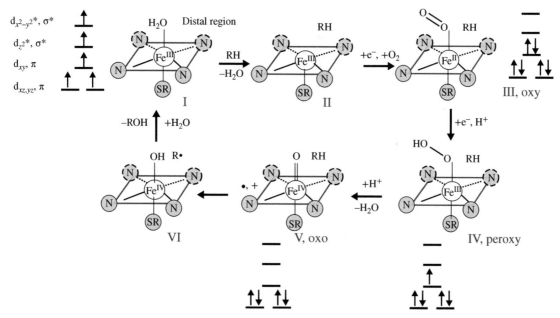

Figure 10.11 *Cytochrome P450 cycle mechanistic details. Fe electron configurations are for approximate* D_{4h} *symmetry; the high spin diagram at the left is for structure II. Electrons at steps* II → III *and* III → IV *come from a redox partner*

considered a **ligand-centered** redox reaction (Equation 10.5) in contrast to the metal-centered reactions in Section 10.4.1.

$$M^{n+}(P) + Ox \rightarrow M^{n+}(P\bullet^+) + Ox^-; \quad then \quad M^{n+}(P\bullet^+) + RH \rightarrow M^{n+}(P)H + R\bullet \quad (10.5)$$

or

$$M^{n+}(P) + Ox \rightarrow M^{n+}(P\bullet^+) + Ox^-; \quad then \quad M^{n+}(P\bullet^+) \rightarrow M^{n+1}(P)$$

Other heme–Fe enzymes include the peroxidases, which have similar mechanistic details that start with structure **IV** in Figure 10.11. Peroxidases convert H_2O_2 to H_2O while oxidizing an organic substrate (e.g., glutathione or GSH in Equation 10.6). Other metals are used in peroxidase enzymes including Vanadium in the haloperoxidases.

$$2GSH + H_2O_2 + 2e^- + 2H^+ \rightarrow GS\text{-}SG + 2H_2O \quad (10.6)$$

10.6.2　Conversion of NH_4^+ to NO_3^- (Nitrification or Aerobic Ammonium Oxidation)

Other monooxidases use Cu blue proteins similar to hemocyanin and Cu–heme groups to catalyze the oxidation of alkanes, nitrogen species, and NH_4^+ in the first two-electron transfer step [24, 29]. Ammonium monooxygenase is a Cu–Fe **metalloenzyme** that archaea and bacteria use to convert ammonia to NH_2OH as the first intermediate along the path to form nitrate. Similar to cytochrome P450, O_2 must be converted to H_2O_2 (Equation 10.4) as the O_2 reaction with ammonia to form NH_2OH is unfavorable thermodynamically

Table 10.2   *Half reactions (equations from Table 2.2), voltages and enzymes for the three reactions resulting in nitrate formation (nitrification)*

Half reaction (Table 2.2 equation)		mV (pH 7)	Enzyme
$NH_4^+ \rightarrow NH_2OH$	(N6c)	-1137	Fe/Cu ammonium monooxygenase
$NH_2OH \rightarrow NO_2^-$	(N19)	$+60$	P460 heme hydroxylamine oxidoreductase
$NO_2^- \rightarrow NO_3^-$	(N1)	$+55$	Mo nitrite oxidoreductase

(Equations 10.7a and 10.7b) whereas the H_2O_2 reaction is favorable (Equations 10.7c and 10.7d).

$$NH_4^+ + O_2 \rightarrow NH_2OH + H_2O_2 + H^+ \quad -707 \text{ mV} \quad [\text{N6c–O2 Table 2.2}] \tag{10.7a}$$

$$NH_3 + O_2 + H_2O \rightarrow NH_2OH + H_2O_2 \quad -722 \text{ mV} \quad [\text{N6b–O2 Table 2.2}] \tag{10.7b}$$

$$NH_4^+ + H_2O_2 \rightarrow NH_2OH + H_2O + H^+ \quad 209 \text{ mV} \quad [\text{N6c–O3 Table 2.2}] \tag{10.7c}$$

$$NH_3 + H_2O_2 \rightarrow NH_2OH + H_2O \quad 275 \text{ mV} \quad [\text{N6b–O3 Table 2.2}] \tag{10.7d}$$

The four-electron oxidation of NH_2OH to NO_2^- is mediated by a P460–heme known as hydroxylamine oxidoreductase, and the oxidation of NO_2^- to NO_3^- is performed by an MoS complex in the enzyme system known as nitrite oxidoreductase (Table 10.2). Thus, three distinct enzyme systems are used for the eight-electron oxidation of NH_4^+ to NO_3^- (**nitrification**), and the key step is the monooxygenase converting O_2 to H_2O_2 to oxidize NH_4^+ to NH_2OH. Section 12.7 discusses anaerobic ammonium oxidation (**anammox**) in which a similar process occurs when NO and NH_4^+ react to form N_2.

10.7   Oxygen Production in Photosynthesis

Thus far, the discussion has centered on O_2 as an oxidant and its conversion to water as in Equations 10.3a–10.3d, or O_2 binding with metals for O_2 transport within organisms. The reverse process for Equations 10.3a–10.3d is the oxidation and splitting of water to form O_2 during **photosynthesis** (Equation 10.8), which is a complex process that continues to be elucidated [30, 31]. In the water oxidation reaction, an $[Mn_3O_4Ca]Mn$ complex (where brackets indicate a cluster) known as the **oxygen-evolving complex** (OEC) in photosystem center II (PSII) stores oxidizing equivalents for the four-electron oxidation of two water molecules to form O_2. Equation 10.3d shows that the first electron transfer (half-reaction) in H_2O oxidation has a significant thermodynamic barrier of -2.12 V.

$$2H_2O \rightarrow O_2 + 4H^+ + 4e^- \tag{10.8}$$

As noted above, oxidation of Mn^{2+} is not a thermodynamically favorable process at neutral pH. During photosynthesis, there are several transformations involving electrons and hydrogen ions that use **light** harvesting from chlorophyll-*a* to oxidize Mn. During the redox transformations, the $[Mn_3O_4Ca]Mn$ cluster (Figure 10.12a) contains a different number of Mn(III,IV) ions (Figure 10.12b) and possibly a Mn(V) ion with a Mn = O bond. Also, it has a distorted $[Mn_3O_4Ca]$ cube as Mn(III) ions exhibit **Jahn–Teller** distortion. There are O bridging atoms between the metal ions so that all metal ions are **six coordinate**, and these O atoms may become protonated during photosynthesis. Interestingly, Ca^{2+} and Mn^{3+} are labile metal ions (Table 9.1), but Mn^{4+} tends to be inert and has shorter bonds. The Mn^{3+} lability is important in the first part

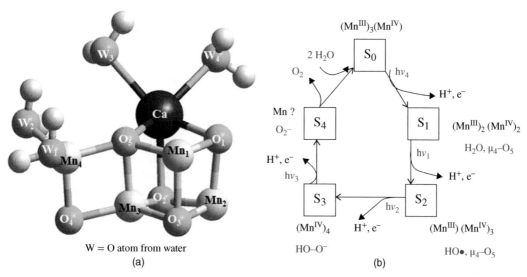

Figure 10.12 *(a) Structure and numbering system representing the [Mn$_3$O$_4$Ca]Mn oxygen-evolving complex produced using Hyperchem™ version 8.0.10. The Mn and Ca ions are bound by other protein residues (not shown) including histidine, and bridging carboxyl groups of aspartate and glutamate. (b) The Kok cycle with the probable Mn oxidation states and intermediate oxygen species leading to O$_2$ formation*

of the process and the Mn^{4+} inertness later in the process. The electrons and hydrogen ions produced in PSII are ultimately used to reduce CO_2 to produce carbohydrates via the Calvin Benson cycle. Once formed, the carbohydrates are precursors for the formation of other molecules needed by the organism; thus, energy from sunlight is stored as chemical energy.

Figure 10.12b shows the Kok cycle, which indicates the five proposed redox states (S_0 through S_4) for the Mn ions in each S state. These states have been studied with dark acclimation followed by light flashes. The S_0 state is the most reduced state, but the S_1 redox state is the native dark state, which has 2 Mn^{3+} and 2 Mn^{4+} ions (both in the cage) and where $\mu_4 - O_5$ forms in the OEC (an open cluster exists at S_0 and becomes a closed cluster at S_1). After dark acclimation and providing the first three light flashes, O_2 formation is highest. O_2 is formed again at every fourth subsequent flash, but production decreases (O_2 is produced at each S state when there is full light). The OEC is near a membrane containing chlorophyll-a, and in the dark, no absorption of a photon occurs at the chlorophyll-a in the PSII reaction center (P_{680} for the absorption peak maximum at 680 nm). When exposed to flashes of light, P_{680} is oxidized leaving an electron hole that oxidizes tyrosine (an intermediate donor molecule) in a protein that is part of the reaction center. Once tyrosine is oxidized, it oxidizes one of the Mn ions in the OEC, which in turn oxidizes water. The entire water oxidation process occurs when four successive photons are absorbed by the reaction center so that there are four consecutive Mn photo-oxidations between states $S_0 - S_1$, $S_1 - S_2$, $S_2 - S_3$, and $S_3 - S_4$ resulting in the loss of electrons and hydrogen ions during each photo-oxidation step (the duration of each flash of light occurs over the range of 30–300 μs and increases with S redox state number). Although two water molecules are bound to the OEC at the S_0 state, the formation of the O–O bond appears to first occur at the S_3 state [32, 33]. At S_3, all Mn ions are +4 so that the Mn distances are at their shortest and ligand dissociation is less. Once the four oxidizing equivalents are formed at S_4, the spontaneous formation (∼1 ms) and release of O_2 occurs, leading to the S_4 to S_0 transformation. During the entire process, water molecule 2 bound to Mn_4 or water molecule 3 bound to

Ca moves to O_5 that binds with Mn_1 to form the O–O bond prior to O_2 release [33]. The Mn oxidation states in the S_4 state are not precisely known at this time, but a Mn(V) oxidation state has been suggested. Because the metal and other redox centers are stabilized or embedded in the protein, the reorganization energy λ from Equation 9.76 is ~0.5 eV, and electron transfer is fast.

At present, the energies of the Mn oxidation state transformations are not well documented. Because Mn^{4+} is reduced to Mn^{3+} in the OEC by water, a possible thermodynamic way to look at these S state conversions is to look at the energetics of the MnO_2 (Mn^{4+}) to $MnOOH$ (Mn^{3+}) reduction coupled with the water oxidation reactions using the equations from Chapter 2 (Table 2.2; reaction Mn4 – reactions O6,O7,O8,O9a). Equations 10.9 through 10.13 represent possible S redox transitions shown in Figure 10.12b. The last reaction ($S_4 \rightarrow S_0$) is thermodynamically spontaneous. The other four are not, so require light activation. The reactions which form ·OH are the steps that require the most energy (-1.57 V), and in the $S_0 \rightarrow S_1$ step, the ·OH becomes the μ_4-O_5 bridged species. The unfavorable energetics of these full reactions (Equations 10.9–10.12) are compensated for by the oxidation of P_{680} (photon of 1.82 eV), which has a redox difference of $+1.7$ V when the phaeophytin electron acceptor is considered [34].

$$MnO_2 + H_2O \rightarrow MnOOH + \cdot OH \text{ (as } \mu_4 - O_5) \qquad \Delta E = -1.57 \text{ V} \quad S_0 \rightarrow S_1 \qquad (10.9)$$

$$MnO_2 + H_2O \rightarrow MnOOH + \cdot OH \qquad \Delta E = -1.57 \text{ V} \quad S_1 \rightarrow S_2 \qquad (10.10)$$

$$MnO_2 + \cdot OH + OH^- \text{(as } \mu_4 - O_5) \rightarrow MnOOH + HO_2^- \quad \Delta E = -0.03 \text{ V} \quad S_2 \rightarrow S_3 \qquad (10.11)$$

$$MnO_2 + HO_2^- \rightarrow MnOOH + O_2^- \qquad \Delta E = -0.35 \text{ V} \quad S_3 \rightarrow S_4 \qquad (10.12)$$

$$MnO_2 + O_2^- + H^+ \rightarrow MnOOH + O_2 \qquad \Delta E = +0.71 \text{ V} \quad S_4 \rightarrow S_0 \qquad (10.13)$$

References

1. Boukhalfa, H. and Crumbliss, A. L. (2002) Chemical aspects of siderophore mediated iron transport. *BioMetals*, **15**, 325–339.
2. Raymond, K. N. and Carrano, C. J. (1979) Coordination chemistry and microbial iron transport. *Accounts of Chemical Research* **12**, 183–190.
3. Batinić-Haberle, I., Spasojević, I., Hambright, P., Benov, L., Crumbliss, A. L. and Fridovich, I. (1999) Relationship among redox potentials, proton dissociation constants of pyrrolic nitrogens, and in vivo and in vitro superoxide dismutating activities of manganese(III) and iron(III) water-soluble porphyrins. *Inorganic Chemistry*, **38**, 4011–4022.
4. Stumm, W. and Sulzberger, B. (1992) The cycling of iron in natural environments: considerations based on laboratory studies of heterogeneous redox processes. *Geochimica Cosmochimica Acta*, **56**, 3233–3257.
5. Williams, R. J. P. and Frausto da Silva, J. J. R. (2006) *The Chemistry of Evolution: The Development of Our Ecosystem*, Clarendon Press, Oxford, pp. 481.
6. Hille, R., Hall, J. and Basu, P. (2014) The mononuclear molybdenum enzymes. *Chemical Reviews*, **114**, 3963–4038.
7. Stumm W. and Morgan J. J. (1996) *Aquatic Chemistry* (3rd ed.), John Wiley and sons, Inc., New York, pp. 1022.
8. Hem, J. D. and Lind, C. J. (1983) Nonequilibrium models for predicting forms of precipitated manganese oxides. *Geochimica Cosmochimica Acta*, **47**, 2037–2046.
9. Murray, J. W., Dillard, J. G., Giovanoli, R., Moers, H. and Stumm, W. (1985) Oxidation of Mn(II): initial mineralogy, oxidation state and ageing. *Geochimica Cosmochimica Acta*, **49**, 463–470.

10. Millero F.J., Sotonlongo S., Izaguirre M. (1987) The oxidation kinetics of Fe(II) in seawater. *Geochimica Cosmochimica Acta*, **51**, 793–801.

11. Morgan J. J. (2005) Kinetics of reaction between O_2 and Mn(II) species in aqueous solution. *Geochimica Cosmochimica Acta*, **69**, 35–48.

12. Luther, G. W. (2005) Manganese(II) oxidation and Mn(IV) reduction in the environment—two-one-electron transfer steps versus a single two-electron step. *Geomicrobiology Journal*, **22**, 195–203.

13. Wehrli B. (1990) Redox reactions of metal ions at mineral surfaces, In *Aquatic Chemical Kinetics*, Chapter 11 (ed. W. Stumm). John Wiley and Sons, New York, pp. 311–336.

14. Murphy, S. A., Solomon, B. M., Meng, S., Copeland, J. M., Shaw, T. J. and Ferry, J. L. (2014) Geochemical production of reactive oxygen species from biogeochemically reduced Fe. *Environmental Science and Technology*, **48**, 3815–3821.

15. Rosso, K. J., Morgan, J. J. (2002) Outer-sphere electron transfer kinetics of metal ion oxidation by molecular oxygen. *Geochimica Cosmochimica Acta*, **66**, 4223–4233.

16. Duckworth O. W. and Sposito G. (2005) Siderophore—Manganese(III) interactions. I. Air oxidation of manganese(II) promoted by desferrioxamine B. *Environmental Science and Technology*, **39**, 6037–6044.

17. Luther G. W., Kostka J. E., Church T. M., Sulzberger B. and Stumm, W. (1992) Seasonal iron cycling in the salt marsh sedimentary environment: the importance of ligand complexes with Fe(II) and Fe(III) in the dissolution of Fe(III) minerals and pyrite, respectively. *Marine Chemistry*, **40**, 81–103.

18. Wallar B. J. and Lipscomb J. D. (1996) Dioxygen activation by enzymes containing binuclear non-heme iron clusters. *Chemical Reviews*, **96**, 2625–2657.

19. Tebo B. M., Bargar J. R., Clement B. G., Dick G. J., Murray K. J., Parker D., Verity R. and Webb S. M. (2004) Biogenic manganese oxides: properties and mechanisms of formation. *Annual Reviews of Earth and Planetary Science*, **32**, 287–328.

20. Luther, III, G. W. (2010) The role of one and two electron transfer reactions in forming thermodynamically unstable intermediates as barriers in multi-electron redox reactions. *Aquatic Geochemistry*, **16**, 395–420.

21. Barnese, K, Gralla, E. B., Cabelli, D. E. and Valentine, J. S. (2008) Manganous phosphate acts as a superoxide dismutase. *Journal of the American Chemical Society*, **13**, 4604–4606.

22. Shikama, K. (1998) The molecular mechanism of autoxidation for myoglobin and hemoglobin: a venerable puzzle. *Chemical Reviews*, **98**, 1357–1373.

23. Flores, J. F., Charles R. Fisher, Carney, S. L., Green, B. N., Freytag, J. K., Schaeffer, S. W. and Royer, Jr., W. E. (2005) Sulfide binding is mediated by zinc ions discovered in the crystal structure of a hydrothermal vent tubeworm hemoglobin. *Proceedings of the National Academy of Sciences USA*, **102**, 2713–2718.

24. Solomon, E. I., Heppner, D. E., Johnston, E. M., Ginsbach, J. W., Cirera, J., Qayyum, M., Kieber-Emmons, M. T., Kjaergaard, C. H., Hadt, R.G. and Tian, L. (2014) Copper active sites in biology. *Chemical Reviews*, **114**, 3659–3853.

25. Kryatov, S. V. and Rybak-Akimova, E. V. (2005) Kinetics and mechanisms of formation and reactivity of non-heme iron oxygen intermediates. *Chemical Reviews*, **105**, 2175–2226.

26. Beal E. J., House C. H. and Orphan V. J. (2009) Manganese- and iron-dependent marine methane oxidation. *Science*, **325**. 184–187.

27. Poulos, T. L. (2014) Heme enzyme structure and function. *Chemical Reviews*, **114**, 3919–3962.

28. Ener, M. E., Lee, Y.-T., Winkler, J. R., Gray, H. B. and Cheruzel, L. (2010) Photooxidation of cytochrome P450-BM3. *Proceedings of the National Academy of Sciences USA*, **104**, 18783–18786.

29. Maia, L. B. and Moura, J. J. G. (2014) How biology handles nitrite. *Chemical Reviews*, **114**, 5273–5357.

30. Kärkäs, M. D., Verho, O., Johnston, E. V. and Åkermark, B. (2014) Artificial photosynthesis: molecular systems for catalytic water oxidation. *Chemical Reviews*, **114**, 11863–12001.

31. Yano, J. and Yachandra, V. (2014) Mn_4Ca cluster in photosynthesis: where and how water is oxidized to dioxygen. *Chemical Reviews*, **114**, 4175–4205.
32. Cox, N., Retegan, M., Neese, F., Pantazis, D. A., Boussac, A. and Lubitz, W. (2014) Electronic structure of the oxygen evolving complex in photosystem II prior to O-O bond formation. *Science*, **345**, 804–808.
33. Suga, M., Akita, F., Hirata, K., Ueno, G., Murakami, H., Nakajima, Y., Shimizu, T., Yamashita, K., Yamamoto, M., Ago, H. and Shen, J-R. (2015) Native structure of photosystem II at 1.95 A° resolution viewed by femtosecond X-ray pulses. *Nature*, **517**, 99–105.
34. Falkowski, P. G. and Raven, J. A. (2007) *Aquatic Photosynthesis* (2nd ed.), Princeton University Press, Princeton, NJ. pp. 484.

11

Solid Phase Iron and Manganese Oxidants and Reductants

11.1 Introduction

When Fe and Mn are oxidized in the environment, metal (oxy)hydroxides [$MnOOH$, $FeOOH$, $Fe(OH)_3$] and oxides (MnO_2, Fe_2O_3) form. These are susceptible to reduction by a variety of natural reductants including sulfide and organic compounds, and the reactions are pH dependent. In environmental systems, the most powerful and abundant oxidant after O_2 is MnO_2 (Equation 1.21). Sulfide is used to describe metal oxide and (oxy)hydroxide reduction (other reductants undergo similar pathways). In natural sediments and stratified waters, the iron cycle would appear to be limited to one-electron transfer reactions whereas the manganese cycle can undergo one- or two-electron transfer reactions [1]. However, these are solids with bands of orbitals so two-electron transfers are possible. Figure 11.1 shows aspects of these environmental cycles that were noted in Section 1.8.2 and in Figure 1.13, which showed the overlap of dissolved oxidants and reductants due to molecular diffusion.

In some natural systems (e.g., the Black Sea and sediments), Mn is the dominant metal at the oxic/sub-oxic (OMZ)/anoxic transition zones (Section 1.8.2), but Fe is also important in most systems. The reaction of upward diffusing Mn^{2+} and Fe^{2+} with downward diffusing O_2 leads to ROS, and the reactions of oxidized Mn and Fe with upward diffusing H_2S lead to intermediate oxidation state sulfur compounds (S_x^{2-}, S_8, $S_2O_3^{2-}$). During these latter reactions, Fe(II) forms iron sulfide phases (FeS, FeS_2) with either H_2S or S_x^{2-} (see Section 12.6) whereas Mn does not. Lastly, the reaction of Fe(II) compounds with Mn(III,IV) phases is also favorable and facile.

11.2 Reduction of Solid MnO_2 and $Fe(OH)_3$ by Sulfide

In Section 5.4.8, the oxidation of H_2S by O_2 was discussed, and the one-electron transfer was shown to be thermodynamically unfavorable. The oxidation of sulfide by oxidized iron and manganese phases is now discussed, and they are important reactions serving to recycle solid phase iron and manganese back to the dissolved state. These are more complicated reactions as the solids have a band of orbitals to accept electrons

Inorganic Chemistry for Geochemistry and Environmental Sciences: Fundamentals and Applications, First Edition. George W. Luther, III.
© 2016 John Wiley & Sons, Ltd. Published 2016 by John Wiley & Sons, Ltd.
Companion Website: www.wiley.com/go/luther/inorganic

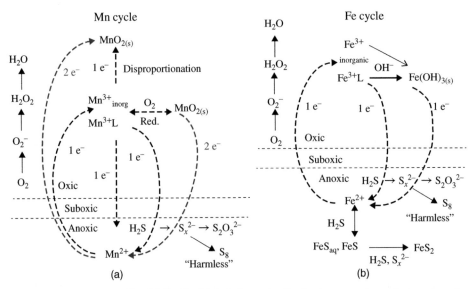

Figure 11.1 *(a) Mn, O_2, H_2S cycle. (b) Fe, O_2, H_2S, FeS_x cycle. Dashed arrows show electron transfer reactions for the metals; blue arrows indicate two-electron transfer reactions. On oxidation of M^{2+} to M^{3+}, MOOH solids or metal–ligand complexes ($M^{3+}L$) can form as products*

for reduction (conduction band) rather than discrete orbitals as for O_2. They also have a band of orbitals (valence band) to donate electrons to form inner sphere complexes. Also, sulfide exists in solution as H_2S and/or HS^- (pK_{a1} in freshwater equals 7.0; 6.55 in seawater). With this as the starting point, the calculations in Figure 11.2 show that the two-electron transfer reactions of $Fe(OH)_3$ with H_2S and HS^- to form $S(0)$ are thermodynamically favorable from low pH to nearly 9–10; the $pH_{(ZPC)}$ for many Fe(III) solids ranges from 5.4 to 8.0.

Calculations show that there is no apparent thermodynamic barrier for the two-electron transfer reactions of MnO_2 with H_2S and HS^-; the $pH_{(ZPC)}$ for δMnO_2 is 4.6. Two-electron transfer steps to form $S(0)$ are chosen as one-electron transfer steps have a thermodynamic barrier in forming the product $HS\bullet$ [1] similar to O_2 reacting with HS^- (Section 5.4.8). As the metal oxide solid phases have a band of orbitals, two-electron transfer from the sulfide species should be facile. However, the kinetics [2, 3] of these reactions show a clear increase and then decrease with pH (Figure 11.2); the peak reaction rate occurs at a pH of 5.2 for MnO_2 and 6.3 for $Fe(OH)_3$. Understanding the kinetics of these reactions requires more detailed knowledge of the chemical speciation of the reactants. The nature of the oxidized metal is now discussed.

11.2.1 Fe(III) and Mn(IV) Electron Configurations

The electron configurations for the high spin octahedral metal ions are $t_{2g}^3 e_g^{*2}$ for Fe(III) and t_{2g}^3 for Mn(IV). Fe(III) in $Fe(OH)_3$ ($pH_{ZPC} \sim 5.4$) is a labile metal cation. At lower pH, the surface iron atoms can be represented as $>Fe-OH$ or $>Fe-OH_2^+$. The OH^- or H_2O can dissociate from the Fe(III) so that a sulfide species could attack in an inner sphere process and transfer two electrons into the conduction band of the solid. In contrast, Mn(IV) has an inert electron configuration so that OH^- or H_2O will not readily dissociate from Mn. The E_{HOMO} value for the valence band in MnO_2 is -5 eV and attributed to the oxide 2p orbitals [4] with a band gap of 0.25 eV (Table 6.1); the conduction band is -4.75 eV. In Appendix 5.2, the E_{HOMO}

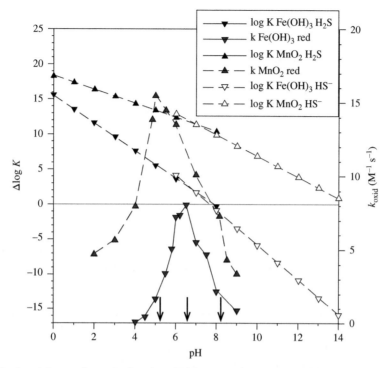

Figure 11.2 *Calculated thermodynamic data from Table 2.2 with experimental kinetics data for the oxidation of Fe(OH)$_3$ and MnO$_2$ by H$_2$S in seawater [2, 3] to form dissolved Fe(II) and Mn(II) as well as S(0) as S$_8$. Favorable reactions are those above the horizontal line. The arrows indicate from left to right the peak kinetics for MnO$_2$ and Fe(OH)$_3$, and the calculated pH when the Fe(OH)$_3$ reaction becomes unfavorable*

values and orbitals of H$_2$S and HS$^-$ are -10.47 eV (1b$_2$ orbital) and -2.32 eV (1π orbital), respectively, and HS$^-$ is the better reductant or electron donor. These are formally p$_y$ (π) orbitals from the S atom so an outer sphere process is possible to t_{2g} orbitals for Fe(III) but not Mn(IV). The p$_y$ orbital can also attack Fe and Mn on the bond axis (e_g) in an inner sphere process. The $\Delta S^{\neq}$ is -14 J K^{-1} mol^{-1} for Mn((IV) [5] indicating an associative step prior to electron transfer. Because of the reasonably fast second-order reaction kinetics, these are surface-controlled and inner sphere electron transfer reactions.

11.2.2 MnO$_2$ Reaction with Sulfide

Because these are redox reactions, a second-order rate law (Equation 11.1), which is first-order in the total concentration of each reactant, can be written as in the example of the MnO$_2$ reaction.

$$-\frac{d}{dt}[H_2S]_T = \text{Rate} = k_2[MnO_2]_T[H_2S]_T \tag{11.1}$$

To better understand the system's reactivity, the speciation of the aqueous H$_2$S system (Figure 7.1) and the surface speciation of the δMnO$_2$ surface over pH need to be determined. Surface MnO speciation is governed by the pH$_{zpc}$ expression below (see Table 7.9).

$$>Mn - OH \rightarrow MnO^- + H^+ \qquad {}^*K^S{}_{a2} = 10^{-4.6}$$

At the pH_{zpc} of 4.6, there are two dominant Mn species, $> Mn - OH$ and MnO^-. [At extremely low pH, the surface MnO_2 species would be $> Mn - OH_2{}^+$] Both $> Mn - OH$ and MnO^- may react with H_2S and HS^- so that there would be a total of four precursor complexes (between the reaction arrows in the following equations) with surface reaction rate constants k_1', k_2', k_3', and k_4'.

$$> Mn - OH + H_2S \Rightarrow > Mn - OH \cdot H_2S \rightarrow Mn - OSH + 2H^+ \qquad k_1'$$

$$> Mn - OH + HS^- \Rightarrow > Mn - OH \cdot HS^- \rightarrow Mn - OSH + H^+ \qquad k_2'$$

$$> MnO^- + H_2S \Rightarrow > MnO^- \cdot H_2S \rightarrow Mn - OSH + H^+ \qquad k_3'$$

$$> MnO^- + HS^- \Rightarrow > MnO^- \cdot HS^- \rightarrow Mn - OSH \qquad k_4'$$

The overall rate law with Mn surface and H_2S speciation is given in Equation 11.2.

$$-\frac{d}{dt}[H_2S]_T = k_1'[> MnOH \cdot H_2S] + k_2'[> MnOH \cdot HS] + k_3'[MnO \cdot H_2S] + k_4'[MnO \cdot HS] \qquad (11.2)$$

Multiplying the fraction of each MnO_2 species by each H_2S species at each pH and plotting versus pH predicts the fraction of each precursor complex over pH. The major surface precursor species, which correlates well with the observed second-order rate constants [2] in Figure 11.3, is $> MnO^- \cdot H_2S$ with the rate constant k_3'. Also, a minor amount of the precursor complex, $> Mn - OH \cdot HS^-$, exists in the same pH region but contributes a negligible amount to k_2. The precursor complex $> MnO^- \cdot HS^-$ has two negative species, which can repel in the transition state, and increases in concentration as k_2 decreases with increasing pH so does not contribute to the reaction rate. Likewise, the precursor complex $> MnOH \cdot H_2S$ decreases in concentration with increasing pH to pH 5 as k_2 increases so does not contribute to the reaction rate.

The thermodynamic speciation calculations in Figure 11.3 trend with the kinetics data, which show that the reactions slow down significantly above and below pH 5.6. However, Figure 11.2 showed that there was no

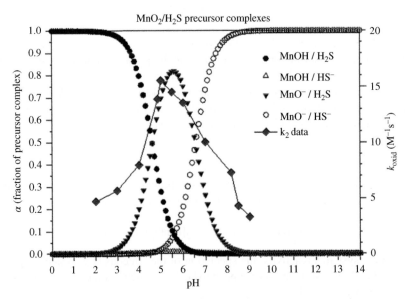

Figure 11.3 *Predicted fraction of the four MnO_2 and H_2S surface precursor complexes occurring at each pH along with the second-order rate constant data, k_2 [2]*

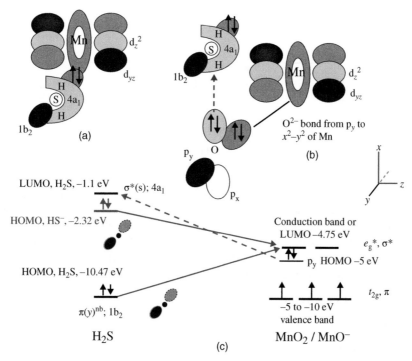

Figure 11.4 *(a) Electron transfer from H_2S $1b_2$ (HOMO) to d_{z^2} Mn(IV). (b) Formation of the $>MnO^-$ • H_2S precursor complex; p_y O atom on MnO^- interacts with $4a_1$ H_2S LUMO; after H^+ transfer from H_2S to MnO^-, HS^- can donate two electrons to the MnO^- e_g^* LUMOs. (c) Orbital energies; solid arrows indicate direct electron transfer from sulfide species to Mn(IV) and dashed arrow represents possible interaction of MnO surface with H_2S prior to H^+ transfer*

thermodynamic barrier to the reaction over all pH, yet the reaction occurs over a limited pH range with a peak at pH = 5.6. These data indicate that the transition state should be affected by pH. Although the speciation calculations indicate that several possible surface precursor complexes form, they do not indicate the type of bonding interactions that occur for each in the transition state.

The speciation calculations in Figure 11.3 predict that the most reactive inner sphere surface complex is $> MnO^-$ • H_2S and Figure 11.4 gives the framework for the formation of the complex and redox electron transfer. As the MnO_2 e_g^* conduction band is -4.75 eV, the stable HOMO of H_2S ($E_{HOMO} = -10.47$ eV; $1b_2$) can donate electrons directly to Mn (d_{z^2}) in MnO_2 (within the 6 eV uphill barrier limit; Figure 11.4a and c) if there is a vacant (or defect) surface site at Mn, but HS^- ($E_{HOMO} = -2.32$ eV) can donate even better (2.43 eV downhill). The oxidation of H_2S or HS^- results in S(0) formation, which can lead to polysulfide and/or S_8 as identifiable products [5]. Electron transfer to Mn atoms in the solid could go to surface or interior Mn atoms as electron transfer depends on which atoms make up the lower orbitals of the conduction band. For example, two different Mn(IV) ions could each accept an electron to form two Mn(III) ions or one Mn(IV) could accept two electrons to form one Mn(II). Once dissolution occurs more vacant Mn sites are available for sulfide species to attack and continue the redox reaction.

There is an alternate inner sphere redox possibility (Figure 11.4b and c). E_{LUMO} for H_2S is -1.1 eV so the oxide on the $>MnO^-$ surface ($E_{HOMO} = -5$ eV; also the t_{2g} band ranges from -5 to -10 eV) can donate electrons to H_2S to form a weak surface complex (3.9 eV uphill barrier). Once formed, H^+ transfer to an

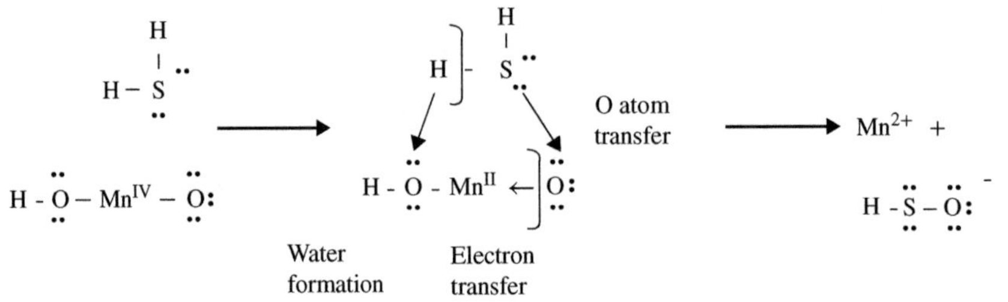

Figure 11.5 *Lewis structure schematic of the* $>$ MnO$^-$ • H$_2$S *surface complex followed by* H$^+$ *and O atom (two–electron) transfer. Brackets indicate bond breakage. Formation of* H$_2$O *may lead to dissociation from* Mn^{2+}

oxide ion on the surface would produce HS$^-$ (HSAB interaction Section 7.8.3), which could displace OH$^-$. Direct bonding between Mn-SH would then permit two-electron transfer into the conduction band of MnO$_2$ (Figure 11.4a) leading to polysulfide and S$_8$. Alternately, a S–O bond can form with O atom transfer from MnO$_2$ to sulfide (Figure 11.5). Continued reaction of HSO$^-$ should lead to sulfite, thiosulfate, and sulfate, which have been identified as products [5], as in the following reactions (Equations 11.3a–11.3d). Note that the electrons as products in these half reactions are transferred into the MnO$_2$ conduction band to form Mn^{2+}.

$$HSO^- + H_2O \rightarrow HSO_2{}^- + 2H^+ + 2e^- \tag{11.3a}$$

$$HSO_2{}^- + H_2O \rightarrow HSO_3{}^- + 2H^+ + 2e^- \quad (HSO_2{}^- \text{ is not a stable species}) \tag{11.3b}$$

$$HSO_3{}^- + H_2O \rightarrow SO_4{}^{2-} + 3H^+ + 2e^- \tag{11.3c}$$

or

$$HSO_3{}^- + S^0 \rightarrow S_2O_3{}^{2-} + H^+ \tag{11.3d}$$

Above a pH of 7, the precursor complex, [$>$MnO$^-$] • [HS$^-$] for $k_4{}'$, dominates, and the second-order rate constant decreases significantly. In the transition state, both species have negative charge and would repel each other so that formation of the precursor complex is hindered.

Below a pH of 4, the predicted precursor complex, [$>$Mn $-$ OH] • [H$_2$S] for $k_1{}'$, is the dominant species and corresponds to the other region with smaller second-order rate constants. In the transition state, both species have neutral character so there is no electrostatic driving force. However, the oxide ion surface binds H$^+$ (E_{LUMO}) better than S in H$_2$S (HSAB; Chapter 7) so binding to S is not expected. There is evidence that surface $>$Mn $-$ OH(H)$^+$ [6] exists at pH < 3 and H$_3$S$^+$ exists at pH ≤ 5 [7]. Thus, in the transition state, both species have positive charge and would repel each other so that formation of the precursor complex is hindered.

11.2.3   Fe(OH)$_3$ Reaction with Sulfide

Similar kinetic data and calculations have shown that clusters or nanoparticles of MnO$_2$ are also important environmental oxidants of thiols [5] and nitrite [8]. These data indicate that the transition state should vary with increasing pH as the pH$_{(ZPC)}$ of any metal oxide phase changes from positive to negative at the same time that H$_2$S (or any acid) loses H$^+$ to form its conjugate base. The behavior mentioned above for MnO$_2$ also occurs when Fe(OH)$_3$ reacts with H$_2$S, and a second-order rate law (Equation 11.4) can be written similar to Equation 11.1. The Fe(III) surface has two acid dissociation constants as given below.

$$-\frac{d}{dt}[H_2S]_T = k_2[Fe(OH)_3]_T[H_2S]_T \tag{11.4}$$

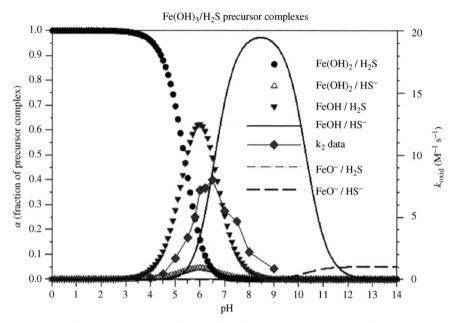

Figure 11.6 *Predicted fractions of four* $Fe(OH)_3$ *and* H_2S *surface precursor complexes occurring at each pH along with the second-order rate constant data,* k_2 *[3]*

$$> Fe(OH)_2{}^+ \rightarrow Fe(OH) + H^+ \qquad *K^S_{a1} = 10^{-5.4}$$

$$> Fe(OH) \rightarrow FeO^- + H^+ \qquad *K^S_{a2} = 10^{-10.3}$$

The kinetic data for the $Fe(OH)_3$ plus H_2S reaction from Figure 11.2 are replotted with the speciation data for H_2S species and the possible precursor complexes in Equation 11.5 (note the similarity to Equation 11.2) and Figure 11.6.

$$-\frac{d}{dt}[H_2S]_T = k'_1[> FeOH_2 \bullet H_2S] + k'_2[> FeOH_2 \bullet HS] + k'_3[FeOH \bullet H_2S] + k'_4[FeOH \bullet HS] \quad (11.5)$$

Note that the $Fe(OH)_3$ and FeO^- surface precursor complexes are unimportant. The most important predicted surface precursor complex (by a factor of 10) is FeOH • H_2S followed by $Fe(OH)_2$ • HS^-. These two predicted complexes account for the kinetic data. Unlike the MnO_2/H_2S reaction, S_8 is the dominant oxidation product [3]. Because of the lability of Fe(III), a likely mechanism is the dissociation of H_2O and/or OH^- from Fe(III), followed by formation of Fe^{III}-SH complexes and then electron transfer from HS^- to the t_{2g} conduction band of $Fe(OH)_3$.

For H_2S to give two electrons to an Fe(III) phase would require that each H_2S donate the electrons to a t_{2g} band containing several orbitals that are partially occupied so that two Fe(III) ions would accept the electrons. Although Fe(III) is a labile ion, the slower kinetics for oxidation of H_2S by Fe(III) versus Mn(IV) are likely due to (1) the lower number of reactive surface precursor complexes at the pH of the transition state ($\alpha_T = 0.8$ in Figure 11.3 and 0.7 in Figure 11.6) and (2) the partially filled t_{2g} conduction band (LUMO) for Fe(III) and empty e_g* conduction band for Mn(IV). Thus, there is less negative charge in the Mn(IV) conduction band.

11.3   Pyrite, FeS₂, Oxidation

Fe^{2+} formed during sulfide oxidation of Fe(III) minerals reacts with sulfide, which leads to the formation of the iron sulfides, FeS and FeS_2, which are described in Section 12.6. Pyrite is the most stable and insoluble iron sulfide mineral, but when exposed to O_2 and other oxidants, it oxidizes. Understanding the oxidation of pyrite is an important environmental problem in iron and sulfur recycling as it occurs in freshwater and marine systems as well as **acid mine** areas. Available data on the oxidation of pyrite indicate that the rate of oxidation with O_2 is 100 times less than that with soluble Fe(III) at a pH ≤ 2.2. At higher pH (>5), the rates become similar. Kinetic data indicate that a surface precursor complex must form prior to electron transfer. Thus, Fe(III) solubility is critical, and at higher pH, dissolved Fe(III)-organic complexes can enhance the rate of pyrite oxidation [9, 10]. The following overall stoichiometric reactions (Equations 11.6a–11.6d) are often used to characterize pyrite oxidation at low pH. Hydrogen ions are a major product of pyrite oxidation, and oxygen isotope data indicate that the oxygen atoms in sulfate ion come from water, which results in the release of H^+. Thus, a dramatic decrease in pH occurs in many environmental systems, and microbial Fe(II) oxidation is necessary to sustain the oxidation. For example, in acid mine areas where pyrite is a constituent in coal, the pH can actually be near or below zero [11].

$$FeS_2(s) + 7/2\, O_2 + H_2O \rightarrow Fe^{2+} + 2SO_4^{2-} + 2H^+ \qquad \text{aerobic} \qquad (11.6a)$$

$$Fe^{2+} + 1/2\, O_2 + 2H^+ \rightarrow Fe^{3+} + H_2O \qquad Fe^{3+}\,\text{formation} \qquad (11.6b)$$

$$Fe^{3+} + 3H_2O \rightarrow Fe(OH)_3(s) + 3H^+ \qquad Fe^{3+}\,\text{solubility} \qquad (11.6c)$$

$$FeS_2(s) + 14Fe^{3+} + 8H_2O \rightarrow 15Fe^{2+} + 2SO_4^{2-} + 16H^+ \qquad \text{anaerobic} \qquad (11.6d)$$

Thiosulfate has been shown to be one of the major intermediate products during pyrite oxidation near pH 7. When Fe(III) is the oxidant and is in excess concentration to pyrite, the following equations (Equations 11.7a–11.7c) represent the complete oxidation of pyrite. **Sulfite and polythionates** have also been detected as minor products or intermediates in solution.

$$3H_2O + FeS_2 + 6Fe^{3+} \rightarrow 7Fe^{2+} + 6H^+ + S_2O_3^{2-} \qquad (11.7a)$$

$$5H_2O + S_2O_3^{2-} + 8Fe^{3+} \rightarrow 8Fe^{2+} + 10H^+ + 2SO_4^{2-} \qquad (11.7b)$$

$$FeS_2 + 8H_2O + 14Fe^{3+} \rightarrow 15Fe^{2+} + 16H^+ + 2SO_4^{2-}\,(\textbf{NET}) \qquad (11.7c)$$

11.3.1   Pyrite Reacting with O_2

Despite the complexity of pyrite oxidation, the initial phase of the reaction of nanoparticulate FeS_2 with O_2 at seawater pH has been shown to be first order in each reactant [12].

$$-\frac{d}{dt}[FeS_2]_T = k_2[FeS_2][O_2] \qquad (11.8)$$

To understand pyrite reactivity [13], it is important to review its crystal structure and orbitals. Figure 11.7a shows that pyrite has a sodium chloride lattice with the Fe^{2+} ions octahedrally coordinated to six sulfur atoms from the disulfide ions, S_2^{2-} (formally persulfide ions), and each S atom is coordinated to three Fe^{2+} ions. The center of the S–S bond would be analogous to the Cl site in NaCl (see NaCl in Figure 6.14). The Fe–S_A–S_B bonding arrangement is angular and not linear where A indicates the S atom bound to Fe_A, and B indicates the terminal S atom (although it is formally bound to one Fe_B, it is pointing toward the solution so can be

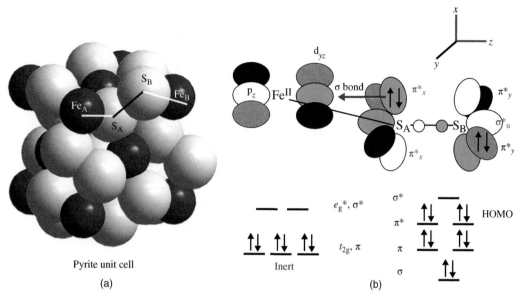

Figure 11.7 *(a) FeS$_2$ unit cell indicating the relative positions of Fe^{2+} and S$_2^{2-}$ (see NaCl Figure 6.14). (b) Frontier molecular orbitals for Fe^{2+} and S$_2^{2-}$ (persulfide is analogous to peroxide); black lines indicate the Fe–S–S bond angle shown in (a). Models created with the Hyperchem™ program (version 8.0.10)*

considered terminal). On the crystal surface, S$_B$ gives pyrite its low pH$_{zpc}$ of 1.4 (Table 7.9), but within the crystal, S$_B$ is bound to three other Fe^{2+} ions. Fe^{2+} has a low spin t_{2g}^6 configuration (Figure 11.7b) so there is no dissociation of the S$_2^{2-}$ ligand from the Fe^{2+} (low solubility) because of high MOSE (CFSE). One of the π^* HOMO orbitals (π_x^* in Figure 11.7b) from S$_2^{2-}$ forms a σ bond with one of the sp^{3}d^2 orbitals from Fe^{2+}; the other π_y^* HOMO can interact with chemical species in solution. Because both π^* orbitals are filled, S$_2^{2-}$ is formally a σ donor (like NH$_3$, see Section 8.6.1) to Fe^{2+}. This discussion is simplified as each one of these sets of orbitals interacts with similar orbitals in the crystal to form a band of orbitals.

Using this information and Figure 11.8, the surface reaction of pyrite with O$_2$ is now described [13].

O$_2$ cannot bind to the pyrite surface with a normal two-electron bond in an inner sphere process because of its partially occupied π^* SOMO orbitals. This is similar to the reaction of O$_2$ with HS$^-$ (Section 5.4.8). However, O$_2$ is a neutral molecule so can approach and possibly chemisorb to the pyrite surface. An outer sphere electron transfer process is possible because the O$_2$ acceptor SOMOs have the same symmetry as the S$_2^{2-}$ donor HOMOs. The energy gap between these SOMO and HOMO orbitals is about 3 eV so there is a possible kinetic barrier to fast electron transfer. The loss of an electron from the S$_2^{2-}$ HOMO (about 3.5 eV) results in an increase of its bond order from 1.0 to 1.5 whereas the O$_2$ bond order decreases from 2.0 to 1.5. The strengthening of the S–S bond leads to S$_2$O$_3^{2-}$ as a metastable product (see next section).

11.3.2 Pyrite Reacting with Soluble Fe(III)

Figure 11.9a shows a possible surface reaction for the first electron transfer from the S$_2^{2-}$ in pyrite with soluble Fe(III) to form Fe^{2+}. Fe(III) (as [Fe(H$_2$O)$_6$]$^{3+}$ at low pH) is a labile ion with a $t_{2g}^3 e_g^{*2}$ electron configuration; thus, H$_2$O or another ligand can dissociate from the Fe(III) so it becomes unsaturated (**EAN** = 15). To achieve EAN = 17, two electrons on S$_B$ in S$_2^{2-}$ from the π_y^* HOMO (pH$_{ZPC}$ = 1.4) can donate to the vacant p$_z$ (or d$_{z^2}$)

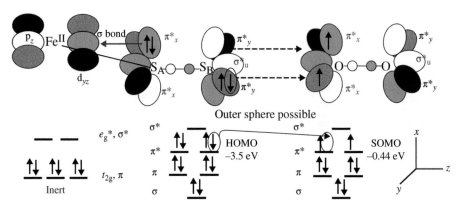

Figure 11.8   *Pyrite and O_2 frontier molecular orbitals describing a possible outer sphere electron transfer process (dashed arrows). Energy level diagrams for Fe(II), S_2^{2-}, and O_2 show electron transfer*

orbital of Fe(III) forming a surface complex on pyrite. Thus, the S_2^{2-} acts as a bridge between pyrite–Fe^{2+} and Fe(III) in solution. This new σ bond begins the weakening of the Fe^{2+} bond with the S_A atom from S_2^{2-} in pyrite.

Although an outer sphere electron transfer process ($\pi_x^* \rightarrow d_{xz}$ or $\pi_y^* \rightarrow d_{yz}$) is possible without the formation of a surface Fe(III)-S_B bond, the formation of an angular σ bond brings the oxidant and reductant closer together for easier and faster **inner sphere** electron transfer between a π^* HOMO from S_2^{2-} and a t_{2g} orbital from Fe(III). Thus, there should be less reorganization energy (λ) in the transition state. Because of the vacant orbital from ligand dissociation on Fe(III), there is no significant HOMO–LUMO energy gap between pyrite and Fe(III). As noted in the reaction with O_2, the S–S and O–O bond orders both become 1.5. During this process, *free radical cations* are formed on the surface of pyrite as in Figure 11.9b, and then the S_B atom binds with oxygen atoms from water. Formally H^+ ions are lost to solution as water attacks the pyrite cation (Fe–S_A–S_B^{2+}) allowing the oxide ion, O^{2-}, to bind the terminal S_B. The reaction lowers the pH to < 2 in acid mine areas undergoing oxidation. To produce thiosulfate, $[S_2O_3^{2-}]$, six Fe(III) ions are necessary to fully break the bond between Fe^{2+} and S_2^{2-} as indicated in the following reactions (Equations 11.9a–11.9d).

$$FeS_2 + 2Fe^{3+} + H_2O \rightarrow 2Fe^{2+} + 2H^+ + FeSSO \tag{11.9a}$$

$$FeSSO + 2Fe^{3+} + H_2O \rightarrow 2Fe^{2+} + 2H^+ + FeSSO_2 \tag{11.9b}$$

$$FeSSO_2 + 2Fe^{3+} + H_2O \rightarrow 2\ Fe^{2+} + 2\ H^+ + FeSSO_3 \tag{11.9c}$$

$$FeSSO_3 \rightarrow Fe^{2+} + S_2O_3^{2-} \quad \text{(detachment)} \tag{11.9d}$$

The SSO in FeSSO is isoelectronic with SO_2 and NO_2^- (see Section 5.7.3) so an inner sphere bridge for electron transfer between the pyrite Fe^{2+} and aqueous Fe^{3+} can easily occur. Also, the SSO_2 group in $FeSSO_2$ is isoelectronic with SO_3 and CO_3^{2-} (see Section 5.8) so another inner sphere bridge for electron transfer between the pyrite Fe^{2+} and the Fe^{3+} can easily occur. The SSO and SSO_2 groups have delocalized π systems so can accept t_{2g} electrons from pyrite Fe^{2+} to form π bonds to stabilize FeSSO and $FeSSO_2$ as intermediates during electron transfer to Fe^{3+}.

The S_2^{2-} ligand is unique as it not only bridges two iron atoms but also transfers electrons from its π^* HOMO and acts as a reducing agent in pyrite. The process described allows for the pyritic iron to remain as Fe^{2+} in solution where it can then become oxidized. The strengthening of the S–S bond during the process is

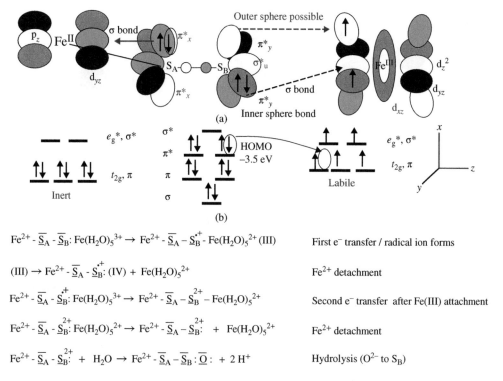

Figure 11.9 (a) FeS$_2$ and Fe(III) frontier molecular orbitals describing a possible inner sphere surface precursor complex (black dashed arrow); energy level diagrams show electron transfer. (b) Plausible reactions for the transfer of the first two electrons from FeS$_2$ to Fe(III)

a possible driving force during the oxidation, and the radical cations (e.g., Fe–S$_A$–S$_B^+$) can lead to electron delocalization in the S$_2^{2-}$ band of orbitals in the solid.

11.3.3 Pyrite Reacting with Dihalogens and Cr^{2+}

Pyrite is also reactive with strong oxidants and reductants [13, 14]. For example, the dihalogens have an unfilled LUMO σ^* orbital so can readily accept electrons to oxidize pyrite in a manner similar to that described for the reaction of HS$^-$ with X$_2$ (Section 5.4.7). In addition, the S$_2^{2-}$ ligand can accept electrons from a reductant such as [Cr(H$_2$O)$_6$]$^{2+}$ to form H$_2$S and [Cr(H$_2$O)$_6$]$^{3+}$ (Equation 11.10), and this reaction is used as a method for the analysis of pyrite.

$$FeS_2 + 2[Cr(H_2O)_6]^{2+} + 4H^+ + 6H_2O \rightarrow 2H_2S + [Fe(H_2O)_6]^{2+} + 2\ [Cr(H_2O)_6]^{3+} \qquad (11.10)$$

[Cr(H$_2$O)$_6$]$^{2+}$ has a $t_{2g}^3 e_g^{*1}$ electron configuration so transfer of the $e_g^{*1}(\sigma^*)$ electron to the σ^* LUMO of S$_2^{2-}$ is a symmetry-allowed outer sphere electron transfer process (Figure 11.10). Because this reaction occurs readily at room temperature, formation of an inner sphere Fe–S–S–Cr precursor complex is likely prior to electron transfer. [Cr(H$_2$O)$_6$]$^{2+}$ is very labile and Cr(II) is a "soft metal" so can react with S$_2^{2-}$ in pyrite. For example, the π_y^* orbital from S$_B$ in S$_2^{2-}$ donates a pair of electrons to the vacant p$_z$ orbital of Cr(II) to form an angular σ bond followed by electron transfer from Cr(II) to pyrite. This new σ bond also begins to weaken

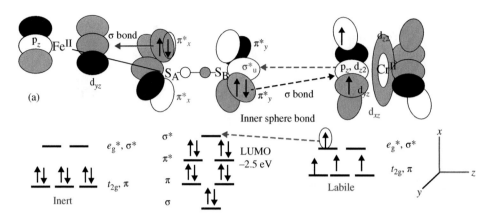

Figure 11.10   *Pyrite and Cr^{2+} frontier molecular orbitals describing a possible surface precursor complex (black dashed arrow) and inner sphere electron transfer process (blue dashed arrow). Energy level diagrams show electron transfer*

the binding of Fe^{2+} with the S_A atom from S_2^{2-} in pyrite. Again, there should be less reorganization energy (λ) in the transition state for electron transfer.

References

1. Luther III, G. W. (2010) The role of one and two electron transfer reactions in forming thermodynamically unstable intermediates as barriers in multi-electron redox reactions. *Aquatic Geochemistry*, **16**, 395–420.
2. Yao, W. and Millero, F. J. (1993) The rate of sulfide oxidation by δMnO_2 in seawater. *Geochimica Cosmochimica Acta*, **57**, 3359–3365.
3. Yao, W. and Millero, F. J. (1996) Oxidation of hydrogen sulfide by hydrous Fe(III) oxides in seawater. *Marine Chemistry*, **52**, 1–16.
4. Audi, A. A. and Sherwood, P. M. A. (2002) Valence-band X-ray photoelectron spectroscopic studies of manganese and its oxides interpreted by cluster and band structure calculations. *Surface and Interface Analysis*, **33**, 274–282. Also see http://www.nist.gov/srd/surface.cfm for HOMO / LUMO data for $Fe(OH)_3$ and MnO_2.
5. Herszage J. and dos Santos Afonso, M. (2003) Mechanism of hydrogen sulfide oxidation by manganese(IV) oxide in aqueous solutions. *Langmuir*, **19**, 9684–9692.
6. Balistrieri, L. S. and Murray, J. W. (1982) The surface chemistry of δMnO_2 in major ion seawater. *Geochimica Cosmochimica Acta*, **46**, 1041–1052.
7. Luther, III, G. W. and Ferdelman, T. G. (1993) Voltammetric characterization of iron(II) sulfide complexes in laboratory solutions and in marine waters and porewaters. *Environmental Science and Technology*, **27**, 1154–1163.
8. Luther G. W. and Popp J. I. (2002) Kinetics of the abiotic reduction of polymeric manganese dioxide by nitrite: an anaerobic nitrification reaction. *Aquatic Geochemistry*, **8**, 15–36.
9. Luther, G. W.,Kostka, J. E., Church, T. M., Sulzberger, B. and Stumm, W. (1992) Seasonal iron cycling in the salt marsh sedimentary environment: the importance of ligand complexes with Fe(II) and Fe(III) in the dissolution of Fe(III) minerals and pyrite, respectively. *Marine Chemistry*, **40**, 81–103.

10. Peiffer, S. and Stubert, I. (1999) The oxidation of pyrite at pH 7 in the presence of reducing and nonreducingFe(III)-chelators. *Geochimica Cosmochimica Acta*, **63**, 3171–3182.

11. Druschel, G. K., Baker, B. J., Gihring, T. M. and Banfield, J. F. (2004) Acid mine drainage biogeochemistry at Iron Mountain, California. *Geochemical Transactions*, **5** (2), 13–32, doi: 10.1063/1.1769131.

12. Gartman, A. and Luther, III, G. W. (2014) Oxidation of synthesized sub-micron pyrite (FeS_2) in seawater. *Geochimica Cosmochimica Acta*, **144**, 96–108.

13. Luther, G. W. (1987) Pyrite oxidation and reduction: molecular orbital theory considerations, *Geochimica Cosmochimica Acta*, **51**, 3193–3199.

14. Luther, III, G. W. 1990. The frontier molecular orbital theory approach in geochemical processes, In *Aquatic Chemical Kinetics*, Chapter 6, (ed. W. Stumm), John Wiley and Sons, New York, pp. 173–198.

12

Metal Sulfides in the Environment and in Bioinorganic Chemistry

12.1 Introduction

The discussion in Chapter 6 centered on the structure of solids or crystals, and Chapter 11 centered on the dissolution of minerals; however, there was no discussion regarding the formation of solids or minerals. In this chapter, the formation of solids via clusters and nanoparticles is presented using metal sulfides (MS) as the starting point. Natural MS clusters or nanoparticles are found in fresh and marine waters, sediments, waste waters, and hydrothermal vent waters; they are typically resistant to oxidation by O_2. In addition, manufactured MS nanoparticles (i.e., quantum dots), which are used for a variety of applications, are released to the environment. The properties of nanoparticles can deviate from the bulk material because of their high surface area and diverse structural features, which affect their reactivity.

Although the free metal ions and hydrogen sulfide can each be toxic to many organisms on the surface of the earth, the metal sulfides are typically not toxic to organisms because of their high thermodynamic stability and low solubility. Also, copper and zinc in sulfides are typically not bioavailable for life processes. However, iron sulfides are used in many biochemical processes, and researchers have shown that the formation of FeS membranes at neutral pH as well as the formation of FeS_2 at lower pH may have been essential pathways for the production of organic matter and the origin of life.

For the purposes of the following discussion, the following definitions with approximate size ranges are provided.

A *complex* is a mononuclear metal coordination compound in which the central metal atom unites one or more ligands; a complex is a dissolved species and has a size typically less than 0.4 nm.

A *cluster* is a polynuclear metal complex containing a discrete number of metal atoms, which are typically bridged by ligands; it behaves as a dissolved species. Size range is from a complex to about 1 nm.

A *nanoparticle (NP)* is a higher order material with a structural characteristic of that of the first condensed or solid phase (see Section 6.4). Size range is from 1 to 100 nm and can pass through filters that are commonly used (200 and 400 nm).

A *solid or mineral* is an infinite lattice of atoms or ions as described in Chapter 6.

Inorganic Chemistry for Geochemistry and Environmental Sciences: Fundamentals and Applications, First Edition. George W. Luther, III.
© 2016 John Wiley & Sons, Ltd. Published 2016 by John Wiley & Sons, Ltd.
Companion Website: www.wiley.com/go/luther/inorganic

12.2 Idealized Molecular Reaction Schemes from Soluble Complexes to ZnS and CuS Solids

The **Ostwald step rule** or the "rule of stages" postulates that the precipitate with the highest solubility will form first in a consecutive precipitation reaction indicating that mineral formation occurs via precursor intermediates that are molecular. The reaction of the free divalent metal ions of the first transition series (Fe^{2+}, Co^{2+}, Ni^{2+}, Cu^{2+}, Zn^{2+}) at pH 7–8 with sulfide shows an example of the complexity of the reaction process even though it is rather rapid (<1–100 ms) [1–3]. The process can be broken into three major reactions (Equations 12.1–12.3) as in the following reaction sequence for Zn^{2+} (Cu^{2+} is similar). Several principles outlined in Chapters 8 and 9 and data in Table 12.1 are used to show that metal sulfide mineral formation is an entropy-driven process and an example of **self-assembly** reactions.

$$3[Zn(H_2O)_6]^+ + 3HS^- \rightarrow 3[Zn(H_2O)_5(SH)]^+ + 3H_2O; \ HS^- \text{ substitution; Eigen–Wilkins} \qquad (12.1)$$

$$3[Zn(H_2O)_5(SH)]^+ \rightarrow Zn_3S_3(H_2O)_6 + 3H^+ + 9H_2O; \text{ ring formation}; H^+, H_2O \text{ loss}; \ \Delta S \text{ favored} \qquad (12.2)$$

$$5Zn_3S_3(H_2O)_6 + 3HS^- \rightarrow 3[Zn_4S_6(H_2O)_4]^{4-} + 3Zn(H_2O)_6^{2+} + 3H^+; \text{ cluster forms}; \ \Delta H \text{ favored} \qquad (12.3)$$

For pH 7–8, there is no significant hydrolysis for these free ions (Table 12.1). Thus, these metal ions are formally six coordinate, $[M(H_2O)_6]^{2+}$ (there is evidence for Cu^{2+} being five coordinate), and in the MS mineral, the metal ions become four coordinate. Hydrogen sulfide exists as H_2S and HS^- and also becomes four coordinate in the mineral. Overall, there are significant intra- and intermolecular arrangements occurring during the transformation of the reactants to the products. In order to form the products, a series of reactions or steps must occur between the metal and sulfide species to form **covalent bonds**, which give ZnS and CuS minerals their inherent insolubility.

The first step is the nucleophilic displacement of water by HS^- (Equation 12.1) to form a **covalent** Zn–S bond in $[Zn(H_2O)_5(SH)]^+$. This step is fast for all the free metal ions listed in Table 12.1 (except for Ni^{2+}) because of their fast water exchange rates, and Equation 12.1 is an example of an **Eigen–Wilkins** mechanism. For Cu^{2+} and Zn^{2+} (also Ag^+, Cd^{2+} and Pb^{2+}), Equation 12.1 is a diffusion-controlled reaction

Table 12.1 *Molecular orbital stabilization energy for octahedral geometry in units of βS_σ^2 from Table 8.2 (maximum stability is 12 βS_σ^2 for the d^0 to d^3 cases). The smaller βS_σ^2 values indicate a tendency toward tetrahedral geometry over octahedral. Water exchange is k_{-w} and the negative log of the hydrolysis constant is pK_H*

Ion		Octahedral configuration	MOSE (βS_σ^2)	$k_{-w} (s^{-1})$	pK_H
Mn^{2+}	d^5	$t_{2g}^3 e_g^{*2}$	6.00	2.15×10^7	10.70
Fe^{2+}	d^6	$t_{2g}^4 e_g^{*2}$	6.00	4.4×10^6	10.10
Co^{2+}	d^7	$t_{2g}^5 e_g^{*2}$	6.00	3.2×10^6	9.60
Ni^{2+}	d^8	$t_{2g}^6 e_g^{*2}$	6.00	3.15×10^4	9.40
Cu^{2+}	d^9	$t_{2g}^6 e_g^{*3}$	3.00	8.9×10^9	7.53
Zn^{2+}	d^{10}	$t_{2g}^6 e_g^{*4}$	0	6.0×10^8	9.60
Cd^{2+}	d^{10}	$t_{2g}^6 e_g^{*4}$	0	3.0×10^8	10.08
Pb^{2+}	$s^2 d^{10}$	$t_{2g}^6 e_g^{*4}$	0	7.0×10^9	7.71

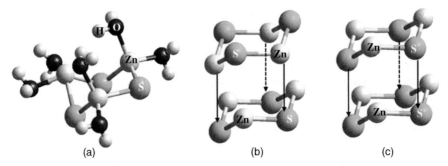

(a)   (b)   (c)

Figure 12.1   *(a) $Zn_3S_3(H_2O)_6$ ring. (b) Zn and S atoms on different rings overlap to form Zn–S bonds. (c) S atoms on different rings overlap to form S–S bonds. The waters coordinated to Zn are omitted for clarity in b and c. Models created with the Hyperchem™ program (version 8.0.10)*

as $k_f > 10^8$ $M^{-1}s^{-1}$. $[Zn(H_2O)_5(SH)]^+$ can then polymerize to form cyclic six-membered $Zn_3S_3(H_2O)_6$ (Equation 12.2), which results in four coordination for Zn, two coordination for S and alternating Zn–S–Zn–S bonds as shown in Figure 12.1a. Equation 12.2 also shows H^+ as well as water loss, which results in an increase in **entropy** and therefore a decrease in ΔG (see Equations 9.8 and 9.9, also Equation 7.6). The loss of MOSE (Table 12.1) for Zn^{2+} and Cu^{2+} is another driving force for the change in coordination from six to four. The choice of $Zn_3S_3(H_2O)_6$ as a polymeric building block is due to the geometrical arrangement of Zn and S in the mineral (see Figures 6.13 and 6.14). The Zn_3S_3 ring is structurally similar to cyclohexane, and can then further polymerize so that the coordination of the sulfide changes to four prior to precipitation as a solid (**enthalpy favored process**). Once this occurs ZnS exhibits the diamond structure as the four S^{2-} ions provide an octet of electrons around tetrahedral Zn^{2+}. Figures 12.1b and c show two possible ways for Zn_3S_3 rings to overlap or cross-link completely to form a larger dimeric cluster Zn_6S_6 (examples of **Ostwald ripening**). Figure 12.1b shows bonding overlap between Zn and S whereas Figure 12.1c shows bonding overlap between only S atoms. In both cases, the bonding can be viewed as the symmetry-allowed overlap of p orbitals between the atoms on one ring with the p orbitals of the atoms on the other. Figure 12.1c is not possible for the ZnS system because this would require oxidation of two $S(-2)$ ions to form S_2^{2-}. There is only one statistical opportunity for collision in these two cases.

The probability of more collisions between two Zn_3S_3 rings increases to eight when two Zn atoms on one ring can overlap with two S atoms on the other ring (another example of **Ostwald ripening**); this will enhance the reaction rate for cluster, nanoparticle, and mineral formation. Figure 12.2a shows the overlap of two Zn_3S_3 rings in the presence of excess HS^-, which would be likely in anoxic basins and sediments. The HS^- provides another way to cross-link the Zn_3S_3 rings through Zn atoms on the two rings to form $[Zn_4S_6(H_2O)_4]^{4-}$. In Equation 12.3 and Figure 12.2a, $[Zn_4S_6(H_2O)_4]^{4-}$ is chosen for cluster formation because it gives the adamantine-like structure shown in Figure 6.13 for sphalerite, and because it has been found in laboratory and field studies [4].The overall process results in one Zn_3S_3 ring being capped by a ZnS_3 group with H^+ and Zn^{2+} being released to solution. In the absence of **capping agents** such as thiol (RSH) compounds, the reaction will lead to nanoparticles and eventually minerals depending upon the concentration of the zinc and sulfide in solution.

The same reactivity pattern is available for Cu^{2+}; however, it is also reduced by HS^- to Cu^+ once the Cu_3S_3 rings or other clusters form [5, 6]. The one-electron transfer reaction of HS^- (π) with Cu^{2+} ($t_{2g}^6 e_g^{*3}$) is **symmetry forbidden** ($\pi \rightarrow \sigma$ or e_g^*) as an **outer sphere process**, and is thermodynamically unfavorable

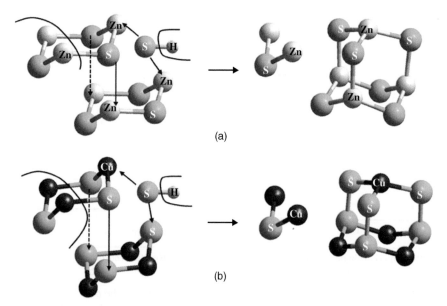

(a)

(b)

Figure 12.2   *(a) Overlap of metal atoms on one ring with the sulfur atoms on the second ring in the presence of excess HS$^-$ to form [Zn$_4$S$_6$(H$_2$O)$_4$]$^{4-}$. (b) Overlap between sulfur atoms on both rings to form the Cu$_4$S$_6$ species. The waters coordinated to Zn and Cu are omitted for clarity*

[Cu1, S3 in Table 2.2 (p$\varepsilon = 2.72 - 18.26 = -15.5$)]. However, HS$^-$ donates an electron pair as a ligand to Cu^{2+}. Once a cluster forms, a two-electron inner sphere transfer from HS$^-$ to two Cu^{2+} is favorable at pH $= 7$ [Cu1, S2 in Table 2.2 (p$\varepsilon = 2.72 - (-1.06 - (0.5*7)) = 7.28$)] through the **delocalized band of orbitals** in the cluster. At this point, S$_2{}^{2-}$ is formed during the redox reaction. Figure 12.2b shows the cross-linking of two Cu$_3$S$_3$ rings in the presence of excess HS$^-$; a similar adamantane-like structure results but S–S bonds form between the two rings because of redox. Both Cu–S and S–S bonding between planes (similar to that found in both of the adamantane-type structures in Figure 12.2) are found in the mineral covellite (CuS) [7, 8].The structure with S–S bonding shows more planar character (closer to benzene) in the Cu$_3$S$_3$ rings and some of the Cu atoms in covellite have trigonal bipyramidal five coordination. Cu$_3$S$_3$ and Cu$_4$S$_6$ stoichiometries have been detected by mass spectrometry in laboratory and field samples [2, 4].

The energetics of these reactions along the reaction coordinate is summarized in Figure 12.3. Figure 12.3a describes the reaction of the complexes to form the mineral as the final product, and Figure 12.3b describes the formation of a stable nanoparticle. In both figures, there is very little activation energy when going from the complex ions to the first M$_3$S$_3$ ring as well as going from the M$_3$S$_3$ ring to the larger nanoparticle because the formation of neutral six-membered rings allows for more collisions in the transition state as described above. Neutral rings are excellent building blocks and react quickly with each other as well as with positive or negative ions (e.g., [Zn$_4$S$_6$(H$_2$O)$_4$]$^{4-}$). Sulfur becomes four coordinate on formation of higher order clusters, nanoparticles, and minerals, which completes the conversion of solution species to the covalent mineral. These initial entropy-driven reactions (Equation 12.2) are examples of self-assembly reactions, which are rapid, as there is no thermodynamic barrier or significant kinetic barrier to reactivity. The main difference between Figures 12.3a and b is that ΔG_r for the nanoparticle is more stable than that for the mineral.

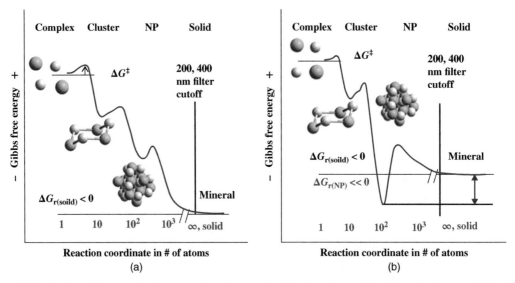

Figure 12.3   (a) Reaction coordinate in number of atoms for the formation of the mineral via metastable clusters or nanoparticles. (b) Reaction coordinate for the formation of a stable nanoparticle over the mineral; the double arrow notes that the NP is more stable than the mineral

12.3   Nanoparticle Size and Filtration

It is important to note the cutoff diameter for filters, and the number of atoms found in clusters and nanoparticles in Figure 12.3. For example, a PbS nanoparticle (cubic NaCl structure) of 10^3 total Pb and S ions with a length of 1.515 nm on each edge (see Table 6.4 for ionic radii and Figure 6.21) has a cubic diameter of 2.62 nm and volume of 3.48 nm^3. This particle size is two orders of magnitude smaller than the 200 or 400 nm diameter Nuclepore™ filters commonly used in environmental sampling (see Appendix Figure 12A.1).

12.4   Ostwald Ripening versus Oriented Attachment

Nanoparticles can react with the basic cluster building block via **Ostwald ripening** as above and with each other in a process termed **oriented attachment**. The growth of nanoparticles by these two methods has been described mathematically. Growth by **Ostwald ripening** can be described by the Lifshiz–Slyozov–Wagner model [3, 9] (Equation 12.4), which is represented schematically in Figure 12.4a. Note that smaller clusters in this example react with a PbS nanoparticle A (model with 64 atoms) to form a slightly larger nanoparticle B.

$$d = at^{1/n} + d_0 \tag{12.4}$$

Here, t is time, a is a temperature-dependent kinetic constant, d is the average particle diameter, and d_0 is the diameter at $t = 0$. The exponent n will have a value between 2 and 4, depending on what controls nanoparticle growth. When $n = 2$, growth is controlled by surface diffusion at the solid/liquid interface; when $n = 3$, growth is controlled by volume diffusion of ions in solution; and when $n = 4$, growth is controlled by the dissolution kinetics of the initially formed species.

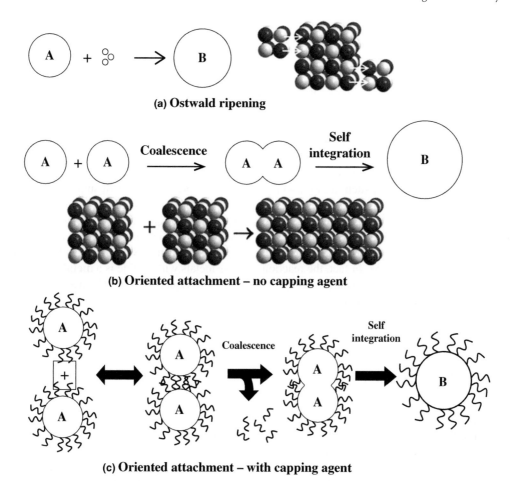

Figure 12.4   *(a) Nanoparticle A reacts with several clusters to form nanoparticle B with no doubling in size (white arrows show covalent bonding alignment). (b) Doubling of A without any capping agent to form B. (c) Doubling of A with capping agents to form B*

In addition to Ostwald ripening, nanoparticle growth can also occur by a mechanism of coalescence when growth occurs as two nanoparticles come together to form larger species (**schematic in Figure** 12.4b). First, two nanoparticles form a "floc," which is held together by weak van der Waals interactions or electrostatic forces. Then, the two nanoparticles coalesce (Figure 12.4b). In the absence of surface adsorbed species or capping agents, the reaction rate can be <1 *s*. If there are capping agents, these need to be eliminated via M–L dissociation before coalescing occurs (Figure 12.4c). Capping agents and increased ionic strength increase the reaction rate to ≫1 *s* [10]. These mechanisms are referred to as **oriented attachment** because adjoining nanoparticles must adopt a crystallographic orientation that will allow new covalent bonds to be formed for coalescence and self-integration to occur. The models in Figure 12.4b show perfect alignment of two PbS nanoparticles of the same size (64 atoms) reacting to form one nanoparticle of 128 atoms, which is twice the size of the original. There are many more ways to orient the two PbS nanoparticles in Figure 12.4b to form covalent bonds, but these lead to other structures or materials that are less stable. Equation 12.5 shows a simple mathematical model for the doubling of average nanoparticle size when two nanoparticles come together; all the variables are defined as in Equation 12.4 and k is the rate constant for nanoparticle growth [9].

$$d = \frac{d_0 \left(\sqrt[3]{2} kt + 1 \right)}{(kt + 1)} \tag{12.5}$$

12.5   Metal Availability and Detoxification for MS Species

Zinc and copper sulfide nanoparticles are kinetically inert to ligand dissociation, decomposition, and oxidation by O_2. Interestingly, metals such as Ag^+, which react even more strongly with sulfide than Zn^{2+} and Cu^{2+}, can displace these metal ions to form AgS nanoparticles as in Equation 12.6 [2]. The order of thermodynamic stability based on these reactions is $Ag_2S_{NP} > CdS_{NP} > CuS_{NP} > ZnS_{NP}$, which aligns with the inverse order of solubility for these MS minerals. Thus, these metals and others (e.g., Pb^{2+}, Hg^{2+}) bound to sulfide should not be available to organisms for uptake and use as indicated by Williams and da Silva [11], who used solubility criteria alone. In fact, the reaction of these metals with sulfide is a **metal detoxification** process as documented for *Daphnia magna* using the chemistry in Equation 12.6 [12]. Metal binding with sulfide is also a way to detoxify sulfide.

$$ZnS_{NP} + 2Ag^+ \rightarrow Ag_2S_{NP} + Zn^{2+} \tag{12.6}$$

12.6   Iron Sulfide Chemistry

12.6.1 FeS_{mack} (Mackinawite)

The reaction of $Fe(H_2O)_6^{2+}$ with HS^- also proceeds via an Eigen–Wilkins mechanism as in Equation 12.1 with $k = 1.1 \times 10^7$ $M^{-1}S^{-1}$ [1, 11]. The first product formed is mackinawite with the stoichiometry FeS $[Fe^{2+}S^{2-}]$. Figure 12.5a shows that this is a tetragonal solid with a spacing of 503 pm (5.03 Å) between the iron layers on the z axis; the other two axes have a distance of 367.4 pm. Within the iron layers, the Fe–Fe bond distance of 259.8 pm is identical to that found in metallic iron, indicating that there is Fe–Fe bonding. Instead of an M_3S_3 ring as shown in the copper and zinc sulfides, the basic building block is a Fe_2S_2 ring. Figure 12.5b shows a 90° rotation (the orientation of atoms Fe_1, Fe_2, S_1 and S_2 are identical in Figures 12.5a and b). The Fe^{2+} ions have almost perfect tetrahedral coordination with four S atoms (Fe_2 in Figure 12.5b) whereas the S atoms are bound asymmetrically to four Fe atoms that are in a plane directly above or below the two S atom layers.

Figure 12.5c shows the structure for a $Fe_2S_2(H_2O)_4$ or $[Fe_2(\mu_2\text{-}S_2)](H_2O)_4$ cluster, which is the simplest possible molecule with a Fe_2S_2 ring. Evidence for it and higher order clusters, which are electroactive and detectable by voltammetry, have been found in freshwater and marine waters [13]. The first solid or condensed phase appears to have the stoichiometry $Fe_{150}S_{150}$. Because Fe^{2+} is high spin, replacement of water by other ligands and the Fe by other cations such as Ni, Mo, W is possible.

Figure 12.5a and b shows that the surface S and Fe atoms are exposed to the solution. The S^{2-} ions can act as nucleophiles (bases), and the unsaturated Fe^{2+} ions can act as Lewis acids to accept a pair of electrons from a ligand and increase the Fe coordination number, or to accept electrons from a reductant such as H_2. Also, Figure 12.5a shows that there is significant spacing between the S atom layers. Because H^+ and M^{+2} ions are typically smaller than 100 pm (0.100 nm, Table 6.4), they can insert between the S atom layers and change the properties of mackinawite.

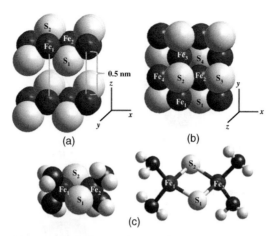

Figure 12.5 *(a) Mackinawite structure with the xz plane in the plane of the paper. (b) Same structure with the xy plane in the plane of the paper. (c) [Fe$_2$(μ_2-S$_2$)(H$_2$O)$_4$ structure (Fe$_2$S$_2$ ferredoxin type). Models created with the Hyperchem™ program (version 8.0.10)*

12.6.2 FeS$_{mack}$ Conversion to Pyrite, FeS$_2$

FeS$_2$ is the principal iron sulfide formed in anoxic waters and sediments. The kinetics and rate laws of FeS$_2$ formation have been described for two major FeS$_2$ forming reactions (1) FeS and polysulfides as reactants (Equations 12.7a,b) and (2) FeS and H$_2$S as reactants (Equations 12.8a,b). The FeS starting material is nanoparticulate mackinawite and acts differently in each reaction; also, it is formally a soluble entity that has been given the formula FeS$_{aq}$ [13]. Although the reactions increase with increasing temperature, the conversion of FeS$_{aq}$ to solid phase mackinawite also increases and prevents optimal kinetics [11].

$$\text{FeS} + \text{S}_x^{2-}(x = 4, 5) \rightarrow \text{FeS}_2 + \text{S}_{(x-1)}^{2-} \tag{12.7a}$$

$$\frac{d}{dt}[\text{FeS}_2] = k_2[\text{FeS}_{aq}][\text{S}_x^{2-}], \text{ simplified form} \tag{12.7b}$$

$$\text{FeS}_{aq} + \text{H}_2\text{S} \rightarrow \text{FeS}_2 + \text{H}_2 \quad \Delta G_f = -33.93 \text{ kJ mol}^{-1} \tag{12.8a}$$

$$\frac{d}{dt}[\text{FeS}_2] = k_2[\text{FeS}_{aq}][\text{H}_2\text{S}] \tag{12.8b}$$

12.6.2.1 Polysulfide Reaction

Linear polysulfide ions (S$_x^{2-}$) form as intermediates during the oxidation of sulfide [13]. The internal S atoms approach zero charge as the chain lengthens and with increasing protonation or binding. Thus, they are a source of S(0), which can react with FeS to form FeS$_2$. Figure 12.6a shows that the starting FeS has similar characteristics to the product FeS$_2$ (the cubic unit cell axes have a length of 541.7 pm compared with the 503 pm Fe–Fe layer distance in FeS$_{mack}$). Looking down the z axis of each, there is a plane of five Fe atoms. Figure 12.6b shows a schematic of Equation 12.7a with S$_5^{2-}$ as the polysulfide. Sulfur isotopic data [13, 14] indicate that the reaction of FeS with linear polysulfides (S$_x^{2-}$) results in both S atoms in FeS$_2$ coming from S$_x^{2-}$. The kinetics for Equation 12.7a indicate that S$_x^{2-}$ attacks Fe^{2+} to form a [FeS(S$_x^{2-}$)] complex as in Figure 12.6b and c. Figure 12.6b shows that there are two bonds formed (dashed arrows) and two bonds broken (two brackets). Because S^{2-} is a strong nucleophile, it can attack the S(0) atoms at the center of the

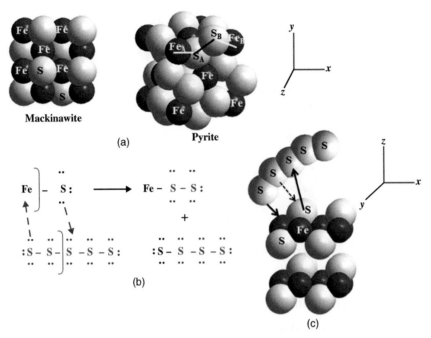

Figure 12.6 (a) Relationship between FeS and FeS$_2$ surfaces in their unit cells. (b) Schematic of reaction in Equation 12.7a; blue color for S atoms from S$_5^{2-}$. (c) Attack of S$_5^{2-}$ on the FeS surface to form an [FeS(S$_x^{2-}$)] surface complex; the dashed arrow indicates that the S$_2^{2-}$ formed slips between Fe atoms. Models created with the Hyperchem™ program (version 8.0.10)

S$_x^{2-}$ chain resulting in the formation of a five-membered FeS$_4$ ring, which can break up to form S$_2^{2-}$ (from S$_x^{2-}$), S$_{(x-1)}^{2-}$ and FeS$_2$. Although an S atom from FeS leaves the surface to form S$_{(x-1)}^{2-}$, the S$_2^{2-}$ slips between the Fe atoms to open up the Fe–Fe bond distance from 260 to 542 pm along the xy plane. The FeS bond distance is 226 pm in FeS with a range from 223–230 pm in FeS$_2$. The Fe(II) changes from high to low spin during the reaction. Overall, FeS acts as a Lewis acid when S$_x^{2-}$ attacks surface Fe^{2+} and as a Lewis base when S^{2-} from FeS attacks S$_x^{2-}$. Thus, both reactants exhibit amphoteric (Lewis acid–base) properties.

12.6.2.2 H$_2$S Reaction

For the reaction of FeS with H$_2$S, sulfur isotopic data indicate that one S atom comes from each reactant [13, 14]. There are only two requirements for this pyrite formation pathway; (1) the reactants must lead to S–S bond formation and weakening of H–S bonds to form H$_2$, and (2) in the intermediate, tetrahedral Fe(II) must undergo a high spin ($e^3t_2^3$) to low spin (t_{2g}^6) octahedral conversion. In this case, FeS acts only as a Lewis base attacking the H$_2$S LUMO (Figure 12.7a), which has a stable energy of -1.1 eV and is delocalized across all three atoms. This nucleophilic attack permits simultaneous S–S bond formation and H–S bond weakening to form H atoms adsorbed to the surface. H–H bond formation is facilitated by the $\sim 93°$ H$_2$S bond angle.

Overall, H$_2$S is the oxidant of FeS, but it is the sulfur and not the Fe in FeS that is oxidized as S$_2^{2-}$ forms. The FeS HOMO energy has been calculated to be $+0.5$ eV [13]. Figure 12.7b shows H$_2$S molecules interacting with the FeS surface to form FeS $\rightarrow$ SH$_2$ intermediates; the dashed arrow indicates slippage of the S atom on FeS to accommodate the extra S atom from H$_2$S to open up the Fe–Fe bond distance in the plane.

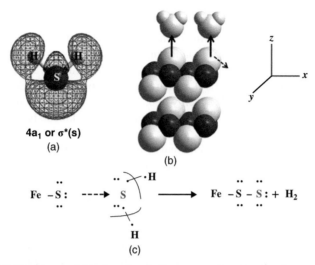

4a₁ or σ*(s)

(a)

(b)

(c)

Figure 12.7 *(a)* H_2S *LUMO; blue and black colors indicate + and − sign for the wave function, respectively. (b)* H_2S *molecules interacting with the FeS surface. (c) Dashed arrow indicates the formation of the FeS → SH₂ intermediate before breakup to products; blue S atom from* H_2S*. Models and orbital created with the Hyperchem™ program (version 8.0.10)*

Figure 12.7c shows a schematic of the reaction with formation of a Lewis acid–base complex (FeS → SH₂) as an intermediate. The H_2 product can remain adsorbed to the surface in another Lewis acid–base complex as its LUMO is −5eV, which is lower in energy than the FeS (+0.5 eV) and FeS₂ (near-3.5 eV) HOMOs. H_2 adsorption on FeS₂ is likely important in the catalytic conversion of CO and CO_2 to small organic compounds at hydrothermal vents (Section 12.6.3).

12.6.2.3 FeS₂ Formation Pathways and pH

The H_2S pathway forms pyrite two orders of magnitude faster than the polysulfide pathway [11, 12]. Figures 12.6c and 12.7b give a possible indication as to why this is so. First, H_2S should be faster because there are less steric factors for H_2S to adsorb onto the FeS surface versus S_x^{2-} ($x > 4$). Thus, more H_2S molecules can attack the same FeS surface area than can S_x^{2-}. Second, there is more ordering needed for the S_x^{2-} to adsorb to the surface for the formation of the five-membered FeS₄ ring. Once formed, the FeS₄ ring reduces the amount of S_x^{2-} that can attack the same FeS surface. Third, the mobility of the S_2^{2-} to slip between the Fe plane of atoms is more facile as less atomic motion is necessary for the H_2S pathway.

Both FeS₂ formation reactions occur, and depending on the pH, occur simultaneously. The S_x^{2-} reaction is fastest at pH > pK_{a2}, which is pH > 6.05 for S_5^{2-} and >6.6 for S_4^{2-}. The H_2S reaction is fastest at pH < 7 (pK_{a1} H_2S ~ 7), whereas at pH > 7, HS⁻ is not a reactant because it is a Lewis base and cannot accept electrons from FeS$_{aq}$. Thus, at pH ≥ 8, the polysulfide pathway is dominant. Pyrite is ubiquitous as it forms in many environments on earth including sediments, stagnant water columns, and hydrothermal vents (see Figure 2.8). In marine sediments and stagnant water columns, the pH is typically > 7.5. In near shore sediments, the pH is nearer 7. In wetlands, salt marsh sediments, and hydrothermal vents, the pH is typically < 6.5. The H_2S pathway seems to be the more important one in a host of environments, and during the early earth period before the Great Oxidation Event, it was likely the dominant pathway as oxidants, which would oxidize H_2S to S_x^{2-} or S_8, were not present in the environment to a great extent.

12.6.3 FeS as a Catalyst in Organic Compound Formation

The reaction of CO and CO_2 with FeS alone or in the presence of H_2S leads to the formation of COS, CS_2, and simple organic sulfur compounds (e.g., CH_3SH) indicating that the FeS system has significant reducing power to activate CO_2 [15]. CH_3SH is an excellent starting material for biochemistry as CO insertion (similar to the oxo process; Section 9.13.1) forms $CH_3(CO)SH$, which is a precursor for acetyl-CoA found in the reverse or reductive citrate cycle [16]. The FeS HOMO energy (+0.5 eV [13]) is similar to the energy of the CO_2 π_u^* LUMO (Section 5.7.2) so direct electron transfer is possible with little kinetic barrier when compared to the HS^- HOMO of −2.32 eV.

Wächtershäuser [17, 18] proposed the hydrothermal synthesis of organic compounds during the inorganic synthesis of FeS_2 as a primary pathway. The FeS/H_2S reaction (Equation 12.8a) is an important source of H atom production on the pyrite surface and eventual H_2 production by inorganic (abiotic) means. Ni has been added to the FeS starting material, and it is possible that a **Fischer–Tropsch** type reaction (see Section 9.13.4) can occur during the formation of FeS_2 as the H_2 (or H atoms) along with the oxidant CO or CO_2 can remain on the Fe(Ni)S or $Fe(Ni)S_2$ surface. As an example, the reaction of H_2 with CO_2 to form formic acid (Equation 12.9) is endergonic but when coupled with Equation 12.8a, the reaction (Equation 12.10) is favorable. Thus, reduction and activation of CO_2 can be accomplished with the FeS system. These reactions would occur in a more acidic ocean because H_2S and not HS^- is the reactant.

$$CO_2 + H_2 \rightarrow HCOOH \qquad\qquad \Delta G_f = +22.07 \ \text{kJ mol}^{-1} \qquad\qquad (12.9)$$

$$CO_2 + FeS_{aq} + H_2S \rightarrow HCOOH + FeS_2 \qquad \Delta G_f = -11.86 \ \text{kJ mol}^{-1} \qquad (12.10)$$

Under basic conditions (as high as pH 11), the oxidation of FeS_{mack} by O_2 as well as other electron acceptors leads to S_8 (and other S species) and Fe^{3+}. This and the Fe–Fe bond distance led Russell [19] to hypothesize that FeS semipermeable membranes doped with Ni^{2+} and/or Co^{2+} could be a primordial source of electrons and catalysis for the formation of organic compounds from inorganic substrates such as CO and CO_2. On one side of the FeS_{mack} membrane, H_2 is oxidized to H^+ with the electrons going into the Fe layers with metallic bonding. These electrons then reduce the oxidant (e.g., organic acids, NO_3^- and CO_2) on the other side of the FeS species. The Fe layers act as an electrical conduit.

12.6.4 FeS as an Electron Transfer Agent in Biochemistry

FeS is an excellent reducing agent as noted in the above reactions; thus, it is not surprising to find that FeS clusters are used as reducing agents in nature [20, 21]. FeS proteins, which are responsible for electron transfer in many biochemical reactions, have embedded FeS clusters including Fe_2S_2 clusters (formally $[Fe_2(\mu_2\text{-}S_2)]$ or [2Fe-2S]) and Fe_4S_4 clusters (formally $[Fe_2(\mu_3\text{-}S_4)]$ or [4Fe-4S]) known as ferredoxins where the brackets indicate the cluster within the protein. Figure 12.5c shows the $[Fe_2S_2](L)_4$ structure with water as ligand. In proteins, the fully oxidized form is $[Fe_2S_2]^{2+}$ with 2 Fe^{3+} ions. Figure 12.8 shows the cubic $[Fe_4S_4](L)_4$ structure with the $[Fe_2S_2]$ ring. L is usually cysteine, but can also be inorganic S, methionine, or histidine. The structural similarity of these FeS clusters to the Fe_2S_2 structural unit in mackinawite is striking.

The ferredoxin clusters have a wide potential range (Table 12.2, Figure 10.1), which can be tuned by changing the ligand bound to Fe. Replacing cysteine (S–Fe bond) with histidine (N–Fe bond) leads to an increase in the reduction potential for the $[Fe_2S_2]$ cluster core (also Table 10.1) whereas a higher cluster oxidation state leads to a higher reduction potential for the $[Fe_4S_4]^{3+}$ cluster.

Fully reduced $[Fe_4S_4]^0$ and fully oxidized $[Fe_4S_4]^{4+}$ clusters have been produced in the laboratory but are unknown in nature [21], so only the +1, +2, and +3 clusters are found in proteins. Each Fe has unpaired electrons, but m_s equals 0 for $[Fe_4S_4]^{2+}$, which is diamagnetic and not active to electron paramagnetic resonance

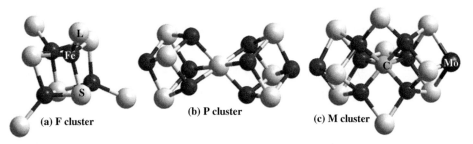

Figure 12.8 *Examples of the* Fe_4S_4 *structure in biochemistry. (a) Ferredoxins. (b) P-cluster in nitrogenase. (c) FeMo-cofactor cluster in nitrogenase. Models created with the Hyperchem™ program (version 8.0.10)*

Table 12.2 *Reduction potentials for some oxidized* $[Fe_nS_n]L_4$ *ferredoxin clusters [20]. L = cysteine unless stated otherwise. HiPIP = high potential ferredoxin*

Cluster	Potential (V)	Redox state of oxidant; Comments
$[Fe_2S_2]^{2+} + e^- \rightarrow [Fe_2S_2]^+$	−0.10 to +0.49	$2Fe^{3+}$ with 2 histidines on 1 Fe; *HiPIP*
$[Fe_2S_2]^{2+} + e^- \rightarrow [Fe_2S_2]^+$	−0.40 to 0	$2Fe^{3+}$
$[3Fe-4S]^+ + e^- \rightarrow [3Fe-4S]^0$	−0.05 to −0.45	$3Fe^{3+}$; pH dependent
$[Fe_4S_4]^{2+} + e^- \rightarrow [Fe_4S_4]^+$	−0.25 to −0.65	$2Fe^{3+}, 2Fe^{2+}$
$[Fe_4S_4]^{3+} + e^- \rightarrow [Fe_4S_4]^{2+}$	0.025 to 0.45	$3Fe^{3+}, 1Fe^{2+}$; *HiPIP*

spectroscopy. The value of m_s is 0.5 for $[Fe_4S_4]^{3+}$ and can be 0.5, 1.5, or 2.5 for $[Fe_4S_4]^+$. Ferredoxins are used to provide electrons in a variety of processes including photosynthesis and P450 cytochrome reactions (e.g., Section 10.6.1). The CO_2/CO couple has a reduction potential of −515 mV at pH = 7, and this is within the potential range of many of the highly reduced $[Fe_4S_4]^+$ ferredoxins, which are effective reducing agents.

12.6.4.1 Assimilatory N_2 Reduction

Figure 12.8b and c shows the two metal clusters in the iron–molybdenum (FeMo) protein called nitrogenase, which is involved with assimilatory reduction of N_2 to NH_3 for amino acid and peptide synthesis. The P-cluster is involved with electron transfer from the Fe protein to the FeMo-cofactor, which is the site where N_2 becomes bound and is reduced. The Fe–Fe bond distances in these rings is similar to that found in mackinawite and Fe metal, which would permit H_2 (or H^-) to bridge across Fe atoms [22]. The extended structures of the P and M clusters permit more bonding of the substrates (N_2 and H_2) with the advantage of having more bridging hydride ions (Section 9.13.4) across the increased number of Fe_2S_2 rings.

Because the thermodynamics of N_2H_2 is unknown, the operative reactions to reduce N_2 to NH_3 are Equations 12.11 and 12.12. The potential for these reactions at pH 7 is also given, and the reduction of N_2 to hydrazine, N_2H_4, is similar to the −515 mV potential of the CO_2/CO couple. Thus, the highly reduced $[Fe_4S_4]^{2+}/[Fe_4S_4]^+$ ferredoxin couple can effect N_2 reduction. There is no barrier to NH_3 formation once hydrazine (N_2H_4) forms.

$$\tfrac{1}{4}(N_2 + 4e^- + 4H^+ \rightarrow N_2H_4) \qquad -511\ \text{mV} \qquad (12.11)$$

$$\tfrac{1}{2}(N_2H_4 + 2e^- + 2H^+ \rightarrow 2NH_3) \qquad 523\,\text{mV} \qquad (12.12)$$

If the N species become protonated, the hydrazine reduction reactions (Equations 12.13 and 12.14) are now more unfavorable, and out of the range of the ferredoxins' reducing power. Thus, reduction of H^+ to H_2 must be completely efficient at the catalytic site.

$$\tfrac{1}{4}\,(N_2 + 4e^- + 5H^+ \rightarrow N_2H_5^+) \qquad -745\ \text{mV} \tag{12.13}$$

$$\tfrac{1}{2}\,(N_2H_5^+ + 2e^- + 2H^+ \rightarrow NH_4^+) \qquad 652\ \text{mV} \tag{12.14}$$

12.7 More on the Nitrogen Cycle (Nitrate Reduction, Denitrification, and Anammox)

The nitrogen cycle is one of the most important element cycles in nature and conversion from ammonia to nitrate involves eight electrons. The following inorganic compounds and ions represent the nine oxidation states of nitrogen.

-3	-2	-1	0	$+1$	$+2$	$+3$	$+4$	$+5$
NH_4^+	N_2H_4	NH_2OH	N_2	N_2O	NO	NO_2^-	NO_2	NO_3^-

Figure 12.9 summarizes the chemical species involved with the major processes of the nitrogen cycle. NH_4^+ oxidation reactions are on the left, and on the right are reduction reactions to form $NH_4^+(NH_3)$ or N_2.

Once N_2 is fixed as NH_3 in an organism, NH_3 is used to form amino acids, peptides, RNA, and DNA, which comprise the soft parts of the organism. NO_3^- can also be assimilated by organisms and reduced to NH_3 (see Table 12.3). The **assimilatory** reduction (assimilatory ammonification) of NO_3^- has been described as a two-step process. However, plants appear to have a three-step process; the third reaction in Table 12.3 is a four-electron transfer process and NH_2OH appears to be an intermediate in the process. The potentials for these processes are accessible with the **metalloenzymes** indicated. **Dissimilatory** nitrate reduction to NH_4^+ occurs during organic matter decomposition when O_2 is $<5\ \mu M$ (Section 1.8.2). The overall reactions are similar to those for assimilatory reduction of NO_3^- in Table 12.3, but the enzyme at step 2 is different.

NO_3^- and NH_4^+ are called fixed-N as they are soluble entities that are not lost to the atmosphere so can be recycled leading to eutrophication. However, nitrogen can be lost from an aquatic ecosystem as the gases

N₂ fixation

nitrification

$$N_2 \rightarrow \ H_2NNH_2 \ \rightarrow NH_3$$

$$NH_4^+ \ \rightarrow \ NH_2OH \ \rightarrow \ NO_2^- \ \rightarrow \ NO_3^-$$

Assimilatory / Dissimilatory
NO₃⁻ reduction (ammonification)

anammox

$$NO_3^- \rightarrow \ NO_2^- \ \rightarrow \ NH_2OH \ \rightarrow \ NH_4^+$$

$$NH_4^+ \ \rightarrow \ H_2NNH_2 \ \leftarrow \ NO \ \leftarrow NO_2^- \ \leftarrow NO_3^-$$

denitrification

$$\downarrow$$

$$NO_3^- \ \rightarrow \ NO_2^- \ \rightarrow \ NO \ \rightarrow \ N_2O \ \rightarrow \ N_2$$

$$N_2$$

Figure 12.9 *The chemical species formed during the major processes in the nitrogen cycle (gases are in blue).* NH_2OH *(blue italics) is a possible intermediate in plant assimilatory* NO_3^- *reduction*

Table 12.3   *Half reactions (equations from Table 2.2), voltages, and enzymes for eventual NH_3 formation during assimilatory nitrate reduction reactions*

Half reaction (Table 2.2 equation)	mV (pH 7)	Enzyme
$NO_3^- \rightarrow NO_2^-$ (N1)	+422	Mo nitrate reductase
$NO_2^- \rightarrow NH_4^+$ (N23)	+340	Siroheme nitrite reductase and Fe/S
Possible plant pathway after N1		
$NO_2^- \rightarrow NH_2OH$ (N19)	−60	Cytochrome-c-Fe heme nitrite reductase
$NH_2OH \rightarrow NH_4^+$ (N6c)	+1137	Octaheme hydroxylamine reductase

Table 12.4   *Reduction half reactions (equations from Table 2.2), their voltages, and enzymes for the four denitrification reactions*

Half reaction (Table 2.2 equation)	mV (pH 7)	Enzyme
$NO_3^- \rightarrow NO_2^-$ (N1)	+55	Mo nitrate reductase
$NO_2^- \rightarrow NO$ (N13)	+346	Fe heme/Cu nitrite reductase
$NO \rightarrow N_2O$ (N15)	+1169	Fe heme/non-heme Fe nitric oxide reductase
$N_2O \rightarrow N_2$ (N4)	+1352	Fe, Cu nitrous oxide reductase

N_2O and N_2. During organic matter decomposition with O_2 as electron acceptor, NH_4^+ is released to solution and oxidized to NO_3^- as shown in Section 10.6.2. Because many processes occur under low or zero O_2 conditions, NO_3^- becomes the electron acceptor for dissimilatory organic matter decomposition, leading to N_2 (denitrification) or NH_4^+ formation (ammonification) (see Section 1.8.2). Table 12.4 shows that the half reactions for each electron transfer step are quite favorable so ferredoxins are not required as reductants (see Figure 10.1).

Structural studies on NO_2^- reduction to NO using a Cu nitrite reductase indicate that Cu binds across one of the N–O bonds (bidentate or side on bonding) in NO_2^- [23] as shown in Figure 8.4c. The hapto nomenclature from Section 9.10 is used as in [(η^2-NO_2)Cu]. This type of bonding facilitates breaking of the other N–O bond in NO_2^- and formation of NO.

NO formation is also important in anaerobic ammonium oxidation (**anammox**) [24], which leads to N_2 as product. As in denitrification, NO forms from reduction of NO_2^-, which reacts with NH_4^+ via cytochrome-c-heme hydrazine synthase to form N_2H_4 (Equation 12.15; formally a **comproportionation**-type reaction). The hydrazine intermediate then forms N_2 readily via cytochrome-c-heme hydrazine oxidoreductase (Equation 12.16). The reaction between NO and NH_4^+ alone (Equations 12.17a–12.17c) is unfavorable (−427 mV; Equation 12.17c). Thus, the enzyme needs to supply hydrogen ions and electrons to affect NO reduction (much like O_2 to H_2O_2 in cytochrome P450 in Section 10.6.1), which then leads to NH_4^+ oxidation (the unfavorable step in the process). Formally, a three-electron reduction of NO with addition of two H^+ as in Equation 12.15 would give NH_2O^- (deprotonated hydroxylamine) as an intermediate, which would react with NH_4^+.

$$NO + NH_4^+ + 2H^+ + 3e^- \rightarrow N_2H_4 + H_2O \qquad \text{+109 mV; overall with enzyme} \qquad (12.15)$$

$$N_2H_4 \rightarrow N_2 + 4H^+ + 4e^- \qquad \text{+712 mV; reverse eq. N7b in Table 2.2} \qquad (12.16)$$

$$NH_4^+ \rightarrow \tfrac{1}{2}N_2H_4 + 2H^+ + e^- \qquad \text{−679 mV; reverse eq. N5b in Table 2.2} \qquad (12.17a)$$

$$\tfrac{1}{4}NO + H^+ + e^- \rightarrow \tfrac{1}{8}N_2H_4 + \tfrac{1}{4}H_2O \qquad\qquad +252 \text{ mV; eq. N20 in Table 2.2} \qquad (12.17b)$$

$$NH_4^+ + \tfrac{1}{4}NO \rightarrow \tfrac{5}{8}N_2H_4 + H^+ + \tfrac{1}{4}H_2O \qquad\qquad -427 \text{ mV; overall} \qquad (12.17c)$$

Both denitrification and anammox lead to loss of fixed nitrogen mainly as N_2 gas to the atmosphere from the water body where the organisms reside. NH_2OH and H_2NNH_2 also react with Mn(III,IV) to form N_2O and N_2, respectively. N_2O is a major greenhouse gas and is produced in many OMZs (Section 1.8.2), and its presence indicates that denitrification is occurring, but not anammox.

APPENDIX 12.1

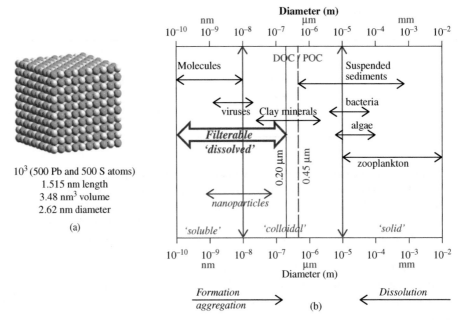

Figure 12A.1   *(a) Model of a PbS nanoparticle with 500 Pb and 500 S atoms. Model created with the Hyperchem™ program (version 8.0.10). (b) Operationally defined size ranges of natural materials found in waters, and what size ranges are considered "soluble", "colloidal" and "solid". On the left, molecules, viruses and nanoparticles as well as some colloids (with a size range between the blue vertical arrows) and clays can pass through 200 or 450 nm filters. The cutoff for dissolved organic carbon (DOC) and particulate organic carbon (POC) is commonly defined as 450 nm.*

References

1. Rickard, D. (1995) Kinetics of FeS precipitation: Part I. competing reaction mechanisms. *Geochimica et Cosmochimica Acta*, **59**, 4367–4379.
2. Luther, III, G. W. and Rickard, D. T. (2005) Metal sulfide cluster complexes and their biogeochemical importance in the environment. *Journal of Nanoparticle Research*, **7**, 389–407.
3. Tiemann, M., Marlow, F., Hartikainen, J., Weiss, O. and Linden, M. (2008) Ripening effects in ZnS nanoparticle growth. *Journal of Physical Chemistry C*, **112**, 1463–1467.

4. Rozan, T. F., Lassman, M. E., Ridge, D. P. and Luther, III, G. W. (2000) Evidence for Fe, Cu and Zn complexation as multinuclear sulfide clusters in oxic river waters. *Nature*, **406**, 879–882.

5. Luther, III, G. W., Theberge, S. M., Rozan, T. F., Rickard, D., Rowlands, C. C. and Oldroyd, A. (2002) Aqueous copper sulfide clusters as intermediates during copper sulfide formation. *Environmental Science and Technology*, **36**, 394–402.

6. Luther III, G. W. (2010) The role of one and two electron transfer reactions in forming thermodynamically unstable intermediates as barriers in multi-electron redox reactions. *Aquatic Geochemistry*, **16**, 395–420.

7. Wells, A. F. (1986). *Structural Inorganic Chemistry* (5th ed.), Clarendon Press, Oxford. pp. 1382.

8. Evans, H. T. and Konnert, J. A. (1976) Crystal structure refinement of covellite. *American Mineralogist*, **61**, 996–1000.

9. Huang, F., Zhang, H. Z. and Banfield, J. F. (2003) Two-stage crystal-growth kinetics observed during hydrothermal coarsening of nanocrystalline ZnS. *Nano Letters*, **3**, 373–378.

10. Mullaugh, K. M. and Luther III, G. W. (2011) Growth kinetics and long term stability of CdS nanoparticles in aqueous solution under ambient conditions. *Journal of Nanoparticle Research*, **13**, 393–404.

11. Williams, R. J. P. and Frausto da Silva, J. J. R. (2006) *The Chemistry of Evolution: The Development of Our Ecosystem*, Clarendon Press, Oxford, pp. 481.

12. Bianchini, A., Bowles, K. C., Brauner, C. J., Gorsuch, J. W., Kramer, J. R. and Wood, C. M. (2002) Evaluation of the effect of reactive sulfide on the acute toxicity of silver(I) to *Daphnia magna*, Part 2. Toxicity results. *Environmental and Toxicological Chemistry*, **21**, 1294–1300.

13. Rickard, D. and Luther, III G. W. (2007) Chemistry of iron sulfides. *Chemical Reviews*, **107**, 514–562.

14. Butler, I. B., Böttcher, M. E., Rickard, D. and Oldroyd, A. (2004) Sulfur isotope partitioning during experimental formation of pyrite via the polysulfide and hydrogen sulfide pathways: implications for the interpretation of sedimentary and hydrothermal pyrite isotope records. *Earth and Planetary Science Letters*, **228**, 495–509.

15. Heinen, W. and Lauwers, A. M. (1996) Organic sulfur compounds resulting from the interaction of iron sulfide, hydrogen sulfide and carbon dioxide in an anaerobic aqueous environment. *Origins of Life and Evolution of Biospheres*, **26**, 131–150.

16. Huber, C. and Wächtershäuser, G. (1997) Activated acetic acid by carbon fixation on (Fe,Ni)S under primordial conditions. *Science*, **276**, 245–247.

17. Wächtershäuser, G. (1988) Before enzymes and templates: theory of surface metabolism. *Microbiological reviews*, **52**, 452–484.

18. Huber, C. and Wächtershäuser, G. (1998) Peptides by activation of amino acids with CO on (Ni,Fe)S surfaces: implications for the origin of life. *Science*, **281**, 670–672.

19. Russell, M. J., Nitschke, W. and Branscomb, E. (2013) The inevitable journey to being. *Philosophical Transactions of the Royal Society B*, **368**, 20120254.

20. Liu, J., Chakraborty, S., Hosseinzadeh, P., Yu, Y., Tian, S., Petrik, I., Bhagi, A. and Lu, Y. (2014) Metalloproteins containing cytochrome, iron–sulfur, or copper redox centers. *Chemical Reviews*, **114**, 4366–4469.

21. Lee, S. C., Lo, W. and Holm, R. G. (2014) Developments in the biomimetic chemistry of cubane-type and higher nuclearity iron-sulfur clusters. *Chemical Reviews*, **114**, 3579–3600.

22. Hoffman, B. M., Lukoyanov, D., Yang, Z.-H., Dean, D. R.and Seefeldt, L. C. (2014). Mechanism of nitrogen fixation by nitrogenase: the next stage. *Chemical Reviews*, **114**, 4041–4062.

23. Solomon, E. I., Heppner, D. E., Johnston, E. M., Ginsbach, J. W., Cirera, J., Qayyum, M., Kieber-Emmons, M. T., Kjaergaard, C. H., Hadt, R.G. and Tian, L. (2014) Copper active sites in biology. *Chemical Reviews*, **114**, 3659–3853.

24. Maia, L. B. and Moura, J. J. G. (2014) How biology handles nitrite. *Chemical Reviews*, **114**, 5273–5357.

13

Kinetics and Thermodynamics of Metal Uptake by Organisms

13.1 Introduction

Metals such as manganese, iron, copper, and zinc are needed by organisms for enzyme function. At the base of the food chain are phytoplankton, which live in river, lake, or near-shore environments that are replete with these metals. However, in the center of the ocean, phytoplankton need to uptake these metals, which are at nanomolar levels or lower. Thus, oceanic phytoplankton have developed specialized uptake strategies to take the metal from the solution environment into the cell. However, too much metal uptake can be "toxic" to an organism. **Toxicity** can occur when one metal such as cadmium (1) enters an organism at high concentrations (**bioaccumulates**) due to anthropogenic effects, and thus replaces a necessary metal or (2) enters an organism even at low levels where it can outcompete for the ligands within the cell that normally bind the necessary metal.

In natural waters, metals are bound to organic ligands, which normally are in excess to the total metal concentration. There is a variety of natural metal binding ligands that are derived from a variety of sources – either as direct excretions from an organism or as degradation products on cell lysis or death. These include natural humic material, which is not well-defined structurally. Diatoms and plants produce porphyrins and phytochelatins (RSH; thiol groups). Bacteria produce **siderophores** (a term for iron bearing ligands) to acquire iron, but can also bind a host of other metals. The functional groups include hydroxamate, catecholate, and β-hydroxyaspartate bidentate functional groups with up to three bidentate groups in a molecule. Freshwater and marine mussels produce peptides and proteins containing catecholate groups. Figure 13.1 (also Figure 8.1) shows some specific structural features for these natural ligands, which can be obtained from isolation of excreted material in laboratory cultures and sometimes by laboratory syntheses. Based on the structure of these natural ligands, chemists have also used known ligands or synthesized model ligands that contain similar functional groups including carboxylic acid, amine, hydroxyl, catecholate, hydroxamate, phosphate, and sulfhydryl (thiol) functional groups in order to mimic natural conditions in the laboratory.

Inorganic Chemistry for Geochemistry and Environmental Sciences: Fundamentals and Applications, First Edition. George W. Luther, III.
© 2016 John Wiley & Sons, Ltd. Published 2016 by John Wiley & Sons, Ltd.
Companion Website: www.wiley.com/go/luther/inorganic

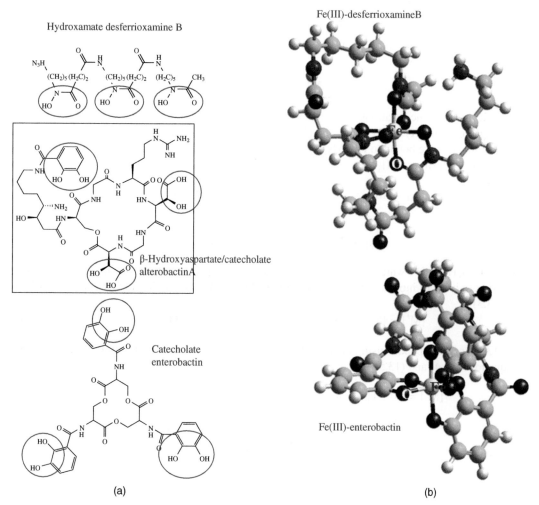

Figure 13.1 (a) Two structures for select siderophores with six ligating oxygen atoms showing the hydroxamate, catecholate and β-hydroxyaspartate bidentate functional groups. (b) Three dimensional models when bound to Fe(III). Models created with the Hyperchem™ program (version 8.0.10)

13.1.1 Conditional Metal–Ligand Stability Constants

As these natural or synthetic metal–ligand complexes approach the surface of the cell, a protein transporter site (another ligand) must then acquire the metal from the metal–ligand complex in solution via an associative or dissociative pathway. The uptake of metals is related to the strength of these metal–ligand binding interactions, which can be described by the kinetic and thermodynamic expressions for metal–ligand complexes in Chapter 9. There are three principal ways to report the value of a stability constant (recall Equation 7.23 and Equations 9.5–9.6), which are discussed before describing metal uptake.

Because the metal concentration (c_M or $[M]_T$) is the parameter measured in natural water by metal-specific techniques (e.g., inductively coupled plasma mass spectroscopy, atomic absorption spectroscopy, voltammetry, ion selective electrodes), metal complexation is expressed as the **conditional equilibrium constant**, $K_{\text{cond M'L'}}$ (Equation (13.1)), which is also equal to the ratio of the rate constant of complex formation, k_f, to the rate constant of complex dissociation, k_d.

$$K_{\text{cond M'L'}} = \frac{[ML]}{[M'][L']} = \frac{k_f}{k_d} \qquad (13.1)$$

Here M' and L' are the concentrations of the metal and ligand that are not bound to each other. M' is defined as all the inorganic forms (X) of M (where X = chloride, sulfate, carbonate, hydroxide, etc.) and L' is defined as all ligand forms that are bound to H^+, Mg^{2+}, Ca^{2+}, etc. These are related to the total metal $[M]_T$ and $[L]_T$ where Equation (13.2) shows $[M]_T$.

$$c_M = [M]_T = [M^{n+}] + [ML] + [MCl^+] + [MSO_4] + \cdots$$

or

$$c_M = [M]_T = [M^{n+}] + [ML] + \Sigma[MX]_i \qquad (13.2)$$

The free metal $M[H_2O]_6{}^{n+}$, which is abbreviated as $[M^{n+}]$ or $[M]_{\text{free}}$ to indicate its activity $\{M\}$, plus the metal bound to only other inorganic ligands (chloride, sulfate, carbonate, hydroxide, etc.) equals $[M']$ as in Equation (13.3).

$$[M'] = [M]_T - [ML] = [M^{n+}] + [MCl^+] + [MSO_4] + \cdots \qquad (13.3)$$

Equation (13.3) can be expanded using the equilibrium expressions for each metal–inorganic ligand complex to give Equations (13.4) and (13.5).

$$[M'] = [M^{n+}]\{1 + K_{MCl}[Cl^-] + K_{MSO4}[SO_4{}^{2-}] + \cdots\} \qquad (13.4)$$

$$[M'] = [M^{n+}] + [M^{n+}] \sum K_{MXi}[X]_i = [M^{n+}](1 + \sum K_{MXi}[X]_i) \qquad (13.5)$$

The **fraction of a free metal**, α_M, in solution *without the organic ligand* is given by Equation (13.6).

$$\alpha_M = \frac{[M^{n+}]}{[M']} \qquad (13.6)$$

On substitution of the equilibria for all inorganic forms of M (Equation (13.5)) into Equation (13.6), α_M is expanded into Equation (13.7).

$$\alpha_M = \frac{1}{1 + \sum K_{MXi}[X]_i} \qquad (13.7)$$

The reciprocal of Equation (13.7) is known as the **inorganic side reaction coefficient** for M' (given the symbol $\alpha_{M'}$) as in Equation (13.8), and is frequently used in the literature.

$$\alpha_{M'} = \frac{[M']}{[M^{n+}]} \qquad (13.8)$$

The same procedure is also used to express the **fraction of free ligand, α_L, and the side reaction coefficient for L' (given the symbol $\alpha_{L'}$)**. Unfortunately, in natural waters, the ligand is normally unknown because there are several ligands that can bind the metal; thus, the conditional constant from Equation (13.1) can only be corrected for the free metal ion. A measure of the total ligand concentration can be determined by metal titration and other experiments [1–5]. Substituting from Equation (13.8) for M' into Equation (13.1) leads to Equation (13.9).

$$K_{\text{cond M'L'}} = \frac{[ML]}{[M'][L']} = \frac{[ML]}{[M^{n+}]\alpha_{M'}[L']} \tag{13.9}$$

Expanding Equation (13.9) for the correction of all inorganic ligands gives a new conditional stability constant known as $K_{\text{cond ML'}}$, which is related to $K_{\text{cond M'L'}}$ by Equation (13.10). These two stability constants are those normally reported in the environmental literature so it is important that the reader be able to discriminate between them easily. Also, $K_{\text{cond ML'}} > K_{\text{cond M'L'}}$.

$$(\alpha_{M'})K_{\text{cond M'L'}} = \frac{[ML]}{[M^{n+}][L']} = K_{\text{cond ML'}} \tag{13.10}$$

13.1.2 Thermodynamic Metal–Ligand Stability Constants

When the ligand is known as in laboratory experiments, similar corrections can be made for the free ligand concentration as its α_L (and $\alpha_{L'}$) for complexes with H^+, Ca^{2+}, Mg^{2+}, etc. can be calculated. Thus, the activities, expressed as { }, of both the metal and ligand are known, and the thermodynamic constant can be calculated as in Equation (13.11); $K_{\text{therm}} > K_{\text{cond ML'}} > K_{\text{cond M'L'}}$.

$$K_{\text{therm}} = \beta = \frac{\{ML\}}{\{M^{n+}\}\{L^{n-}\}} = K_{\text{cond M'L'}}(\alpha_{M'})(\alpha_{L'}) = \frac{k_f}{k_d}(\alpha_{M'})(\alpha_{L'}) \tag{13.11}$$

The resulting K_{therm} is then a pH- and solution species-independent constant as all solution conditions are properly specified. However, in environmental samples, the interactions of H^+, Ca^{2+}, and Mg^{2+} with the ligand are almost always unknown, so only conditional stability constants ($K_{\text{cond ML'}}$ or $K_{\text{cond M'L'}}$) are reported. Table 13.1 shows the conditional and thermodynamic constants for M^{2+}–EDTA or $[M(EDTA)]^{2-}$ complexes determined in seawater by stripping voltammetry techniques; these constants are several orders of magnitude higher than those for the Group II cations (see Table 7.6). Table 13.2 shows the conditional and thermodynamic constants for several natural ligands bound to Fe(III) in seawater. Note that the side reaction coefficient for Fe' in seawater is over eight orders of magnitude larger than that for the divalent cations because of Fe(III) hydrolysis.

Equation (13.11) can also been expressed using the free metal fraction, α_M, to give Equation (13.12).

$$K_{\text{therm}} = \beta = \frac{[ML]\alpha_{ML}}{[M']\alpha_M[L']\alpha_L} \tag{13.12}$$

Table 13.1 *Relationship between the conditional stability constants ($K_{\text{cond M'L'}}$ and $K_{\text{cond ML'}}$) determined in seawater (pH = 8.0) and the ACTUAL thermodynamic stability constant (log K_{therm}) for several metals bound with EDTA and. The log of $\alpha_{EDTA'}$ equals 8.0 [1–4]*

Complex	log $K_{\text{cond M'L'}}$	log $\alpha_{M'}$	log $K_{\text{cond ML'}}$	log K_{therm}
Zn-EDTA	7.9	0.32	8.22	16.3
Cu-EDTA	8.6	1.38	9.98	17.94
Cd-EDTA	7.7	1.55	9.25	16.5
Pb-EDTA	8.6	1.54	10.14	18.0

Table 13.2 *Relationship between the conditional stability constants and the thermodynamic stability constant for Fe(III)-complexes bound to several natural ligands in seawater modified from [5]. The log of $\alpha_{Fe'}$ in seawater equals 10.0*

Complex	$\log k_f$	$\log k_d$	$\log K_{\text{cond M'L'}}$	$\log K_{\text{cond ML'}}$	$\log K_{\text{therm}}$
Fe-ferrichrome (hydroxamate)	5.56	−7.30	12.9	22.9	29.1
Fe-desferrioxamine-B (hydroxamate)	6.29	−5.82	12.1	22.1	31.0
Fe-alterobactin-A	5.58	−6.76	12.3	22.3	51
Fe-protoporphyrin-IX	5.79	−6.15	11.9	21.9	–

For a complex with more than one metal bound to more than one ligand combining to form a cluster (M_xL_y), Equation (13.11) is expressed as Equation (13.13).

$$K_{\text{therm}} = \beta = \frac{\{M_xL_y\}}{\{M^{n+}\}^x \{L^{n-}\}^y} = K_{\text{cond M'L'}} (\alpha_{M'})^x (\alpha_{L'})^y \tag{13.13}$$

13.2 Metal Uptake Pathways

There are multiple ways in which single cell organisms can uptake or assimilate ions, nutrients, and metals. Figure 13.2 shows a schematic of a cell with several transport proteins (P) for the transport of elements across a membrane, which has hydrophobic components extending out into solution. There are different membrane transport proteins for cations and anions, and the focus is primarily on metal ions. Figure 13.2 demonstrates several different cation possibilities. First, the Group I ions (Na, K) exist as hexaaqua complexes, which dissociate at the cell surface to form M^+P complexes (these are discussed in detail in the next section). Second, divalent cations including Mn^{2+}, Fe^{2+}, Co^{2+}, Ni^{2+}, Cu^{2+}, Zn^{2+}, Cd^{2+}, and Pb^{2+} exist in nature mainly as metal ligand complexes ML or $[M(L)]^{2+}$, which can dissociate at the surface to give metal chloro complexes (as in seawater) or the free ion, which in turn forms $M^{2+}P$ complexes. Third, higher oxidation state cations such as Fe^{3+} and Mn^{3+} also exist as metal–ligand complexes, but because of significant hydrolysis, any dissociation would result in hydroxide or oxyhydroxide species rather than the free ion prior to forming $M'P$. The term M' is used to distinguish the free metal ion from metal complexes with inorganic ligands such as the halides, carbonate, and hydroxide. Thus, the transport protein must have the capability of outcompeting the inorganic ions for Fe^{3+} so that iron can be transported into the cell. For all cases, the dissociation or partial dissociation of a metal–ligand complex in solution is important to form the surface MP complex, and the kinetics of this process is discussed in Section 13.2.2. Once MP forms, the metal is transferred into the cell (M_{cell}) with a rate constant of k_{in} and goes to a center, C_1, where it can perform its biochemical function.

For the monovalent and divalent cations, Figure 13.2 shows that the free ion normally is formed prior to formation of an MP complex. For higher oxidation state cations, the metal–ligand complex can be reduced in solution or at the cell surface by chemical reagents or in solution by photochemical pathways. For Fe^{3+}, the result would be faster uptake of the resulting Fe^{2+} as the thermodynamic stability constants for Fe(II)L complexes are twenty log units lower than those reported in Table 13.2 for Fe(III)L complexes [6]. In early Earth history prior to the Great Oxidation Event, iron existed predominantly as Fe^{2+} so the uptake of iron would not have required a more specialized membrane transport protein to bind and acquire Fe^{3+}.

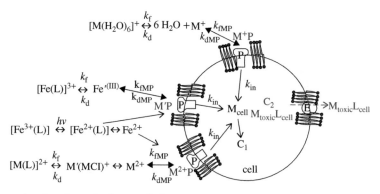

Figure 13.2   *Schematic of a phytoplankton cell showing membrane transport proteins (P) for metal ions along with the reactions and rate constants for $[M(H_2O)_6]^{n+}$, ML, MP and M_{cell}. C_2, E (efflux site) and $M_{toxic}L_{cell}$ are discussed in Section 13.2.2.3*

13.2.1   Ion Channels for Potassium

There are specific **cation membrane transport proteins** for Na^+ and K^+ (also NH_4^+ and Ca^{2+}), which are major ions in natural waters and are required for many biological functions including electrical conduction and cell signaling. These cations pass through membrane transporter or ion channels based on a chemical gradient across the membrane. Because they have fast water exchange rates and form weak complexes in solution, their ligands dissociate readily so that the rate of cation conduction is fast, on the order of diffusion control. To enter the pore and pass through ion channels, the $M(H_2O)_6^+$ complexes lose all six water molecules and any ion pair partners such as chloride so the metal ions can complex the O atoms from carbonyl groups in peptides within the transporter channel (Figure 13.3). The stability constants, K_{therm}, of these ligand complexes with K^+ in the channel are about 10^2, which is larger than that for simple ion pairing in solution. These metal–ligand complexes in the channel are also labile to dissociation so movement is fast. To ensure this cation movement, the channel size appears to be thermodynamically tuned to the ionic radius of the specific ion much like ions pack in a crystal more efficiently; the only difference is that there are no atoms or ions between K^+ in the channel resulting in Coulombic repulsion between adjacent K^+ ions that aid their movement through the channel [7]. The ionic radii for Na^+ and K^+ are 116 and 152 pm for six coordinate complexes, respectively. Thus, selectivity may be achieved with the proper pore or channel size (called a selectivity filter). Channels or compounds that are metal selective are called **ionophores**.

As in the case of Na^+, the smaller size and higher charge of the transition and post transition metals normally prevent them from entering the K^+ membrane transport protein. In addition, the transition and post transition metals are lower in concentration in natural waters and have significantly stronger complexes (are also inert or slowly dissociating) with anions and ligands in solution; thus, the free ion is in low concentration. Thus, metals such as Hg and Ag cannot pass through these channels. Unfortunately, thallium(I) ($r_+ = 164$ pm) can pass through these channels along with K^+, and once inside the cell can oxidize to Tl(III), which binds well with internal cell ligands leading to toxic side effects for the cell. Pb^{2+} has an ionic radius of 133 pm so may also be able to pass through the K^+ channel.

The synthesis of crown ethers, which are cyclic oligomers of $-CH_2CH_2O-$, showed that cyclic complexes with a specific ring size can lead to better Group I-ligand binding, which encourages the correct ion to enter the cavity (in this case K^+). Figure 13.4 shows the structure of the **crown ether**, 18-crown-6, which has

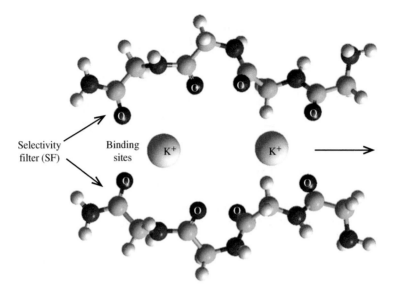

Figure 13.3 *Depiction of K$^+$ ions passing through a portion of a membrane transport protein at the cell surface. The tetrapeptides are bound to membranes above and below (not shown). Models created with the Hyperchem™ program (version 8.0.10)*

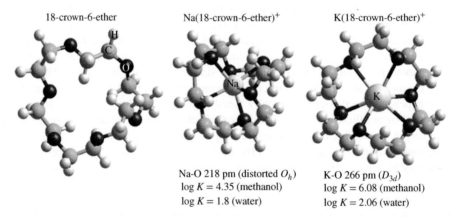

Figure 13.4 *The structure of 18-crown-6-ether without metal binding, bound to Na$^+$ and bound to K$^+$. Models created with the Hyperchem™ program (version 8.0.10)*

6-CH$_2$CH$_2$O- groups. The number 18 refers to the total number of structural atoms (C and O) in the molecule, and the number 6 refers to the total number of oxygen atoms, which can bind to Na$^+$ and K$^+$. The Na$^+$ complex has a distorted octahedral geometry, and the crown ether is no longer open but partially buckled in order to bind with the Na$^+$. The K$^+$ complex has an overall D_{3d} symmetry and has an open structure with a cavity or hole for the K$^+$ to sit in. The thermodynamic stability constants for these complexes vary with the solvent. As shown in Figure 13.4, the K$^+$ complex is more stable than the Na$^+$ complex by almost 2 log units in methanol, but is only more stable by 0.2 log units in water [8]. Also, the stability constants for each complex are 3–4 log units more stable in methanol than in water. As shown in Chapter 9 and Equations (13.14) and (13.15), the stability constant is a function of the formation and dissociation rate constants. For ions to be mobile through the membrane transport protein and to carry charge, these rate constants must be high because dissociation and formation must occur frequently for the cations to get through the channel for ion conduction. For the K^+-18-crown-6 ether complex in water, k_f is 4.8×10^8 $M^{-1}s^{-1}$ and k_d is 3.7×10^6 s^{-1} so transport of the K^+ could occur quickly as these complexes are very labile.

Cl^-, SO_4^{2-}, and HPO_4^{2-} are major anions that cross the membrane via **anion membrane transport proteins**. Cl^- is required for balancing the cation charge. HPO_4^{2-} is a nutrient required for adenosine triphosphate (ATP), adenosine diphosphate (ADP), DNA, and RNA as well as the hard parts or the structural material for bone and teeth. Isoelectronic anions such as $HAsO_4^{2-}$ can also pass through the phosphate membrane transport protein, resulting in toxic effects to organisms.

Because H_2S is toxic to many organisms, SO_4^{2-} is assimilated via ion channels for reduction to sulfide, and its fast incorporation into cysteine and glutathione as well as ferredoxins. Thiosulfate, SSO_3^{2-}, can also enter and pass through the SO_4^{2-} channel. However, if Ag^+ is bound to thiosulfate, silver enters the organism, leading to toxic side effects [9]. Other isoelectronic metal anions such as MoO_4^{2-} and CrO_4^{2-} can also pass through an SO_4^{2-} channel with possible toxic effects for Cr. Interestingly, MoO_4^{2-} and WO_4^{2-} are actually used as inhibitors of sulfate reduction because they cannot be easily reduced.

Neutral and nonpolar species (H_2S, NH_3) can diffuse directly across the bilayer membranes due to their lipid solubility. Unfortunately, these and metal species ($HgCl_2$ and Hg^0) can pass across the hydrophobic component of the membrane causing toxic effects [10].

13.2.2 Metal Uptake by Cells via Ligands on Membranes

The transition and post transition metals exist in the environment at trace or ultra-trace levels (see Figure 1.9), and also have significantly stronger complexes with anions and ligands in solution than Group I metals. Thus, the free ion is in low concentration. The membrane has more specialized membrane transporter protein sites with ligands that can outcompete the metal–ligand complex in solution. The thermodynamics and kinetics of transition metal complexes can be used to make quantitative predictions of metal uptake at the cell surface.

Figure 13.2 shows a schematic of metal transport across a cell membrane when free metal cations, inorganic metal ion complexes, and metal–ligand complexes are all present. This model relies on the thermodynamics and kinetics of free metal or M' in solution as well as MP formation on the cell surface and transfer into the cell, M_{cell}, to make quantitative predictions of metal uptake. The student should correlate the discussion below with the metal–ligand kinetics Section 9.2.5. Although associative reactions can occur, they can be broken down into two or more elementary steps that involve dissociation of the metal–ligand complex.

Outside the cell, ML dissociation is a key step as many complexes do not dissociate rapidly based on Equations (13.14) and (13.15). Thus organic complexation in solution can exert a major influence on metal uptake rates. $K_{cond\,M'L'}$ in Equation (13.15) is the same conditional constant $K_{M'L'}$ from Equation (13.1) without any correction for the side reaction coefficients for the metal and the ligand as discussed above. Thus, $K_{cond\,M'L'}$ is dependent upon the conditions of the experiment such as ionic strength, pH, etc. It is also related to the formation and dissociation rate constants (see Table 13.2 for data on selected Fe(III)–ligand complexes).

$$M' + L' \underset{k_d}{\overset{k_f}{\rightleftharpoons}} ML \tag{13.14}$$

$$K_{cond\,M'L'} = \frac{[ML]}{[M'][L']} = \frac{k_f}{k_d} \tag{13.15}$$

At the cell surface, the dissociated metal reacts with the metal transporter protein, P, which is represented by Equation (13.16). Note the notation to discriminate ligands (L) in solution and on the cell surface (P).

$$M + P \underset{k_{dMP}}{\overset{k_{fMP}}{\rightleftharpoons}} MP \xrightarrow{k_{in}} M_{cell} \tag{13.16}$$

For metal uptake to be effective, k_f of MP must always be greater than k_d of MP as shown in the K^+ channel example above. The rate of metal uptake (V) by the cell can be expressed as Equations (13.17) and (13.18) (similar to 9.43).

$$MP \xrightarrow{k_{in}} M_{cell} \quad \text{rate limiting step} \tag{13.17}$$

$$\frac{d[M_{cell}]}{dt} = V = k_{in}[MP] \tag{13.18}$$

Applying the **steady-state approximation** for an associative reaction to form the cell surface complex MP gives the rate expression Equation (13.19), which is similar to Equation 9.44.

$$\text{rate} = \frac{d[MP]}{dt} = 0 = -k_{in}[MP] + k_{fMP}[M][P] - k_{dMP}[MP] \tag{13.19}$$

Solving for [MP] gives Equation (13.20).

$$[MP] = \frac{k_{fMP}}{k_{in} + k_{dMP}}[M][P] \tag{13.20}$$

Rearranging leads to the ***conditional constant*** for MP formation, K_{MP}, in Equation (13.21) also known as K_{aff} or the affinity constant for the metal with the membrane binding site (P). When $k_{in} \ll k_{dMP}$, then $K_{MP} = k_{fMP}/k_{dMP}$. In many cases, $k_{in} \gg k_{dMP}$.

$$K_{MP} = \frac{[MP]}{[M][P]} = \frac{k_{fMP}}{k_{in} + k_{dMP}} = K_{aff} \tag{13.21}$$

The reciprocal or inverse of Equation (13.21) is the MP dissociation constant, which is also known as the **Michaelis–Menten half saturation constant**, K_M. Equation (13.20) is now written as Equation (13.22).

$$[MP] = K_{aff}[M][P] \quad \text{or} \quad [MP] = \frac{M[P]}{K_M} \tag{13.22}$$

At any time, $P = P_T - MP$ and on substituting this expression for P in Equation (13.22) and rearranging gives Equations (13.23) and (13.24).

$$[MP](K_M + M) = M[P_T] \tag{13.23}$$

or

$$[MP] = \frac{M[P_T]}{K_M + M} \tag{13.24}$$

The rate of formation of the product, M_{cell}, can be rewritten by substituting Equation (13.24) for [MP] into Equation (13.18). The result is the complete ***associative rate*** expression for a metal reacting with P in Equation (13.25). V is the uptake rate and is dependent upon the concentration of both the metal and the membrane transporter protein.

$$\frac{d[M_{cell}]}{dt} = V = k_{in}[MP] = \frac{k_{in}M[P_T]}{K_M + M} \tag{13.25}$$

The **maximum uptake rate** (V_{MAX}) is related to the total concentration of P, $[P_T]$, and k_{in} as in Equation (13.26). Substituting Equation (13.26) into Equation (13.25) gives Equation (13.27), which is the **Michaelis–Menten** equation. At (V_{MAX}), the rate constant, k_{in}, is also called **the turnover number**, which is the maximum number of metal ions in solution converted to produce metal ions in the cell per membrane transporter protein site per second.

$$V_{MAX} = k_{in} \ [P]_T \tag{13.26}$$

$$\frac{d[M_{cell}]}{dt} = V = k_{in}[MP] = \frac{V_{MAX}(M)}{K_M + M} \tag{13.27}$$

On rearrangement and plotting [M] versus ([M]/V) for data obtained on increasing [M], V_{MAX} (the slope) and K_M (the intercept) can be obtained (Equation (13.28)).

$$[M] = \frac{V_{MAX}(M)}{V} - K_M \tag{13.28}$$

Dividing Equation (13.27) through by K_M gives an alternate form of it as shown in Equation (13.29).

$$V = \frac{V_{MAX}[M]K_{aff}}{1 + [M]K_{aff}} \tag{13.29}$$

[M] is evaluated by calculation from the stability constant with known [M] and [L] and/or by analytical determination of samples (environmental or laboratory incubations).

13.2.2.1   Metal Uptake Limiting Cases

1. There are two limiting cases for the uptake of the metal as discussed in the stoichiometric mechanisms for an associative mechanism (S_N2 limiting) in **Section 9.2.5.3**. Equation 9.48 described the k_{obs} for an activation-controlled reaction. For the **first case of $k_{in} \ll k_{dMP}$** ($k_{dMP} \gg k_{in}$), Equation (13.16) leads to k_{obs} (Equation (13.30)) and to V (Equation (13.31)), which is similar to Equation 9.46.

$$k_{obs} = k_{in} K_{MP} \tag{13.30}$$

$$V = k_{obs}[M][P] = k_{in} K_{MP} [M][P] \tag{13.31}$$

At equilibrium of M with L and P, $[M^{x+}]_{free}$ depends on binding with L external to the cell. The uptake rate, V, is under **thermodynamic or equilibrium ML control** (depends on the k_d of ML) and is slow compared to all complexation and dissociation reactions. This case is found for metals with high ligand exchange rates (water exchange rate constants are a guide) and no significant hydrolysis. Examples include the discussion on the K^+ channel above as well as many essential divalent cations such as Mn^{2+}, Co^{2+}, Cu^{2+}, and Zn^{2+} and the more toxic metals, Cd^{2+} and Pb^{2+}.

2. **For $k_{dMP} \ll k_{in}$ (see Equations 9.46 and 9.47 that transform to Equations (13.32) and (13.33)),** V depends on the k_f of [M]′ (free plus labile inorganic complexes) with P or MP in Equation (13.16) so is under **kinetic control (for formation of MP)**. This case is important for biolimiting trace elements (Figure 1.8) and especially for those metals with higher oxidation states. Metals or ML complexes with very small k_d values such as Fe(III)L complexes are in this category.

$$k_{obs} = k_{fMP} \tag{13.32}$$

$$V = k_{obs}[M][P] = k_{fMP}[M][P] \tag{13.33}$$

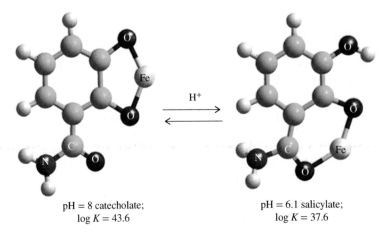

pH = 8 catecholate;
log *K* = 43.6

pH = 6.1 salicylate;
log *K* = 37.6

Figure 13.5 *Catecholate bonding to Fe(III) on the left slips to salicylate on decreasing pH. Models created with the Hyperchem™ program (version 8.0.10)*

For very low values of k_d, dissociation of the ML complex is rate limiting, and Equation (13.15) can be arranged to give a limiting uptake rate as in Equation (13.34).

$$V = k_d[ML] = k_f[M][L] \qquad (13.34)$$

As noted in Figure 13.2, another way to enhance the uptake of a high oxidation state metal such as Fe is to reduce the Fe(III)–ligand complex to Fe(II), which has a significantly lower thermodynamic stability constant and a faster water exchange rate. In the light, π donor complexes can undergo internal photoredox as in Equation (13.35). Lowering the pH can also decrease the FeL stability constant by several orders of magnitude as in the case of catechol binding to Fe(III) in alterobactin-B (Figure 13.5). Here, Fe(III) binding slips to the salicylate binding mode [11].

$$Fe^{3+}(L) + h\nu \rightarrow Fe^{2+}(L^{\bullet+}) \qquad (13.35)$$

13.2.2.2   Example Phytoplankton Uptake Data

Because the oceanic concentrations of trace metals are nanomolar or lower, trace metals are limiting nutrients (sometimes referred to as micronutrients) for the growth of phytoplankton. To measure the cellular uptake rate of a metal at these levels, researchers commonly use radiolabeled isotopes of the metal along with the addition of the metal at natural environmental levels. The isotope is added at ultra-trace levels to a culture containing an organism of interest so as not to affect organism function. Over a period of time, aliquots of the culture are filtered, and then washed with seawater or an appropriate reagent to remove traces of the isotope from the surface of the cell. The filter is then counted to determine the amount of the isotope and the metal in the cell, from which *V* can be calculated. Figure 13.6 shows the use of the equations to describe iron uptake for two different phytoplankton species. The cellular uptake rate, *V*, increases with increasing Fe′ concentration and asymptotically approaches its maximum value, V_{MAX}. At this point, all the membrane proteins sites are bound to Fe′. The Michaelis–Menten constant, K_M, is the Fe′ concentration at which the uptake rate, *V*, is at half maximum (V_{MAX}). The smaller the K_M value (the inverse indicates higher affinity of P for Fe′), the faster the rate approaches V_{MAX}. In both examples of Figure 13.6, V_{MAX} is reached quickly and does not decrease

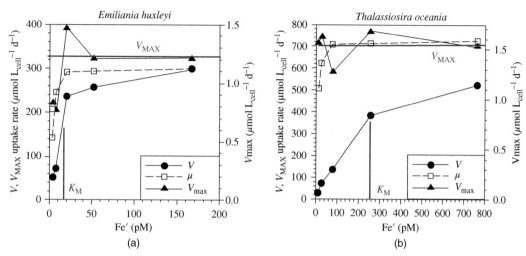

Figure 13.6 *Fe uptake data for two different oceanic phytoplankton; data from [12]. (a)* Emiliania huxleyi. *(b)* Thalassiosira oceania

with increasing Fe′ concentration. However, the K_M values are different for each organism, indicating that the organisms would likely compete for the available iron (and other nutrients) in seawater when both are present.

The specific growth rate (μ) is the uptake rate normalized to the concentration of M in the cell as in Equation (13.36).

$$\mu = \frac{V_{Fe}}{[M]_{cell}} \tag{13.36}$$

13.2.2.3 Competitive Interactions of Metals at the Membrane Transporter Site, P

The first row transition metals from Mn^{2+} to Zn^{2+} have ionic radii descending in size from 97 to 83 pm, and their stability constants increase on going across the row according to the **Irving–Williams** stability relationship (Section 7.8.1). Thus, these kinetic equations can also be expanded to describe the competition of two different metal ions for the same membrane transporter protein site for M^{2+} ions [13]. Competition can be antagonistic, leading to a toxic metal such as Cd^{2+} ($r_+ = 109$ pm) or Cu^{2+} (at high levels; $r_+ = 87$ pm) replacing a metal needed for biochemical function such as Mn^{2+} ($r_+ = 97$ pm) or can be favorable.

In the antagonistic case, organisms have developed positive and negative feedback systems (Figure 13.2), which permit the release or efflux of a toxic metal from the organism. When site C_1 is bound to Mn but more importantly a toxic metal such as Cd, a negative feedback signal is sent to decrease the number of active membrane P sites, thus decreasing V_{MAX}. A second site, C_2, when bound to Cd can initiate its complexation with an internal ligand like a phytochelatin (RSH compound), which binds Cd^{2+} more strongly than the required metal, Mn^{2+}, resulting in the toxic metal–ligand complex ($M_{toxic}L_{cell}$) being excreted from the cell at efflux site, E.

In the synergistic favorable case, a metal considered toxic, Cd^{2+}, may in fact be able to replace the required metal, Zn^{2+} ($r_+ = 88$ pm), which may be lower in concentration than the toxic metal. Here Cd and Zn are in the same family of the periodic table and have similar chemistry. Certain organisms have shown that they can use Cd to replace Zn when Zn is lower in concentration than Cd in seawater [14, 15].

13.2.3 Evaluation of k_f, k_d, and $K_{cond\,M'L'}$ from Laboratory and Natural Samples

Two experiments can determine these parameters. First, the formation of an M'L' complex (Equations (13.37a), (13.37b)) can be accomplished by adding *excess* M' to a solution with L' in the medium of interest and determining M' over time. Figure 13.7a shows a plot of the loss of [M'] over time as M'L' forms. In this case, M' is detectable but the complex M'L' is not. The integrated rate expression, Equation (13.38), can be used to calculate the second order rate constant k_f from a plot of t versus the right side of Equation (13.38), which gives a straight line. Figure 13.7a is an example of the reaction rate for an oxidized metal such as Fe' with a siderophore. To give perspective, K^+ has a second-order rate constant, k_f, of 4.8×10^8 $M^{-1}s^{-1}$ when reacting with 18-crown-6-ether. At nanomolar K^+ levels, the reaction would be complete within 1.5 s. Because K^+ is at tens of millimolar in cells, the reaction would be complete in less than a microsecond.

$$M' + L' \xrightarrow{k_f} M'L' \tag{13.37a}$$

$$\frac{d}{dt}[M'L'] = k_f[M'][L'] \tag{13.37b}$$

$$k_f t = \frac{1}{[M'_0] - [L'_0]} \times \ln\left\{\frac{[L'_0]([M'_0] - [M'L'])}{([L'_0] - [M'L'])[M'_0]}\right\} \tag{13.38}$$

Similarly, a competitive ligand, Y, can be used to assess k_d as shown in the following discussion. The method is a ligand exchange reaction (Equation (13.39)) for the known (or unknown) M'L' complex with *excess* Y ligand. The reaction is performed under pseudo-first-order conditions, and the greater the excess of Y, the faster the reaction. A plot of M'Y, which is measurable, versus time is shown in Figure 13.7b. For the

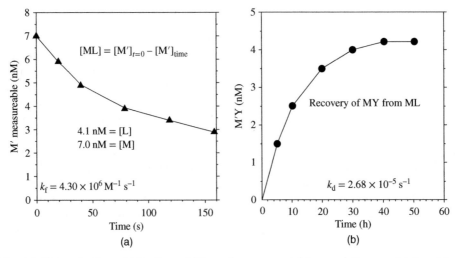

Figure 13.7 (a) Determination of M' after addition of excess metal to a solution containing 4.1 nM ligand. (b) The recovery of M' from M'L' with another ligand, Y (using Equation 13.48)

membrane transporter protein on a cell surface, Y becomes P.

$$M'L' + Y \xrightarrow{k_{obs}} M'Y + L' \tag{13.39}$$

The rate expression for Equation (13.39) is given in Equation (13.40).

$$-\frac{d}{dt}[M'L'] = \frac{d}{dt}[M'Y] = k_{obs}[M'L'][Y] \tag{13.40}$$

Integrating this expression results in Equation (13.41).

$$\ln[M'L'] = -k_{obs}[Y]t \tag{13.41}$$

The dissociation rate constant of an $M'L'$ complex can be determined using the steady-state approximation from the associative reaction (Equation (13.39)), which can be broken into *two elementary reaction steps*: First, the *dissociation* of the $M'L'$ complex to form M' (Equation (13.42), is the inverse of K_{cond} for $Mn'L'$ in Equation (13.15)), where M' represents all inorganic forms of M in the solution of interest at ambient pH; and second, the *reaction* of M' with Y (Equation (13.43)):

$$M'L' \underset{k_f}{\overset{k_d}{\rightleftharpoons}} M' + L' \tag{13.42}$$

$$M' + Y \xrightarrow{k_2} M'Y \tag{13.43}$$

The rate expression for Equation (13.43) is Equation (13.44).

$$\frac{d}{dt}[M'Y] = k_2[M'][Y] \tag{13.44}$$

The M' concentration will always be very small and, thus, applying the *steady-state approximation* to Equations (13.42) and (13.43) yields Equation (13.45):

$$\frac{d}{dt}[M'] = 0 = k_d[M'L'] - k_f[M'][L'] - k_2[M'][Y] \tag{13.45}$$

Rearranging and solving for M' gives Equation (13.46).

$$[M'] = \frac{k_d[M'L']}{k_f[L] + k_2[Y]} \tag{13.46}$$

Substituting $[M']$ from Equation (13.46) into Equation (13.44) gives Equation (13.47).

$$\frac{d}{dt}[M'(Y)] = -\frac{d}{dt}[M'L] = \frac{k_d[M'L']k_2[Y]}{k_f[L'] + k_2[Y]} \tag{13.47}$$

When the $M'L'$ complexes react completely with *excess* Y so that $k_2[Y] \gg k_f[L']$, Equation (13.47) reduces to Equation (13.48).

$$\frac{d}{dt}[M'(Y)] = -\frac{d}{dt}[M'L'] = k_d[M'L'] \tag{13.48}$$

Integrating Equation (13.48) gives Equation (13.49).

$$\ln[M'L'] = -k_d t = -k_{obs}[Y]t \tag{13.49}$$

A plot of ln [M$'$L$'$] versus time gives k_d. The half-life for the Mn$'$L$'$ complex dissociation (an estimate of its residence time) is given in Equation (13.50).

$$t_{1/2} = \frac{0.693}{k_d} = \frac{0.693}{k_{obs}[Y]} \tag{13.50}$$

$K_{cond\,M'L'}$ is now evaluated from k_f and k_d using Equation (13.15).

References

1. Coale, K. H. and Bruland, K. W. (1988) Copper complexation in the North East Pacific. *Limnology and Oceanography*, **33**, 1084–1101.
2. Bruland, K. W. (1989) Complexation of zinc by natural organic ligands in the central North Pacific. *Limnology and Oceanography*, **34**, 176–198.
3. Capodaglio, G., Coale, K. H. and Bruland, K. W. (1990) Lead speciation in surface waters of the eastern North Pacific. *Marine Chemistry*, **29**, 221–233.
4. Bruland, K. W. (1992) Complexation of cadmium by natural organic ligands in the central North Pacific. *Limnology and Oceanography*, **37**, 1008–1017.
5. Witter, A. E., Hutchins, D. A., Butler, A. and Luther, III, G. W. (2000) Determination of conditional stability constants and kinetic constants for strong Fe-binding ligands in seawater, *Marine Chemistry*, **69**, 1–17.
6. Spasojević, I., Armstrong, S. K., Brickman, T. J. and Crumbliss, A. L. (1999) Electrochemical behavior of the Fe(III) complexes of the cyclic hydroxmate siderophores Alcaligin and Desferrioxamine E. *Inorganic Chemistry*, **38**, 449–454.
7. Köpfer, D. A., Song, C., Greune, T., Sheldrick, G. M., Zachariae, U. and de Groot, B. L. (2014) Ion permeation in K$^+$ channels occurs by direct coulomb knock-on. *Science*, **346**, 352–355.
8. Gokel, G. W., Leevy, W. M. and Weber, M. E. (2004) Crown ethers: sensors for ions in molecular scaffolds for materials and biological models. *Chemical Reviews*, **104**, 2723–2750.
9. Fortin, C. and Campbell, P. G. C. (2001) Thiosulfate enhances silver uptake by a green alga: role of anion transporters in metal uptake. *Environmental Science and Technology*, **35**, 2214–2218.
10. Mason, R. P., Reinfelder, J. R. and Morel, F. M. M. (1996) Uptake, toxicity and trophic transfer of mercury in a coastal diatom. *Environmental Science and Technology*, **30**, 1835–1845.
11. Lewis, B. L., Holt, P. D., Taylor, S. W., Wilhelm, S. W., Trick, C. G., Butler, A. and Luther III, G. W. (1995) Voltammetric estimation of iron(III) thermodynamic stability constants for catecholate siderophores isolated from marine bacteria and cyanobacteria. *Marine Chemistry*, **50**, 179–188.
12. Sunda, W. G. and Huntsman, S. A. (1995) Iron uptake and growth limitation in oceanic and coastal phytoplankton. *Marine Chemistry*, **50**, 189–206.
13. Sunda, W. G. and Huntsman, S. A. (1998) Processes regulating cellular metal accumulation and physiological effects: phytoplankton as model systems. *The Science of the Total Environment*, **219**, 165–181.
14. Morel, F. M. M. (2013) The oceanic cadmium cycle: biological mistake or utilization? *Proceedings of the National Academy of Sciences of the United States of America*, **110**, E1877.
15. Hormer, T. J., Lee, R. B. Y., Henderson, G. M. and Rickaby, R. E. M. (2013) Nonspecific uptake and homeostasis drive the oceanic cadmium cycle. *Proceedings of the National Academy of Sciences of the United States of America*, **110**, 2500–2505.

Index

Inorganic Chemistry for Geochemistry and Environmental Sciences: Fundamentals and Applications, First Edition. George W. Luther, III.
© 2016 John Wiley & Sons, Ltd. Published 2016 by John Wiley & Sons, Ltd.
Companion Website: www.wiley.com/go/luther/inorganic

The Elements

Atomic number	Name	Symbol	Mass (g mol^{-1})	Atomic number	Name	Symbol	Mass (g mol^{-1})
1	Hydrogen	H	1.008	32	Germanium	Ge	72.61
2	Helium	He	4.003	33	Arsenic	As	74.92
3	Lithium	Li	6.941	34	Selenium	Se	78.96
4	Beryllium	Be	9.012	35	Bromine	Br	79.90
5	Boron	B	10.81	36	Krypton	Kr	83.80
6	Carbon	C	12.01	37	Rubidium	Rb	85.47
7	Nitrogen	N	14.01	38	Strontium	Sr	87.62
8	Oxygen	O	16.00	39	Yttrium	Y	88.91
9	Fluorine	F	19.00	40	Zirconium	Zr	91.22
10	Neon	Ne	20.18	41	Niobium	Nb	92.91
11	Sodium	Na	22.99	42	Molybdenum	Mo	95.94
12	Magnesium	Mg	24.30	43	Technetium	Tc	98.91
13	Aluminum	Al	26.98	44	Ruthenium	Ru	101.1
14	Silicon	Si	28.09	45	Rhodium	Rh	102.9
15	Phosphorus	P	30.97	46	Palladium	Pd	106.4
16	Sulfur	S	32.07	47	Silver	Ag	107.9
17	Chlorine	Cl	35.45	48	Cadmium	Cd	112.4
18	Argon	Ar	39.95	49	Indium	In	114.8
19	Potassium	K	39.10	50	Tin	Sn	118.7
20	Calcium	Ca	40.08	51	Antimony	Sb	121.8
21	Scandium	Sc	44.96	52	Tellurium	Te	127.6
22	Titanium	Ti	47.87	53	Iodine	I	126.9
23	Vanadium	V	50.94	54	Xenon	Xe	131.3
24	Chromium	Cr	52.00	55	Cesium	Cs	132.9
25	Manganese	Mn	54.94	56	Barium	Ba	137.3
26	Iron	Fe	55.85	57	Lanthanum	La	138.9
27	Cobalt	Co	58.93	58	Cerium	Ce	140.1
28	Nickel	Ni	58.69	59	Praseodymium	Pr	140.9
29	Copper	Cu	63.55	60	Neodymium	Nd	144.2
30	Zinc	Zn	65.39	61	Promethium	Pm	145
31	Gallium	Ga	69.72	62	Samarium	Sm	150.4

Atomic number	Name	Symbol	Mass (g mol^{-1})	Atomic number	Name	Symbol	Mass (g mol^{-1})
63	Europium	Eu	152.0	91	Protactinium	Pa	231.0
64	Gadolinium	Gd	157.2	92	Uranium	U	238.0
65	Terbium	Tb	158.9	93	Neptunium	Np	237.0
66	Dysprosium	Dy	162.5	94	Plutonium	Pu	244
67	Holium	Ho	164.9	95	Americium	Am	243
68	Erbium	Er	167.3	96	Curium	Cm	247
69	Thulium	Tm	168.9	97	Berkelium	Bk	247
70	Ytterbium	Yb	173.0	98	Californium	Cf	251
71	Lutetium	Lu	175.0	99	Einsteinium	Es	252
72	Hafnium	Hf	178.5	100	Fermium	Fm	257
73	Tantalum	Ta	180.9	101	Mendelevium	Md	258
74	Tungsten	W	183.8	102	Nobelium	No	259
75	Rhenium	Ru	186.2	103	Lawrencium	Lw	262
76	Osmium	Os	190.2	104	Rutherfordium	Rf	261
77	Iridium	Ir	192.2	105	Dubnium	Db	262
78	Platinum	Pt	195.1	106	Seaborgium	Sg	266
79	Gold	Au	197.0	107	Bohrium	Bh	264
80	Mercury	Hg	200.6	108	Hassium	Hs	269
81	Thallium	Tl	204.4	109	Meitnerium	Mt	268
82	Lead	Pb	207.2	110	Darmstadtium	Ds	271
83	Bismuth	Bi	209.0	111	Roentgenium	Rg	272
84	Polonium	Po	209	112	Copernicium	Cn	277
85	Astatine	At	210	113	Ununtrium	Uut	
86	Radon	Rn	222	114	Flerovium	Fl	289
87	Francium	Fr	223.0	115	Ununpentium	Uup	
88	Radium	Ra	226.0	116	Livermorium	Lv	293
89	Actinium	Ac	227.0	117	Ununseptium	Uus	
90	Thorium	Th	232.0	118	Ununoctium	Uuo	

Printed and bound by CPI Group (UK) Ltd, Croydon, CR0 4YY

27/10/2024

14580301-0004